THE FACTS ON FILE ENCYCLOPEDIA OF SCIENCE, TECHNOLOGY, AND SOCIETY

Editorial Board and Contributors

CONTRIBUTORS

A.H.S. Ann Helton Stromberg, Pitzer College
A.J. Alan Jones, Pitzer College
A.L. Arthur Layton
A.M. Arthur Molella, National Museum of American History, Smithsonian Institution
A.R. Ami Radunskaya, Pomona College
A.S. Adam Serchuk
A.Wa. Albert Wachtel, Pitzer College
A.W. Al Wurth, Lehigh University
A.Z. Andrew Zanella, The Joint Sciences Center of the Claremont Colleges
B.H. Blair Haworth, Duke University
B.L. Barbara Lawrence, Eastern Illinois University
B.N. Bern Nagengast
B.D.S. Brett D. Steele, UCLA
B.E.S. Bruce E. Seely, Michigan Technological University
B.S. Bruce Sinclair, Georgia Institute of Technology
C.D.D. C. Douglas Dalgleish, Arizona State University
C.H. Charles Haines
C.I. Christine Ilgen
C.S. Carlene Stephens, National Museum of American History, Smithsonian Institution
D. F. Deborah Fitzgerald, Massacusetts Institute of Technology
D.A.B. Debra A. Brooks, Rancho Santiago College—Orange Campus
D.B. Dipannita Basu, Pitzer College
D.C.J. Donald C. Jackson, Lafayette College
D.J.W. Deborah Jean Warner National Museum of American History, Smithsonian Institution
D.K. Dennis Keith
D.R.M. David R. Mets, School of Advanced Airpower Studies, Air University
E.W.C. Edward W. Constant II, Carnegie-Mellon University
E.C.R. Elwin C. Robison, Kent State Universitry
E.K. Ellen Koch
E.L. Edward Layton, University of Minnesota
G.K. Gregory Kunkle, Lehigh University
G.T.S. G. Terry Sharer, National Museum of American History, Smithsonian Institution
H.M. Helmut Maier, The University of the Ruhr
H.W. Harvey Wichman, Claremont McKenna College
J.A.G. Joanne Abel Goldman, University of Northern Iowa
J.C. Jonathan Coopersmith, Texas A&M University
J.C.B. Joseph C. Bigott
J.D.H. J.D. Hunley, NASA Dryden Flight Research Center
J.D.S. John D. Sullivan, Pitzer College
J.F. Jodi Frost, University of Minnesota
J.G. Judith Grimshaw
J.H. Jim Hoste, Pitzer College
J.K.S. John K. Smith, Lehigh University
J.M. John Mountcastle, U.S. Army Center for Military History
J.N.G. J. Nadine Gelberg, Pennsylvania State University
J.S. JoAnne Simson
J.V.G. Judith V. Grabiner, Pitzer College
J.W. James Williams, DeAnza College
K.D. Kelly DeVries, Loyola College
K.G. Karl Gurcke
L. H. Lamont Hempel, Claremont Graduate University
L. S. Larry Schweikart, University of Dayton
L.E.M. Lynne E. Miller
L.G. Lawrence Gross, Universitry of Massachusetts, Lowell
L.S. Lisa Sullivan, Harvey Mudd College
M.A. Marsha Ackermann, Michigan State University
M.B. Madeleine Brink
M.O.B. Morton O. Beckner, Pomona College
M.F. Mark Freeman
M.H.R. Mark H. Rose, Florida Atlantic University
M.J.N. Michael J. Neufeld, National Air and Space Museum, Smithsonian Institution
M.L. Mark Levinson, University of Washington
M.M. Marguerite Malakoff, Harvey Mudd College
M.R. Martin Reuss, U.S. Army Corps of Engineers
M.R.W. Michael R. Williams, University of Calgary
M.S. Monica Smith, Lemelson Center for the Study of Invention and Innovation, Smithsonian Institution
M.S.K. Michael S. Klein
M.W. Michael Woodcock, Pitzer College
M.W.B. Molly W. Berger, Case Western Reserve University
M.W.R. Matthew W. Roth, University of Southern California
N.C. Newton Copp, The Joint Sciences Center of the Claremont Colleges
N.P. Noel Price
P. C. Paul Ceruzzi, National Air and Space Museum, Smithsonian Institution
P.E.H. P. Edward Haley, Claremont McKenna College
P.F. Paul Faulstich, Pitzer College
R.F.H. Richard F. Hirsh, Virginia Polytechnic Institute
R.G.G. Roger G. Gilbertson
R.H.E. Richard H. Elderkin, Pomona College
R.K. Ramunas Kondratus, National Museum of American History, Smithsonian Institution
R.L.B. Regina Lee Blaszczyk, Boston University
R.O. Richard Olson, Harvey Mudd College
R.C.P Robert C. Post, National Museum of American History, Smithsonian Institution
R.Q.H. Richard Q. Hofacker, Jr.
S.E.W. Sara E. Wermiel, Massachusetts Institute of Technology
S.H.C. Stephen H. Cutcliffe, Lehigh University
S.J.D. Steven J. Dick, U.S. Naval Observatory
S.L. Steven Lubar, National Museum of American History, Smithsonian Institution
S.N. Stephen Naftilan, The Joint Sciences Center of the Claremont Colleges
T.B. Tad Beckman, Harvey Mudd College
T.H. Thomas Helliwell, Harvey Mudd College
T.I. Thomas Ilgen, Pitzer College
T.S. Tim Stroshane
T.S.R. Terry Reynolds, Michigan Technological University
W. B. William Banks, Pomona College
W.R. Wayne Rasmussen

THE FACTS ON FILE ENCYCLOPEDIA OF SCIENCE, TECHNOLOGY, AND SOCIETY

VOLUME 2 E–N

Rudi Volti

Facts On File, Inc.

The Facts On File Encyclopedia of Science, Technology, and Society

For information contact:

Facts On File, Inc.
11 Penn Plaza
New York, NY 10001

Library of Congress Cataloging-in-Publication Data

Volti, Rudi.
The facts on file encyclopedia of science, technology, and society / Rudi Volti.
p. cm.
Includes bibliographical references and index.
ISBN 0-8160-3459-1 (v.1)
ISBN 0-8160-3460-5 (v.2)
ISBN 0-8160-3461-3 (v.3)
ISBN 0-8160-3123-1 (set) (alk. paper)
1. Science—Encyclopedias. 2. Technology—Encyclopedias. 3. Science—Social aspects—Encyclopedias. 4. Technology—Social aspects—Encyclopedias. I. Title.
Q121.V65 1998
503—dc21 98–39014

Facts On File books are available at special discounts when purchased in bulk quantities for businesses, associations, institutions or sales promotions. Please call our Special Sales Department in New York at 212/967-8800 or 800/322-8755.

You can find Facts On File on the World Wide Web at **http://www.factsonfile.com**

Text design by A Good Thing, Inc.
Cover design by Cathy Rincon

Printed in the United States of America.

VB AGT 10 9 8 7 6 5 4 3 2

This book is printed on acid-free paper.

THE FACTS ON FILE ENCYCLOPEDIA OF SCIENCE, TECHNOLOGY, AND SOCIETY

E

E. coli

The bacterium *Echerichia coli*, or *E. coli*, is a normal inhabitant of the human intestine. Some strains also can be pathogens, causing food poisoning, especially from raw or undercooked ground beef. A fast-food chain gained notoriety in 1994 when several people became ill and a few died after eating hamburgers contaminated with a deadly strain of *E. coli*.

Cells are classified on the basis of a cellular nucleus, or *karyon*. *E. Coli* is a *prokaryote*: like other bacteria, it contains no cell nucleus, and all of its DNA resides in a single chromosome. Yet it has served well as a model for higher, *eukaryotic* organisms in which DNA is organized into chromosomes that are compartmentalized in a nucleus. In eukaryotes, a higher level of complexity is imposed on the same basic processes that take place in *E. coli*.

The molecular biology of *E. coli* is the best understood of any organism because it has been the model organism for research in biochemistry and bacterial genetics. At the same time, *E. coli* is of great importance for scientific research because it is the most convenient host for the propagation of recombinant DNA. As such, *E. coli* can serve as a kind of factory for the production of important pharmaceuticals. Recombinant proteins of commercial value, including human insulin and human growth factor, produced in *E. coli*, now have reached the market and others are likely to follow.—C.I.

See also BIOTECHNOLOGY; DNA

Earthquake-Resistant Building Standards

Earthquakes are among the most dangerous and destructive of natural occurrences. Consequently, attempts to develop designs for buildings to help them withstand earthquakes have occupied the attention of structural engineers for well over a century. In the United States it was the 1906 San Francisco earthquake that provided earthquake engineering its first real impetus. The loss of approximately 1,000 lives and one-half to $1 billion in damages devastated the city. The earthquake led to the establishment of the Structural Association of San Francisco, later to become the Structural Engineers Association of California (SEAOC), and the creation of the Seismological Society of America. Due to the particularly earthquake-prone nature of California, the SEAOC's recommendations became the basis for most earthquake-resistant building codes.

Earthquakes in Santa Barbara in 1925 and in Long Beach in 1933, both of which measured 6.3 on the Richter scale, reinforced the need for improved building designs. The development of the first strong-motion accelerographs in the early 1930s allowed earthquake engineers to visualize ground shaking in terms of amplitude, frequency, and duration. These two earthquakes also revealed that some of the largest damage problems occurred with unreinforced brick buildings constructed with sand-lime mortar, a type of construction then typical of many school buildings. Within a month the California legislature passed the

Field Act, which required enhanced lateral force design standards for all future school construction. This act has proved to be extremely important, for during subsequent earthquakes, many pre-1933 schools suffered severe damage, while those erected or subsequently modified fared far better. Although somewhat weaker in terms of standards, the 1933 Riley Act incorporated similar requirements for nonschool buildings and has similarly proved its worth. At the same time, individual cities such as Los Angeles began incorporating mandatory, rather than merely suggested, earthquake provisions into their building codes.

The SEAOC formed a Seismology Committee in 1957 whose resulting Recommended Lateral Force Requirements report has subsequently been incorporated into numerous building codes around the world, including the U.S. Uniform Building Code (UBC) in 1961. The philosophy that underlies the requirements is not one of absolute 100 percent, risk-free protection, which is never possible, but rather the "reasonable" protection of life and property, which recognizes the likelihood of "some structural as well as nonstructural damage."

The 1971 San Fernando earthquake led to major advances in the understanding of aseismic design. Measuring 6.4 on the Richter scale, it resulted in 65 deaths and over $550 million in property damage. Of these deaths, 47 resulted from the collapse of several pre-1933 unreinforced masonry hospital buildings. The SEAOC in 1974, prompted by the San Fernando quake, revised its seismic code recommendations, which were in turn incorporated into the 1976 edition of the UBC. Included in the enhanced requirements were considerations for the "seismicity" of the area in which a building was located, the primary usage or occupancy of a building, and a factor that accounts for differences in underlying soils. At this time, the availability of computers and their associated modeling capabilities allowed designers to incorporate and code writers to require dynamic analysis of irregularly shaped buildings with differences in stiffness between adjacent sections. It was this development of computer modeling techniques that also made it possible to construct high-rise buildings in particularly earthquake-prone zones, such as Los Angeles and in Japan, which completed its first real skyscrapers only in the late 1960s, including the 40-story Tokyo World Trade Center Building in 1970.

The 6.9 Loma Prieta, Calif., earthquake resulted in substantial loss of life (63) and, up to that point, caused the most property damage ($7.1 billion) from an earthquake in the United States. Among the most vulnerable buildings were those with "weak" first floors, such as three- to four-story residences with numerous garage door openings at ground level. Damage was especially severe where extensive liquefaction of the soil occurred. This problem has led to increased attention being paid to the development of soil liquefaction maps, which, it is to be hoped, will lead to better-informed land use planning decisions.

Less than 5 years later, on Jan. 17, 1994, the most destructive earthquake in the United States to date struck Northridge, Calif. At 6.7 on the Richter scale, the Northridge earthquake was slightly smaller than the Loma Prieta temblor, yet strong ground motion caused it to be the most destructive earthquake ever in the United States in terms of property damage, with losses amounting to over $20 billion. Almost miraculously, only 57 people lost their lives. What was most surprising, at least from an engineering perspective, was the extensive damage done to structures built according to recent design codes and thought to be earthquake resistant.

Seismic recordings revealed very dramatic and rapid ground movement—both laterally and vertically—with some ground movements on the order of .6 m (2 ft), and then back again within a period of 5 seconds. Such movements are far beyond the limits anticipated in existing building codes and suggest that a tall structure, even if built to the most stringent codes, might not be able to handle such movement and stress. Fortunately, Northridge had few tall buildings or structures older than 20 years. Had downtown Los Angeles been at the epicenter, it is conceivable that some large skyscrapers might have fallen.

The Northridge quake revealed a number of design problems, one of the more serious being three-story, wooden-framed apartment buildings with garages and open carports underneath that weakened the structures' ability to resist lateral forces. According to one estimate, approximately half the projected $20 billion in total damage was due to the poor performance of wooden-framed buildings. Much of the destruction can be attributed to slipshod construction practices and inadequate field inspection, as well as to unsound design. Of all the damage, however, the most disconcerting was the failure of welded steel moment-frame (WSMF) structures previously thought to be very safe seismically. Subsequent examination of such buildings has revealed that more than 120 had cracks in the connecting joints where the beams are welded to the supporting columns. Although none failed catastrophically, it was an open question whether they could survive another such earthquake. The problem is compounded by the difficulty of detecting existing damage, because the structural steel is often covered by architectural finishing and fireproofing materials. An especially troubling aspect of the discovery of cracked welds is that WSMF construction was used in more than half of all new commercial buildings erected worldwide in 1994.

An important lesson to be learned from all earthquakes, but especially from the unexpected damage occasioned by the Northridge earthquake, is that we must not become complacent about earthquake engineering and the design of buildings. This lesson is no less applicable outside California, where most of the attention has been focused. The East Coast, although less seismically active, contains many high-rise buildings that could collapse in the event of a major earthquake. The priority of other public issues, the immediacy of costs coupled with uncertain benefits, and the invisibility of public safety, all work against the adoption and enforcement of seismic building codes. Still, it is only a matter of time before the next quake will occur, and we must guard against a false sense of security.—S.C.

See also BUILDING CONSTRUCTION, STEEL-FRAME; CONCRETE; EARTHQUAKES; MATHEMATICAL MODELING; RICHTER SCALE; RISK EVALUATION

Further Reading: "On Shaky Ground: A History of Earthquake Resistant Building Design Codes and Safety Standards in the United States in the Twentieth Century," *Bulletin of Science, Technology, and Society*, vol. 16, no. 5 (1996).

Earthquakes

The Earth's surface contains many faults, breaks in the crust where movement occurs. When this happens, energy is released through the Earth and is felt as an earthquake. Earthquakes occur in many locations and are almost always associated with tectonic plate boundaries. Few natural phenomena are feared as much as an earthquake, although large earthquakes are the exception, not the rule. Every day thousands of earthquakes occur and go unnoticed, or are felt and cause little to no damage.

What distinguishes earthquakes is how energy moves through the Earth after it has been released by movement along a fault. There are two major types of energy released by an earthquake, body waves and surface waves. Body waves result from energy moving through a substance. The two major types of body waves are P waves (primary waves) and S waves (secondary waves). P waves are compressional, and the particle motion within the wave is back-and-forth in the same direction as the direction of the wave propagation. S waves are shear waves, and the particle motion within the wave is back-and-forth in a direction perpendicular to the direction of the wave propagation. P waves are generally felt as jolts; S waves are felt as shaking.

The second way that earthquake energy moves is as surface waves, which move along surfaces or interfaces within the Earth rather than straight through the rock. Surface waves are felt as a rolling motion and frequently cause the most structural damage during an earthquake.

Seismic energy of all wave types occurs during an earthquake. What is actually felt is partly determined by a number of factors other than the type of wave. The energy released as seismic waves is weakened, bent, and filtered as it radiates out from the earthquake origin. The four major factors that affect how the seismic waves change are: (1) the distance to the earthquake origin, (2) the size of the earthquake, (3) the rock type that the waves travel through, and (4) the rock type present at the location where the earthquake is felt.

When an earthquake occurs, the ground breaks. This break, or fault, may be new, but more likely the movement is on a preexisting fault. Also, the fault may not continue to break all the way to the surface of the Earth. The origin of the earthquake is the focus or "hypocenter" of the earthquake and occurs at some underground location on the fault. The closer an observer is to the focus, the stronger the earthquake feels, since the energy weakens as it moves away from the focus. The location at the surface of the Earth directly above the focus is the "epicenter," and frequently the epicenter is mistakenly called the *origin* of the earthquake. Even though it is not the origin of the earthquake, it is the location from which a large earthquake receives its name. For example, the epicenter of the 1906 San Francisco earthquake occurred near the city of San Francisco. This earthquake was very strong and was felt as far north as Oregon and as far south as Los Angeles. Even though the earthquake was felt far away, the earthquake did little damage outside the greater San Francisco area since the energy weakened with distance from the focus.

The second factor to consider is the size of the earthquake. There are three major ways to measure size: by the damage caused, the ground motion as recorded by machines, or the amount of energy released. In general, the more energy released during an earthquake, the bigger the earthquake, the more the ground moves, and the more damage occurs. Unfortunately, measuring energy release is comparatively difficult. It is much easier to measure damage and ground motion.

Damage, or earthquake intensity, is the historical way to measure earthquakes. After an earthquake occurs, the damage to structures and landforms is compiled and mapped. The intensity of the earthquake is determined by comparing the damage against the descriptive values of an intensity scale; with more damage, the higher the number on the scale and the greater the intensity. Measurements of intensity are most important to people who build and design structures and to people who are responsible for disaster planning. They are not as important to seismologists because they are not the best way to compare different earthquakes.

The second way to measure the size of an earthquake is by measuring the magnitude or ground motion caused during the earthquake. Seismometers are machines used to measure ground motion. Some are designed to measure the velocity of the ground motion, others measure ground acceleration. The record of the ground motion is a seismogram; these are studied to determine the types, amounts, and timing of seismic waves. From studying seismograms, the location and the size of the earthquake and the structure of the Earth between the earthquake and the seismometers can be determined.

The Richter scale is the most common magnitude scale. Earthquake energy is recorded as a series of waves, and the maximum amplitude or wave height is used to determine size. It works best for local rather than distant earthquakes. However, the ease of having just one number to describe the size of an earthquake eventually led to modifications of the scale for use with distant earthquakes. All of the magnitude scales are based on wave amplitude corrected for the distance to the focus of the earthquake. They are logarithmic, so an increase of 1 means a 10 times increase in ground motion; e.g., the ground motion is 10 times greater for a magnitude-6 earthquake than for a magnitude-5 earthquake. Seismometers are much better at measuring earthquakes than are human beings. Except in very special circumstances, people do not usually feel earthquakes unless they are at least a magnitude 2.

Seismograms are also useful because they can be used to measure and compare the size of earthquakes that occur far away, even in unpopulated areas such as the sea floor. Also, they are used to determine whether the energy moving through the earth is from an earthquake or from a man-made source such as a nuclear explosion. Having recordings of the seismic waves generated by past earthquakes also makes it easier to design structures that could withstand similar earthquakes when they reoccur in that area.

The third way to measure the size of an earthquake is to measure the energy released by the earthquake, the "seismic moment." Energy is a measure of the work that can be done; in this case, the work is the movement on the fault. To measure the seismic moment, it is necessary to determine how much work was done, usually by measuring the amount of earth moved. While this should be the best way to compare earthquakes, it is difficult because not all faults break through to the surface of the Earth, and not all earthquakes occur in areas where the length, width, and height of the faulted surface can be measured.

When possible, both the seismic moment and the magnitude are measured, and an empirical relationship between the two has developed. A magnitude difference of one unit roughly corresponds to a 30-fold energy difference. Therefore, the energy released by a magnitude 6 earthquake is the same as the energy released by 30 magnitude-5 earthquakes. Given this relationship, earthquake magnitudes are now sometimes stated in terms of the "moment magnitude scale," a measure of the seismic moment derived from the magnitude. Since the moment magnitude is energy dependent, it is the preferred description for scientific use.

The next factor that helps determine what is felt during an earthquake is the type of rock between the focus and the seismometer. Earthquake energy moves through rock at different speeds. P waves are called *primary waves* and S waves are called *secondary waves* because P waves are always faster than S waves. On a seismogram, the P-wave energy arrives first, the S-wave energy is second, and the surface wave energy arrives last. The difference in speed between P waves and S waves has been measured in laboratory experiments on different rocks and has been confirmed by measuring the differences in P-wave and S-wave arrival times when energy has been released by man-made forces such as quarry blasts and nuclear explosions. The time difference between them can therefore be used to calculate the distance from the seismometer to the focus of the earthquake. This is how earthquakes are located; seismograms from many stations in the area are used to triangulate on the focus, and a preliminary location of both the focus and epicenter can usually be reported within minutes after an earthquake.

As the seismic waves travel through the Earth, they may be altered by changes in rock type or properties. When they reach a boundary between rock types, some waves pass through unchanged, and some may be changed. An example of such a change occurs when an S wave reaches a liquid, such as a magma chamber. While P waves move through the liquid rock, the S waves convert to either P waves or surface waves. S waves are shear waves and cannot move through liquids. Also, seismic waves can change speed or direction as they move across a boundary. They may be bent, and either reflected back or refracted through at a different angle as they move across the boundary. If the seismograms from many stations are studied together, these changes in the speed, direction, and type of seismic energy can be mapped. Seismic waves sometimes are channeled along different paths and do not travel out uniformly. For example, during the 1906 San Francisco earthquake, much of the energy was channeled north and south of the focus by the internal structuring of the rocks under California.

The last factor to be considered is the type of rock at the location where the earthquake is felt. Surface rock type will have a major influence on earthquake damage. The less stable the surface, the more the Earth moves, and the greater the possibility of structural damage. Unconsolidated sediment, which is frequently found in beach areas, presents a special problem for engineers and builders. People want to live near a beach, but during an earthquake the ground may actually liquefy and become incapable of supporting the weight of structures. Similar problems arise when structures are built on hillsides. The view may be spectacular, but if the rock and soil near the surface are not well consolidated, earthquakes can cause landslides. Structures may experience essentially the same seismic energy during an earthquake, but if they are built on different rock types the effects of the ground motion can be dramatically different.

Earthquakes are a natural phenomena like the weather. Weather prediction is currently far easier, because meteorologists have a understanding of the properties of the atmosphere and which properties are the most useful for purposes of prediction. There are many different properties that can be measured within the Earth, but seismologists do not completely understand how all of these properties change before and during earthquakes and which are the most important for prediction. People have been keeping records of atmospheric change for centuries; seismograms have been in existence only since the early part of this century. It is easy to tell *where* an earthquake is most likely to occur, but it is difficult to tell exactly *when*. To be sure, some earthquakes have been predicted. Chinese seismologists predicted a magnitude-7.4 earthquake in Liaoning province in February 1976, and many lives were saved. However, using the same techniques, they failed to predict the magnitude-7.6 earthquake that occurred 5 months later in Tangshan and killed more than 250,000 people. Until earthquake prediction techniques are improved, people who live in earthquake-prone areas should base their safety on preparedness rather than prediction.—D.A.B.

See also EARTHQUAKE-RESISTANT BUILDING STANDARDS, LOGARITHM; PLATE TECTONICS; RICHTER SCALE; WEATHER FORECASTING

Further Reading: Bruce A. Bolt, *Earthquakes*, 1988.

Ecology

Ecology is the scientific study of the interrelationships between plants, animals, and the environment. The word *ecology* is derived from the Greek *oikos*, meaning "household" or "home," combined with the suffix *logy*, meaning "the study of." Thus, ecology has developed into the scientific study of interdependent communities of organisms and their environments. The word was coined in 1869 by the German zoologist, Ernst Haeckel (1834–1919), who applied the term *oekologie* to the "relationships of animals with both the inorganic and organic environments." In recent years scientists have come to regard ecology as the interdisciplinary study of Earth's life-support systems. Its principles have proved useful in broad aspects of the related fields of natural resource management, conservation biology, forestry, agriculture, and pollution control. As such, ecology has developed into one of the most popular and most important aspects of biology.

Ecology evolved from the natural history of the Greeks and emerged as an academic discipline in the late 19th century in Europe and America. The historical roots of ecology lie in diverse disciplines, including plant and animal physiology, oceanography, and evolution. Darwin's theory of evolution, for example, was essentially ecological; he postulated that those organisms that adapt to their environment will be most likely to survive and reproduce. Because of its reliance on measurement and mathematics, ecology has occasionally been called "scientific natural history."

Until recently, ecology lacked a solid conceptual base. Contemporary ecology, however, now focuses on the related concepts of *community* and *ecosystem*. An ecosystem consists of interacting organisms, mineral cycling, energy flow, population dynamics, and all aspects of a given natural environment. They function by maintaining a flow of energy and nutrients within the biotic (living) and abiotic (nonliving) components of the community. Ecosystems tend towards maturity, or stability, and develop in the direction of increasing complexity. This change from a less complex to a more complex state is called *succession.*

Through succession, an ecosystem that has been disturbed through fire, flood, or other disaster tends eventually to restore itself. The first species to reestablish in a disturbed area are called *pioneers*. They are opportunistic species with proficient means of dispersal and high reproductive capabilities. Lichens, grasses, and other herbaceous species are the most common pioneers. They begin to heal the disturbed zone by increasing the organic content of the area, creating microenvironments, and releasing nitrogen-fixing bacteria into the soil. Through succession, conditions are made increasingly suitable for new organisms that use less energy for reproduction and more energy for sustenance. These species eventually bring the natural system to a point of stability, referred to as a "climax community."

Communities are assemblages of organisms living together; the study of groups of organisms is called *community ecology*. Some communities are tiny, such as those composed of invertebrates inhabiting a decomposing log. Other communities are defined on a macro level and may be as extensive as an entire forest ecosystem. The largest natural communities are called *biomes*; they comprise enormous geographic areas. The major global biomes are arctic tundras, coniferous forests, deciduous forests, grasslands, deserts, and tropical rainforests. While biomes are generally defined by the predominance of characteristic plant species, animals that are notably associated with them also contribute to their distinctiveness. Ecology has demonstrated, however, that rather than discrete ecological communities, there exists a continuum of communities that blend together. Communities have various organic structures, including vertical stratification, diurnal patterns, and seasonal variation.

The number and variety of species within a community is called *species diversity*, or *biodiversity*. Biodiversity has two components: richness and evenness. If the number of different species inhabiting an ecosystem is great, the community is said to have rich diversity. However, species are not always equally represented within an ecosystem. If a few species prevail, the diversity is considered uneven. The most ecologically stable communities are those constituting a complex food web made up of many species, each relatively abundant.

One of the principal aims of ecological research is to understand how organisms retain and recycle the chemical features of an ecosystem. Retention and recycling are carried out in many efficient ways, such as the way bacteria and fungi break down organic matter into its mineral components, thereby increasing the rate at which nutrients are returned to the soil. A second functional aspect of ecosystems is energy flow. Energy in the form of food is transferred through a series of organisms, creating a food web in which there is more energy passing through the plants than there is in the herbivores that eat them, and still less in their predators. A third major ecological function of ecosystems is population regulation. Herbivores, for instance, serve to regulate plant populations, and predators and parasites function partly as regulators of organisms they consume.

From its inception, ecology has stressed the holistic study of the structure and function of nature. This emphasis, however, has not always been practiced. In the mid-20th century, for example, plant and animal ecology developed separately until biologists began to emphasize the interrelation of floral and faunal communities as a biotic whole. Modern ecology is variously divided into terrestrial ecology, marine ecology, and freshwater ecology (limnology), or into community ecology, ecosystem ecology, and population ecology. Specialized subfields include paleoecology, ecoclimatology, and physiological ecology.

The major functional unit of the ecosystem is population. The study of population dynamics developed during the mid-1900s, fueled by the early 19th-century insights of Thomas Malthus, who called attention to the incompatibility of an expanding human population with the capability of Earth to supply food. Contemporary population ecology is the study of processes that influence the demographics of plant and animal populations. A major component of population ecology is population genetics, which deals with the operation of genes in natural populations.

Ecology came of age in 1942 with the development of the trophic-dynamic concept, which details the operative flow of energy through an ecosystem vis-à-vis food consumption. Ecologists commonly classify organisms according to their function in an ecosystem; the ecological function of an organism constitutes its niche. Green plants (called *producers*) manufacture their own food from carbon dioxide, water, minerals, and sunlight; they are incapable of metabolically synthesizing food and must obtain it externally. The advanced study of energy flow and nutrient cycling, prevalent in contemporary ecology, was an outgrowth of techniques that enabled ecologists to identify, trace, and measure the movement of energy and nutrients through ecosystems.

The term *ecology* is infused with divers meanings, some principally scientific, some more philosophical, and some functioning as social or political movements. Between the mid-1800s and the mid-1900s, wilderness advocates—most notably Henry David Thoreau, John Muir, and Aldo Leopold—began to reexamine the relationship between culture and nature. These thinkers brought ethical, arcadian, and philosophical concerns to ecology. Responding to the mechanistic orientation of classical science, they reasoned that human beings are sentient elements in the evolutionary process and thus are obliged to evaluate their actions from a reflexive and ethical standpoint. The normative ecologies formulated by these men have greatly influenced the orientations of contemporary environmental movements.

In recent decades, "human ecology" has added a behavioral science orientation to scientific ecology. Focusing on the interactions between people and natural systems, human ecology is largely concerned with how *Homo sapiens* adapts to the environment.

In the 1970s, a movement known as "deep ecology" emerged that calls for a new paradigm to replace the dominant rationalistic paradigm of the past 300 years. It advocated for fundamental changes in human relations with natural systems, and proposed a new ecological ethic. Deep ecology draws on basic concepts from the science of ecology—complexity, diversity, and symbiosis—to clarify the place of our species within nature.

Another recent development, "social ecology," analyzes the technologies and the political and social institutions that people use to mediate between themselves and nonhuman nature. It proposes new patterns of production and offers theories that explain the social causes of environmental problems and alternative ways to address them. Prevailing theories of social ecology employ the approach of Karl Marx and Friedrich Engels to ecology and society.

One of the most pronounced trends in scientific ecology is the increased use of mathematical modeling involving the use of computers. The methods of computer science and applied mathematics have encouraged a new stage in the evolution of ecology known as *systems ecology*, which concentrates on input and output analysis and the organization and function of ecosystems. Systems ecology utilizes theoretical analysis to study the disruption of ecological systems and the dynamics of their reconstruction.

Systems ecology in turn has stimulated the rise of applied ecology, which is concerned with the application of ecological principles to the management of natural resources, agricultural production, and environmental pollution. Late 20th-century ecologists are increasingly involved with finding solutions to problems created by increased population, patterns of consumption, energy usage, and ecosystem destruction.—P.F.

See also EVOLUTION; FOOD CHAIN; MATHEMATICAL MODELING; NATURAL SELECTION; PARADIGM; PHOTOSYNTHESIS

Further Reading: Eugene P. Odum, *Ecology and Our Endangered Life-Support Systems*, 1993.

Ecosystem

An ecosystem is an interacting assemblage of living (biotic) and nonliving (abiotic) natural components. It is a dynamic complex of plant, animal, fungal, and microorganism communities and their associated environment interacting as an ecological whole. The term *ecosystem* was coined in 1935 by the English botanist Arthur Tansley, who conceived of the diverse components of nature working together as a whole complex.

The biotic components of ecosystems include producers (green plants), consumers (animals), and decomposers (microorganisms). The nonliving components include solar energy, temperature, air, water, and chemical elements. The type, quantity, and variation of these components determine what plants and animals can exist in any given natural system. In a healthy ecosystem, matter is recycled and never depleted. Living organisms are in balance with nonliving constituents, and they do not drain their resources.

The biotic and abiotic parts of an ecosystem are connected by a constant exchange of materials through nutrient cycling powered by energy to food energy by means of photosynthesis. The consumers are animals that feed on producers and other consumers; herbivores are the primary consumers, and carnivores are secondary consumers because they eat the animals that eat the plants. Decomposers constitute the largest and perhaps most important biotic component of ecosystems. They include the bacteria and fungi that break down detritus, converting its complex molecules into simpler molecules, which are then used as nutrient by producers to carry on photosynthesis. In this way, all matter in nature is recycled.

A characteristic of ecosystems is that energy and matter flow through them. Energy, however, flows through the system in one direction and dissipates, whereas matter cycles through the system and is used over and over. Consider how forests cycle materials. Like all ecosystems, forests require nutrients—especially oxygen, carbon, hydrogen, and nitrogen—which are gathered, retained, and recycled by trees and other producers. Bacteria convert nitrogen into compounds that are absorbed by the root systems. Rainfall, too, supplies a host of nutrients to plants. As stored water evaporates from plants, more moisture and nutrients are drawn upwards from the soil into the plants' leaves, where they combine with carbon dioxide from the air to produce carbohydrates through photosynthesis. In all forests, fungi, earthworms, termites, and other decomposers break down organic debris, cycling nutrients back into the soil. In a well-ordered ecosystem, such biochemical cycles keep matter circulating between biotic and abiotic components.

Inputs and outputs, then, are important components of the ecosystem concept. Ecosystems are open systems, wherein things are constantly entering and leaving, even though their general appearance and basic function may remain relatively constant. Energy is a necessary input, with the sun being the ultimate energy source for the biosphere. Energy also flows out of the system in the form of heat and organic matter.

Ecosystems depend on a constant transfer of energy, which is transferred from solar radiation in a sequence from green plants to consuming animals to microorganisms. The sequence by which energy is transferred from sunlight, to green plant, to herbivore, to carnivore is known as the *food chain*. The food pyramid represents the total amount of energy that flows along a food chain; in any given ecosystem, plants are more common than herbivores, and herbivores are more common than carnivores. A food chain or food web (a system of linked food chains) represents the relationship of who eats whom and illustrates the interweaving of many different food chains.

The biosphere is a mosaic of countless ecosystems, which may be divided into two broad categories: aquatic ecosystems and land ecosystems. Terrestrial and aquatic ecosystems are complementary types, populated by different kinds of organisms. On land, the predominant producers are usually rooted plants, while in open-water ecosystems the producers are microscopic suspended plants called *phytoplankton*.

Ecosystems share similar structure and function throughout the world, even though their component parts may be structurally different. Ecologically similar species, known as *ecological equivalents*, are found in different parts of the globe where the physical environment is similar. The grasslands of temperate Australia are composed of different plant and animal species than the grasslands of North America, but the distinct ecosystems function in similar ways. Australia's grazing kangaroos, for example, are the ecological equivalents of North America's bison and antelope, and occupy ecologically similar habitats. A habitat is the place where a species is found, and an ecological niche is the biological role of an organism in its community. Thus, the kangaroo and bison, although not closely related genetically, occupy similar niches within their respective habitats.

The biosphere is characterized by a series of gradients of physical characteristics; an example is the temperature gradient from a mountain peak to a valley. Frequently, conditions and the organisms adapted to them change gradually along a gradient, but often there are points of abrupt change. The area at the boundary between two distinct ecosystems is known as an *ecotone*, as, for example, the intertidal zones of a shoreline. An ecotone is not simply a boundary but involves the active interaction between two or more ecosystems. This results in the ecotone having characteristics that are unique and nonexistent in the adjoining ecosystems.

Ecosystems exist in space and evolve with discernible stages of development over time. Left undisturbed, an expansive coral reef may last thousands of years. Conversely, a tiny microbial community may last less than 12 hours. Knowing the dimensions and characteristics of an ecological community is vital to understanding how it works and how it may respond to changing conditions.

Ecosystems respond in different ways to human impacts, and some are more vulnerable to perturbation than others. Generally, complex ecosystems are more stable than simple ones. Natural systems (e.g., saltwater estuaries), which are generally more diverse than artificial systems (e.g., agricultural fields), tend to be most stable. Likewise, tropical rain forests have been thought of as more diverse and, consequently, more stable, than less complex temperate forest communities. In the same way, the relatively simple communities of islands have proved to be highly vulnerable to anthropogenic disturbance. However, doubt has been expressed as to whether the classic concept of the causal linkage between complexity and stability is consistently valid. There is, for example, increasing awareness of the susceptibility of tropical forest ecosystems to human disturbance.

Some ecosystems tend to be vulnerable, while others display the property of elasticity and are able to recover from damage. Lakes, as an example, are closed systems and act as natural traps. They are more vulnerable to disturbance than are rivers (which continually receive new inputs) or oceans (which are much larger). In general, small, isolated systems tend to possess low elasticity, while larger, more complex systems tend toward greater elasticity.

Two important properties of elasticity are resilience (the number of times a system can recover from stress), and inertia (the ability to retain systemic characteristics). Despite their low productivity, desert ecosystems display high resilience and inertia. The flora and fauna of desert environments are accustomed to variable environmental conditions, including day and night temperature extremes. Most desert organisms evolved to a pattern of brief favorable environmental conditions alternating with long periods of stress. They have preadapted resilience to tolerate extreme conditions; they have the ability for rapid and frequent recovery. Additionally, desert plants have various delay and trigger mechanisms, and desert animals have flexible and opportunistic eating habits.

Human beings have a tendency to simplify ecosystems by making them less diverse. Resource extraction, introduced species, development, habitat reduction, and agriculture have impacted ecosystems by decreasing their biological diversity and reducing their cycling of organic matter. Agriculture, for example, creates artificial ecosystems that are energy inefficient within the law of thermodynamics. Simplifying ecosystems usually requires a large input of energy and causes the system to become unstable and vulnerable. The entire system is weakened, not just its component parts. Modern conservation biology has responded by emphasizing conservation of the ecosystem rather than conservation of individual species.—P.F.

See also COMPOSTING; CONSERVATION OF ENERGY; DEFORESTATION; ECOLOGY; ENTROPY; FOOD CHAIN; MONOCULTURE; PHOTOSYNTHESIS; SPECIES EXTINCTION

Further Reading: Eugene P. Odum, *Ecology and Our Endangered Life-Support Systems*, 1993.

Electric Chair

In the late 19th century, electrocution began to be used to execute criminals convicted of a capital offense. The first execution by electrocution occurred in Auburn State Prison in Buffalo, N.Y., on Aug. 6, 1890. On that day, William Kemmler, who had been convicted of murdering his girlfriend with an ax, was executed by a lethal application of electricity. The execution did not proceed smoothly, as the executioners did not know how long the current had to be applied. The first jolt failed to kill Kemmler, so a second one had to be applied. The execution was not a pretty sight; Kemmler's chest heaved, his mouth foamed, and his hair and skin burned as the current ran through him.

At the time of Kemmler's execution, the use of the electric chair was a macabre feature of an ongoing debate about electrical technology: the "battle of the currents" that pitted alternating current against direct current. Due to the ease with which it could be transmitted over long distances, many inventors and entrepreneurs favored alternating current. Direct current, however, enjoyed the support of America's most eminent inventor, Thomas Edison. Never comfortable with abstraction and mathematical formulations, Edison found alternating current difficult to comprehend. Moreover, he had invested a considerable amount of human and physical capital in the development of direct current, and he was determined to make it the foundation of the electrical industry.

One of the arguments that Edison used in support of direct current was that it was safer than alternating current. To emphasize the danger of AC, on several occasions Edison used 1,000 volts of alternating current to publicly electrocute cats and dogs. In a dig at the major manufacturer of AC equipment, Edison liked to refer to the electrocution process as being "Westinghoused." He also helped to design an electrocution apparatus that was unsuccessfully used in 1892.

During the early 20th century, the evident superiority of alternating current put an end to the debate. By this time the electric chair had become well entrenched as the preferred means of execution, although it failed

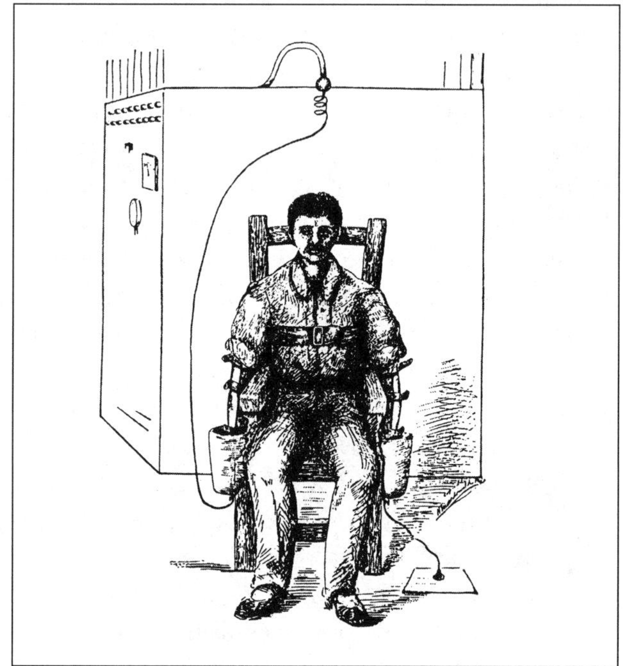

The execution of Charles MacElvaine in 1892. The apparatus, which had been designed by Thomas Edison, performed poorly (two separate charges of 1,600 volts were required), and it was never used again (courtesy T. Bernstein).

to live up to its billing as a rapid and painless method of execution. Beginning in 1924, poison gas began to be used for execution, but the electric chair continued to be used in many states. From 1971 to 1995, of the 313 executions conducted in the United States, 121 were by electrocution. By the mid-1990s, however, other means were being more widely employed. In 1996, of the 45 executions conducted in the United States, 7 were by electrocution, 36 used lethal injection, and hanging and a firing squad accounted for 1 each. In 1997, an execution in Florida that resulted in the condemned man's head being engulfed in flame focused attention on the shortcomings of the electric chair. At the same time, however, the electric chair's deficiencies from a humanitarian point of view may be reckoned as virtues by those who think that capital punishment serves as a deterrent to the committing of capital offenses.

See also ALTERNATING CURRENT; DIRECT CURRENT; EXECUTION BY LETHAL INJECTION; GAS CHAMBER; TRANSFORMER

Further Reading: James E. Penrose, "Inventing Electrocution," *American Heritage of Invention and Technology*, vol. 9, no. 4 (Spring 1994): 34–44.

Electric Meter

Broadly speaking, an electric meter is an instrument that measures electrical current, voltage, power (wattage), or all three. More narrowly, to most consumers of electricity a meter is a device that records their electrical consumption so that a monthly bill can be prepared. This kind of meter has been an integral part of the electric-power industry virtually since its inception, for without some way of determining how much electricity is consumed by individual consumers, there was no way of determining charges or operating in accordance with changing demands.

One of the first meters used for commercial purposes was invented by Frank Julian Sprague (1857–1935). Best known for his pioneering work with electric street railways, in 1882 Sprague installed a meter in a Boston building to complement a 15-hp electric motor used for powering a freight elevator. A year later, Oliver Shallenberger of Westinghouse invented an induction meter for the measurement of alternating current, a crucial element in what was to become the standard means of transmitting electricity. Also in that year, Edward Weston (1850–1936) designed an electric meter based on principles still employed today: a permanent magnet and movable coil attached to a needle that recorded the flow of current. The next major innovation did not come until 1971, when the first solid-state meter for the measurement of large amounts of electricity was introduced.

It is paradoxical that the distribution of electricity, an enterprise characterized by a high level of technological sophistication, still has to rely on a small army of meter readers to record and transmit electrical consumption. This situation may change as a result of the introduction of solid-state electronic meters that can automatically relay the amount of electricity consumed to a central office. Meters of this sort may also play an important role in reducing the cost of electrical power for individual consumers by recording electrical usage over brief periods of time. This would allow suppliers of electricity to charge different rates depending on the time of use so that more could be charged during times of peak usage—typically 2 to 6 P.M.—and less at other times. Accordingly, a residential consumer could use the meter in conjunction with a computerized control to turn off air conditioning during periods of peak consumption and correspondingly high rates. An industrial consumer could also increase electrical usage during times when rates are lower, and vice versa. These meters, which are now being field tested, cost about $100, but the initial cost would be recouped by lower electrical bills in a fairly short period of time.

See also ALTERNATING CURRENT; STREETCAR

Electric Motor—see Motor, Electric

Electric-Power Blackouts

In its early years, the electric-power distribution system in the United States was limited in geographic scope. Power for cities and their surrounding areas was generated and consumed entirely within these localities. As the electrical industry grew, however, large regions were connected into extensive powerline systems, and today electrical energy flows great distances from point of generation to point of consumption. The energy flows through a complex grid of powerlines, which ties together generating sites with regional power companies and their customers. Such grids can cover very large regions, such as the one that covers the western United States from the Dakotas to the Pacific Ocean, and from Canada to Southern California.

A major problem for such extensive power grids is the occurrence of power outages or blackouts. Such events have a variety of causes, including equipment malfunction or failure, accidents, and stress to the system from extremely high energy demand. The consequences of such power blackouts are technical, economic, and social. When a power system experiences a blackout, engineers are faced with two sets of problems: first, identifying the cause of the blackout and repairing the damage, and then getting the system back into operation and restoring power to consumers.

Power outages can disrupt the functioning of business and government, resulting in lost productivity, sales, and service to customers and citizens. In addition to losses and disruption during the blackout, business and government face additional costs when power is restored. Often, equipment must be reset or recalibrated, and operating procedures must be reestablished. To provide one example, during a blackout in Southern California during the weekend of Aug. 10, 1996, many electric stoplights ceased to function properly. When power was restored, the stop signals defaulted to a constant blinking red. They subsequently had to be reset, and that process involved a cost to the cities involved. Business faced a similar set of problems as operations resumed after the blackout. In a world that is increasingly dependent on digital computers, a blackout can cause major disruptions as work and data are lost. The resumption of power delivery can also be a problem, for a sudden voltage spike can do serious damage to computer components.

The social consequences of blackouts can be extensive and can range from spoiled food in home freezers to looting and violence when the blackout occurs at night. In the 1996 Southern California blackout, the abnormal operation of traffic signals did not cause undue difficulties. In general, motorists seemed to adjust to the situation; at most intersections, cooperative behavior was evident, and they were not the scenes of accidents. In contrast, during the New York blackout of 1977, there was extensive looting and violence at night as the absence of power and lights seemed to lead to a breakdown of the social order. In this instance, the costs of the blackout included losses and damage to property, along with the added costs of deploying police and fire personnel to cope with the disruptions.

A closer look at the recent blackout in California illustrates the context in which these blackouts occur. The map, adapted from the *New York Times*, presents a picture of the Western power grid. As can be seen, the grid consists of a large number of power lines and power stations. An important part of this grid is the Pacific Intertie, which links the Northwest (where a great deal of energy is generated) with California, which has high energy demand. On this map, this grid

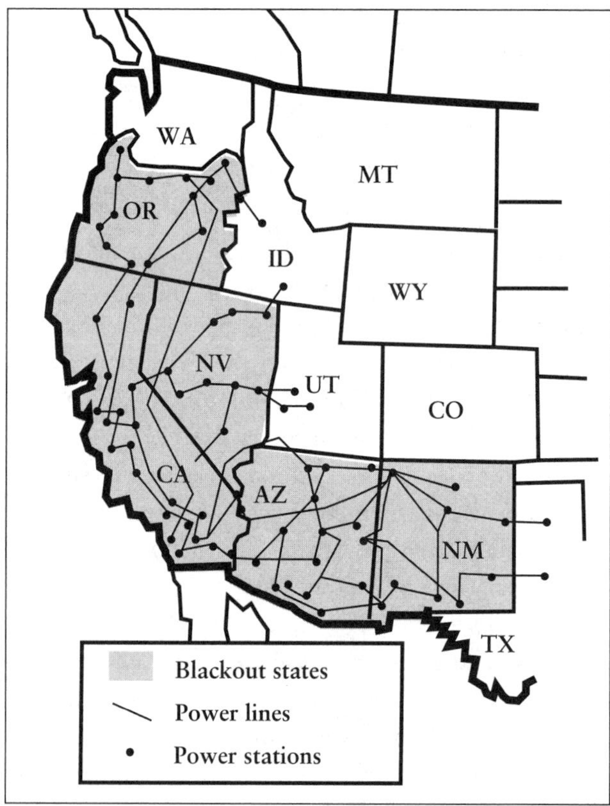

Area affected by the August 1996 power blackout, showing the main power lines and power stations (from *The New York Times,* Aug. 19, 1996).

gives the appearance of a well-integrated, easily managed system. But reality departs substantially from this simple picture. There are a number of separate governmental agencies at the federal and state level that play a role either in generating or consuming energy, or both. Private utilities also play an important role as consumers of energy from the grid. In addition, there is no single management entity that governs the entire grid. Rather, there are a myriad of agencies that manage parts of the grid and coordinate their activities with other elements of the grid, either contractually or through less formal arrangements.

The map highlights the Big Eddy line, where it is believed that the blackout of August 1996 began. It appears that some powerlines in this area sagged as a result of summer heat, apparently brushed against some trees, and began to short-circuit and shut down. As the flow of electricity was disrupted, the situation was further complicated by the failure of the Federal Power Authority in Oregon to warn a power company in Northern California of the problem. Other power lines began to shut down, and a hydroelectric station on the Columbia River followed suit. On this particular day, a dam further up-river might have provided support for the system, but it happened to be shut down to allow salmon to make their spawning runs. The result was, as the map indicates, a blackout over a major portion of the West.

The consequences of this blackout were relatively mild, whereas the 1977 New York blackout, because it occurred at night and literally shut down the entire city, had much more serious consequences. In the future, we can expect to see many such blackouts, from mild to serious, as demand for energy increases. Many states are considering deregulation of their power industries, and the increased competition that is likely to result from such deregulation might exacerbate the consequences of blackouts. Blackouts, like earthquakes and violent storms, are a part of life; it is important for business, governments, and individuals to plan for their occurrence even though they are quite unpredictable.—J.D.S.

See also ALTERNATING CURRENT

Further Reading: Clinton J. Andrews, ed., *Regulating Regional Power Systems*, 1995. Robert Curvin and Bruce Porter, *Blackout Looting! New York City, July 13, 1977*, 1979.

Electric Starter

At the beginning of the 20th century, the best source of power for automobiles was still undetermined; steam, electricity, and the gasoline-fueled internal combustion engine all had their adherents. Yet within a decade the internal combustion engine had emerged as the clear winner, powering the vast majority of automobiles on the road. Even so, it suffered from a number of shortcomings, not the least of which was the difficulty encountered in getting it started. A hefty swing on a crank was required, and if the operator forgot to retard the spark, the engine could kick back, occasionally causing serious injury. To surmount this problem, automobile manufacturers tried a number of starting devices based on springs, compressed air, and mechanical linkages, but none was markedly successful.

One of the victims of hand cranking was Byron Carter, himself an automobile manufacturer, who suffered a broken jaw when the crank kicked back. The gangrene that ensued took his life. A friend of his, Henry Leland (1843–1932), the chief executive of the Cadillac Motor Company, was determined to avoid such tragedies in the future, and he set about engaging his engineering staff in the design of an effective self-starter: An electric motor seemed to hold the most promise, but engineering calculations seemed to indicate that the cranking of an engine would require a motor of excessive size. Leland's company was part of General Motors, which also encompassed the Dayton Engineering Laboratories Company (Delco), on whose staff was an engineer named Charles F. Kettering (1876–1958). Before coming to Delco, Kettering worked for the National Cash register Company, where he had designed an electrically powered cash register that required only a small motor. Kettering realized that an automobile starter motor could be small because it did not have to bear a constant load: It would be required to supply only brief bursts of power. This principle was embodied in the starter he designed for Cadillac, which was first offered on Cadillac's 1912 models. During the following year, the electric starter was adopted by virtually every manufacturer, augmented by an improved starter drive that had been designed by Vincent Bendix (1882–1945).

The electric starter made operating an automobile

easier for drivers, who in the past had sweated and cursed while trying to get their cars started. It also opened up a new market by making driving feasible for people who were physically or mentally blocked from hand cranking. The electric car, already in decline, received a coup de grâce; an easy-to-start gasoline-powered automobile had so many advantages that cars with alternative forms of propulsion saw their markets shrivel into insignificance. The electric starter did not create today's automobile-based society, but without it the automobile would not be the nearly universal mode of transportation that it is today.

See also AUTOMOBILE; AUTOMOBILE, ELECTRIC; ENGINE, FOUR-STROKE

Further Reading: James Flink, *The Automobile Age*, 1988.

Electrical-Power Systems

By the 1890s, Thomas Edison's direct-current (DC) electricity technology had begun to retreat before the alternating current (AC) system devised by George Westinghouse and others, which proved more efficient at transmitting power across long distances. Nevertheless, Edison's fame justly derives from envisioning and constructing a complete technological system. From the 1882 debut of Edison's original Pearl Street Generating Station, built to serve and attract the patronage of New York City's Wall Street financiers, American electrification has proceeded by means of a centralized, multilayered system that integrated technology, management technique, government policy, and rapid growth in consumption. Only recently have events challenged the centralized nature of the system's physical layout and control structure.

Most entrepreneurs in the late 19th-century conceived of electricity as a luxury product with which hotels, department stores, and wealthy homes could power lights, elevators, and novelty items. However, Chicago utility manager (and former secretary to Edison) Samuel Insull (1859–1938) saw that large centrally located generators offered lower costs per unit of power. Starting in the early 1900s, he encouraged widespread and growing use of electricity, allowing him to achieve unprecedented economies of scale with newly introduced steam turbines, which powered the generators that produced electricity. Insull also concluded that because of the capital-intensive nature of the business, one firm in any given area served the public more cheaply than two or more. Thus, he tried to eliminate competition and the costly duplication of facilities. Finally, Insull noted that the total expenses of a central station remained nearly constant, no matter how much electricity it produced. He attributed this phenomenon to utilities' need for plants big enough to serve maximum load, although much of the plants' capacity stood idle most of the time. Insull came to believe that the secret to profits lay in attracting customers whose demand peaks each occurred at different times of the day and year. During the next 75 years, the electric utility industry pursued a strategy based on Insull's insights.

A thicket of overhead wires at turn-of-the-century at Broadway and John Street in New York City, 1890 (from R. Burlingame, *March of the Iron Men*, 1938).

Government regulators came to play an important—though increasingly symbolic—role in the expanding system. As early as 1907, policymakers in some states came to agree that electric monopolies could serve the public more efficiently than multiple competing firms. But to prevent the growing enterprises from demanding exorbitant fees for their increasingly pervasive commodity, states created regulatory commissions to oversee utility operations. Often, the utilities collaborated in this effort. Insull himself believed that regulation would prevent market entry by potential competitors, and that it would legitimize the power industry in the eyes of suspicious citizens, who at that time were pondering the advantages of municipally owned electric systems. As the electric companies provided a growing customer base with ever-cheaper power, however, many regulators took a passive role, prompting occasional charges that they had been "captured" by the interests they were intended to monitor.

Although electric utilities in some regions, notably California, soon extended service to rural users, firms in most areas concentrated on profitable urban markets, where densely packed customers could share the fixed costs of distribution equipment. Most American farmers did not receive electric service until President Franklin Roosevelt's New Deal created the federally owned and operated Tennessee Valley Authority (1933) and the Rural Electrification Administration (1935), which extended low-interest loans to rural cooperatives for the creation of their own distribution networks.

In the mid-1960s, the American electric system appeared immensely successful. Nearly all citizens enjoyed dependable electric service. Between the 1890s and 1967, real rates fell from about 96 to 2 cents per kilowatt-hour (in inflation-adjusted 1967 cents). Industrial customers largely abandoned the on-site "cogeneration" of electricity and steam for heating that had been common early in the century. Most Americans patriotically associated cheap electric power with unmatched industry and a rising standard of living.

Soon thereafter, the rosy tableau disintegrated. Electrical-equipment manufacturers reached plateaus of size and thermal efficiency in the design of new generating units. Although the cantankerous machines failed to provide the expected improvements in total system efficiency, utilities continued their growth-oriented marketing strategies, pushing rates up rather than down. Then, in the 1970s, political turmoil in the Middle East disrupted supplies of oil, causing the price of all energy resources to skyrocket. As electricity rates soared, irate consumers allied themselves with newly influential environmentalists, who objected to the breakneck pace of power plant construction, as well as to the impact of thermal- and nuclear-generating technology on the nation's ecosystem.

These unprecedented stresses forced utility managers to cede part of their control over the system. Environmental and consumer advocates proved unexpectedly successful at staking out a role in the regulatory process. Long-quiescent state regulators reasserted themselves, and federal regulators pressed for inclusion as well. Equally significant, policymakers questioned the economic logic of the utility companies' unchallenged control over the physical system. The Public Utility Regulatory Policies Act (PURPA), passed by Congress in 1978 to boost energy efficiency, required utilities to purchase electricity from industrial cogenerators and from entrepreneurs using natural gas turbines and renewable resources such as wind and biomass. In a manner unforeseen by its creators, PURPA effectively deregulated the generation sector of the electric system.

In the mid-1990s, the system remains in transition. Electric utilities will probably persist as one type of supplier in a competitive market. Less clear is the extent to which market principles will come to govern the transmission and distribution functions of the system. Federal and state regulators may soon force utilities to rent their transmission grids to any independent generator able to negotiate a contract with a potential customer and pay the utility a reasonable usage fee. And as small-scale energy technologies such as fuel cells reach viability, the nation may see increasing decentralized generation, not connected to the grid at all. In sum, while no clear trend has yet emerged, centralization and control remain key concepts to understanding the nation's electric system.—A.H.S. and R.F.H.

See also ALTERNATING CURRENT; COGENERATION; DIRECT CURRENT; FUEL CELL; INSECTICIDES; TENNESSEE VALLEY AUTHORITY; TURBINE, STEAM

Further Reading: Richard F. Hirsh, *Technology and Transformation in the Electric Utility Industry*, 1989.

Electrification, Rural

In the early 1920s, rural America began to emerge from its long isolation. Joining a massive cultural shift, farm families increasingly craved the electric conveniences enjoyed by their city cousins, particularly lights and radio receivers. But the tide of electrification rose unevenly across America. In 1907, 25 years after Thomas Edison opened his first central generating station, 8 percent of urban homes and a minuscule but uncounted fraction of farms enjoyed electric service. By 1920, the figures rose to 47.4 percent of urban homes and 1.6 percent of farms; by 1930, the figures were 84.8 percent and 10.4 percent, respectively. In California, due to a variety of factors, utilities electrified 60 percent of the state's farms by the mid-1930s. But in the rest of the nation, the burgeoning electric utility industry ignored the rural market.

Economic and demographic conditions made utility managers reluctant to enter rural markets. Building a long transmission line and an extensive distribution network to serve lonely farms earned utilities less profit than urban electrification, where aggregations of closely grouped customers shared fixed distribution costs. Moreover, few managers believed that cash-poor farmers would buy enough electricity to pay back their company's expense. And reflecting the business ideology of the day, managers believed that private enterprise bore no responsibility to serve those unable to pay.

In any case, with less than half of city homes served in 1920, utilities had hardly saturated the urban market. In the meantime, some communities formed cooperatives; by 1923, 31 groups of farmers in 9 states had pooled resources to buy power from local utilities and distribute it through their own network of lines. In a few cases, the co-ops even built their own generating facilities. Other farmers turned to wind and water power, along with gasoline-powered generators, to produce their own electricity. Most, however, continued to rise with the sun, depending on kerosene for light.

In 1928, Governor Gifford Pinchot of Pennsylvania published evidence that six financial empires used complex holding companies to control two-thirds of the nation's generating capacity, allowing them to raise electricity rates artificially high. To check these webs of private enterprise, Pinchot urged public takeover of American electric capacity, especially in rural areas. His stand echoed that of "Progressive" presidential candidate Robert M. La Follette, who in 1924 had advocated transferring social control from unregulated business interests to the government. For Pinchot and other social reformers, rural electrification became a

Linesmen stringing wires for the Rural Electrification Administration (from C. W. Pursell Jr., ed., *Technology in America*, 1982).

tool with which to enlist government in their war on the monopolies.

Nevertheless, the farms remained dark until President Franklin Roosevelt, who viewed the lack of electric service as the denial of a basic right, propounded two types of federal intervention. In 1933, Congress created the Tennessee Valley Authority (TVA) to furnish cheap hydroelectric power and fertilizer for the Tennessee River Valley and to make navigable the region's rivers. Second, as part of the Federal Emergency Relief Act of 1935, Congress formed the Rural Electrification Administration (REA), which disbursed low-interest government loans to nascent rural electrical cooperatives.

REA cooperatives encountered various obstacles. Many managers at established for-profit utilities determined to block the emergence of competitors within their territory. Some, on learning that a cooperative had applied for a federal loan, convinced state regulators that the utility had, in fact, been about to enter that area, and that the proposed cooperative infringed on its franchise. In other cases, utilities quickly ran a line through the locale, picking off the most promising customers. These "spite lines" often made the cooperative economically impractical and stranded most area farms without electricity.

By 1959, however, the REA had extended loans at 2 percent interest (which by then exceeded the government's cost of money) to 1,026 rural cooperatives serving 4.3 million consumers, then 95.9 percent of American farms enjoyed electric service. To the surprise of those who doubted that farmers would constitute a profitable market for electricity, the TVA proved that a combination of cheap power and promotional campaigns could indeed stimulate rural demand. Many farms sported not only the electric lights, stoves, and refrigerators common in urban homes but also electric milking machines, milk coolers, incubators, brooders, grinders, hoists, irrigation pumps, hay dryers, and other devices. Electrification initiated complex changes in family structure and the nature of work, but almost all affected applauded its arrival.

Midcentury critics of the REA denounced it as unwarranted government interference in the market and even as "creeping socialism." Its defenders retorted that it had succeeded in extending electric service when private utilities refused, and they noted that most cooperatives had faithfully repaid their government loans, making the process cost-free to taxpayers. The REA still exists in the 1990s, with its mission expanded to include rural telephone cooperatives. Modern critics note that the interest on REA loans (set at 5 percent in 1973, although "hardship cases" still qualified for 2 percent loans), while cheaper than commercially available loans, are also well below the government's cost of money. Since all co-ops in areas once classified as "rural" qualify for REA loans, regardless of their actual financial need, such critics charge that the agency now often uses tax dollars to subsidize the operations of large, well-capitalized firms. Although the REA did good work once, the argument goes, the original socioeconomic distinction between rural and urban has lost its pith, and the agency persists only through effective political lobbying.—A.H.S. and R.F.H.

See also HYDROELECTRIC POWER; TENNESSEE VALLEY AUTHORITY

Further Reading: David Nye, *Electrifying America: Social Meanings of a New Technology 1880–1940*, 1990.

Electrocardiogram

An electrocardiogram (often abbreviated as ECG or EKG) is a record of voltage changes produced by heart muscle contractions. The machine that is used to obtain this record is known as an *electrocardiograph*. The history of EKGs goes back more than 100 years. In 1887, Augustus Waller, a British physician, used an electrometer to record electrical currents produced by heartbeats. The electrometer used by Waller was delicate and slow to respond, and was rarely used in clinical practice. A more practical electrocardiograph was devised during the first years of the 20th century by Willem Einthoven (1860–1927), a Dutch physiologist. The 275-kg (600-lb) unit consisted of a quartz filament coated with silver, the ends of which were placed between the poles of an electromagnet. Wire leads at each end of the filament ran to corresponding liquid-filled electrode jars. When the patient immersed his or her arms in the jar, the electric current produced by the heart caused a deviation of the filament. Its slight movement was then recorded as the filament cast its shadow on a moving photographic plate. By the late 1920s, the weight of these devices had been brought down to 14 kg (30 lb). Electrocardiographs were also improved by using electronic amplifiers to increase the signal received.

Today, the usual procedure for obtaining an EKG is to attach electrical leads to the patient's extremities and chest. One end of each lead is placed on the surface of the skin, using an electrolytic compound to increase electrical conductivity; the other lead goes to the electrocardiograph. The EKG that is obtained can then be scrutinized for signs of abnormality. By examining the height, form, and duration of the wave patterns that appear on the electrocardiogram, a trained physician can diagnose specific heart disorders, such as myocardial infarcts (regions where the heart muscle has died) or decreased oxygen supply to the heart. Some EKGs are obtained while the patient runs or walks on a treadmill; this allows the discovery of abnormalities that occur when the body is under physical stress.

The development of solid-state electronics has paved the way for the miniaturization of electrocardiographs, which in turn has increased their value by making it possible for them to be used in a variety of circumstances. Some electrocardiographs can be worn and monitored continuously over a 24-hour period in order to give warning of an imminent heart attack. It is also possible to transmit electrocardiogram data through telephone lines to a central station, allowing an expert cardiologist to monitor a patient who may be thousands of miles away.

The success of the electrocardiograph has been dependent on the development of clear standards for what constitutes normality or abnormality in an EKG. Early electrocardiographs were of limited use until a number of autopsies could be performed in order to determine clear connections between the anatomical record and the tracings produced by the machines. Even today, an EKG record may contain considerable ambiguity. Although a skilled cardiologist can learn much about the condition of a patient's heart by scrutinizing them, EKGs are best used in conjunction with other diagnostic procedures.

See also AMPLIFIER; ELECTROMAGNET; SEMICONDUCTOR

Further Reading: Stanley Joel Reiser, *Medicine and the Reign of Technology*, 1978.

Electroencephalogram

An electroencephalogram (EEG) is a recording of the electrical activity of the brain. The process of making EEGs is known as *electroencephalography*. Electrical activity in the brain was discovered by the German physician Joannes Berger (1873–1941) in the 1920s. Berger made his discovery through the use of a string galvanometer (a device for the measurement of tiny amounts of electrical current), which was invented by Willem Einthoven and subsequently used by him for recording the first electrocardiograms. After placing electrodes on the skull of his young son and hooking them up to the galvanometer, Berger was able to detect two dominant signs of electrical activity, which he called *alpha waves* and *beta waves*. Berger's findings drew little attention, until a decade later two British neurophysiologists, Edgar Adrian (1889–1977) and B. H. C. Matthews, used electroencephalography in their own research.

During the late 1930s, electroencephalograms began to be used for the diagnosis of seizures. Their value was further demonstrated in the years that followed, when EEGs were used for the diagnosis of other brain disorders. Electroencephalograms are also widely used for the study of normal brain functioning. For example, a great deal of sleep research has been based on electroencephalography. The study of the resulting EEGs has shown that sleep is not an even process but one marked by many changes in brain functioning.

In identifying specific brain waves, electroencephalography still makes use of Berger's original terminology. Alpha rhythm is defined as waves with a frequency of 8 to 12 Hz. Beta rhythms are faster, having a frequency of 14 to 30 Hz. In addition, there are theta waves (4–7 Hz), which are found during drowsiness and light sleep, and delta waves (0.5–4 Hz) that are present during deep sleep. Delta waves are also predominant when a person is in a comatose state. A set of "flat" brain waves is indicative of brain death and is often used as the legal definition of death itself.

The use of computers and color graphic displays in conjunction with electroencephalography has made it possible to map the distribution of brain wave frequencies in particular areas of the brain. This has been

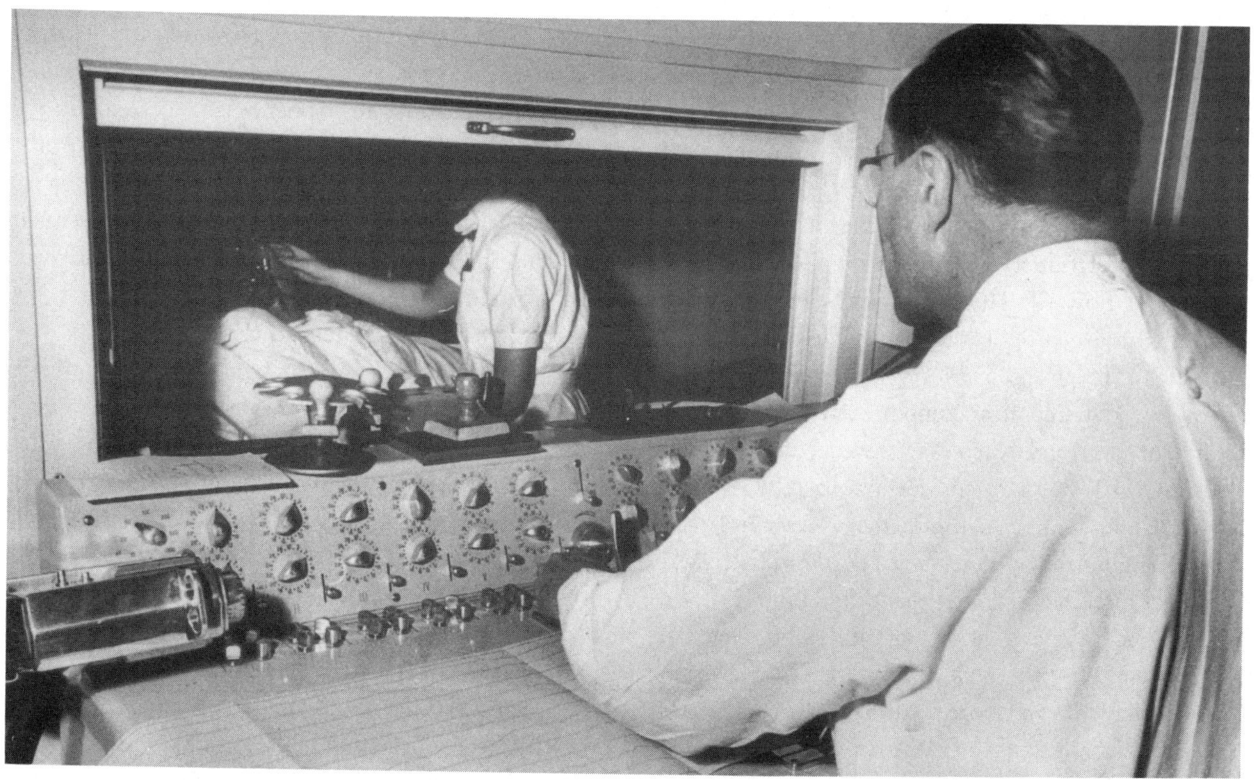

Monitoring an electroencephalograph (courtesy National Library of Medicine).

very useful for the diagnosis of psychiatric disorders, learning difficulties, and dementia. At the same time, however, electroencephalograms are not perfect diagnostic tools; some epileptics, for example, show normal EEGs after many trials. The presence of Alzheimer's disease, which causes severe brain malfunctions in many older people, cannot be positively identified by reading EEGs. In fact, positive proof of Alzheimer's disease can be obtained only through a direct examination of brain tissue.

See also BRAIN DEATH; ELECTROCARDIOGRAM

Electrolysis

Alessandro Volta's (1745–1827) publication in 1800 of the design of the first battery opened up many new possibilities for scientific research. Shortly after Volta's article reached England, Anthony Carlisle (1768–1840) and William Nicholson (1753–1815) constructed their own battery and used it to decompose water into hydrogen and oxygen. In 1807, another British Scientist, Humphrey Davy (1778–1829), passed an electrical current through fused potash (potassium carbonate, K_2CO_3) and fused soda (sodium carbonate, Na_2CO_3). The result was the separation of metallic potassium and sodium, the first isolation of these elements. A year later, he used a more complicated procedure to produce the first samples of elemental calcium, barium, strontium, and magnesium, although at the time he was not certain that they were elements.

In 1833, Davy's former assistant, Michael Faraday (1791–1867), pursued more experiments in decomposition by electricity, giving the name *electrolysis* to the process. In conjunction with William Whewell (1794–1866), he coined the other key words used to denote the major components of electrolytic processes: electrode (a conductor through which an electrical current enters or leaves), electrolyte (a substance or solution through which the current passes), anode (a positively charged electrode), and cathode (a negatively charged electrode). They also were responsible for the word *ion*, although their conception differed from the modern one.

Through his experiments, Faraday showed that the amount of each material released was proportional to the total quantity of electricity that had been used (that is, to the product of the current and the time during which it flowed). The quantity of each substance released was also proportional to its chemical equivalent weight. These discoveries helped to advance the general idea that chemical compounds were formed as a result of electrical reactions between different elements. In 1881, Herman von Helmholtz (1821–1894) recognized, as Faraday had not, that the second relationship implies the atomic character of electricity. As Helmholtz hypothesized, if elements and molecules are distinguished by their atomic or molecular weights, and if equal amounts of electricity liberate masses of material that are proportional to their molecular weights, a given amount of electricity must be associated with each particle of the dissociated material.

In addition to its importance to the advancement of scientific knowledge, electrolysis is the basis for a number of important industrial processes, such as the extraction of chlorine and alkali from brine, the purification of metals, and the electroplating of metals. In electroplating, the object to be plated is made the cathode and is bathed in a metallic salt solution; when current is passed through the solution, the metal collects in a film on the surface of the cathode. Should hydrogen become an important fuel in the future, it will be obtained through the large-scale electrolysis of water.—R.O.

See also BATTERY; CAPACITOR; ELECTROPLATING; HYDROGEN

Electromagnet

An electromagnet uses the flow of electrical current to produce a magnetic field. Since at least the 18th century, it was suspected that electricity and magnetism were related, as evidenced by Benjamin Franklin's discovery that sewing needles could be magnetized by an electrical discharge from a Leyden jar. In 1820, Hans Christian Oersted (1777–1851) made a fundamental discovery when he found that the needle of a compass could be moved by an electric current. This indicated that an electric current produces a magnet field. Further experiments into electromagnetic phenomena were conducted in France by André Marie Ampère (1775–1836) and Dominique Arago (1786–1853), and in Germany by Johann Schweigger (1179–1857). Schweigger's particular contribution, achieved a year after Oersted's discovery, was to produce a strong magnetic field by passing current through a coil of wire that had been wrapped 100 times around a frame. In England, William Sturgeon (1783–1850) built what is considered the first electromagnet, the description of which was first published in 1824. He formed an electromagnet by bending a foot-long piece of soft iron in the shape of a horseshoe, covered it with varnish for insulation, and then wound 16 turns of copper wire around it. When a voltaic cell passed a current through the wire, the electromagnet held a weight of 4.1 kg (9 lb). In the United States, Joseph Henry (1797–1878) was able to produce a more powerful electromagnet in 1828 by keeping the turns of wire at right angles to the bar around which they were wound. Since the wires now touched, they had to be insulated; this was accomplished by wrapping the wire with silk. Henry continued to work on electromagnets, and by 1832 he had constructed one capable of holding 1,587 kg (3,500 lb), a record that held for many years.

Subsequent electrical experiments and their application to useful devices made great use of the fact that an electric current induces a magnetic field. To take one important example, this principle underlies the operation of electric motors. Electromagnets themselves have found many applications. Their most mundane use is for picking up scrap iron, while at the other end of the spectrum, they are essential to the functioning of particle accelerators used for high-energy physics. In between, electromagnets find many commercial applications in solenoids, relays, and clutches.

See also BATTERY; CAPACITOR; MOTOR, ELECTRIC; PARTICLE ACCELERATOR

Electromagnetic Induction

Electromagnetic induction is the production of electricity from magnetism or, more specifically, the production of an electromotive force (emf) in a circuit as a result of a changing magnetic field around the circuit. The discovery of the electromagnetic inductive force made possible the invention of the generator, and it can be said without exaggeration that the generator's production of powerful electric currents initiated the modern electrical age.

In 1820, the Danish physicist Hans Christian Oersted (1777–1851) announced the discovery of electromagnetism, the production of a magnetic force around a current-carrying wire. This inspired scientists to look for the reciprocal effect, the production of an electric current by the application of magnetic force. Motivating them in this search was a sense of symmetry in nature and the idea that electricity and magnetism may be fundamentally the same force. A common experimental setup was to juxtapose a current-carrying wire (the primary) with another wire (the secondary) in which a second current was to be induced. This did not produce the hoped-for effect because these experiments used only steady currents—and therefore steady magnetic fields—that do not produce inductive effects. Induction requires a changing magnetic field.

The discovery of electromagnetic induction was made independently in 1831 by Michael Faraday (1791–1867), a British scientist at the Royal Institution in London, and Joseph Henry (1797–1878), an American who later became the first secretary of the Smithsonian Institution. Having published his results first, Faraday is credited with the discovery, although there is anecdotal evidence that Henry was the first to produce electromagnetic induction.

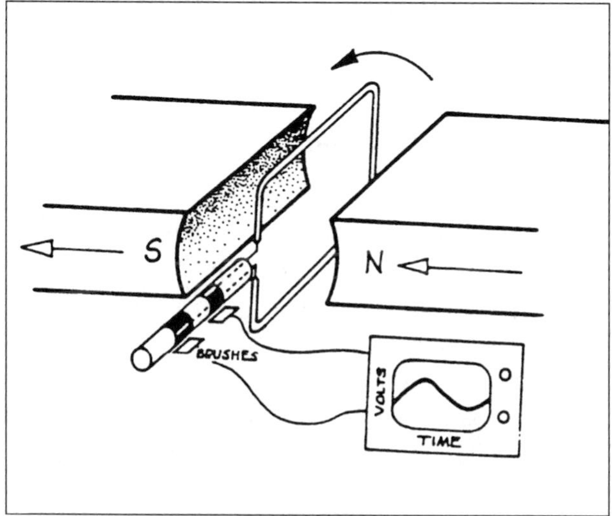

The principle of electromagnetic induction: An alternating electric current is produced when the wire coil rotates within a magnetic field (from Newton H. Copp and Andrew W. Zanella, *Discovery, Innovation and Risk,* 1993; used with permission).

After realizing that it was the change in magnetic force that produced the current, Faraday generated a current in a wire in two ways: (1) by changing the current in an adjacent wire, and (2) by moving a magnet relative to the wire. These two methods became the basis for the modern theory of electromagnetic induction. Both Faraday and Henry amplified the effect by winding their wires into coils, with and without an iron core. To explain the need for a changing magnetic force as well as the pattern of those forces, Faraday introduced his theory of the electromagnetic field. Faraday's field concept revolutionized the understanding of electricity and magnetism and eventually led to a general explanatory framework, not only for electricity and magnetism but also for gravity and light. In terms of field theory, we may say that a current is induced in a circuit whenever it is exposed to a changing magnetic field or flux, which is produced either by varying the intensity of the field or by relative motion of the field and the circuit.

The production of a current in one wire by another as described above is known as *mutual induction.* Joseph Henry is credited as sole discoverer of the closely related phenomenon of *self-induction*; a varying current in a wire generates a changing magnetic field, which in turn induces a voltage in the same wire. In recognition of his contributions, the unit of inductance was named the *henry*.

Working independently of Faraday and Henry and simultaneously duplicating many of their findings, the German-Russian investigator H. F. E. Lenz (1804–1865) arrived at a law for determining the direction of an induced current. Applying the principle of the conservation of energy, Lenz's law posited that the direction of an induced current is such as to oppose the cause producing it. For example, if the south pole of a magnet is moved toward a coil, the induced current in the coil will be in such a direction as to make the end of the coil nearest the magnet also a south pole, and the two

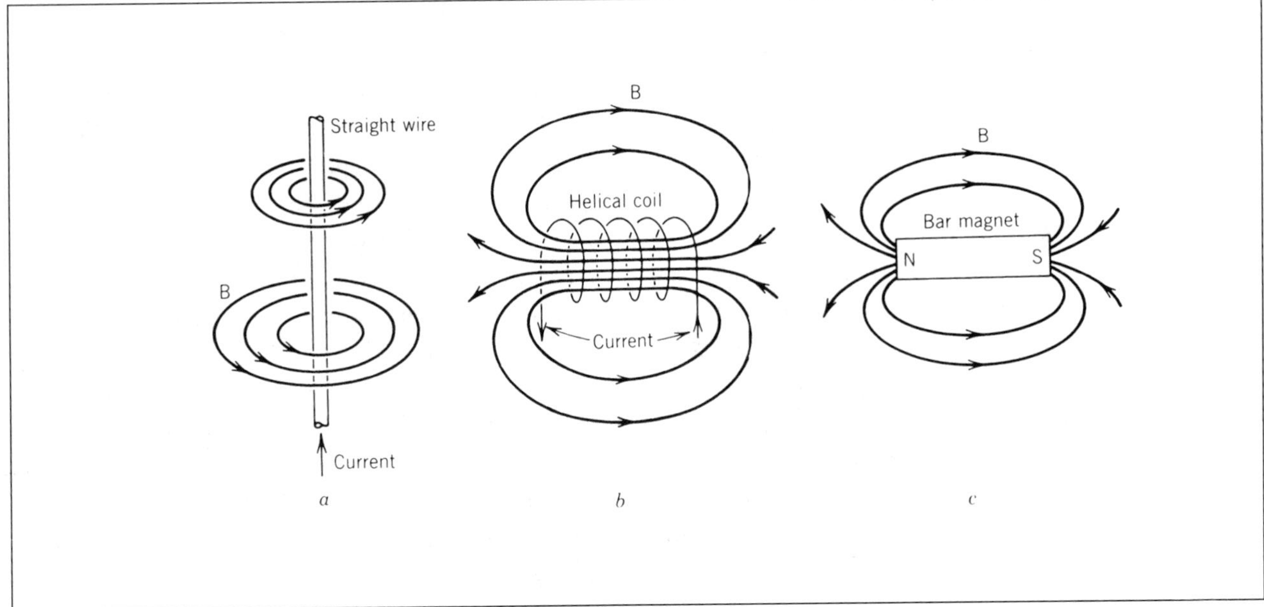

Magnetic fields surrounding (a) a straight wire, (b) a helical coil of wire, and (c) a bar magnet (copyright © N. Spielberg and B. Anderson, *Seven Ideas that Shook the Universe*, 1987; reprinted by permission John Wiley & Sons, Inc.).

south poles will repel each other.

In the case of mutual induction, i.e., when a changing current in a primary circuit induces a voltage in a neighboring circuit, the induced secondary emf, Σ, is proportional to the rate of change of the primary circuit:

$$\Sigma = M\, \Delta L / \Delta t$$

where M is a constant called mutual inductance, L is current, and t is time. If Σ is in volts and $\Delta L/\Delta t$ is amps/sec, M is in henrys.

For the case of a coil or solenoid, the magnitude of emf induced in a coil is proportional to the change of the magnetic flux, $\Delta\Phi/\Delta t$ linked with the coil, and to the number of turns, N, in the coil:

$$\Sigma = -N\, \Delta\Phi / \Delta t$$

The minus sign indicates that the induced emf opposes the cause that produces it (by Lenz's law). The emf is in volts if $\Delta\Phi/\Delta t$ is in webers/sec.

Although Faraday and Henry were interested in the pursuit of electrical research for its own sake, their discoveries in electromagnetic induction revolutionized technology, leading not only to the generator but also to such ubiquitous devices of the electrical era as the induction coil and the transformer.—A.M.

See also ALTERNATING CURRENT; ALTERNATORS AND GENERATORS; DIRECT CURRENT; ELECTROMAGNET; GRAVITY; ENERGY, CONSERVATION OF; TRANSFORMER

Further Reading: L. Pearce Williams, *Michael Faraday: A Biography*, 1965. Gerald Holton, *Introduction to Concepts and Theories in Physical Science*, 2d ed., 1973.

Electron

Toward the end of the 19th century, most physicists believed that all matter was made up of atoms that, by their very nature, were indivisible. That view was challenged in 1897, when J. J. Thomson's (1856–1940) experiments with cathode rays indicated the presence of subatomic particles that were deflected by electric and magnetic fields. By measuring the amount of the deflection, Thomson was able to determine the velocity as well as the charge-to-mass ratio of these particles (which he originally called *corpuscles*). Since the charge-to-mass ratio was much larger than that of an ion of hydrogen, the lightest atom, Thomson inferred that the particle either was much lighter than the smallest atom or that it had a much greater charge. Through an ingenious set of experiments, Thomson was able to determine that the corpuscle had the same charge as the ions found in salt solutions. It thus became evident that the particle was much smaller than the smallest atom, and hence that it was a component part of all atoms.

This particle soon acquired the name *electron*, a word first used in 1894 by George Johnstone Stoney (1826–1911) for the charge of individual ions in an electrolytic solution. It was the Dutch physicist Hendrik Antoon Lorentz (1853–1928) who first suggested that *electron* be used in place of corpuscles.

It is now generally agreed that an electron is a true elementary particle, i.e., one that is not composed of even smaller particles, as is the case with the other major constituents of atoms, protons and neutrons. An electron is much lighter than a proton or neutron; with a mass of 9.1×10^{-31} kg, it has 1/1,836 the mass of a proton. The electron's charge is the smallest ever detected in isolation, 1.6×10^{19} coulombs. At the same time, however, its attractive force to positively charged particles is much greater than gravitational attraction.

One useful model of the atom has its electrons arrayed around the nucleus in a series of concentric orbits known as *shells*. As first hypothesized by Niels Bohr (1885–1962) in 1913, as an electron moves from an outer shell to an inner one, it gives off energy in the form of a photon. When it moves from an inner to an outer shell, it absorbs energy. Electrons in the outer shells of atoms also are the basis of the chemical bonds that unite atoms to make molecules. Molecules form when the outer electrons of two or more atoms share a common shell.

The use of the term *orbit* to describe the movement of electrons around an atom's nucleus implies that electrons move around the nucleus like planets orbiting the sun. This is adequate as a first approximation, but the reality is considerably more complex. In fact, in regard to its motion, an electron is as much a wave as it is a particle, a theoretical conceptualization first postulated by the French physicist, Louis DeBroglie (1892–1987). The mathematics necessary for an adequate understanding of this wave/particle duality was first done by Paul A. M. Dirac (1902–1984) and by Erwin Schroedinger (1887–1961). Further complicating matters is the Heisenberg Uncertainty Principle, which posits that the path of a single electron moving around a nucleus is unknowable. However, the average paths of large numbers of electrons can be described with a high degree of accuracy through the use of statistical quantum mechanics.

Dirac's mathematical description of the electron pointed to the existence of positively charged electrons, or positrons. In 1932, Carl Anderson (1905–1991) was able to confirm their existence by examining tracks made in a cloud chamber. Physicists generally believe that electrons and positrons emerged in the first moments following the Big Bang that created our universe. When an electron encounters a positron, the two annihilate each other, and both are converted into energy. Electrons may also disappear through a process known as *electron capture*, whereby an electron combines with a proton to form a neutron in the nucleus of an atom. Conversely, electrons are created when the atomic nuclei of some atoms decay. In this circumstance, the liberated electron is also known as a *beta particle*.

Although interest in the electron and its properties was originally confined to physicists, it soon became apparent that electrons could form the basis of many practical devices. The discovery of the electron put electrical technology on a firmer scientific basis, making it evident that electricity was not a "subtle fluid," but the flow of electrons through a conductor. Beginning with the thermoionic tube early in the 20th century, the manipulation of electrons was turned to commercial advantage. Today, the scientific discoveries that began at the end of the 19th century are the basis of the vast field known as *electronics*.

See also atomic theory; antimatter; beta particle; Big Bang theory; cathode-ray tube; chemical bonding; cloud chamber; gravity; Heisenberg Uncertainty Principle; proton; neutron; quantum theory and quantum mechanics; thermoionic (vacuum) tube

Electronic Mail (e-mail)

Electronic mail, or e-mail for short, is a computer-based system for the transmission and receipt of messages. In the early days of computing, messages were sent to and from people who worked at the same mainframe computer. In the 1970s, electronic messages began to go beyond individual computers when some research laboratories, defense contractors, and military staff were linked together by a government-sponsored computer network known as ARPAnet. ARPAnet (which stands for Advanced Research Projects Agency Network) was originally intended to serve as a means of transferring data, but it was soon realized that ordinary communications also could be sent over it.

Messages and data could be electronically transmitted because in the 1960s computer specialists had created *time sharing*. This gave computers the ability to handle outputs and inputs to and from more than one user at a time. Due to time sharing, the senders or recipients of electronic-mail messages did not have to wait for their turn to use the computer. The computer could accommodate several users at once, making the transmission of electronic mail virtually instantaneous. In 1976, the first commercial e-mail service, OnTyme, was established. It achieved only limited success because few firms and individuals had computers. Within a few years, however, the diffusion of personal computers greatly expanded the number of people who could tie into a network that supported e-mail. In 1984, the development a standard for exchanging e-mail, the Simple Mail Transfer Protocol, ensured that all systems would be able to talk to one another. As computer networks rapidly expanded in the 1990s, large numbers of people gained access to e-mail. By the mid-1990s, an estimated 30 to 40 million people in more than 160 countries were able to send and receive electronic messages.

Unlike regular mail (which is derisively dubbed "snail mail" by e-mail enthusiasts), e-mail is not routed through some kind of electronic "post office." As is the case with the Internet as a whole, electronic-mail messages do not travel along a set route, and they are not routed by centralized distribution facilities. This in part is a reflection of the culture of e-mail's creators. Computer networks are creations of people known for their independence and antiauthority spirit; today's e-mail technology reflects their successful effort to create a communications system impervious to central control.

Communicating via e-mail has many advantages. One of them is speed; it takes only a few seconds for an e-mail message to reach a computer located on the Internet backbone, and a few hours for it to reach a satellite node. An e-mail message can be sent to a thousand people as easily as it can be sent to a single party. The software that supports e-mail usually includes an automatic mailing list creator. Once a list is made up, a keystroke or the click of a mouse can send a message to every address on the list. This is not an unmixed blessing, however. The ease of sending messages to large numbers of people has resulted in electronic mailboxes being filled by messages of little or no interest, and even the emergence of annoying "junk e-mail." To forestall this problem, many organizations employ e-mail systems that do not allow a message to be sent to large groups of people by means of a single command.

An advantage of e-mail is that it is quite easy to reply to a message: Once the response has been typed up, it too can be sent with a keystroke or the click of a mouse. Many users of e-mail find that they tend to reply to e-mail messages more quickly than they do to regular letters. This is due to the relative ease of responding via e-mail; eliminated is the need to print out the response, look up an address, write it on an envelope, affix a stamp, and then take the letter to a mail depository.

The most common use for e-mail is the sending of letters and other verbal information, but e-mail is capable of much more. E-mail can be used for the transmission of anything that can be digitized, i.e., coded in binary digits. Raw data, mathematical formulas, engineering drawings, pictures, and even sound can all be sent via e-mail. However, users of e-mail should know that where privacy is concerned, e-mail messages are more like postcards than letters. Everything that is sent or received through e-mail may be read by a third party, sometimes with unfortunate consequences for the sender or recipient. More than one employee has been fired because his or her boss intercepted an e-mail message that the employee thought would remain private.

The speed and versatility of e-mail make it a formidable competitor for traditional communications systems. Conventional postal delivery is particularly vulnerable. From 1988 to 1994, the volume of mail delivered in the United States went up by 5 percent, but business-to-business mail dropped 33 percent due to fax machines, electronic funds transfers to financial institutions, and e-mail.

E-mail has provided many benefits for its users; seasoned users of e-mail sometimes wonder how they ever got along without it. But as is the case with information technologies in general, not everyone has benefited to the same extent. The rapid development of e-mail has been paralleled by the polarization of society into electronic "haves" and "have nots." Given the importance of the ability to communicate rapidly and to gain access to information, being cut off from e-mail will be a severe disadvantage. For this reason, one Rand Corporation study included a recommendation that the U.S. federal government provide as much as $1 billion to ensure that everyone in America has access to e-mail.

See also BINARY DIGIT; BYTE; FAX MACHINE; COMPUTER NETWORK; INTERNET; MAINFRAME COMPUTER; MOUSE; PERSONAL COMPUTER

Elevator

At the end of the first quarter of the 19th century, elevators were beginning to increase the profitability of buildings through the vertical transportation of goods and, eventually, people. Hoists, based on pulley systems, had been in use since ancient times for lifting heavy loads, in mining and quarrying, and for loading ships. Men, animals, and waterwheels powered these early hoists. The use of steam power directed attention towards steam-assisted vertical transport. Indeed, the modern history of the elevator begins with the addition of steam power to hoisting systems. In the 1830s, in mills, factories, and other commercial buildings, power from steam engines was transmitted to platform hoists through a system of belts. However, the absence of safety braking mechanisms, combined with the tendency for ropes and cables to stretch and break, led to deadly accidents that occurred with enough frequency to preclude the use of these hoists for passenger travel.

At New York City's 1853 Crystal Palace Exhibition, a celebration of American technological ingenuity, Elisha Graves Otis (1811–1861) introduced a new safety device for hoists and elevators. To demonstrate its reliability, each day Otis stood high on a hoisted platform and had the supporting ropes cut, much to the horror of the fair's onlookers. Otis's invention won the attention of factory owners, especially after Otis developed a compact steam engine that allowed the elevator's location to become independent of a factory's main driveshaft and belts. In 1857, Otis installed what was most likely the first public passenger elevator in the elegant five-story E. V. Haughwout store in New York City. The elevator enabled the store's wealthy clients to shop without having to climb the stairs from floor to floor.

By the 1850s, the palace style of commercial architecture predominated in New York City and other large American urban centers, resulting in large, block buildings of seven or eight stories. Otis Tufts, a Boston mechanical engineer, introduced the vertical screw elevator into the Fifth Avenue Hotel (New York, 1859) and the Continental Hotel (Philadelphia, 1861). This elevator, which consisted of an ornate passenger car centered on a steam-powered iron screw extending through the height of the building, eliminated the threat posed by suspensions ropes and pulleys. Even though it was safer, it proved to be slow, inefficient, expensive to operate, and difficult to repair. And despite the great height of these buildings, proprietors featured the elevator more as a luxurious appointment rather than as a means of conveyance.

This changed when architects for the Equitable Life Assurance Building (New York City, 1868–1870) designed a structure 40 m (130 ft) tall that depended on an elevator to make the upper stories rentable, thus increasing the value of the land and the building's profitability. From this point on, elevator design advanced rapidly. Structural building techniques developed through the 1880s and 1890s that resulted in the steel-framed skyscraper, a building that utterly depended on fast, reliable, and safe elevators to function properly.

Several different types of elevators were developed during this time. The water-balance elevator, similar to systems used in funicular railways, employed a water-filled steel container to counterbalance the weight of the elevator car. A steam-powered pump filled and emptied the container as the elevator car rose or descended. Drum elevators were secured by a cable equal in length to the distance the elevator needed to travel. The elevator car rose or fell as the cable wound itself around a large drum turned by a steam engine located under the elevator shaft. This system proved impractical in taller buildings because of the enormous size needed for the drums. The hydraulic elevator came into use in the later 1870s. This elevator required a parallel shaft to house a vertical cylinder containing a piston, or room in the basement for a horizontal cylinder. Rods attached to the piston passed through the ends of the cylinder and were connected to a pulley system that magnified the movement of the elevator car. Water pressure, either from city water mains or steam engines, acted on water enclosed in the cylinder, moving the piston and the car with it.

Beginning in the 1880s, engineers began to experiment with elevator designs that used electric motors; these were mostly applied to systems similar to the ones using steam power. At the end of the 19th century, electrically powered systems did not yet offer clear advantages over hydraulic systems. However, in 1903 the Otis Elevator Company introduced traction elevators that completely eliminated complicated pulley systems, enabled the motors to be mounted at the top of the shaft, and often utilized electric power purchased from central plants, thus clearing out the huge amount of basement space formerly devoted to elevator equipment. By 1906, nearly 90 percent of new elevator installations were of this type.

The history of the elevator is entwined with that of the skyscraper and the increasing density of urban centers. As skyscrapers became complete cities within cities, vertical transportation became analogous to surface transportation links like railroads and boulevards. The engineer challenged the architect as the ranking principal designer. The speed at which one could travel the vertical distance of a tall building became the determinant of profitability. And as the height of buildings surpassed the human capacity for stair climbing, the elevator took control of one's passage through time and space.—M.W.B.

See also MOTOR, ELECTRIC; PULLEYS; SKYSCRAPERS; STEAM ENGINE

Further Reading: Cecil D. Elliott, *Technics and Architecture, The Development of Materials and Systems for Buildings*, 1992.

Emissions Controls, Automotive

Automotive emissions are a primary source of air pollution in general, and photochemical smog in particular. The internal combustion engines that power virtually all automobiles emit water vapor, carbon dioxide, carbon monoxide, unburned hydrocarbons, and oxides of nitrogen (NO_X) in the course of their operation. In addition to its exhaust emissions, an automobile with no pollution equipment also vents pollutants from its crankcase and gasoline tank. An effective pollution-reduction program has to address emissions from all of these sources, but tailpipe emissions are of the greatest importance.

The control of automotive emissions had become a major issue by the 1960s, when the air quality in some parts of the United States, most notably the Los Angeles basin, had become intolerable. As part of a general effort to reduce air pollution, the state of California initiated a number of laws and programs to reduce automotive emissions. The first of these mandated the installation of positive crankcase ventilation (PCV) valves on all 1963 model cars in order to prevent the escape of hydrocarbons from engine crankcases. The state then set limits on exhaust emissions for all new cars, beginning with the 1966 model year. By the time this had gone into effect, the federal government also was involved with automotive emission control. The first piece of legislation directly targeting emissions, the Motor Vehicle Air Pollution and Control Act of 1965, simply allowed the secretary of Health, Education, and Welfare (HEW) to set emissions standards, but later in the same year Congress required HEW to do so. This resulted in the setting of emissions standards that replicated those already established in California. Beginning with the 1968 model year, federal standards set hydrocarbon emissions at no more than 275 parts per million (ppm), and put the acceptable level of carbon monoxide at 1.5 percent of total emissions. By the 1970 model year, these had to drop to 180 ppm and 1.0 percent, respectively. The Clean Air Act, passed in 1970, set lower limits, and amendments to the act that were passed in 1977 lowered them even more. Another set of amendments passed in 1990 gave the automobile industry its current emissions standards: Beginning with the 1994 model year, carbon monoxide has been limited to 3.4 grams per mile, hydrocarbons to .25 grams per mile, and NO_X to .4 grams per mile. Also included is a requirement that emissions controls have to perform acceptably for 10 years after a car is manufactured.

The reduction of automotive emissions has required the development of a number of new technologies: fuel injection, computerized engine management systems, exhaust gas recirculation (EGR) devices, and catalytic converters. Emissions also have been lowered through the use of reformulated gasoline, while the use of catalytic converters made it necessary to remove the lead from gasoline. The result of all these technological fixes has been an impressive reduction in automotive emissions compared to cars produced in 1960: a 96 percent decrease in emissions of carbon monoxide and hydrocarbons, and a 76 percent decrease in oxides of nitrogen.

Emissions-control devices have made a major contribution to cleaner air, but their effectiveness depends on their working properly. As was noted above, federal law required that antipollution devices had to maintain their effectiveness for 10 years, but this mandate is of value only if there is some way to ensure that cars remain in compliance. Although emissions standards are set by the federal government, enforcement of these standards is the responsibility of individual states. The federal government, however, is able to retain some control over the process by threatening to withhold federal highway funds from states deemed to have inadequate testing procedures.

The 1990 amendments to the Clean Air Act required that the testing procedures had to simulate actual driving conditions. The actual place where the procedure was performed was left up to the individual states. The federal government has favored a network of central facilities that do nothing but emissions testing, but this has been strongly resisted by owners of gas stations and repair shops who perform emissions testing as an adjunct to their other operations. The problem with this arrangement is that a large number of facilities makes it difficult to inspect the equipment and personnel performing the tests. There is also an inherent conflict of interest when enterprises are in the business of making repairs in addition to conducting emissions tests.

Disagreements concerning the nature of the testing facility may eventually be rendered moot by the development of procedures that test emissions while cars are actually on the road. The most promising of these is the remote sensing system developed by Donald Stedman, a chemistry professor at the University of Denver. Stedman's remote sensing device uses an infrared light that is beamed across a roadway. The

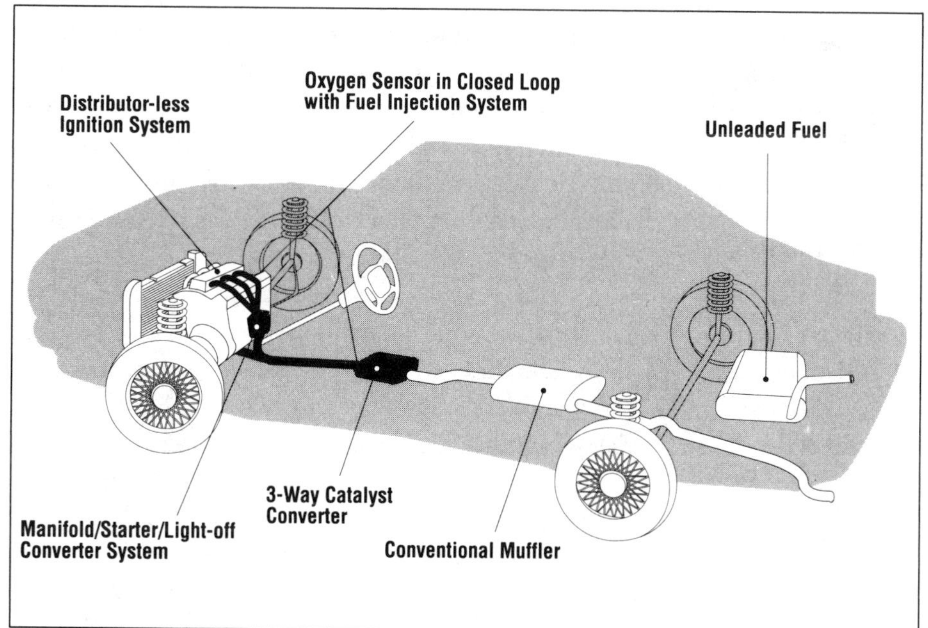

The key elements of an automobile's emission-control system (courtesy Manufacturers of Emission Controls Association).

beam passes through the exhaust of a car traveling on the roadway and strikes a computerized detector located on the other side. This provides an instantaneous measurement of the concentrations of carbon monoxide, carbon dioxide, and hydrocarbon in the exhaust. Equipment for the measurement of NO_X is still under development.

Remote sensing can be used to test the tailpipe emissions of more than 1,000 cars in an hour, at a cost of about 50 cents a car. It is not as precise as conventional testing procedures, but this may not be much of a drawback, for there is substantial evidence that a relatively small number of cars are responsible for a disproportionate share of automotive emissions. A study conducted by the National Research Council reported in 1991 that 50 percent of the ozone-forming emissions from mobile sources are produced by fewer than 10 percent of the vehicles in operation. Getting these "gross polluters" into compliance (or, if this is not possible, off the road) may be the most cost-effective way of reducing automotive emissions.

Remote sensing of tailpipe emissions has not met with universal acclaim. Questions remain about the accuracy and reliability of remote sensing. For example, it does not distinguish between a fully warmed-up engine and one that has just started up—the period when emissions are at their peak. Concerns have also been voiced about the "Big Brother" aspect of remote sensing, since it entails taking pictures of license plates so the owners of high-emissions cars can be notified.

See also CATALYTIC CONVERTER; ENGINE, FOUR-STROKE; FUEL INJECTION; GASOLINE; INFRARED RADIATION; INTERSTATE HIGHWAY SYSTEM; NATIONAL SCIENCE FOUNDATION; OCTANE NUMBER; SMOG, PHOTOCHEMICAL

Empiricism and Scientific Method

In the 17th century, epochal breakthroughs in astronomy, mechanics, biology, and other fields cumulatively produced the Scientific Revolution. No less important than the discoveries themselves was the development of new methods of inquiry, for this period marked the decline of Aristotelian science and the ascendance of modern empiricism. For the former, the method of science was clear: Knowledge of the world was based on deduction from higher principles. In contrast, empiricism taught that knowledge comes out of experience, and that we can only experience particulars not principles. Principles might exist, but they had to be derived through induction, i.e., by accumulating knowledge of particulars, and eventually arriving at a general principle or set of principles.

Since the postrevolutionary history of science seems to have demonstrated a great deal of success in producing reliable information about both principles and particulars, modern science seems to be an effective inductive tool. But why should this be so? The traditional assumption has been that modern science is in possession of a distinctive and highly effective method of advancing our understanding of the world. If this method is properly followed, reliable knowledge will be gained regarding fundamental principles, and will provide an understanding of particular events that have not been observed in experience.

Beginning with Frances Bacon (1561–1626) in the 16th century, various scientists and philosophers tried to capture the essential logic of the scientific method of induction; however, all attempts ultimately have failed to reveal a hard-and-fast method of obtaining general physical principles from particular experiences and observations of nature. The last powerful attempt was put forward by the Logical Positivists from the 1920s onward, but the problem has not been solved. Indeed, today, the strongest methodological position in empirical science makes heavy use of the calculus of probabilities. While observations of experienced particulars cannot deliver knowledge of general principles with complete certainty, most will agree that it is possible to calculate probabilities for realizing statements of general principles, and that this offers a reliable means of differentiating statements that we should trust from those that we should not.

Although it does not accurately characterize the nature of scientific inquiry, many introductory texts in the sciences still cling to a Baconian version of scientific method. According to this rather naive notion of empiricism, all science proceeds by collecting the relevant observations, sifting through them to discover generalizations, and concluding with the statement of laws of nature. Centuries of philosophical criticism, however, have made it clear that we rarely know what is truly relevant until much later, that observation is a complex process always involving selection and exclusion, that generalizations are suggested more often than they are discovered, and that there are no laws of nature of this kind. In reality, we go to nature more often to support our existing beliefs and designs than to discover virgin ideas.

What, then, is our best account of the success that science seems to have had? In recent decades, largely stimulated by Thomas Kuhn's book *The Structure of Scientific Revolutions* (1962), various philosophers have attempted to uncover the reasons for this success through a consideration of social factors, in particular the ways in which institutions direct scientific activities. Kuhn's explanation suggested that scientific disciplines are alternately dominated by periods of radicalism and conservatism. In the era of conservatism, the field is governed by a "paradigm" that provides a common approach to specific problems within a field of study. The central paradigm (or set of paradigms) is defended by teaching, editing, funding, and reviewing policies that uphold the intellectual status quo. This gives the field the maximum time in which to develop the fullest appreciation of paradigm-related ideas and observations. It is, however, inevitable that the process of development will uncover both general and specific inadequacies of a given paradigmatic approach. A natural period of distress and defensiveness ensues, eventually giving way to the emergence of a new, radical point of view that promises new solutions to intellectual problems and rewards scientists with new successes. Thus, according to Kuhn's social theory of science, a science advances only after scientists have spent extensive periods of time gathering everything they can, all the while tolerating a certain amount of troublesome information that threatens overall coherence. Nevertheless, revolutionary progress is still allowed in the long run, and usually happens through the emergence of new, younger leaders who do not share the same commitments to the past.

What we see here is not "method" but "process." Moreover, it is a process that is strongly affected by external factors. To be sure, individuals pursue their activities empirically, critically, and methodically, but science itself moves socially and institutionally. We do not, in fact, throw a theory out because we find a counterexample; instead, we work around the problem so that we can develop the largest successful domain of the theory's application—hence the impression that scientific method is extremely successful. Still, it must always be kept in mind that achieving a scientific success is not the same thing as apprehending the truth.—T.B.

See also EXPERIMENT; HISTORIES OF SCIENCE, INTERNAL AND EXTERNAL; HYPOTHESIS; INDUCTIVE REASONING; MECHANICS, NEWTONIAN; PARADIGM; SCIENTIFIC METHODOLOGY; SCIENTIFIC THEORY

Enamel

In the paint industry, the term *enamel* denotes glossy pigmented varnishes that are highly resistant to abrasion, atmospheric corrosion, and many chemicals. Typically, automobiles and household appliances are painted with enamel. But the term also has an older definition, namely coatings that appear glossy or vitreous (glasslike) upon drying—either for purposes of utility or ornament. Such coatings have been applied to many materials, including porcelain and ceramics as well as metal. So-called "vitreous enamels" are composed chiefly of quartz, feldspar, clay, soda, and borax, with borates or saltpeter as a flux, to which color and opacity are imparted by mixing in metallic oxides. This mass is fused, reduced to a powder called *frit*, washed, applied, and heated in a furnace until the enamel melts and adheres.

Historically, enameling was an ancient art in both Asia and Europe, with focal points in England under Roman and Anglo-Saxon rule, Celtic Ireland, and Byzantium after the decline of the Roman Empire. In medieval times, enameling was part of the goldsmith's art, and it continued to thrive in Renaissance Italy and in France around Limoges. Among specific types of enamelware, the term *majolica* denotes a marblelike finish made by mixing enamels of different colors. Cloisonné is a type of enamel with colored cells separated by thin metal (traditionally, gold) strips. (Cloisonné is now imitated in plastic under such names as Enameloid.) A *japan finish* is the name given for black baking enamels, expensive to apply by hand but extremely durable.

The art of enameling declined after the 17th century, then revived at the beginning of the 19th. One persistent technical problem concerned the differential expansion rate of iron and enamel, which often resulted in chipping. During World War II, however, new vitreous enamels were developed for use on the exhaust systems of aircraft engines; these could withstand high temperatures without peeling or chipping. Vitreous enamels are now routinely applied to the insides of steel tanks and to road signs, cookware, and siding for buildings.—R.C.P.

See also PAINT; PORCELAIN

Endangered Species Act

A major environmental concern is the loss of entire species of animals and plants. Some ecologists believe that if present trends continue, as many as half the Earth's species could disappear by the end of the present century. In the United States, one response to this problem is the Endangered Species Act. Signed into law by President Richard Nixon in 1973, the law forbids the destruction of plants and animals designated as "endangered" or "threatened," as well as the habitats in which they live. The Endangered Species Act has contributed to the preservation of many species, but at the same time it has been bitterly opposed by some farmers, ranchers, loggers, and others whose livelihood it may affect. One of the most contentious issues centered on the protection of the habitat of the northern spotted owl from the effects of logging in the Pacific Northwest. Loggers challenged the act in the courts, claiming that it did not apply to private property, but in 1995 the U.S. Supreme Court upheld it by a 6 to 3 margin.

As of January 1993, more than 758 species in the United States have been officially designated as endangered or threatened. An endangered species is one that is determined to be in danger of extinction, while the threatened designation pertains to a species likely to be listed as endangered in the near future. The listing of a species can begin with petitions from individuals and organizations; final determination is the responsibility of the U.S. Fish and Wildlife Service, or the National Marine Fisheries Service for ocean species. On the list were 65 species of mammals, 86 species of birds, and 34 species of reptiles, as well as 44 species of clams, 19 species of snails, and 11 species of crustaceans.

With a large number of species designated as threatened or endangered, numerous conflicts have arisen between landowners and the federal government. Landowners whose activities may upset the habitat of a designated species must obtain a permit from the Fish and Game Service contingent on the submission of a plan that mitigates the adverse effects of these activities. Land held by the federal government and projects involving federal money also require prior consultations with the Fish and Game Service to ensure that the project does not jeopardize the survival of a species. In most cases an accommodation is reached; between 1987 and 1992 the Fish and Game Service conducted about 97,000 consultations concerning possible harm to endangered species, but only 54 of the proposed activities were either withdrawn or rejected.

At the same time, however, private landowners on occasion have had to bear considerable expenses while creating a conservation plan to protect a particular species; in some cases the process may take several years and the expenditure of hundreds of thousands of dollars. A fair amount of time may pass before the issue is resolved, during which time the land is not producing any income. As a result of these problems, some landowners have taken measures to rid their land of an endangered species before their presence is known to the outside world. For example, some farmers have kept in continuous cultivation formerly fallow land that had been the home of an endangered animal. In cases like this, the act has the paradoxical consequence of harming the habitats of the plants and animals it is supposed to protect.

As an alternative to dealing with matters on a case-by-case basis, in a few instances federal and state officials have worked in conjunction with private developers to set up regional sanctuaries for endangered plants and animals, thereby taking the pressure off private landholdings. Some economists have suggested the use of tradable credits similar to pollution credits. These would give credits to landowners who preserve or improve habitats. These credits could then be sold to developers and businesses whose activities lead to habitat modification or destruction. It has also been argued that the government should financially compensate landowners for losses incurred as a result of the act, but in a time of shrinking governmental budgets, this seems unlikely.

See also POLLUTION CHARGES AND CREDITS; SPECIES EXTINCTION

Energy Efficiency

In general, efficiency is the ratio of output to input. In regard to energy, efficiency is the ratio of energy output to the energy that went in. For example, an undershot waterwheel is turned by the kinetic energy of the running water that strikes its blades: If 10 percent of the water's kinetic energy emerges at the output shaft of the wheel, the waterwheel has an efficiency of 10 percent. Any energy conversion process, such as burning a fuel to produce mechanical motion, results in a loss of energy and a concomitant increase in heat; an energy efficiency of 100 percent is a physical impossibility.

Energy efficiency is not necessarily the same thing as economic efficiency, which relates energy output to the *cost* of the energy input. One kind of machine may use energy more efficiently than another one, but the energy it uses may be more expensive than the energy used by the less energy-efficient machine. Under these circumstances the machine that is less energy efficient but cheaper to run will likely be used.

Most of the energy generated and used in the world today is supplied by heat engines such as internal combustion engines and steam turbines. The most important measure of their efficiency is *thermal efficiency*, which is determined by how much of the energy contained in the fuels they burn is converted into usable energy. Over time, these engines reached progressively higher levels of thermal efficiency before reaching a plateau. The first Newcomen steam engines worked at about 0.5 percent efficiency; subsequent detail improvements in the 18th century moved the figure up to a bit over 1 percent. James Watt's invention of the separate condenser resulted in a large improvement in steam engines. Watt's early engines operated at about 2.7 per cent efficiency. By the second half of the 19th century, the introduction of high-pressure steam engines, along with compounding, the use of superheated steam, and other improvements, pushed the best stationary steam engines close to 25 percent efficiency. Steam locomotives, however, only achieved about half this level of efficiency.

The efficiency of steam engines is limited by the relatively low temperatures achievable with steam. A basic thermodynamic principle stipulates that the thermal efficiency of any kind of heat engine is determined by the following equation:

$$\frac{T_1 - T_2}{T_1}$$

where T_1 is the temperature at the beginning of a cycle, and T_2 is the temperature at the end of a cycle. Thus, all other things being equal, the higher temperature at the beginning, the greater the efficiency. This condition is better met by an internal combustion engine, with which hotter temperatures can be produced inside the combustion chamber than are possible with steam. Light, compact internal combustion engines, such as those found under the hood of a car, typically operate at close to 30 percent efficiency. A diesel engine is even better, operating at more than 30 percent efficiency.

It is also possible (and useful) to determine the energy efficiency of devices that use energy. An ordinary incandescent lightbulb converts only about 1.5 percent of its electrical input into light; the remainder is lost as heat. A fluorescent light is much better, operating at close to 20 percent efficiency.

As noted above, useful energy is lost every time one form of energy is converted into another form, with a consequent loss of efficiency. To take one important example, the production and transmission of electrical energy entails a number of conversion processes, each one of which entails some loss of energy: when a fossil fuel is burned to produce steam, when the steam is used to power a turbine, when the turbine runs an electrical generator, and when the electricity moves through high-tension lines. It is therefore important to analyze any energy-producing technology in terms of *net energy*: the usable energy that remains after all of the energy losses have been subtracted from the energy supplied by the original source. When this is done, many energy-producing processes turn out to be quite inefficient. The energy supplied by the food we eat is gained at the cost of very substantial energy inputs; on average, the production of one calorie of food energy requires nine calories of fossil fuel energy. The solar energy that reaches the Earth is free, but given present technologies the energy costs incurred in producing equipment to gather, convert, and transmit solar energy are probably greater than the amount of energy actually produced. The word *probably* has to be used here, for determining net energy is necessarily an inexact science. Calculating the net energy return of a particular technology always requires a fair number of assumptions concerning which energy inputs should be included and how to estimate their size.

See also CARNOT CYCLE; CONSERVATION OF ENERGY; ENGINE, DIESEL; ENGINE, FOUR-STROKE; ENTROPY; LIGHTS, FLUORESCENT; SEPARATE CONDENSER; STEAM ENGINE; TURBINE, STEAM; WATERWHEEL

Energy Intensity

Energy intensity measures the cost of producing something, not in terms of money but of energy. For example, aluminum is an extremely expensive material in terms of energy used, requiring 227 to 342 megajoules of energy for the production of a kilogram of aluminum from bauxite ore. By contrast, only 20 to 25 megajoules of energy are required to produce a kilogram of iron from raw ore. Largely for this reason, iron is used much more extensively than aluminum, even though deposits of aluminum ore are far more extensive than deposits of iron ore.

While energy is required for the production of goods and services, it is also necessary for the production of energy itself. Coal has to be dug out of the ground, hydroelectric power stations have to be built and maintained, and so on. Sources of energy differ in the energy costs they incur. Oil, especially when it is extracted from the rich fields of the Middle East, is exceedingly cheap in energy terms; extraction costs are.5 to 5 kilojoules per kilogram of oil. Coal is more energy intensive, requiring between 4 kilojoules and 4 megajoules per kilogram. At the extreme end of the scale, when viewed only as a source of energy the food we eat is quite expensive, needing on the order of nine units of fossil fuel energy for each unit of energy actually consumed.

As with single organisms, national economies cannot survive for long without regular infusions of energy, and growing organisms and economies may need more energy than static ones. Much of the economic and technological advance of previous centuries was the result of using more and more energy. Steam engines, internal combustion engines, and electric motors played central roles as economic growth and energy use increased in unison. Had this pattern continued, energy might have been the limiting factor in the pursuit of economic growth. But by the second decade of the 20th century, economic growth and energy consumption were "decoupled" as output grew without concomitant increases in energy usage. There are several reasons for this. First, energy was used more efficiently due to the introduction of new production technologies. Second, more energy-efficient sources of fuel were being used, as oil and natural gas began to replace wood and coal. Finally, the structure of the economy began to change, as economic activities increasingly centered on the provision of services rather than the production of industrial goods.

These trends were reinforced by the energy crises of the 1970s, which motivated a more efficient use of energy. As a result, from 1972 to 1985 the U.S. economy grew by 40 percent, yet the amount of energy consumption hardly increased, and energy intensity fell by approximately 28 percent. A modern economy still uses vast (and increasing) amounts of energy, but it gets more in return for the energy it uses.

See also ENERGY EFFICIENCY; FUELS, FOSSIL

Energy, Conservation of

Gaining a clear understanding of the nature of energy was one of the key contributions of 19th-century physics. The process through which this occurred entailed a combination of careful experimentation and the precise definition of key terms. As a first step, energy can be defined as the ability to do work. In this context, work is defined as the application of force over a distance. It is also important to differentiate two types of energy: kinetic energy and potential energy. Kinetic energy is the energy of a moving object, like that of a cannonball after it has been fired. In the formulation used by physicists today, $E = {}^1/_2\ mv^2$, where E equals energy, m equals the mass of the object, and v equals its velocity. Potential energy is the ability to acquire kinetic energy; a cannonball resting on the top of a wall has the potential to gain kinetic energy by falling off that wall and being accelerated by the force of gravity.

The world would be a strange place if energy manifested itself in only these two forms. Consider a ball located in a U-shaped channel. From its place at the top of one leg of the U, the ball would then roll downward, and the potential energy would be converted to kinetic energy. The kinetic energy would cause it to roll up the other leg of the U until it reached the top, whereupon all of its kinetic energy would have been converted to potential energy. It would then roll toward the bottom and then start the process all over again, beginning another cycle of a never-ending process. Such things don't happen in the real world for a simple reason: Some of the energy is converted to heat. As the ball moves, its contact with the surface it rides on produces friction, and more friction is generated as it moves through the air. Friction produces heat, and this heat represents the system's loss of kinetic and potential energy.

Although the conversion of kinetic and potential energy into heat is clearly understood today, many years passed before this concept was fully understood. Much of the impetus for the study of heat and energy came as a result of trying to understand and perfect the steam engine during the first half of the 19th century. One person who gave serious thought to the matter was a young French artillery officer named Nicholas Leonard Sadi Carnot (1796–1832). In thinking about what it would take to produce a perfectly efficient steam engine, Carnot determined that in an ideal gas engine, thermal efficiency is dependent on the initial temperature of the gas and the extent to which it "falls" to a lower temperature at the conclusion of its working cycle. Carnot came to the conclusion that the power of a steam engine was derived from heat, and that the steam was in effect a way of making use of that heat. When he first formulated his theory, Carnot operated under the assumption that heat was an actual substance known as *caloric fluid* or simply *caloric*. But before his premature death from cholera, he began to have doubts about the existence of caloric. He was not the only one; other scientists had to wonder how friction could produce the substance that gave heat. After all, a friction-producing process, such as rubbing one's hands together, generated heat even though there seemed to be no source of the heat-producing caloric fluid.

The ability of friction to produce heat was duly noted by Benjamin Thompson, Count Rumford (1753–1814), who in 1798 performed experiments with the boring of cannon barrels. Rumford noted that friction seemed to produce an inexhaustible supply of heat, for heat was produced as long as the cannon barrel was being bored. The connection between motion and the generation of heat was pursued further by James Joule (1818–1889) in the 1840s. In a classic scientific experiment, Joule used a paddlewheel turning inside a baffled, water-filled cylinder to show through careful and precise measurement that motion was a source of heat. Moreover, since energy was required to turn the paddlewheels, it could be inferred that work could produce heat. By arranging to have the paddlewheel turned by a falling weight, Joule was even able to determine exactly how much mechanical work was required to produce a given amount of heat.

In addition to demonstrating that heat was a form of energy, Joule was able to supply a fundamental insight into the nature of heat. It was not a substance like caloric; rather, it was the product of the motion of the water. This came to be known as the *kinetic theory of heat*, and it proved a powerful basis for explaining how heat was generated. It also went a long way towards explaining why the application of heat caused the volume of gases to expand, while the withdrawal of heat caused their volume to diminish.

At this point, the determination that heat is a form of energy provided a secure basis for the idea that energy is always conserved. Even so, it took a while for the concept to take hold. One of the first to directly state the idea was Julius Robert Mayer (1812–1878), a German ship's surgeon, who came to the idea in the course of thinking about the physiological process of respiration. He published a pamphlet outlining his ideas, but it drew little attention, and when it did, the response was to ridicule the idea. Far more successful was William Thomson, Lord Kelvin (1824–1907), whose stature as an eminent physicist gave him an attentive and respectful audience. Although he initially adhered to the caloric theory, Kelvin became a strong advocate of Joule, the kinetic theory of heat, and the concept that energy is conserved. At about this time, a 26-year-old German named Hermann von Helmholtz (1821–1894) also argued for the conservation of energy. In a paper delivered in 1847, Helmholtz put forth his belief that all forms of energy—heat, light, electricity, and so on—were the result of motion and therefore could be considered as kinetic energy. A change of energy from one form to another was simply the transference of motion, and it necessarily was conserved.

The idea that energy is conserved is now known as the First Law of Thermodynamics. One of the practical consequences of this law is that any effort to create a device that releases more energy than it receives is doomed to failure. However, there is nothing in the First Law that prevents the construction of a device that runs forever with no additional inputs of energy after it gets started. But this too is impossible because heat is a special form of energy, a fact that began to be grasped a few years after the concept of the conservation of energy was fully understood.

See also CALORIC; CARNOT CYCLE; ENERGY EFFICIENCY; ENTROPY; HEAT, KINETIC THEORY OF; PERPETUAL MOTION MACHINES; STEAM ENGINE

Energy, Measures of

As 19th-century physicists began to gain a clear understanding of the nature of energy, their advancing knowledge was reflected in the creation of a standard set of units pertaining to energy and associated physical phenomena.

In the late 18th century, before scientists turned their attention to energy, James Watt (1736–1819) was faced with the need to rate the power of the steam engines he built. The primary use for these early engines was pumping water out of coal mines, a task that hitherto had been done with horses. Quite naturally, Watt began to think of his engines in terms of the number of horses they could replace. In 1782, Watt ascertained that a typical working horse was capable of walking 3 feet (.9 m) per second while pulling 180 pounds (81.6 kg). From this, he was able to calculate that the power exerted by a horse was 32,400 foot-pounds per minute (a year later he rounded this figure up to 33,000 foot-pounds per minute). This figure became the definition of 1 horsepower (hp). In fact, most horses are not capable of this output over a sustained period. Watt may have deliberately overstated the power of a horse (and therefore underestimated the power of his steam engines relative to the power of a horse) in order to ensure that his customers were satisfied with the engines they leased from him.

Power, be it exerted by horses, steam engines, or other devices, is the rate at which energy is used. The standard unit of energy is the joule (abbreviated as J), named after James Prescott Joule (1818–1889), the British pioneer in the study of energy. Energy, in turn, is force exerted over a distance. The standard unit of force is the newton, named after the British physicist Isaac Newton (1642–1727), and abbreviated as N. One newton is defined as a force that accelerates a 1-kilogram mass at a rate of 1 meter per second squared ($1m/sec^2$). Since energy is defined as the exertion of force over a distance, 1 joule is a force of 1 newton applied over a distance of 1 meter. Another commonly used unit of energy is the erg, which is defined as a force of 1 dyne (a force able to accelerate 1 gram at a rate of 1 centimeter per second squared ($1cm/sec^2$) acting over a distance of 1 centimeter. One erg equals 10^{-7} J. Power, the rate at which energy is used, is given in watts (abbreviated as W). One watt is defined as 1 joule per second (1J/sec). It takes about 745.7 watts to equal 1 horsepower.

Before these units were adopted as standards, a number of other units were commonly used in science and technology. One of them is the British thermal unit, or Btu. It is defined as the amount of heat energy required to raise the temperature of 1 pound of water 1 degree Fahrenheit. (1°F). One Btu equals 1,055 joules. A similar heat-based definition of energy is the calorie (or calory), defined as the amount of heat energy required to raise the temperature of 1 cubic centimeter of water 1 degree centigrade (1°C). One calorie equals about 4.2 joules. Calories are a prime concern of people on diets, but what are really involved are kilocalories, where 1 kilocalorie equals 1,000 calories.

As is the case with calories and kilocalories, most units of energy and power can easily be expressed as multiples of 10 through the use of specified prefixes. Some of the most commonly used are as follows:

Number	Prefix	Symbol	Scientific notation
thousand	kilo	k	10^3
million	mega	M	10^6
billion	giga	G	10^9
trillion	tera	T	10^{12}

Some idea of energy output and energy consumption can be gained by looking at some specific examples. As was noted above, horses do not generate 1 horsepower, but human beings are considerably weaker; most people in good health are capable of a sustained output of between $\frac{1}{10}$ and $\frac{1}{5}$ of a horsepower. The typical car has an engine that puts out about 140 hp or 104 kW. A modern mainline diesel-electric locomotive produces 2,600 kW (3,487 hp). A large thermal power station has an output of up to 2 million kilowatts or 2,000 MW. The five Saturn rocket engines used for the moon program collectively put out 2,600 MW or 3,500,000,000 hp. For those not impressed by these figures, it can be noted that the world's annual consumption of commercially produced energy has been estimated to be 9,500,000 MW or 9.5 TW.

See also APOLLO PROJECT; CONSERVATION OF ENERGY; ENTROPY; STEAM ENGINE

Further Reading: Vaclav Smil, *Energy in World History*, 1994.

Engine, Diesel

A type of internal combustion engine, the diesel engine has been employed for more applications than any other power source. Diesel engines are used to power machinery and transportation equipment of every description: trucks, locomotives, farm tractors, ships, earth-moving equipment, oil rigs, electrical generators, refrigerators, air compressors, and more. The ubiquity of the diesel is the result of its ruggedness, economy, and inherent efficiency.

The diesel operates in a manner similar to other internal combustion engines, but there are some important differences. Unlike the spark-ignited four-stroke internal combustion engine, in a diesel engine the air-fuel mixture is not combusted by firing a spark at the proper time. Instead, the compression of air brought into the cylinder raises temperatures to 700° to 900°C (1,290°–1,650°F), causing the fuel to ignite after it has been injected into the combustion chamber. The air is compressed to such an extent because the diesel engines have compression ratios (the ratio of the volume of the cylinder when the piston is at the bottom of its stroke to the volume when it is at the top) of 14:1 to 25:1 (spark-ignited engines typically have compression ratios of around 9:1).

Ironically, this crucial aspect of the diesel engine's operating cycle was of minor importance to its inventor, while the scientific principle on which he based the design proved to be irrelevant. The inventor of the diesel engine was Rudolf Diesel (1858–1913), an engineer with a brilliant academic record and a thorough grounding in science and mathematics. One important benefit of his scientific training was his deep appreciation of the principles of energy conversion first outlined by Sadi Carnot. Diesel's goal was to create an engine that acted in accordance with the principles set down by Carnot. In an ideal Carnot cycle, combustion occurs isothermally, that is, with no increase in temperature as the piston moves away from the place where heat is applied. To build an engine with this feature seemed to be a practical impossibility, but Diesel thought it could be done by having the engine reach its highest temperature through the compression of air alone. Fuel would be injected at a controlled rate after the air had been compressed. After ignition of the air-fuel mixture, the movement of the piston would increase the volume of the cylinder, causing a fall in temperature. This temperature drop would cancel out the increase in temperature caused by the combustion of the fuel. The result would be isothermal combustion, fulfilling the requirements of the Carnot cycle. Unfortunately, the critics were right, and no real-world device could be built to Diesel's design. Not only were the required pressures beyond the capability of contemporary engineering practice, it was also the case that the work of compressing the air required much larger amounts of fuel than Diesel had envisaged. Accordingly, the combustion of the fuel resulted in a steep rise in temperature that obviated the isothermal principle.

While Diesel's efforts to build an engine based on the Carnot cycle were in vain, it was still advantageous to build an engine that operated at high pressures, for the greater the compression of the air-fuel mixture, the greater the amount of power delivered after ignition. But the realization of this potential advantage did not come easily. The most vexing problem lay in getting the right amount of fuel into the combustion chamber at the right moment. The problem was compounded by the use of powdered coal as a fuel. Switching to a liquid fuel helped, but many years were to pass before effective fuel-delivery mechanisms were designed and built.

Although Diesel patented an engine based on his ideas in 1892, it was no more than a paper invention. After a great deal of mental and physical exertion, a working one-cylinder engine was built and put on exhibition in 1897. With a thermal efficiency of 26 percent, it vindicated Diesel's efforts, but it was neither reliable nor economic. The first commercial application of a diesel engine came a year later, when a two-cylinder engine generating 60 hp was installed at St. Louis. This was a start, but a long time would elapse before diesel engines were employed in significant numbers. By the second decade of the 20th century, diesel engines were beginning to be used to propel ships, including most of the submarines used in World War I.

The years after World War I saw continued development of the diesel engine as a stationary power plant and as the power source for a variety of vehicles. The economical operation of the diesel engine was underscored in 1930 when one powered an automobile that was then driven from Indiana to New York City for $1.38 worth of fuel. In the late 1930s, diesel-powered locomotives, which had made their appearance in the mid-1920s, began to pose a serious threat to steam locomotives. Growing numbers of trucks were being equipped with diesel engines at this time, although diesel-powered automobiles did not appear in significant numbers until the 1950s. Diesel-powered cars enjoyed a brief vogue when the energy shocks of the 1970s put a premium on low fuel consumption, something diesels excelled at. By this time, many diesel engines had gained more power through the use of superchargers and turbo-superchargers, devices that first had been used with diesel aircraft engines in the 1930s. Supercharging is especially important to the operation of the two-stroke diesel engines that are widely used in commercial vehicles.

The diesel engine continues to be a popular power plant due to a long service life and the economical operation that comes from a thermal efficiency that can be as high as 40 percent (spark-ignited internal combustion engines manage a thermal efficiency of only 25–32 percent). Diesel engines also run more economically at lower speeds than spark-ignited engines because they bring more air into the combustion chamber, and this air absorbs a great deal of the heat generated by the combustion of fuel. This absorption of heat results in lower operating temperatures, and less heat rejection from the engine means that less of the fuel's energy is wasted as excess heat.

At the same time, however, diesel engines have some drawbacks. High internal pressures require strong construction, and this means more weight than a spark-ignited engine of similar power output. Diesels also tend to produce considerable noise and vibration. Finally, anyone who has found themselves behind a diesel-powered vehicle is aware of their sooty exhaust. Diesel exhaust contributes only moderately to photochemical smog, but the particulates spewed out the exhaust pipe are a serious pollution problem in many places.

See also CARNOT CYCLE; ENGINE, FOUR-STROKE; ENGINE, TWO-STROKE; FUEL INJECTION; LOCOMOTIVE, DIESEL; SMOG, PHOTOCHEMICAL; SUBMARINES; SUPERCHARGER; TRUCKS; TURBOCHARGER

Engine, Four-Stroke

A four-stroke internal combustion engine is a type of heat engine—that is, a device that uses heat to produce power. It is called an internal combustion engine because the heat is generated inside the engine itself; this distinguishes it from a steam engine, which generates heat in a separate boiler.

A conventional four-stroke engine consists of a piston moving with a reciprocating (up and down)

motion inside a cylinder. At one end of the cylinder the piston is connected to a crankshaft that converts the piston's reciprocating motion into a rotary motion. At the other end of the cylinder is a combustion chamber in which a mixture of air and atomized fuel is burned.

This burning process is often described as an explosion, but in an engine that is operating properly the combustion produces a fast-moving flame front. When the fuel-air mixture burns explosively, the result is a hole in the piston crown and other engine damage.

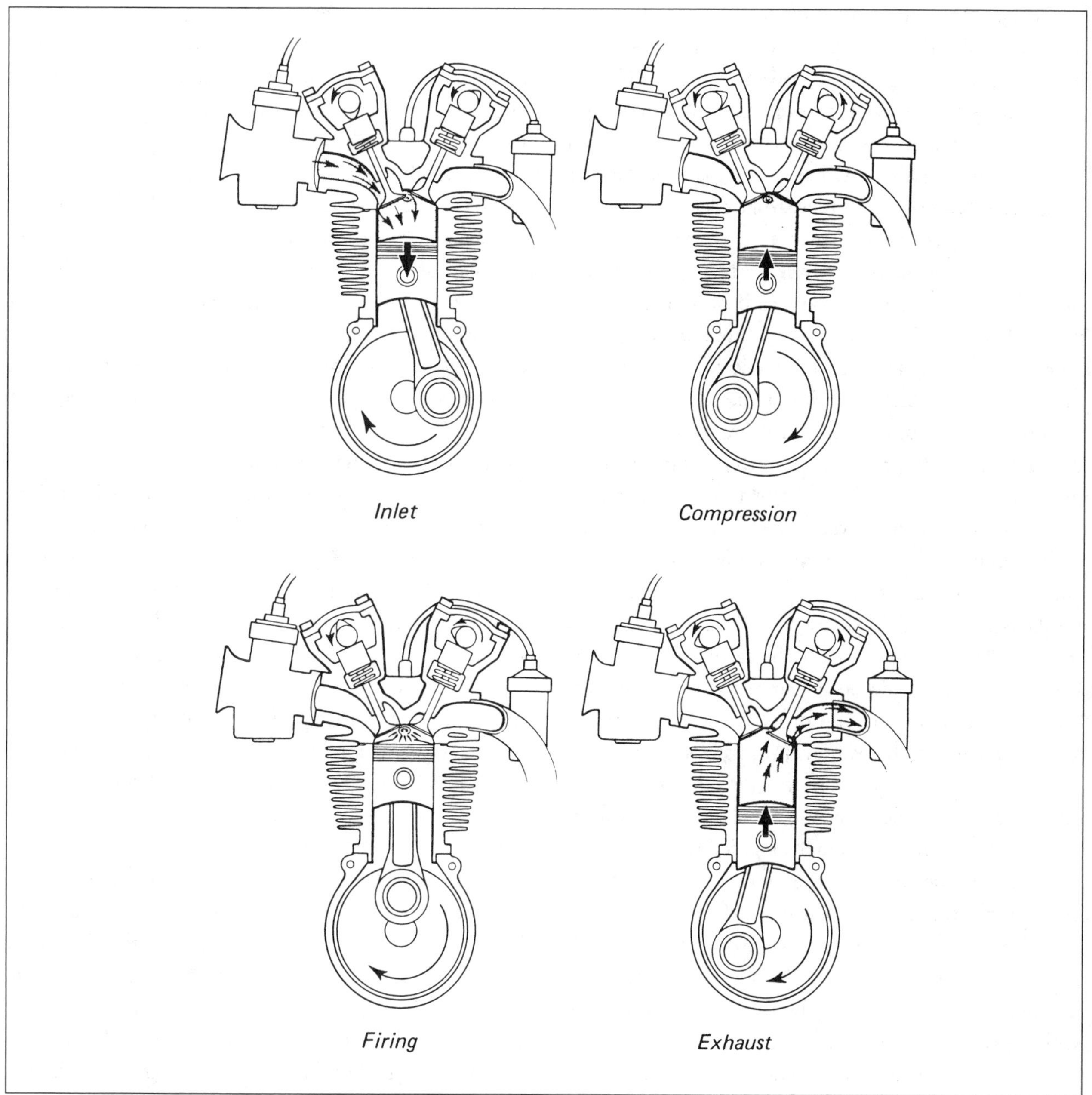

The four phases of a four-stroke engine's operating cycle. During the *inlet* phase, the air-fuel mixture is sucked into the combustion chamber as the piston descends. The piston then moves up, *compressing* the mixture. The *firing* of the spark plug ignites the mixture and forces the piston down. After the completion of the power stroke, the upward movement of the piston forces the *exhaust* gases out. Note that two revolutions of the crankshaft are required for each power stroke (from Vic Willoughby, *Back to Basics,* 1981).

The air-fuel mixture is brought into the combustion chamber by one or more intake valves, which are opened by a camshaft, either directly or through an arrangement of pushrods and rocker arms. An exhaust valve (or valves) is opened after the end of a combustion stroke to allow the products of combustion to leave the combustion chamber.

The four-stroke engine is the most common kind of internal combustion engine, but not the only one—other internal combustion engines operate under different principles. A two-stroke engine, as the name implies, requires only two strokes to complete a cycle; a Wankel rotary engine does away with pistons altogether. A diesel engine uses both two- and four-stroke cycles.

"Four stroke" refers to the four strokes made by the piston in the course of one cycle of the engine's operation. The first stroke is a downward movement of the piston that sucks in the air-fuel mixture that is supplied by a carburetor or through a fuel injection system. The mixture is then compressed during the second stroke as the piston moves upwards. After being compressed, the mixture is ignited by a spark, at which point the pressure produced by the flame front begins to move the piston downwards, applying power to the crankshaft. The final stroke occurs as the piston moves upwards as the exhaust valve opens, scavenging the spent combustion products through the exhaust port. It should be noted that the opening and closing of the valves, as well as the firing of the spark, do not occur at the exact moment that the piston is at the top of its stroke. The time required to complete the various processes, such as filling the cylinder, make it necessary to open and close valves and fire a spark while the piston is still moving upwards. Also, as the engine's rotational speed increases, it is necessary to fire the spark at an even earlier moment; this advancing of the spark is done automatically by the ignition distributor. In recent years, some automobile engines have been equipped with mechanisms that automatically change the valve timing in accordance with the rotational speed of the engine.

Efforts to design and build an internal combustion engine go back to the 17th century, when a few inventors attempted to build engines fueled by gunpowder. This was a reasonable venture, as a gun is simply a kind of one-stroke engine, but their efforts were in vain. In the 19th century, many inventors worked on a variety of internal combustion engines, some of them attaining a modest degree of success. In 1858, Christian Reithmann (1818–1909) was using a gas engine to power machine tools in his workshop. Even more significant were the engines designed and built in France by Etienne Lenoir (1822–1900). These engines produced as much as 12 horsepower and had a higher thermal efficiency than steam engines of equivalent size. Lenoir's engine used a two-stroke cycle and was double-acting; a mixture of air and illuminating gas entered while the piston was beginning to move away from the end of the cylinder. It was then ignited by a spark, which caused the piston to travel the remaining distance. The process was repeated at the opposite end of the cylinder so that the power stroke also served to exhaust the spent mixture on the other side of the piston. The engine thus provided two power strokes per crankshaft revolution, but total power was limited because the air-gas mixture was not compressed. The lack of compression diminished power and efficiency, but over a 5-year period, Lenoir was able to sell as many as 400 engines for industrial uses.

Another noncompression engine was made in Germany by Nikolaus Otto (1832–1891) and his partner Eugen Langen (1833–1895). Patented in England in 1863, it proved even more efficient than the Lenoir engine, even though it used a complicated arrangement of ratchets, friction clutches, and moving wedges to control the movement of the piston and transmit power. Otto's greatest contribution, however, was his design of another internal combustion engine based on principles still in use today, the four-stroke cycle described above. Otto at first viewed the need for a four-stroke cycle as a temporary expedient until something better could be devised. As it turned out, the Otto's four-stroke cycle was incorporated in the great majority of internal combustion engines. Otto and Langen made a good deal of money selling industrial engines built to this design, even though they lost patent rights to the four-stroke cycle, for in 1862 Alphonse Beau de Rochas (1815–1891) had outlined a similar plan in an obscure publication. By 1885, 45,000 engines with a total capacity of more than 300,000 kW (400,000 hp) had left the Otto and Langen factory in Deutz, Germany.

This factory also provided on-the-job training for two men, Wilhelm Maybach (1847–1929) and Gottlieb

Daimler (1834–1900), who went on to turn the large stationary Otto engine into something suitable for powering vehicles. Daimler installed such an engine in a two-wheeler in 1885, and put one in a four-wheeled vehicle a year later. Another German, Karl Benz (1844–1929), used a .6 kW (.8 hp) four-stroke engine to power a three-wheeled carriage in 1885. In both cases, a significant breakthrough was the use of gasoline instead of illuminating gas as a fuel. The use of liquid fuel was greatly facilitated by Maybach's invention in 1893 of a carburetor that used a small jet to spray atomized fuel into the air stream entering the combustion chamber.

By 1900, a Daimler car was powered by an engine that produced 26 kW (35 hp) and weighed only 225 kg (496 lb). Subsequent developments took the four-stroke engine to a level that its early developers could never have envisaged. Although the basic four-stroke cycle remained, power and efficiency were greatly enhanced by design features like overhead valves, higher compression ratios, and balanced crankshafts, along with vastly improved ancillaries such as carburetors and ignition systems. In the second decade of the 20th century the electric starter made its appearance, ending one of the greatest drawbacks of cars powered by internal combustion engines, the need to manually crank the engine to get it started. Beginning in the late 1920s, a few cars were fitted with superchargers; turbochargers began to be used in the 1960s. These devices further increased power, albeit at the cost of greater complexity.

The history of the four-stroke engine illustrates how sustained development can transform a particular technology. In its basic features, the design pioneered by Nikolaus Otto is hardly an example of engineering elegance. The purpose of the engine is to produce rotary motion, yet this is accomplished by having the pistons move up and down rapidly, reversing themselves thousands of times each minute. Also, unlike a two-stroke engine, a four-stroke produces power only on every other downward stroke of the piston. But however awkward the basic operation of the four-stroke engine seems to be, there is no question that it works, and that it works well. While the first automobile engines produced less than 1 horsepower and had a rotational speed of no more than 600 revolutions per minute (rpm), a modern passenger car engine displacing 3 liters (181 cubic inches) puts out close to 150 kW (200 hp) and can turn more than 6,000 rpm. All this is accomplished with minimal maintenance and operator attention. Given proper care, a modern engine is capable of propelling a car a distance of more than 160,000 km (100,000 mi).

No single breakthrough converted the Otto industrial engine into the modern four-stroke engine. The vast improvement in the power and durability of modern internal combustion engines has been the result of a multitude of technological advances, many of them quite small. Some have been the result of advances in basic knowledge, such as a better understanding of the dynamics of gas flow or the nature of vibration. Improved fuels have allowed higher compression ratios and more efficient combustion. Better materials, everything from aluminum to cast iron, have allowed higher rotational speeds, and hence more power. Sophisticated lubricants have made major contributions to the longevity of engines. All of these developments have been combined with advances in manufacturing technologies that have made possible the production of great quantities of engines with no significant deviations from the original design parameters.

Although four-stroke internal combustion engines provide a reliable source of power for hundreds of millions of cars and small trucks, they also have been an important source of environmental degradation. The exhaust products of these engines are major contributors to air pollution and possible global warming. The cyclical nature of the engine's operation and the resultant rapid heating and cooling of surfaces make the control of emissions especially difficult. Exhaust-borne pollutants have been significantly reduced through better engine management technologies (computerized control of air-fuel mixtures and spark timing), fuel injection, and the installation of catalytic converters. All of these are subject to wear and tear, so maintaining a relatively clean exhaust requires appropriate procedures for inspection and repair. The four-stroke internal combustion engine has been an essential part of transportation on land, sea, and air, but overcoming its negative consequences will continue to entail significant individual and collective costs.

See also CARBURETOR; CATALYTIC CONVERTER; CLIMATE CHANGE; ENGINE, DIESEL; ENGINE, TWO-STROKE; ENGINE, WANKEL ROTARY; FUEL INJECTION; GASOLINE;

LIGHTING, GAS; OCTANE NUMBER; SELF-STARTER, AUTOMOBILE; SMOG, PHOTOCHEMICAL; STEAM ENGINE; SUPERCHARGER; TURBINE, STEAM; TURBOCHARGER

Engine, Stirling

Engines that convert heat into power fall into two general categories: external combustion and internal combustion. A steam engine is an example of the former, while a gasoline engine is an example of the latter. A Stirling engine is an external combustion engine, but one that works on principles quite different from those exhibited by steam engines. Invented in 1816 by Robert Stirling, a Scottish clergyman, the engine that bears his name has been little used during the 20th century. It may, however, become a significant source of power in the 21st century.

The operation of a Stirling engine is a bit complicated because a number of things occur simultaneously. The simplest Stirling engines, which were employed during the 19th century, typically consisted of two concentric pistons, a power piston, and a displacer piston. Both pistons moved inside a cylinder that was filled with a gas (usually air) that served as a working fluid. The cylinder was heated at one end and cooled at the other. The displacer piston moved back and forth, alternately moving the gas to the hot and cold ends of the cylinder. The cyclic temperature changes produced by moving gas caused pressure changes, which in turn drove the power piston. An important part of the engine that Stirling devised was a porous heat exchanger that checked the loss of heat from the gas as it was moved back and forth between the hot and cold areas of the cylinder. Located between the pistons, a typical regenerator was a porous, conductive material such as steel wool; the generator was located between the pistons. When the engine was in operation, hot air gave up much of its heat as it passed through the heat exchanger on its way to the engine's cold region. Conversely, cold air picked up heat as it passed through the heat exchanger while moving towards the cold region.

The Stirling engine found a number of applications during the 19th century. Engines with outputs of up to 26 kW (35 hp) had as their main task the powering of pumps, while others were used for butter churns, crushers, and grinders. Although they were more efficient than the steam engines of the time, burned-out cylinders were a common problem. Toward the end of the 19th century, the rapid development of internal combustion engines and electric motors made the Stirling engine a less-attractive proposition.

Modern Stirling engines have been created by the Philips Research Laboratories in the Netherlands, and additional work has been done by General Motors and Ford. Research engines have produced as much as 375 kW (500 hp) per cylinder and have achieved thermal efficiencies of 30 to 45 percent, at least as good as present-day diesel engines. In addition to being efficient, Stirling engines are inherently clean running. Combustion takes place continuously in an external chamber, thereby avoiding the cycles of heating and cooling that occur inside internal combustion engines. Moreover, an unlimited supply of air can be provided, so there is no problem with unburned residual gases. External combustion also allows the use of a wide variety of fuels, everything from alcohol to olive oil. A Stirling engine is also quiet and largely free from vibration. Finally, the engine delivers high torque through most of its operating range, obviating the need for a complicated multispeed transmission.

Offsetting the Stirling engine's virtues are several negative aspects. The engine needs a large radiator for adequate cooling, and building a heat exchanger with an acceptably long life remains a challenge. Sealing is also of critical importance, for the gas used as a working medium cannot be readily replenished. Most critically, Sterling engines are very large and heavy for the power they supply.

See also ENERGY EFFICIENCY; ENGINE, FOUR-STROKE; PUMPS; STEAM ENGINE

Further Reading: Graham Walker, "The Stirling Engine," *Scientific American,* vol. 229, no. 2 (Aug. 1973).

Engine, Two-Stroke

In a conventional four-stroke internal combustion engine, the crankshaft makes two complete revolutions for each power stroke. A two-stroke engine requires only one revolution of the crankshaft for each power stroke. At first glance, this appears to be a more efficient way of producing power, and in fact some of the first internal combustion engines worked on the two-stroke principle. At the same time, however, the two-stroke engine has a number of shortcomings that have impeded its widespread use. Two-stroke engines are often used to power chain saws, lawnmowers, small boats, motorcycles, and motor scooters, but only rarely have they served as power plants for automobiles.

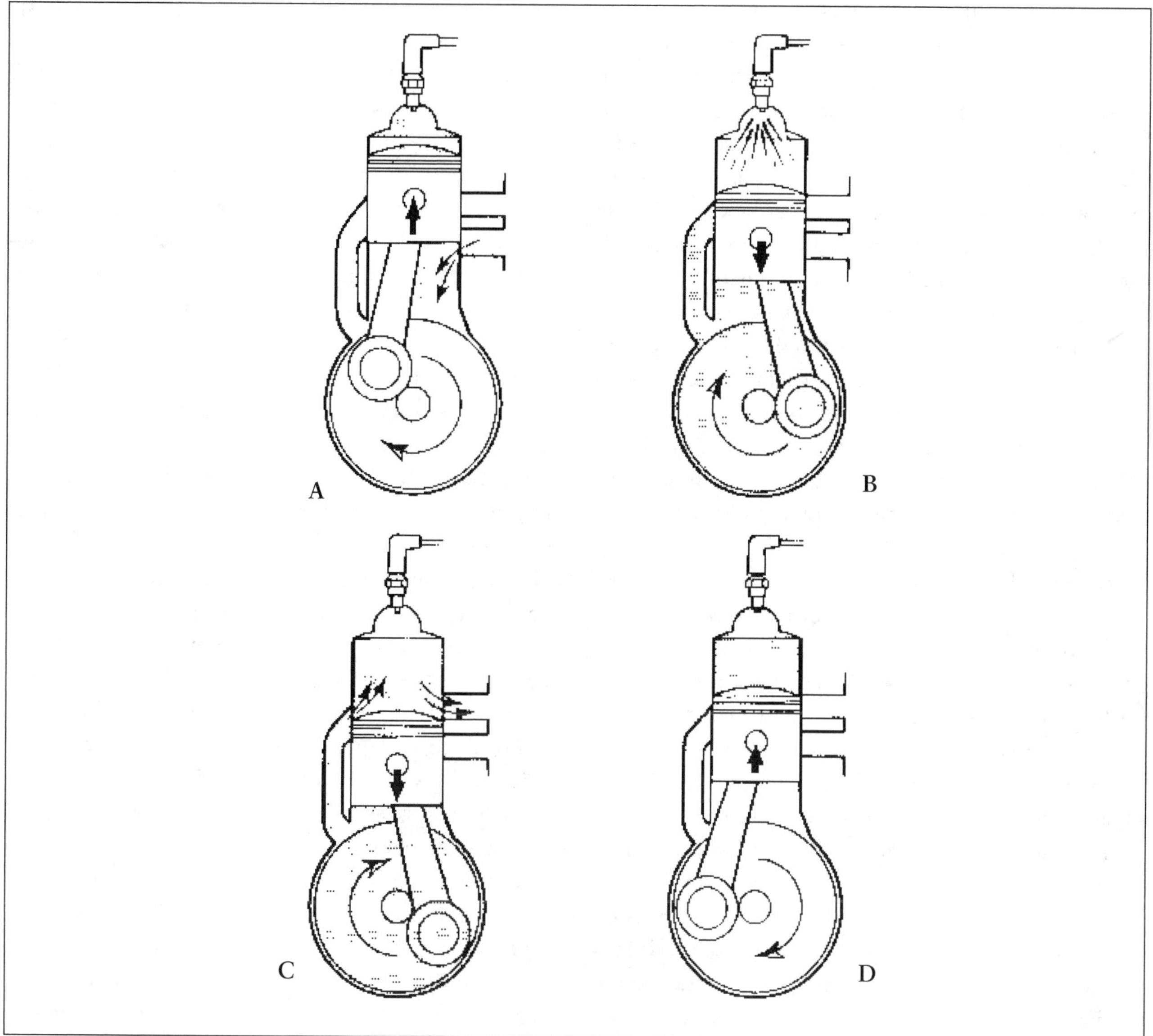

In a two-stroke engine, several things happen at the same time, so the basic engine is simple, but the processes are complex. In *A*, the upward movement of the piston opens up the inlet port, allowing the air-fuel mixture to enter the engine's crankcase. At the same time, an air-fuel mixture that was already in the combustion chamber is compressed by the rising piston. In *B*, the firing of the spark plug ignites the mixture, driving the piston down and producing power. At the same time, the downward movement of this piston closes the inlet port and compresses the air-fuel mixture in the crankcase. In *C*, exhaust gases are blown through the exhaust port, while at the same time a fresh charge enters through the transfer port that connects the crankcase and combustion chamber. In *D*, the rising piston compresses the charge in the combustion chamber, while at the same time closing the transfer and exhaust ports (from Vic Willoughby, *Back to Basics*, 1981).

The first truly practical internal combustion power plant was Nikolaus Otto's four-stroke engine. In 1878, 2 years after Otto's first engine went into operation, Dugald Clark (1854–1932) in Great Britain invented a technically and commercially successful two-stroke engine. Clark's engine used a separate cylinder to compress an air-fuel mixture, which was then admitted to the main cylinder through a one-way valve. The complication of a separate compression cylinder was eliminated when Joseph Day (1855–1946) invented an engine that used the engine's crankcase for this purpose. Day received a British patent for his engine in 1891, while a similar design was patented in the United States by Clark Sintz in 1893. Engines of this sort enjoyed a fair degree of success towards the end of the 19th century, primarily as stationary power plants and for propelling small boats.

The engines invented by Day and Clark set the basic pattern for two-stroke engines. In engines of this type, as a piston moves upwards it uncovers a port through which the air-fuel mixture enters the crankcase. When the piston moves down, it compresses the mixture, which then enters the engine's combustion chamber via a transfer port also opened by the piston. The mixture is then compressed by the piston's upwards movement and is ignited by a spark plug. The expanding gases move the piston downward, producing power and uncovering an exhaust port through which the spent gases escape, helped along by the incoming charge of air and fuel.

Early two-stroke engines used pistons with deflector plates on their crowns. The plates directed the air-fuel mixture upwards, thereby preventing it from being thrown out the exhaust port as it opened. This was an adequate solution, but some of the mixture was still lost. Moreover, the deflector divided the combustion chamber in two, while the piston itself was subject to distortion due to the uneven accumulation of heat. A better means of directing the flow of gases was devised by Adolf Schnürle (1897–1951) in Germany. Schnürle's engines took advantage of the dynamics of moving gases to direct the fresh mixture into the combustion chamber and push the spent gases out the exhaust port. High-performance two-stroke engines also have made good use of tuned exhaust systems that use rebounding pressure waves to keep the fresh mixture from escaping through the exhaust port before it has been burned.

This "loop scavenging" technology continues to the present day, aided by a few refinements, such as the use of reed valves that prevent the incoming charge from being pushed back through the intake port. Also, many two-stroke engines now use oil injection for lubrication, which eliminates the smoky exhaust typical of two-stroke engines that are lubricated by mixing oil with the fuel.

In addition to serving as the basis of some gasoline engines, the two-stroke principle is also used for many diesel engines. In engines of this sort, a supercharger, a turbocharger, or both, force air into the cylinder as the upward movement of the piston opens the intake port. Fuel is squirted into the combustion chamber by an injector after the continued upward movement of the piston has compressed the air. After combustion, the exhaust gases exit through ports opened by the piston or by conventional poppet valves.

Because they have few moving parts, two-stroke engines are light and cheap to manufacture. However, they suffer from several drawbacks. The fact that the engine delivers a power stroke for each rotation of the crankshaft does not make a two-stroke engine twice as powerful as a four-stroke engine of equivalent displacement. Combustion suffers because some of the exhaust gases inevitably mix with the incoming air-fuel charge, and despite the overall effectiveness of loop scavenging, some of this fresh mixture still escapes from the combustion chamber when the exhaust port opens. These defects have been largely overcome through the development of two-stroke engines in which fuel is directly injected into the combustion chamber; the only thing inducted and compressed in the crankcase is air. Engines of this sort combine high power with light weight and compact dimensions. Unfortunately, these engines have not been able to achieve acceptably low levels of emissions of oxides of nitrogen throughout their operating life. Advanced two-stroke engines also suffer from not having been invented and developed by the automobile companies that would be their greatest users.

See also AUTOMOBILE; CHAIN SAW; EMISSIONS CONTROLS, AUTOMOTIVE; ENGINE, DIESEL; ENGINE, FOUR-STROKE; FUEL INJECTION; MOTOR SCOOTER; MOTORCYCLE; NOT-INVENTED-HERE SYNDROME; SUPERCHARGER

Engine, Wankel Rotary

From the early 20th century onward, the four-stroke engine has been the dominant automotive power plant. Its supremacy has not been the result of engineering elegance; in engines of this type the reciprocating (up-and-down) motion of a piston is converted to rotary motion through the use of a connecting rod, crankshaft, and flywheel. Moreover, two complete up-and-down movements of each piston are needed to produce a single power stroke. Not surprisingly, several generations of inventors tried to create an engine that used a rotary rather than reciprocating motion. Beginning in the 1950s, some automobile manufacturers attempted to use turbine engines as automotive power plants, but for a number of reasons they were not successful. In the 1960s, a power plant of a different design, the Wankel engine, became the first rotary engine to be offered in cars sold to the general public.

The Wankel rotary engine consists of rotor, which looks like a triangle with bulging sides, that moves eccentrically inside a housing shaped like an oval pinched at the middle (technically known as an *epitrochoid*). As the rotor turns inside the housing, it sequentially allows an intake port to open, compresses the fuel-air mixture that comes in, ignites it, and exhausts the spent mixture. Each of these phases takes place in a different part of the engine's interior, which changes in shape and volume as the rotor turns. Each revolution of the rotor produces three firing impulses, enhancing the inherent smoothness of rotary motion.

Wankel engines also have the advantage of being smaller and lighter, as well as having fewer components (largely due to the absence of a piston engine's valve train) in comparison with conventional piston engines. However, their manufacture requires the use of sophisticated machining methods for accurately shaping and finishing the epitrochoidal interior of the engine. The engine's inner walls also have to be coated with a nickel-silicone material, a process that further increases the cost of manufacture.

The genesis of the Wankel engine was a long one. Felix Wankel (1902–1988) began to work on a rotary engine in the 1920s and patented his first rotary design in 1934, although it was never produced. In 1954, he used a rotary principle for a successful supercharger used with racing motorcycle engines. In 1957, an engine that rotated both the rotor and housing produced unsatisfactory results. A year later, a new version with a fixed housing showed that Wankel's design could at least be made to work.

Much of Wankel's work was financially supported by the German NSU firm, and in the early 1960s, that company built the first Wankel-powered car offered for sale, albeit in small numbers. Other firms attempted to develop engines using Wankel's principles. The most aggressive was Japan's Toyo Kogyo, whose Wankel-powered Mazda 110S was put on the market in 1967, followed by the R100 a year later. Wankel-powered Mazdas were a commercial success at first. The engines gave them impressive smoothness and excellent acceleration, but their fuel consumption was considerably higher than equivalent piston-engined cars. Many drivers felt that poor fuel economy was an acceptable trade-off, considering their cars' other virtues, but the fuel crisis of 1973– 1974 radically altered the situation. It also had become apparent that problems with the seals at the tips of the rotor produced serious repair expenses. Toyo Kogyo barely survived a disastrous drop of sales and subsequently employed the rotary engine only in its low-volume RX-7 sports car.

During the period prior to the sales collapse of rotary Mazdas, the Wankel engine seemed almost certain to be widely employed. Many car companies announced their intention to use it in large numbers, and backed up their intentions with the payment of licensing fees and the development of prototype engines. Virtually none of these engines were installed in cars sold to the public. The failure of the Wankel engine to live up to expectations shows that apparent engineering sophistication does not guarantee success. The conventional piston engine, a less elegant design, but one with decades of development work behind it, demonstrated formidable staying power. The Wankel engine may enjoy a renaissance in the future, for it is particularly well suited to using hydrogen as a fuel. It is by no means certain, however, that hydrogen will be the fuel of the future and that the Wankel rotary engine will find a home in large numbers of cars and small trucks.

See also ENGINE, FOUR-STROKE; TURBOJET

Engineering Ceramics

There is little resemblance between a chaff-tempered construction brick, a Wedgwood teapot, and the refractory surface coating on the exhaust valves of a modern diesel engine. Yet all are heat-treated nonmetallic materials or, in common parlance, ceramics. Today, the ceramics industry is aiming to compete with engineering resins in a variety of high-tech applications, from high-temperature engines to optical signal processors. Industry prophets are calling ceramics "the polymers of the 21st century."

The resources used for making for ceramics are abundant, and they are easy to work in a plastic state. Moreover, the product retains strength at high temperatures, a prime virtue in engineering applications. Still, as with any material, there are trade-offs. High-tech ceramics are not produced from petrochemicals, as are polymers, but making them is highly energy intensive. They have had some well-publicized applications such as shielding reentry vehicles, but these are not commercially significant. Indeed, aside from insulating integrated-circuit packages and substrates, few commercial markets have been developed beyond traditional automotive applications such as spark plugs insulators and catalytic converters—and the focus of research and development remains in automotive applications.

The great drawback of engineering ceramics is that they are brittle; that is, they deform very little before failing. Also, they are unpredictable: Ceramic parts that are for all practical purposes identical can fail under quite different circumstances. Hence they may appear to have only a limited potential niche in our material culture. Yet it needs be said that in 1920 the same thing appeared true with respect to aluminum, and in 1940 with respect to nylon. The 21st century may see some surprising uses for ceramic materials.—R.C.P.

See also ALUMINUM; NYLON; POLYMERS; SPACE SHUTTLE

Entropy

By the middle of the 19th century the concept of the conservation of energy was well on its way towards establishment as a fundamental physical law. The idea that energy is conserved began to take hold after it was determined that heat is a form of energy, and that some of the energy expended to do work is "lost" in the form of heat. For example, not all of the energy produced by the combustion of fuel will serve to propel an automobile; some of it will show up as heat generated by friction.

Another important rule of physics is that heat flows from hotter objects or areas to colder ones. *Flow* is a bit of a misnomer, for heat is not a fluid, as was once thought. Heat is a manifestation of molecular motion, and the "flow" of heat occurs because the more rapidly moving molecules of a hot object increase the motion of the colder object's molecules. This process moves in only one direction in an isolated system; to get heat to flow from a cold region to a hot one requires the application of energy. A device that uses energy to reverse the normal flow of heat is known as a *heat pump*, and is exemplified by a refrigerator or an air conditioner.

Gaining an understanding of the nature of heat flow was more than an academic concern in the mid-19th century. Significant numbers of steam engines were in operation, yet the basic principles governing their operation were imperfectly understood. It was evident that the development of more efficient steam engines would require an intellectual breakthrough in the emerging science of thermodynamics. That breakthrough—albeit unappreciated at the time—was provided by Nicolas Leonard Sadi Carnot (1796–1832), who was the first to clearly analyze the heat flows that take place in steam and other heat engines.

Carnot's fundamental insights were extended by Rudolph Clausius (1822–1888), a German physicist. According to Clausius, although much of the heat used by an engine is converted into mechanical energy, some of it is inevitably discharged without having done any useful work. In Clausius's mathematical expression, in any process involving heat flow in a closed system, the ratio of the heat content to the temperature of the system always increases. In 1865, Clausius gave this ratio the name *entropy* (based on the Greek word for *transformation*).

Clausius's formulation is the basis of the Second Law of Thermodynamics. This law can be stated in a number of ways, but it essentially says that a heat engine cannot convert into work all of the heat put into it. Some of that heat must be discharged into a lower-temperature reservoir where it will be useless as a source of energy. In other words, the accumulation of "useless" heat represents an increase in the entropy of the system. Another way of looking at this law is to think of entropy as a kind of disorder. An ordered system is one divided between a part of the system that is at a relatively high temperature, and another part of the system that is at a relatively low temperature. It is this ordering that allows heat to flow and work to be done. A system at maximum entropy exists at a single, undifferentiated temperature, and as a result no heat can flow and no work can be done.

A hypothetical device that violates the Second Law of Thermodynamics is known as a "perpetual motion machine of the second kind." One such example is an imaginary ship's engine that is supposed to draw its energy from the heat in the ocean. One premise of this notion is correct; there is quite a lot of heat in the oceans of the world. But the use of that heat as an energy source would require the flow of that heat to a lower-temperature reservoir. Such a reservoir could be produced through the use of refrigeration system, but running it would use more energy than the whole system delivered.

The principle that the entropy of a closed system cannot decrease spontaneously can also be understood on a molecular level. When a vessel contains a gas at a particular temperature, the gas molecules have an average kinetic energy that does not vary. In this sense there is no ordering to the system; the average always remains the same. If, however, these molecules were part of a larger system that contained molecules at a different average level of energy, this difference would make for an ordered system. In such a case the order lies in the differentiation of molecules with one energy level from molecules with another energy level. In this system, work can be done through the movement of high-energy molecules to the part of the system containing the low-energy molecules.

Although every energy conversion device "wastes" energy in the form of heat, some are more wasteful than others. Energy efficiency is defined as amount of energy converted to work relative to the energy taken in. In a heat engine, energy efficiency also can be defined as the ratio of the energy put into the engine in the form of heat to the energy discharged into a lower-temperature reservoir. This can be expressed as a percentage by applying the formula $1 - Q_C/Q_H$, where Q_H is the amount of energy released by the combustion of the fuel, and Q_C is the energy discharged into the lower-temperature reservoir. Because their operation entailed only a modest fall of temperature as work was done, early steam engines were highly inefficient, having a rating of about 0.5 percent. Modern heat engines are far more efficient; a steam turbine operates at an efficiency of around 40 percent, about the same as the efficiency of a diesel engine. But no heat engine can run at 100 percent efficiency; entropy always increases.

The inevitable increase in entropy has implications that go beyond the analysis of heat engines. If the universe is taken to be a very large closed system, it follows that it too must obey the Second Law of Thermodynamics. At some point, after all energy transfer has occurred, it will reach a state where there are no temperature differences—in other words, a state of complete disorganization. This situation, which was first described by Clausius, came to be known as "the heat death of the universe." Although the prospect of the universe slowly running down is a depressing one, it is open to challenge. In the first place, it is not certain that the universe is a closed system. Second, the Second Law of Thermodynamics has been deduced from our experiences with a tiny portion of the universe, and not from first principles. Whether or not it and other physical laws can be extrapolated to the universe as a whole remains unproved.

See also AIR CONDITIONING; CALORIC; CARNOT CYCLE; ENERGY, CONSERVATION OF; ENERGY EFFICIENCY; ENGINE, DIESEL; KINETIC THEORY OF HEAT; PERPETUAL MOTION MACHINES; STEAM ENGINE; REFRIGERATOR; STEAM TURBINE

Further Reading: P. W. Atkins, *The Second Law: Energy, Chaos, and Form*, 1984.

Environmental Impact Report

Environmental impact reports (EIRs) are meant to disclose significant environmental impacts to decision-makers and the general public. These reports allow an informed review of development projects that require governmental approvals or permits. In official use, they are used only for evaluation of proposed projects; they are never conducted on past projects.

EIRs are authorized under "full disclosure" laws such as the National Environmental Policy Act (NEPA, passed by the U.S. Congress in 1969), and the California Environmental Quality Act (CEQA, passed by the California Legislature in 1970). Other states, including Washington, Minnesota, Massachusetts, Hawaii, and Oregon, have similar laws. CEQA is probably the most restrictive such statute in the United States, and was amended subsequently to streamline the requirements for compliance. (Analogous requirements of NEPA are noted in parentheses.)

EIRs, and the full-disclosure laws governing their application, resulted from early political victories of the environmental movement in the 1960s and 1970s. In California, passage of CEQA was prompted in part by public outrage over permit processes that allowed massive urban and suburban sprawl throughout Southern California and the Santa Clara Valley after World War II, widespread concern for important aquatic resources such as San Francisco and Suisun Bays, the construction and operation of the enormous California State Water Project after 1968, bitter battles between highway bureaucrats and neighborhood preservationists over San Francisco's Embarcadero Freeway and other transportation projects, the Santa Barbara offshore oil spill of 1970, and the preference of the electric utility industry for nuclear power to satisfy growing consumer demand for electricity.

The EIR process involves several steps. First, a project is examined to determine if it is exempt from CEQA (NEPA). If it is not exempt, the project is subjected to an Initial Study (Environmental Assessment). If no significant (adverse) impacts are identified, a Negative Declaration (Finding of No Significant Impact, or FONSI) is issued for the project, declaring that the project has no significant environmental impacts.

However, if the Initial Study finds potentially significant environmental impacts reasonably attributable to the project, then CEQA requires that an EIR (Environmental Impact Statement, or EIS) must be prepared. EIRs are "scoped," often through consultation with affected members of the public, to indicate the range of issue areas to be studied; a full description of the proposed project is provided for analysis; a description of the existing social, urban, and/or environmental conditions is developed so that the significant cause and effect interactions of the proposed project with the environment can be analyzed; and mitigation measures are proposed. Under CEQA, all significant impacts must be mitigated so that they are no longer considered adverse (NEPA does not require that impacts be mitigated). In 1989, the California Legislature added a requirement that mitigation measures must also be monitored and reported over time.

Alternatives to the proposed project are developed, and their impacts compared with the proposed project's impacts, including whether any alternatives could achieve the same desired effects as the proposed project. Different sites and a "no project alternative" must be among the alternatives, in addition to different sizes or configurations of the project. An EIR is then released for public review, often for no less than 45 days, during which time public comment is taken. Responses are composed and a final EIR, with responses to comments, is released. Prior to project approval, the local agency responsible for approval must certify the EIR as adequate. Once the EIR is certified, then the project may be approved.

Teams preparing EIRs are often interdisciplinary, since potential problems may cross traditional professional boundaries. The specific environmental topics examined in an initial study suggest the need for an interdisciplinary approach: earth, air, water, biological resources, noise, land use, natural resources, risk of hazards (toxic or other hazards), population distribution and growth, housing, transportation public services, energy consumption, utilities, public health, aesthetics, recreation, and cultural resources (e.g., areas of archaeological interest).

New projects may have controversial cumulative and growth-inducing impacts as well. These impacts must also be analyzed in EIRs. Cumulative impacts may be individually small and mitigated, such as a short freeway bypass around a rural community. But combined with the impacts of other related projects (e.g., a new commercial center along the bypass's frontage at an off-ramp), significant adverse effects may still result (e.g., bypassing a community's downtown may starve it of through traffic that provided patrons for local businesses.

Growth-inducing impacts occur when a proposed project would expand the capacity of urban systems to handle development: sewer line extensions; the setting of local governmental boundaries; construction of reservoirs, roads, and transportation facilities; power facilities; or new land subdivisions that create developable land parcels.

A consultant industry has arisen to professionalize preparation of EIRs through use of standardized methods. Assuring that EIRs are adequate methodologically, however, does not guarantee that substantive debate on a project's real merits occurs. Neighborhood activists may hide racist views of affordable housing projects behind EIR issues (e.g., too many young school children, not enough school facilities); or developers may try to bury opposition under the weight of reams of expert analysis. EIRs thus raise barriers to entry for some individuals and firms seeking to develop new projects due to the cost of hiring consultants to produce the reports.

The legal standard for full-disclosure documents such as EIRs is "adequacy." Preparation and dissemination of the EIR has to follow a clear public notice process and contain specific types of data. If followed, however, the project may still be approved despite significant environmental effects. CEQA allows local jurisdictions to make findings of "overriding considerations" that justify approval of the project despite its expected environmental problems.

In the final analysis, CEQA requires of local governments a respect for truth. EIRs are overlaid atop a community's values; the EIR process imposes no coherent environmental protection regulation or ecological ethic because it needs only to ensure that due diligence was followed by local government in approving a new development project. EIRs inform people about environmental impacts, and what can be done about them. But their consciences must still guide their actions.—T.S.

Environmental Protection Agency

The Environmental Protection Agency (EPA) was established as an independent agency within the executive branch of the American government in late 1970 and charged with implementing federal laws designed to protect the environment. Its creation coincided with growing popular and bipartisan concern about environmental degradation of the country's air, water, and other natural resources. That concern produced comprehensive environmental legislation that is now EPA's responsibility to enforce. The agency administers nine comprehensive environmental protection laws that include the Clean Air Act, the Clean Water Act, the Toxic Substances Control Act (TSCA), the Resource Conservation and Recovery Act (RCRA), and the Comprehensive Environmental Compensation and Liability Act ("Superfund").

EPA was purposely designed to avoid some of the problems of earlier American regulatory agencies, particularly their tendency to be "captured" by the industries they were supposed to be regulating, or to succumb to bureaucratic malaise after the early political enthusiasm for their activities faded away. In particular, by merging 15 existing programs formerly managed by five departments or councils, EPA was positioned to regulate all of industry rather focusing on a particular industry. Rather than becoming an agency within a larger bureaucratic department like the Food and Drug Administration or the Occupational Safety and Health Administration, the EPA has been an independent and very public executive agency headed by an administrator reporting directly to the President. Moreover, the environmental legislation for which the EPA is responsible differed from earlier regulatory laws by limiting agency discretion through the use of clearly specified environmental objectives and inflexible deadlines.

The EPA administrator is assisted by a deputy and nine assistant administrators who manage specific environmental programs and direct other functions. EPA has 10 regional offices across the country that work closely with state and local authorities to ensure that federal environmental laws are enforced. Scientific and technical expertise is provided by the Office of Research and Development within the agency.

The EPA can be credited with contributing to a number of environmental improvements over the past 25 years. Air pollution has declined; lakes and rivers are cleaner due to better wastewater treatment and diminished runoff from agricultural chemicals; airborne lead has been reduced by 90 percent; fuel economy standards are in place for trucks and autos; dangerous pesticides such as DDT have been banned; dumping of toxic wastes has declined and recycling is undertaken by growing numbers of Americans; and environmental education is an integral part of public school curricula. However, many problems remain. Forty percent of lakes and rivers are still not suitable for fishing or swimming; water contamination regularly threatens large metropolitan areas; one in four Americans lives within 6.5 km (4 mi) of a toxic dump site; asthma and breast cancer are rising; and environmental degradation is a suspected cause of these health problems.

Perhaps most troubling is the fact that public confidence in cost-effective environmental management has eroded. EPA's initial approach was to address inappropriate present and past environmental behavior through a command-and-control approach to regulation. Through research, monitoring, and lengthy processes of risk assessment and cost-benefit analysis, EPA would have established uniform standards for all potential offenders, and then engaged in a comprehensive strategy of enforcement. In a political climate where appropriate scientific and technical information was frequently inadequate or incomplete and where environmental and industrial interests quickly became bitter rivals, environmental regulation became highly contentious, routinely litigious, and, all to frequently, ineffective. The politics of the President and the EPA's chief administrator determined how vigorously environmental protection would be pursued, resulting in wide swings in EPA activity in the 1970s and 1980s.

In the 1990s, there has been talk of reinventing environmental protection, adopting a more flexible and customized approach to pollution abatement, and focusing on prevention rather than present and past behavior. Labeled a "common sense" initiative, the agency is supposed to move away from the pollutant-by-pollutant or crisis-by-crisis approach and craft strategies directed at particular industries or even individual firms. Inflexible standards would be replaced by negotiated arrangements that result in a cleaner environment at less cost to both industry and society as a whole. Command and control gives way to the language of partnership and common enterprise. Much of this change is motivated by the economic realities of the 1990s as American government and industry seek to position the national economy to compete more effectively in a global marketplace. At the same time, most Americans have embraced the cause of a protected environment and would applaud EPA efforts to do so in more cost-effective ways.—T.I.

See also CORPORATE AVERAGE; FOOD AND DRUG ADMINISTRATION; FUEL ECONOMY; INSECTICIDES; LEAD POISONING; OCCUPATIONAL SAFETY AND HEALTH ADMINISTRATION; PHOTOCHEMICAL SMOG; RECYCLING; RISK EVALUATION

Enzymes

An enzyme is a protein that acts as a catalyst (i.e., a substance that speeds up a chemical reaction without itself undergoing a permanent change) in biochemical reactions. Enzymes are essential to the metabolic processes that sustain life; without enzymes, life as we know it could not exist.

The first hint of the existence of enzymes came in 1833, when Anselme Payen (1795–1871), a French chemist, was engaged in the refining of sugar from sugar beets. While working with malt extract, he found a substance that sped up the conversion of starch into sugar. He called it *diastase*, and from that time onwards, enzymes usually have been given names that end in "ase." A year later, while in the course of studying the process of digestion, the German biologist Theodor Schwann (1810–1882) prepared an extract of the lining of the stomach. When he mixed it with hydrochloric acid (which was already known to play a key role in digestion), he found that the mixture dissolved meat far more rapidly than the acid alone. In 1836, he was able to form a precipitate by treating the extract with mercuric chloride. He named it *pepsin*, after the Greek word that meant *to digest*.

After its discovery, pepsin was grouped with a number of substances that were collectively known as *ferments*. In 1876, Willy Kühne (1837–1900) extracted from the secretions of the pancreas another ferment, which he called *trypsin*. Kühne was a vitalist, one of the many 19th-century biologists who held to the belief that certain substances could only be the product of biological activity. In his view, the only true ferments were substances that worked within living cells and produced chemical reactions unique to living things. Pepsin and trypsin therefore did not qualify as true ferments, for they were components of the digestive juices, and these juices did not work at a cellular level. He therefore coined the word *enzyme* for substances like pepsin and trypsin. The word meant "in yeast" in Greek and was derived from the resemblance of these substances to the ferments known to reside in yeast cells.

Subsequent events like the synthesis of urea undermined the vitalist position by showing that certain chemical processes could in fact occur in the absence of living organisms. Still, the vitalists tried to bolster their position by pointing to the apparent organic nature of fermentation. The process could be mimicked in the laboratory, but only through the application of very high temperatures and pressures or the use of powerful solvents. In contrast, when fermentation occurred in natural substances, it took place under normal conditions. In order to lend support to the vitalist view, in 1896 the German chemist Eduard Buchner (1860–1917) ground up yeast until all of the cells could be presumed dead. He then added a thick sugar solution to the yeast in order to protect it from bacterial contamination (much as a high sugar concentration prevents the spoilage of fruit "preserves"). Unexpectedly, carbon dioxide bubbles formed before long, indicating that fermentation was taking place. The substances making this happen were evidently no different from the enzymes involved in intracellular chemical activity. Accordingly, the distinction between ferments and enzymes no longer held, and the latter term began to take precedence.

The understanding of enzymes deepened after Buchner's experiment was taken to the next stage by Arthur Harden (1865–1940), an English biochemist. In 1904, he used dialysis to separate the large and small molecules associated with yeast. He did this by putting yeast extract in a bag and immersing the bag in pure water. The bag had been made from a material that allowed small molecules to pass through it but blocked the passage of larger molecules (materials of this sort are known as "semipermeable membranes"). Harden found that the yeast material remaining in the bag did not allow fermentation to take place. However, when the water in which the bag had been immersed was added to the contents of the bag, fermentation occurred. This indicated that the yeast enzyme consisted of two parts. Subsequent tests showed that the substance in the water was not harmed by boiling, which ruled out its being a protein.

The nonprotein components of enzymes came to be known as *coenzymes*. Their importance became evident in the 1930s, when it was determined that vitamins had molecular structures similar to those of coenzymes. It thus became evident that vitamins are essential to the formation of coenzymes. Some vitamins are manufactured by plants but not animals; consequently, animals have to receive their vitamins with the plants (or the meat of plant-eating animals) they consume. Because it adversely affects the workings of enzymes, a deficiency of vitamins leads to poor cell function and possibly the death of the organism. In addition, some coenzymes and enzymes incorporate metal ions into their molecular structure, and for this reason trace amounts of minerals like iron, copper, and zinc are needed for proper bodily functioning.

The mechanisms of enzymatic action began to be explored in 1913, when Leonor Michaelis (1875–1949) and M. L. Menten, German chemists, applied the principles of chemical kinetics (the study of the rates of reactions) to the operation of enzymes. They found that the reaction rate directly varied with the concentration of the substance being acted upon. Their research culminated in the formulation of the Michaelis-Menten equation, which is still used to determine the affinity of an enzyme for the substance it acts upon. More generally, the work of Michaelis and Menten helped to demystify the working of enzymes by showing that they were subject to the same chemical rules as ordinary substances.

At this time, the molecular structure of enzymes still remained in doubt. Although it had long been speculated that enzymes were proteins, existing tests for proteins provided much contrary evidence, and the belief that enzymes were not proteins held sway until the late 1920s. The "nonprotein" hypothesis was undermined in 1926 when the American chemist James Sumner (1887–1955) extracted the enzyme involved in the conversion of urea to ammonia and carbon dioxide. Sumner was able to precipitate some tiny crystals from this extract, and when he dissolved them in water, he found that the crystals acted like the enzyme itself. Tests made on these crystals indicated that they were proteins. Further evidence that enzymes were proteins was provided by another American chemist, John Northrup (1891–1987). In 1930, he succeeded in crystallizing pepsin, which was followed a few years later by the crystallization of trypsin and chymotrypsin, digestive enzymes secreted by the pancreas. Study of these crystals revealed that they too were proteins.

Like proteins in general, enzymes are large molecules, most of them having molecular weights that range from 10,000 to 500,000 g/mol. The large size of enzyme molecules does not give them their catalytic properties; other very large molecules like starch and cellulose do not act as catalysts. Rather, the catalytic qualities of enzymes are derived from their structures. Enzyme molecules are a linear sequence of various amino acids. The sequence of amino acids in these chains and the three-dimensional shapes into which they fold determine the function of particular enzymes.

Molecules of the substance being chemically converted (known as the *substrate*) bind to a section of the enzyme molecule known as its *active site*. In general, the binding is highly selective; only one particular molecule will bind to a particular enzyme and undergo the reaction catalyzed by that enzyme. Organisms have thousands of different types of enzymes, each one catalyzing reactions with a different substrate. The maintenance of an organism requires the operation of thousands of catalyzed chemical reactions, many of them proceeding in a sequential fashion. Each of these reactions depends on the presence of a particular enzyme. A particular enzyme works with only one substrate because the two molecules are uniquely matched in a way that is analogous to a lock that can be opened only by a key with the proper shape. Alternatively, substrates can "induce" the enzyme into a tight fit.

Other compounds with molecular structures similar to a substrate may inhibit a reaction by attaching to an enzyme, a process known as *competitive inhibition*. This is not always bad for the organism; many enzymatic reactions are governed by the presence of both the enzyme and its inhibitor. The enzyme-inhibiting qualities of certain molecules also are the basis for many antibiotics, which kill bacteria by introducing enzymes that block the actions of other enzymes essential to the life of the bacteria.

See also BACTERIA; FERMENTATION; SUGAR; UREA; VITALISM; VITAMINS

Epidemics in History

Somewhere on Earth—probably in a rain forest—an organism exists unknown to be a pathogen of any species. It may be a virus, or perhaps a bacterium. Whatever the species of animal that harbors it, an equilibrium persists between the parasite and its host as long as the population dynamics remain relatively quiescent. A natural mutation might arise and change the balance of that relationship, but more likely, an environmental alteration, maybe man-made, could begin a cascade of events that results in an epidemic never known before among people.

A bacteriophage of *Corynebacterium diphtheriae* offers an approximation for such a scenario. Ordinarily, *C. diphtheriae* produces only a relatively mild inflammation of the larynx and trachea in children, one of several forms of "croup." However, when a "tox+"-gene expressing phage infects the bacterium, it enhances iron binding and so allows *C. diphtheriae* to produce the powerful exotoxin that results in diphtheria. Ancient writers described diseases that may have been diphtheria, but its name dates from the 19th century when epidemics became frequent and severe. A contributing "environmental alteration" was the rise of public education, which caused increasing numbers of children to drink from a common tin cup, chew on the ends of germ-laden pencils, and breath one another's dust. Diphtheria epidemics, like all others, arose from crowding.

In general, epidemics are seen as diseases produced by some special circumstance usually not present in an affected area. An endemic disease, in contrast, prevails in a locality. Malaria, one of humankind's oldest afflictions, illustrates both. Four species of the protozoan genus *Plasmodium* cause malaria in human beings, but other species are parasites of birds, reptiles, and mammals. Perhaps as long as 30 million years ago, plasmodia infected female *Anopheles* mosquitoes and found both an environment in which to reproduce and a vector to extend its range. Malaria among people may have precipitated when the *Anopheles* migration out of Asia encountered the human migration out of Africa. By the time civilizations developed in the valleys of the Mekong, Yangtze, Indus, Euphrates, and Nile, the disease reigned. It existed in Greece centuries before Christ, reached Central Africa about 2,000 years ago, and arrived in the Americas after Columbus. Today, malaria cases are estimated to exceed 100 million, among a population at risk of about half of all humankind.

Epidemics have burst from the endemic areas when the complex relationship between the parasite, vector, victim, and environment (both physical and socioeconomic) produced the right conditions, especially during wars. Armed with DDT insecticide, the World Health Organization launched a malaria eradication campaign in 1955, which quickly produced dramatic results. Cases in Sri Lanka, for example, fell from 2.8 million in 1946 to 100 in 1961. However, after DDT use was curtailed, malaria incidence returned to near its pre-eradication level. Control now focuses on improving economic conditions, while at the same time economic development, based on timber cutting and mining in tropical forests, prompts new epidemics—a situation that applies equally to other protozoan afflictions, principally the two major *Trypansomas*, African sleeping sickness, and Chagas's disease in South America.

Smallpox vaccination in New Jersey, c. 1880 (courtesy National Library of Medicine).

Most of the great plagues that have pulsed through history have been of bacterial origin: cholera and tuberculosis, typhoid and pneumonia, to name a few. Perhaps the most notorious was bubonic plague, known also as "the Plague of Justinian" around 540, and "the Black Death" of 1346–1353. The causative pathogen, *Yersinia pestis*, primarily infects rats and mice, but fleas engorged on the rodent's blood spread the germ to human beings. Antigen changes of the organism can make it more or less virulent and, in the worst case, allow *Y. pestis* to enter cells more easily, resist the host's immune response more effectively, and produce toxins, resulting in a 60 to near 100 percent fatality rate of those infected in the preantibiotic era. These biological changes probably account for epidemic spikes that occurred during the hundreds of years when plague almost always raged somewhere. Public health measures and antibiotics brought bubonic plague under control in the 20th century, as they did most other bacterial diseases. However, *Y. pestis* retains its largest rodent reservoir in Central Asia and holds various tributaries, as among prairie dogs in the American southwest. Consequently, it awaits an opportunity, such as a major war, for people, rodents, and fleas to come closer together.

Antibiotics have not shielded human beings from viral epidemics, though vaccines offer protection and account for medicine's single, complete success against one epidemic disease: smallpox. The World Health Organization hopes to add poliomyelitis to that list by early in the next century. And, of course, researchers around the world are seeking either a vaccine or a chemotherapy for acquired immune deficiency syndrome, or AIDS. The virus thought to be responsible for AIDS has an enormous propensity for changing its antigens, which, so far, has defeated various control strategies. Influenza A also shares that trait to a degree, and with it produced what has been called the greatest single demographic shock that the human species has ever received: the "Spanish flu" pandemic of 1918–1919.

Virologists now suspect that human and avian influenzas naturally recombined by chance early in 1918. Genes of the avian strain, responsible for enormous virulence, made the human strain much deadlier than it had been in prior pandemics (most recently, in 1889–1890). The first serious outbreak may have occurred among American troops embarking for Europe in World War I. By the fall of 1918, the disease had spread through all the armies, and after the armistice, soldiers carried the illness to their homes all over the world. Spain was not a combatant in the war, so its news coverage was not censored, leading many to believe the "Spanish flu" originated there. In a matter of weeks, somewhere between 20 and 40 million people died worldwide. Researchers in England isolated influenza A in 1933, and vaccines have existed since the 1940s. Each spring, vaccine manufacturers try to anticipate the antigens influenza will present the following fall and winter. Their strategy rests on the gamble that the virus will not make a dramatic antigen shift after the "flu season" begins, leaving too little time to produce a new vaccine. Should the worst-case scenario happen, and coincide with an international event such as the Winter Olympic Games, a pandemic of unmatched proportions could take place. In any event, the crowding that makes all epidemics possible, and a global human population that approaches 6 billion, would lead to the reasonable conclusion that mankind's greatest epidemics lay in the future rather than in the past.—G.T.S.

See also ACQUIRED IMMUNE DEFICIENCY SYNDROME; BACTERIA; DDT; INFLUENZA; POLIO VACCINES; SMALLPOX ERADICATION; VIRUS

Ergonomics and Human Factors

The terms *ergonomics* and *human factors* have come to be used synonymously. They refer to an effort to design objects and environments for safer, more efficient, more effective, and more enjoyable use by human beings. Much of what is meant by these terms takes place at the interface of an object and the user of the object; That is, both the characteristics of the object and the capabilities of the user must be considered at the time an object is being designed. For example, prehistoric people chose triangular wedge-shaped stones with which to chip flint because that shape could be held quite well in the human hand. However, the tomahawk developed by Native Americans was a much more sophisticated chopping device. Attaching a wood handle to the stone made it easier to grasp, protected against frequently mashed fingers, and added leverage to the swing of the ax head.

The term *ergonomics* was coined in England by K. F. H. Murrell in the late 1940s. Derived from the Greek, it literally means, "the science of work." Murrell was one of a group of scientists from different disciplines in the United Kingdom who had been working with human problems in the design of equipment used in World War II. The group gathered after the war with the intention of forming a society "concerned with the human aspects of the working environment." Thus, in 1949, the Ergonomics Society was born in Great Britain.

Similar work was simultaneously underway in other countries as well. In the United States, two Southern California organizations, the Aeromedical Engineering Association of Los Angeles and the Human Engineering Society of San Diego, appointed a joint committee to develop a single organization that would be concerned with the *human factors* in design and engineering. The Human Factors Society was founded in 1957, and about 90 people attended the first meeting. By the mid-1990s, the society had more than 5,000 members and publishes a respected journal called *Human Factors* and a magazine called, *Ergonomics in Design*. In 1992, the Human Factors Society added the term *ergonomics* to its name and became The Human Factors and Ergonomics Society. Eventually, the national organizations that began springing up in various countries united to form the International Ergonomics Association, which held its first congress in 1961. Its purpose is to bring together those interested in "the relations between man and his occupation, equipment, and environment in the widest sense, including work, play, leisure, home, and travel situations."

Persons engaged in human factors or ergonomics activities perform two different types of functions that may roughly be regarded as scientific or professional.

The scientific functions revolve around attempting to determine the capabilities and limitations of human beings. The professional functions are associated with the application of these scientific findings in specific design situations. Human factors workers who are more involved with the science aspect of the discipline are typically associated with universities and research institutions, and they study such matters as the range of human physical and mental capacities, how learning takes place, the nature of complex systems, techniques for measuring human performance, and safety principles. Human factors workers more involved with the professional functions work closely with design and engineering teams in industrial, governmental, or military settings. They are involved in the design of specific products such as office furniture; automobile seats, dashboards, and control consoles; aircraft cockpits; tools; computers; and the interiors of buildings.

The picture of what is done by human factors workers is further complicated by the fact that the scientific and professional functions of ergonomics and human factors are multidisciplinary in nature. The sciences most related to human factors are psychology, biology, sociology, physiology, anthropology, mathematics, and statistics. The professions most relevant to human factors are industrial design, engineering, and architecture. While this multifaceted character of the discipline makes it difficult to give it a tidy definition, it also imparts a vitality and attractiveness to the work done by its members. It is fascinating to consider issues such as cultural expectations, social norms, individual needs, physical capabilities, aesthetics, and economic factors, as well as the physics of the machinery when designing something intended to be used safely, efficiently, and comfortably by human beings.

World War II played a great role in bringing human factors issues into sharp focus. Large numbers of young men were recruited and had to be quickly trained to operate new war machines that ranged from rifles and machine guns to tanks, airplanes, submarines, and radios. There had been airplanes in World War I, but they were relatively slow and simple to operate. For example, the WWI Sopwith Camel had a combined total of 16 controls and instruments. In contrast, the WWII Supermarine Spitfire had 57 combined controls and instruments. In the 20 years that separated these two famous fighter planes, aircraft speeds had increased by 300 percent and the number of combined controls and instruments had increased by 350 percent. As a result, pilots complained of "information overload." The speeds of these airplanes left too little time to monitor many instruments and required responses to occur in complex sequences more rapidly than could sometimes be managed. Terrible accidents were occurring with appalling frequency. Crash investigators often attributed the cause to "pilot error." Angry pilots cried "foul" and attributed the crashes to *design error*. Consequently, some researchers began studying ways to communicate information more effectively than with the traditional round-dial "steam gauge" type of instrument. New techniques were developed to train people to perform complex tasks, and recruiting standards were developed to find people who would be able to benefit from the training. Studies were conducted to develop controls that could be differentiated by feel instead of requiring the pilot to look at them. It wasn't long before the pilot and the airplane were being conceived of as two elements of a complex person-machine system. Thus, systems theory, which had been developing in electronics and mathematics, became a key concept in ergonomics.

Human factors and ergonomics experts play important roles in newly developing technologies. For example, heavily repeated activities such as the finger movements in typing at a computer keyboard for long periods of time can produce wrist pathologies such as carpal tunnel syndrome. Human factors scientists study the etiology of such pathologies, and their professional counterparts attempt to design keyboards, tables, chairs, and arm supports that prevent the pathologies from occurring.—H.W.

See also AIRPLANE; SYSTEMS ANALYSIS

Erosion

Erosion is the general name given to combinations of surface processes that loosen soil and rock from the surface of the Earth and transport the loosened material to other places. The Earth's surface is constantly changing, and erosion has played a significant role in creating the landforms of today. The interaction of the surface, the environment, and the response to any tectonic forces loosen and move material from highland areas to lower areas. Erosion destroys existing landforms; the material eroded is then deposited to help build new landforms in other places.

Weathering is the response of the Earth's surface to a changing environment. It is also a general term for the processes that loosen rocks and soil from the surface of the Earth. Weathering is not the same as erosion; it is part of what occurs during erosion. There are two major types of weathering. Mechanical weathering breaks up the surface material into smaller pieces. It doesn't change the composition of the material. Chemical weathering changes the composition of the material. The surface material is decomposed by the removal and/or addition of elements. The chemical agent that helps in these processes is usually water. Once the surface material is loosened, it can then be lifted and moved. The unconsolidated material being lifted and moved is called *sediment.* This transport of sediment is not the same as erosion. As with weathering, the transport is only part of the process of erosion. The lifting and moving is done by an agent of transport, usually running water, wind, or glaciers. These agents of transport are frequently called "agents of erosion."

Running water ("stream transport") is the most common agent of erosion. Streams transport sediment in solution, in suspension, and along the bottom of the stream. Usually all three types of transport occur simultaneously, but the majority of the load is carried in suspension. Most areas of the Earth are primarily shaped by erosion where both mechanical and chemical weathering occur, and streams transport the sediment. The classic signatures of these areas are rounded, softly contoured landscapes with V-shaped valleys in highland areas and broad, flat-bottomed river valleys in lowland areas.

If the region is arid, less chemical weathering occurs, and sediment is usually transported by either water or wind. Stream transport in these areas is intermittent and usually violent, one prominent example being flash floods. The wind moves sediment on a daily basis. Wind transports sediment along the surface of the Earth, and as with stream transport, most of the load is carried in suspension. Since air is much less dense than water, the sediment carried by the wind has smaller-sized particles. Streams usually transport silt and clay in suspension; the wind moves dust. The wind is not confined to a channel like a stream is, so it does not downcut and develop valleys. Landscapes in arid regions tend to be more vertical and angular since mechanical weathering dominates. While wind is a more important agent of erosion in these areas than in other areas, most of the erosion is still done by water.

Glaciers, or moving sheets of ice, occur in areas of high elevation and/or high latitude and are of limited importance as an erosional agent today. However, there were periods of time where glaciers covered a large percentage of the land on Earth. As erosional agents, glaciers are incredibly efficient, and areas formerly covered by glaciers still show the effects of their passing after hundreds of thousands of years. A glacier picks up or "plucks" sediment from the surface over which it moves. The difference between a stream or the wind lifting sediment and the glacier plucking sediment is that the glacier can move pieces of sediment far larger than anything that can be carried by streams or wind. The material plucked from the surface is then carried along within the ice abrading the surface and even more material is plucked and abraded. Glaciers change the landscape dramatically. In highland areas, the V-shaped valleys caused by streams become U-shaped as the glacier scours out all the loose material. In lowland areas, a glacier will scour down anything in its path, leaving large areas of low relief with gentle contours. Glaciers can also deposit vast amounts of earth in huge ridges known as *moraines*; this is how Cape Cod in Massachusetts was formed.

Erosion—be it by water, wind, or glaciers—constantly changes the surface of the Earth. The landforms of today were formed by the cumulative effect of erosion and of deposition. In terms of geological time, these landforms are only temporary, as erosion is constantly reshaping the surface of the Earth.—D.A.B.

See also PLATE TECTONICS

Further Reading: Charles C. Plummer and David McGeary, *Physical Geology*, 7th ed., 1996.

Escalator

The idea of a moving stairway dates back at least to the mid-19th century. In 1859, Nathan Ames of Massachusetts filed a patent application for an "improvement in stairs." Ames's moving stairway was just a paper invention, but it did include one feature that was incorporated in some subsequent designs: a comb-shaped plate at the top. The tops of the steps were shown as having a series of ridges (known today as *cleats*) that passed through the comb as the steps flattened. This would have allowed passengers to alight easily and would have prevented people and things from getting caught between the steps and the plate. Even so, Ames's escalator would have been difficult to operate and maintain, for electrical power was nonexistent at the time of the invention, and the device was projected to use human or steam power.

Toward the end of the century, the advance of electricity offered a practical source of power. Equally important, by this time, there was a perceived need for moving stairways. More buildings were multistoried, and large structures like department stores and railroad stations required something better than staircases to move people between floors. A successful response to this need was an "inclined elevator" devised by Jesse Wilford Reno and first installed at Coney Island in 1896. After a successful 2-week stint at the Iron Pier, the moving stairs were reassembled on the Brooklyn Bridge, where they helped pedestrians reach an adjacent railroad station. Reno's moving stairway achieved a fair amount of success, even though the tops of the steps were tilted forward at a 25-degree angle, putting the passengers' toes higher than their heels, and requiring them to lean forward while riding. This defect was addressed by a moving stairway invented by George Wheeler. It kept the steps level but lacked the comb at the end, so there was always the danger of things getting caught as the tread disappeared beneath the end plate. Wheeler's machine was bought out by Charles Seeberger, who subsequently gave it the name *escalator*. To partially alleviate the potential hazard at the end of the run, Seeberger's escalators were equipped with a waist-high diagonal arm that was intended to direct passengers away from the treads as they flattened.

During the second decade of the 20th century, the Reno and Seeberger machines were built by the Otis Elevator Company, although the original names were retained for many years. During the 1920s, Otis developed what is in its essentials the modern escalator. This was done by creating a moving stairway that combined the level steps of the Seeberger machine with the comb-and-cleats arrangement of the Reno. The new escalators were well received; 2 years after their introduction, more of them were installed than all of the previous versions put together.

Today's escalators are so common and reliable that most riders give little thought to them. They typically travel at about 30 meters per minute (98 ft/min), and have the capacity to move 4,500 passengers per hour with no crowding. Escalators are so well balanced that the weight of passengers moving downward helps to propel those moving upwards, making the energy costs of running an escalator quite low.

Further Reading: William Worthington, Jr., "Early Risers," *American Heritage of Invention and Technology*, vol. 4, no. 3 (Winter 1989): 40–44.

Eugenics

For centuries, people have debated whether individual capabilities are primarily the result of inborn attributes or environmental influences. In this issue of "nature vs. nurture," eugenics holds firmly to the former, postulating that an individual's physical, mental, and perhaps even his or her moral attributes are largely a matter of heredity. In short, "superior" people beget superior progeny, while "inferior" people beget inferior ones. While many biologists, sociologists, and social philosophers argued that individuals were highly malleable and therefore could be reshaped by the environments in which they lived, eugenicists believed that little could be done to alter what had been genetically determined. This view had important political implications. According to eugenicists, the most efficacious social policies did not strive to improve and uplift the physically and mentally weak. Rather, it was the responsibility of government to prevent these people from breeding more of their own kind (negative eugenics), while encouraging parents with superior characteristics to have larger families (positive eugenics).

The word *eugenics* was coined in 1883 by Francis Galton (1822–1911) in his book *Enquiries into Human Faculty*. The word is derived from the Greek, meaning "well born." Prior to developing a systematic theory of eugenics, Galton compiled statistics on a broad variety of phenomena, including the regional distribution of feminine beauty in England. In 1869, he was the first to apply the normal curve to the apparent distribution of human intelligence. Galton also anticipated later research when he conducted studies of identical twins in an effort to demonstrate the importance of heredity.

Galton was a cousin of Charles Darwin, the founder of the theory of evolution through natural selection. Like many intellectuals of his era, Galton had an interest in Darwin's theories, but the eugenics movement was not a simple outgrowth of evolutionary theory, and in some ways it was at odds with it. But eugenicists did find inspiration in the term "survival of the fittest," an expression that Darwin did not use in his own writings but was nonetheless associated with the theory of evolution.

Eugenics had little impact in the years immediately following the publication of Galton's book, but it gained influence during the latter years of the 19th century, and it was immensely influential during the early decades of the 20th century. By this time, eugenics had gained a large measure of legitimacy within the American academic community. Its respectability can be seen in the financial support provided in 1910 by the Rockefeller and Carnegie foundations for a Eugenics Records Office.

Despite its solid academic credentials, much of the support for eugenics rested on a wobbly empirical base. Anecdotal accounts of inherited mental strength and weakness were of particular importance. One of the most famous of these was provided by H. H. Goddard, the director of research at the Vineland Training School for Feeble-Minded Girls and Boys in New Jersey. Goddard purported to have discovered a family lineage that exhibited low mental capabilities generation after generation. After visiting one family unit of this lineage, which had been given the name Kallikak by Goddard, one of Goddard's assistants reported that "In this house of abject poverty, only one sure prospect was ahead, that it would produce more feeble-minded children with which to clog the wheels of human progress." The ancestor of this unfortunate brood was a normal man who had engaged in an illicit affair with a feeble-minded woman. In contrast, after the same man married a woman with normal faculties, all of their succeeding offspring went on to live meritorious lives. The conclusion was obvious: Feeble-mindedness was a hereditary condition that could be traced to a single defective ancestor.

In the United States, eugenics struck a responsive chord in the early 20th century, a time of high levels of immigration. Many native-born Americans of northern European stock feared the inundation of their country by "inferior" breeds. Eugenics provided the intellectual support for the Immigration Act of 1924, which set strict limits on the entry of people from places other than northern Europe. It was not until 1965 that this law was revised to allow large-scale immigration from other parts of the world.

Native-born men and women also were targeted by eugenics legislation. In the United States, 30 states had at some time laws that mandated the sterilization of people considered to be "hereditary defectives": alcoholics, epileptics, moral "deviants," and men and women of limited intelligence. Most of these laws were not rigidly enforced, but by the middle 1930s, an estimated 20,000 Americans, most of them in California, had been sterilized.

The intellectual foundations of eugenics eventually were undermined as biologists gained a better understanding of how genetic information actually was transmitted across generations. Beginning with the work of Gregor Mendel (1822–1884), biologists came to understand that heredity was not the simple process envisaged by most eugenicists. For one thing, hereditary inheritance entailed the transmission of recessive traits that might manifest themselves only after a number of generations had come and gone. It also became evident that genetically transmitted characteristics—both good and bad ones—may be clustered together. For example, there seems to be a statistical association of artistic creativity and bipolar disorder (a condition formerly known as manic depression). Consequently, attempting to remove a "bad" trait through eugenics may result in the loss of some "good" traits as well. Even if eugenics succeeded in making people smarter, stronger, and generally more capable, mutations would still occur and lead to the reappearance of unwanted traits. Finally, psychologists have become increasingly skeptical of the notion that a complex phenomenon like intelligence was a single trait that could be passed down from parent to child.

While the scientific assumptions of eugenics were being called into question, the motivations of eugenicists also came under suspicion. It was evident that the traits adjudged as superior by eugenicists were precisely the ones assumed to exist in themselves and people of their own ethnic group or social class. One example was Cyril Burt (1883–1971), British psychologist, who found statistical proof that low levels of intelligence were most commonly found among non-Christians, nonwhites, women, and members of the working class. That Burt's prejudices informed his research became evident in 1978, when the American psychologist D. D. Dorfman reviewed Burt's data and found that much of it had been fabricated.

By this time eugenics was already moribund. Not only was it based on questionable science, it had been sullied by its association with Nazism and other racist political movements. Eugenics always had been an integral component of Nazi philosophy, and after Hitler came to power in 1933, it was put into practice. In that year, 56,000 Germans deemed genetically unfit were forced to undergo sterilization. Far worse were the atrocities of the years that followed, when millions of Jews, gypsies, and other "undesirables" were murdered by the Nazis.

In recent years, the success of molecular biology in deciphering the genetic code has led to a renewed interest in the mechanisms of heredity. The developing field of sociobiology also has contributed to a renewed interest in the biological foundations of human behavior. In contrast to times past, however, there is much less certainty regarding the extent to which individual traits are the result of heredity. And, most importantly, advances in molecular biology now hold out the prospect of eventually eliminating or alleviating the damage caused by inherited defects. This emerging science and technology, which sometimes goes by the name *euphenics*, may be more effective than eugenics, and will not be burdened by the objectionable aspects of past efforts to control human reproduction.

See also DNA; EVOLUTION; GENE; INTELLIGENCE, MEASURES OF; GENETICS, MENDELIAN; NATURAL SELECTION; NORMAL CURVE; SOCIOBIOLOGY

Further Reading: Daniel J. Kevles, *In the Name of Eugenics: Genetics and the Uses of Human Heredity*, 1985.

Evolution

Initially a mathematical term associated with the generation of curves through some specified operation, *evolution* was generalized during the 17th century to describe the working out in detail of what is potentially contained in any principle. In the 18th century, its application to the growth of organisms as well as to the change of organisms over time was popularized by Erasmus Darwin (1731–1802), Charles Darwin's grandfather. Finally, during the mid-19th century, evolution became associated primarily with theories suggesting that current species of plants and animals originated through a slow and gradual process of development from earlier and initially very simple forms rather than as the result of some special divine creation or creations.

Evolutionary biological theories can be traced back as far as the Ionian Greek, Anaximander (c. 570 B.C.E.), who believed that human beings had developed out of some form of fish, and to Lucretius, a Roman Poet (c. 55 B.C.E.), who theorized that organisms were merely complicated aggregates of atoms and that some complex aggregates were more stable because they were better adapted to avoid disintegration in their environments. Modern evolutionary theories, however, have their most comprehensive and influential progenitor in the writings of Georges Louis Leclerk, Comte de Buffon (1717–1788), whose 1778 *Epochs of Nature* contained a number of ideas that were incorporated by subsequent evolutionary theorists.

First, Buffon argued that the Earth itself had changed dramatically over long time spans. Given the evidence of its magnetic properties and its density, which approximates that of iron, Buffon argued that the Earth had begun as a molten glob of ferrous material thrown off from the sun. Then, using Newton's laws of cooling, he was able to estimate the time that the Earth had taken to cool to its present temperature at about 100,000 years, or nearly 20 times the then widely accepted age of the Earth calculated from biblical genealogies by Bishop James Ussher in 1650. Some 17th-century natural historians, including the Englishman, John Ray (1627–1705), had already suggested extending the biblical time span to allow time, for example, for the erosion of mountains to produce alluvial plains. Later geological theorizing extended this time into the hundreds of millions of years, but Buffon was the first to offer a calculation for the age of the Earth based on commonly accepted physical principles.

As the Earth gradually cooled, Buffon argued, its surface underwent a series of transformations as it formed a crust on the outside. This crust buckled as it cooled, creating mountains and valleys. Vents of steam trapped inside escaped, and the steam eventually cooled to form oceans, lakes, and rivers. At some point, when the atmosphere was still hot and moist, some small number of kinds of flora and fauna, adapted to survive in that environment, came into existence (probably through the direct act of God). As the environment continued to change, those early plants and animals adapted through some unknown mechanism to fit the new environmental niches, creating the huge variety of species that now inhabit the Earth.

Buffon's works also raised a series of definitional issues that caused problems for evolutionary thinkers well into the 20th century. Most natural historians believed that species were "natural kinds," distinguishable from one another in some essential way and characterized by the fact that members of one species were incapable of producing fertile offspring when mated with members of another species. This notion provided a logical barrier to thinking of any transformation of species based on the inheritance of new characteristics (no matter how they may have come into being) that were to be passed on by any individual. Either the individual in question would still be a member of its initial natural kind, in which case no new species could be produced, or it was definitionally of a "new" species, in which case its offspring, when mated with an "original stock" member, could not be fertile and pass those characteristics on.

Buffon resolved this problem by arguing that "species" were defined by human beings for their own classificatory convenience and that they did not exist in nature at all. Instead, there was a continuous distribution of organisms along many dimensions, and animals relatively close to one another reproduced easily with one another, while those far apart did not. At one level, this strategy avoided the traditional problem of how natural kinds could be transformed, and it seems to be close to a position later adopted by Charles Darwin, but it left open a whole series of problems about why many clearly delimitable and isolatable groups of organisms do seem to exist. This set of problems was resolved to the satisfaction of most biologists only in the mid-20th century by Ernst Mayr and other mathematically inclined students of biological populations.

Finally, one of the central notions incorporated within Buffon's discussions made the development of modern evolutionary theories more difficult than it might otherwise have been, and had long-term implications for how evolutionary theories were applied to the understanding of human racial characteristics. In conceiving of the transformation of an initial distribution of organisms over time, Buffon generally thought of those changes as constituting "degenerations" from an initially superior stock. Thus, for example, he believed that most organisms originated in the temperate regions of the Old World and migrated, with changes, into East Asia, the Western Hemisphere, and Africa. This notion led to a correspondence with Thomas Jefferson (1743–1826) over whether American animals were generally smaller than (and presumably inferior to) their European counterparts. More importantly, it led to claims of superiority on behalf of a presumably original European human race over other "degenerate" human races found in Africa, East Asia, and the Americas.

In his 1809 *Philosophie Zoologique*, Buffon's protégé Jean-Baptiste Lamarck (1744–1829) modified Buffon's theories in several ways. He proposed a process of transformation that moved constantly away from a single simple and inorganic origin toward increasing complexity over time. The process began with the creation of simple chemical compounds in the superheated and compressed early atmosphere of the cooling Earth. It proceeded through the joining of complex molecules into primitive, crystallike growing and self-replicating organisms, and then branched into numerous lines of development, each of which continued to divide much as tree branches continue to divide as they move away from the trunk. Thus Lamarck's transformist doctrines had a progressive rather than a degenerative orientation.

Also central to Lamarck's theory was the notion that in the later stages of the transformational process, organisms intentionally changed themselves to adapt to their environment and that these changes were transmitted to their progeny, creating new species. Thus, for example, some early precursors of the giraffe stretched their necks to reach leafy food sources that were above the reach of other animals. The lengthened neck was passed on and extended even more in the next generation, until the giraffe that we know came into existence.

This notion of willed improvement had tremendous appeal to 19th-century social theorists and moral philosophers, encouraging such thinkers as Herbert Spencer (1820–1903) to view social change as a form of superorganic evolution whose driving force lay in the individual effort of competing individuals.

The most successful theory of biological evolution to emerge in the 19th century was the theory of evolution by natural selection, announced simultaneously by Charles Darwin (1809–1882) and Alfred Russell Wallace (1823–1913) in papers published in 1858 in the *Journal of the Linnaean Society of London* and given its most important expositions in Darwin's *On the Origin of Species by Means of Natural Selection, or the Preservation of Favoured Races in the Struggle for Life* of 1859, and *The Descent of Man* of 1870.

During a stint as unpaid naturalist on the globe encircling H.M.S. *Beagle* from 1831 through 1836, Darwin had made extensive observations of the geographical distribution of plant and animal species, setting himself the goal of discovering how new species come into existence. The theoretical key to Darwin's solution came to him while reading Thomas Malthus's (1766–1834) 1802 *Essay on Population.* Malthus had argued that the unchecked population of human beings would expand geometrically, while the food supply could be increased linearly at best. Given this situation, the population would eventually outstrip its food sources, and weaker members of the population would die, leaving a stronger or more-fit population to propagate. Darwin generalized this argument to all organisms.

Darwin initially posited the continuous and random emergence of small inheritable differences (variations or mutations) among individuals constituting a single species. Next, he argued that if any of these variations better adapts the individual to succeed in nourishing itself and reproducing, then the progeny of the changed organism will also be more successful in the struggle for life. Eventually the individuals carrying the variation will come to dominate the population because their unchanged competitors will die out. Over long periods of time, the accumulated gradual changes may become so great as to result in an entirely new species. When the changes occur in an environment that offers many ecological niches (i.e., many sources of food and opportunities for self-defense), the new variants might well fit better into a different niche than the original population. Accordingly, over time new species may come into existence alongside their older parent species. Finally, as the environment of an initially successful species changes, that species may gradually give rise to new, better-adapted species, or it may become extinct, as better-adapted competitors with radically different lineages appropriate available food or protect themselves more effectively against predators.

In *The Origin of Species*, Darwin was able to present a huge amount of evidence to support his theory of evolution by natural selection from the fossil record, from the artificial selection involved in breeder's experiences, and from the geographical distribution of animal and plant populations. In *The Descent of Man*, he focused on the origins of human beings through the evolutionary process and emphasized sexual selection, moving away from the views of Wallace, who continued to insist on environmental fitness. In *The Expressions of Emotions in Man and Animals* of 1872, Darwin examined the evolution of behavior, insisting that behavioral differences must be manifestations of somatic (physical) variations.

Though many biologists became adherents of Darwinian evolutionary theory within a few years after Darwin's work, the theory has never been without both scientific and religious critics. During the late 19th century, Lord Kelvin (1824–1907) was able to argue on the basis of classical thermodynamic arguments that the age of the Earth was less than 50 million years, a time span much shorter than evolutionary theorists needed. But with the discovery of natural radioactivity as a terrestrial heat source, this problem was resolved. For a long time, few scientists, including Darwin, were comfortable without some satisfactory theory about how inheritable changes are produced and transmitted. Only with the emergence of modern genetics in the 20th century has there been wide agreement on these issues, and modern genetics has produced some changes in classical evolutionary theory. It has become clear, for example, that mutations are not always small, gradual, and continuous. This fact helps to account for a puzzling feature of the evidence concerning evolution that has bothered evolutionary biologists beginning with Darwin. In its classical form, evolutionary theory implied that evidence should exist for the continuous transformation of any progenitor species into one or more later species, but the fossil record generally shows substantial discontinuities. Darwin was able to account for the "missing links" in fossil evidence by arguing that phenomena such as floods, glaciation, and earthquakes produce evidentiary discontinuities, but the existence of discontinuous changes based on the discrete character of genetic mutations also helps to explain the nature of the fossil evidence.

Religious opposition to Darwinism has been widespread, though not universal, ever since the publication of *The Origin of Species*. Evolutionary theory was initially perceived as most threatening to the "argument from design" favored by liberal Christian groups (groups that did not insist on a literal interpretation of the nonprophetic books of the Bible and emphasized the use of human reason in evaluating the plausibility of religious claims). According to this argument, the complexity and nearly perfect adaptation of organisms to their environment was evidence for the existence of an immensely clever creator who had organized the structure of the universe so that it resembled an incredibly complicated and precise timepiece. But evolutionary theory suggested that the universe as we know it is the product of random changes selected over time and not the product of any kind of design. Initially, Darwinian evolution seemed to many religious writers to be consistent with the predestinarian arguments of radical Calvinist and evangelical sects, for the fact that only a few favored variations were chosen, while the vast majority were doomed to extinction, paralleled the notion that only a small and arbitrarily chosen group of people would be among God's elect. But during the 20th century, fundamentalist Christians (who focus on the sufficiency of the Bible to salvation and on its inerrancy if taken literally) have been particularly distressed by the aid and comfort that the materialist emphasis of Darwinian evolutionary theory gives to secular humanist trends in Western culture. As a consequence, they have opposed the exclusive teaching of Darwinian evolution as an account of the origins of species and have developed theories of "scientific creationism" intended to incorporate both scientific evidence and biblical literalism. To date, these theories have provided little challenge to evolutionary theories within the scientific community, but they seem to have immense popularity among lay audiences.—R.O.

See also EVOLUTION, PUNCTUATIONAL MODEL OF; GENETICS, MENDELIAN; GEOLOGICAL EPOCHS AND PERIODS; NATURAL SELECTION; SOCIAL DARWINISM

Further Reading: Peter J. Bowler, *Evolution: The History of an Idea*, rev. ed., 1989.

Evolution, Human

The story of human evolution is grounded in complex interactions among the environment, morphology, and behavior. The data currently available indicate that the earliest hominids, those species directly related to our own evolutionary lineage, arose on the savannas of East Africa approximately 4.5 million years ago (mya). Since then, hominids have gradually undergone dramatic changes: The brain has expanded over 300 percent, the body became better adapted for bipedal locomotion, teeth were modified to accommodate a more omnivorous diet, tool cultures became finer and more varied, and the population expanded and spread, eventually to span the entire globe. While much is known about the pathway of our ancestors, new data constantly arise, and scientists must reconsider the model.

One important new source of information was developed in the late 1960s, primarily by two scientists from the University of California, Vincent Sarich and Alan Wilson. Rather than hunting for the bones and stones of our ancestors, Sarich and Wilson worked in a laboratory, studying the immune responses of living creatures. They found that mixing the blood antibodies of one species with the blood antigens of another causes an immune response, and that the strength of that response was a good indicator of how closely the two species were related. Thus, samples from a horse and a chicken would elicit a minimal reaction, as these two species are only distantly related, while samples from a human being and a chimpanzee elicited a strong reaction, presumably because the genetic material of these species is so similar. A major breakthrough in Sarich's and Wilson's work came when they discovered a strong negative correlation between the strength of the immune response and the length of time since the two species last shared a common ancestor. In effect, Sarich and Wilson had discovered a molecular clock that ticked at a constant rate, so that comparing blood samples of two living animals could indicate how long ago those animals had diverged and begun to evolve along separate lines. Sarich and Wilson stunned the world when, in 1967, they announced that their research suggested that human beings and chimps shared a common ancestor as recently as 5 to 8 mya. At that time, it was commonly believed that these species had been evolving separately for perhaps 20 million years. Sarich's and Wilson's assertion seemed preposterous. Since then, however, techniques for comparing DNA have improved dramatically, and they continue to provide similar results. It is now widely accepted that the hominid lineage probably arose 5 to 8 mya.

At the time of Sarich's and Wilson's seminal work, little fossil data existed to support their conclusions. Now, however, a growing body of evidence points to East Africa as the seat of hominid evolution, and new fossils constantly push back the date of our earliest ancestors. The oldest material is commonly placed in the genus *Australopithecus*. These species are distinctive because their skeletal anatomy reflects the adaptation to bipedal locomotion, the first hallmark of hominids. The bones of the pelvis are short and broad—in contrast with the long, narrow bones of a chimpanzee—which allow attachment for large gluteal muscles and provide a basin for support of the viscera. The knees cant in at an angle, giving a knock-kneed appearance, which focuses the body's weight to a single, low point. The feet are significantly different from a chimpanzee's, with a nonopposable big toe, an enlarged heel bone, and high arches running both longitudinally and transversely, all of which provide a stable platform to support and balance the body. The shift from the arboreal locomotion of a chimpanzee-like ancestor to the bipedal walking of a hominid is clearly written upon the fossil bones of the australopithecines.

Why the first hominids began to walk upright is more difficult to discern. Current models propose that the impetus for this adaptation was a change in the environment. Geological forces led to a gradual cooling and drying of East Africa approximately 15 mya, caus-

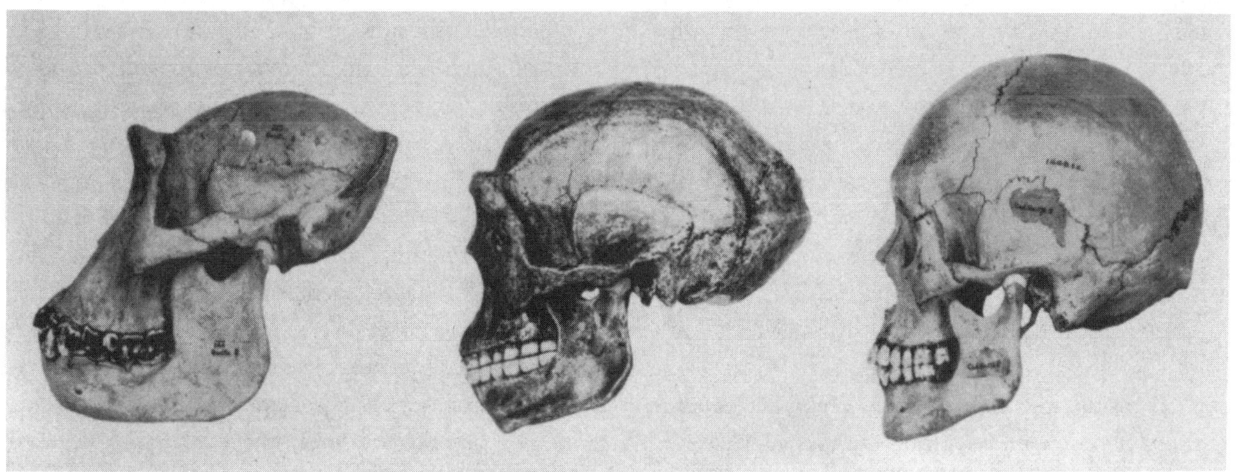

The skull of *Homo erectus* flanked by skulls of a chimpanzee and a modern human (from R. Lewin, *Human Evolution: An Illustrated Introduction*, 1984).

ing the vast forests then present to die off, making way for open woodland and grassland. This new environment, it is proposed, made bipedal walking more advantageous, perhaps for various reasons. Large predators often lay hidden in the grass, and an upright hominid would be more likely to spot such a danger. Upright walking also freed the hands to carry things, such as food; individuals who could carry fruit away from the tree might avoid the intense feeding competition surrounding the food patch. Bipedality is also a very efficient mode of locomotion on the ground; consequently, as trees began to disappear and the resources they provided were more widely dispersed, bipedal hominids could more easily travel between food sources. It is also possible that upright posture helped to cool the body by exposing less surface area to the hot sun. A combination of selective pressures was undoubtedly responsible for the profound anatomical changes associated with bipedality.

The australopithecines were clearly adapted for bipedality, but they retained many apelike characteristics. Fossilized limb bones show that these early hominids had relatively long arms, and long, curved fingers and toes. These features suggest a continued reliance on tree climbing, perhaps to gather food or avoid predators. Australopithecine dental morphology was also somewhat apelike, with large molars for grinding vegetal foods. The oldest species also had long canine teeth, resembling the teeth of a chimpanzee more than a modern human being's teeth. Perhaps the most strikingly nonhuman characteristic, however, is the size of the brain: The earliest australopithecines had a cranial capacity of less than 500 cc, compared with our modern average of 1,400 cc. In contrast with early theories about human evolution, neural expansion did not lead the way; instead, small-brained but upright creatures were the first hominids.

The oldest hominid fossils are currently found in East Africa and date to approximately 4.5 mya. By 3 mya, however, australopithecines also inhabited South Africa. The success of the genus must have led to population expansion, migration into new areas, and eventually a proliferation into several new species. Another major transition in australopithecine evolution occurred approximately 2.5 mya. Again, the environment may have played a major role; based on the array of fossilized animal bones found at both East and South African sites, Elizabeth Vrba has postulated that another burst of atmospheric cooling and drying led to further expansion of the savannas some 3 to 2 mya. Soft, fleshy fruit resources may have become scarce, giving way to harder, less nutritious foods such as roots, tubers, and grass seeds. As the australopithecines began to exploit these new resources, their morphology adapted to accommodate the stresses of heavy chewing. Later australopithecines, those living from 2.5 to 1.0 mya, had massive molar teeth, heavy jaw and facial bones, and bony crests on the tops of their heads for the attachment of large chewing muscles.

It is possible that these "robust australopithecines" were forced into a vegetarian lifestyle not only because of changing resources but also because of competition. Current data indicate that members of our own genus, *Homo*, appeared at about this same time. The earliest species, *Homo habilis*, was distinct from the australopithecines with which it coexisted. Skeletal morphology was similar, but the chewing apparatus suggested an omnivorous diet. In contrast with the robust australopithecines, *Homo habilis* had smaller molars, lighter facial and jaw bones, and no cranial crests, indicating smaller chewing muscles. Furthermore, tiny striations on the molars indicate that this species ate meat in addition to vegetal foods. The hallmark of the genus *Homo*, however, was brain expansion: *H. habilis* had cranial capacities up to 750 cc.

The new morphology of *Homo* is associated with the new behaviors as well. Crude stone tools are found at *Homo habilis* sites in East Africa, along with the bones of butchered animals. It was originally proposed that tool-using *H. habilis* was a proficient hunter; however, more recent analysis has led many to conclude that these hominids procured meat through scavenging the kills of other predators. Concentrations of cultural remains suggest to some scientists that groups of *H. habilis* centered their daily activities around a home base. As the hominid brain grew, culture and social life became more complex. Moreover, as culture and sociality became more important features in hominid survival, they enhanced selective pressures toward increasing brain size. Thus, after 2 million years in which the australopithecines made relatively few evolutionary advancements, *Homo habilis* entered a mutually reinforcing cycle of neural and cultural expansion.

While *Homo habilis* was confined to Africa, its successor, *Homo erectus*, soon expanded its range to include Asia and possibly Europe. The earliest evidence for *H. erectus* comes from East Africa, dating to 1.8 mya. This species is distinct from *H. habilis* in body, brain, and culture. By this stage, the skeletal anatomy had lost virtually all of its apelike characteristics and closely resembled the modern form. Continuing the trend in neural advancement, cranial capacities in *H. erectus* range from 800 to 1,100 cc. And finally, tool cultures associated with this species are more complex, including a wider variety of tool types, more intricate in their manufacture. These behavioral adaptations may have been responsible for *H. erectus*'s success, for this species soon spread into southeast Asia (as Java Man), northern Asia (as Peking Man), and perhaps into Europe as well. They may have been aided by improved hunting techniques, and certainly their lives were changed by the controlled use of fire, the best evidence for which comes from China, dating to roughly 0.5 mya. In addition, *H. erectus* may have led to the demise of the australopithecines, for these hominids became extinct approximately 1.0 mya.

Homo erectus was a fairly static species, changing little in the roughly 1 million years of its existence. However, it gradually gave way to a new species, *Homo sapiens*. As these hominids spread throughout Africa, Asia, and Europe, semi-isolated populations began to develop slightly different morphologies, what we might call *racial variations*, some of which may have been adaptations to their immediate environments. The most well known of these variants is Neanderthal Man. In contrast with the popular view of this species, *Homo sapiens neanderthalensis* was an advanced hominid, highly adapted both morphologically and culturally to the glacial climate in which it lived. The Neanderthals were physically powerful people, short and stocky, and heavily muscled. Their cranial capacities averaged 1,500 cc, larger than the capacities of modern human beings. However, their skulls were long and low, with huge brow ridges blending into sloping foreheads. It is unclear how the differences in cranial shape relate to neural activity, but it is possible that Neanderthals had lower capacities for planning, abstract thought, and communication than do modern human beings. Nevertheless, Neanderthals were sophisticated hominids. They lived in large social groups, often inhabiting caves to protect them from the cold of Europe in the middle of an Ice Age. They hunted cooperatively with refined stone tools. Fossil evidence suggests that they cared for the elderly and infirm members of their groups, and that they buried their dead, possibly the earliest indications of ritual activity.

The Neanderthals existed in Europe and southwest Asia for at least 100,000 years, until they rapidly disappeared, replaced by a newer hominid, anatomically modern *Homo sapiens*. Where and when this subspecies evolved is hotly contested. Some propose that modern human beings evolved in Africa and later spread across the continents, replacing populations of hominids they encountered. Others assert that populations of modern human beings arose separately in different parts of the world, with enough gene flow between groups to maintain a common species. Some support for the former model comes from studies of mitochondrial DNA (mtDNA). Rebecca Cann and Alan Wilson compared the rapidly changing mtDNA of several modern human populations. They came to the conclusion that all human beings living today descend from an ancestral population that lived in Africa approximately 100,000 to 200,000 years ago. While their research and conclusions have been highly controversial, some fossil evidence supports their theory. The earliest known fossils of anatomically modern human beings come from South and East Africa, and are tentatively dated to roughly 100,000 years ago. On the other hand, modern human populations maintain some morphological traits that first appeared in analogous populations of *Homo erectus*, indicating some regional continuity. The advent of modern human beings thus remains unclear.

What is clear is that modern human beings soon expanded to inhabit all of Africa, Asia, Europe, and Australia, and then went on to North and South America. Their success was, again, due to advances in their brains and their cultures. Although the modern human cranial capacity is somewhat less than that of the Neanderthal, neural organization is marked by a huge frontal cortex and considerable cortex convolutions. Associated with this modern brain was an exponential expansion of cultural elements. By 40,000 years ago, the human tool kit included elegant stone blades, stone spear heads hafted on to wooden handles, barbed harpoons, and bone needles with eyes. By 10,000 years ago, human beings were decorating their caves with elaborate paintings, and adorning their bodies with jewelry. Approximately 10,000 years ago, some human populations had become sedentary, and soon they domesticated crops and livestock. Current models of human evolution are based on the complex relationships among environment, morphology, and behavior. However, as new data are discovered, the specifics of these models must continually be revised.—L.E.M.

See also AUSTRALOPITHECINES; DNA; EVOLUTION; GEOLOGICAL EPOCHS AND PERIODS; *Homo habilis*; *Homo sapiens sapiens*; SOCIAL DARWINISM; NEANDERTHALS; NATURAL SELECTION; *Homo erectus*; NEOLITHIC AGRICULTURAL REVOLUTION; PILTDOWN MAN

Evolution, Punctuational Model of

At the core of Charles Darwin's theory of evolution through natural selection is the idea that organisms evolve very slowly through a series of gradual steps. "*Natura non saltum fecit*" ("nature does not make jumps") is a basic assumption of Darwinian evolution. Darwin explained the process:

> . . . As natural selection acts solely by accumulating slight, successive favourable variations, it can produce no great or sudden modification; it can act only by very short and slow steps.

Little was known about the actual mechanisms of heredity in Darwin's lifetime. The science of genetics was launched by Gregor Mendel's investigations into plant heredity and was eventually incorporated into evolutionary theory. The *modern synthesis*, as this coupling came to be known, also embraced the notion of gradualism. As the geneticist Theodosius Dobzhansky (1900–1970) put it:

> . . . species arise gradually by the accumulation of gene differences, ultimately by summation of many mutational steps which may have taken place in different countries and at different times.

Again, gradualism was inherent in evolutionary theory.

Beginning in the 1940s, the accepted model was challenged by an alternative paradigm that postulated the rapid occurrence of evolutionary changes. From this perspective, evolution was not a matter of gradual alterations within an existing species eventually culminating in the emergence of organisms so different that they formed a new species. Rather, evolutionary change occurred as new species rapidly branched off from existing ones. Of course, "rapidly" as it is used here is a relative term; the branching of a new species may take several tens of thousands of years, although some species of freshwater fishes appear to have emerged in less than 10,000 years.

This alternative paradigm goes by a number of labels: "the punctuational model of evolution," "punctuated equilibrium," and "quantum speciation" have all been used. The development of the new model has been the work of a number of scientists. The first were Richard Goldschmidt, a plant geneticist, and J. C. Willis, a plant biogeographer, who in the 1940s posited that very large mutations (macromutation) caused new species to suddenly appear. At about the same time, Ernest Mayr, a biologist trained in ornithology, put forth the claim that some genera of birds had suddenly emerged after small populations had been isolated from their ancestral species. In the early 1970s, Stephen J. Gould, Niles Eldredge, and Steven Stanley argued in support of Mayr's views, and added the terms "punctuational" and "punctuated equilibrium," to the vocabulary of evolution.

By this time, Mayr had already noted that a punctuational model resolved the lack of congruence between evolutionary theory and the fossil record. Darwin himself had ruefully noted that a study of the chronological sequence of fossils gave very little indication of gradual evolution, for it was remarkably lacking in transitional organisms. Darwin attributed this to imperfections in the fossil record, along with the expectation that transitional forms necessarily would be rare. This was a reasonable standpoint at the time he was writing, but more than a century of fossil discovery has not improved the situation greatly. This is not a problem for the punctuational model; it does not require an elaborate sequence of transitional organisms, for speciation (the formation of new species) occurs with relative suddenness.

In the punctuational model, most evolutionary change is the result of rather rapid speciation. The other side of the coin is that established species usually undergo few changes over time. The endurance of species is a problem for the gradualist model of evolution. After all, if natural selection operates on lineages of organisms undergoing a series of small changes, why have some groups of plants and animals gone for millions of years with no apparent significant changes? As punctuationalists are fond of pointing out, some types of organism have endured for so long that they have been labeled "living fossils." For example, lungfishes appear to have remained largely unchanged for 300 million years, while another type of fish, the bowfin, looks about the same today as its did 100 million years ago. As noted above, in the punctuationalist model, evolution occurs through the branching off of new species; living fossils are groups of organisms that have not speciated.

The punctuational model also addresses a thorny problem for evolutionary theory, the very rapid emergence of radically different types of animals. For example, a great variety of mammals appeared over a relatively short span of time in the Earth's history. The first mammals were small, rodentlike creatures, yet within only 12 million years creatures as different as whales and bats had made their appearance. In fact, all major orders of mammals appeared during the first 12 to 15 years of the Cenozoic Era. Timetables like these are difficult to accommodate to a gradualist model of evolution.

In the punctuational model, speciation occurs only when circumstances are favorable. In particular, it is likely to happen only when a small number of individuals reside in an area that is bounded by a geographical feature like a mountain range or a river that seals it off from other environments. With a small, isolated population, a genetic variation (mutation) that arises in a single individual will spread through a larger percentage of the total population than would be the case if the population base is large. With a small population, the genetic variation will not be excessively diluted; rather, it will be augmented as succeeding generations interbreed.

As with conventional views of evolution through natural selection, in the punctuational model, genetic changes endure when environmental circumstances are favorable. A favorable circumstance may be nothing more than the extinction of other, competing species. It is likely, for example, that the number and type of mammals increased rapidly due to the extinction of the dinosaurs. An environment free of competitors also may emerge when geological change creates a lake, an island, or some other place that remains isolated from the surrounding territory. This may not be the end of the matter, however. Geological changes are never-ending, so a species that came into being in a particular environment eventually may be presented with physical pathways that allow it to move into new domains. As a result, species that first had appeared in a limited space may end up covering a large amount of territory.

Models of evolution based on the relatively rapid emergence of new species are not accepted by all evolutionary biologists, and gradualist models are still favored in many quarters. In the years to come, the construction of a more complete fossil record should provide important information about the course of evolution. A better understanding of genetic change and transmission also will be of crucial importance to the development of evolutionary theory.

See also DINOSAURS; EVOLUTION; GENETICS, MENDELIAN; NATURAL SELECTION; PARADIGM

Further Reading: Steven M. Stanley, *The New Evolutionary Timetable: Fossils, Genes, and the Origin of Species*, 1981.

Execution by Lethal Injection

Lethal injection is the most common means of execution in the United States. Of the 38 states that allow capital punishment, 32 either use it as their primary means of execution or give prisoners the option of dying in this way. In 1995, of the 56 prisoners executed in the United States, 49 died by means of lethal injection.

Execution by lethal injection begins when two needles are inserted into the arms of the person to be executed (two needles are used so that if one malfunctions, at least one will be operative). A saline solution is then slowly administered, followed by 5 g of sodium pentothol, a powerful tranquilizer that renders the person unconscious. Death comes as the result of the administration of 50 cc of pancuronium bromide, which brings on muscle paralysis, followed by 50 cc of potassium chloride, which stops the heart. The process takes about 2 minutes.

Although lethal injection is supposed to be painless, observers have reported that some prisoners seem to have been in a great deal of pain, writhing and gasping as the lethal chemicals flowed into them. Simply finding veins in which to insert the needles may be a problem, especially for prisoners who have been heavy intravenous drug users. The use of lethal injection also poses a major dilemma for members of the medical profession. Most states require that a doctor be present at an execution, yet physicians' professional societies forbid direct participation in executions of any sort. This restriction follows from physicians' codes of ethics that prohibit doctors from harming the people they treat. At the same time, however, some doctors have expressed a concern that prison executioners lack the knowledge to always administer lethal injections in a proper manner. Finally, opponents of capital punishment note that while lethal injection may be less painful than hanging or asphyxiation in a gas chamber, its seeming humaneness obscures the fact that it still constitutes state-sponsored killing.

Expanding Universe, The

By the early 19th century, astronomers had begun to get a sense of the immensity of the universe. Some speculated that objects in the night sky that appeared to be clouds of dust were in fact whole galaxies equal in size to our own Milky Way. Twentieth century developments in the construction of telescopes confirmed this speculation. In 1923, the American astronomer Edwin Hubble (1889–1953) trained the 254-cm (100-in.) Mt. Wilson reflecting telescope on a nebula situated in the constellation Andromeda and saw individual stars where only dust had been seen before.

What Hubble saw was a disk-shaped cluster of stars with radiating spiral arms of star clusters. Especially important was his discovery that some of the stars in these clusters exhibited periodic variabilities in their luminosities; that is, the brightness of each star varied over time according to a definite schedule. This observation was of great importance for determining the distances of these stars from the Earth. A few years prior to Hubble's observation, two other American astronomers, Henrietta Swan Leavitt (1868–1921) and Harlow Shapley (1885–1972), observed a group of stars known as *cepheids* and found that for each star there was a close relationship between the period of variation of its luminosity and its absolute luminosity. Consequently, a measurement of a star's variability could be used to determine its absolute luminosity. The absolute luminosity could then be compared to the apparent luminosity, the actual amount of light falling on a telescope's reflecting mirror. Since apparent luminosity depends on both an object's absolute luminosity and its distance from the Earth, when the former is known, the latter can be easily calculated. When Hubble performed a few calculations, he found that the Andromeda nebula was 900,000 light-years from the Earth, 10 times the distance of the farthest known object in our own galaxy. (Later calculations based on better values for the relationship between period and luminosity increased the distance to more than two million light-years.)

Hubble's observations and calculations dramatically increased human understanding of the size of the universe. No less important was the discovery that the universe was getting bigger in the sense that individual galaxies were moving away from one another, and continuously creating more space in the process. The empirical basis for this idea goes back to the 19th century. In 1842, the Austrian physicist Johann Christian Doppler (1803–1853) noted that the wavelengths of sound and light change as their sources move in relationship to an observer. For example, the sound of a car's horn increases in pitch (that is, it has a shorter wavelength) as the car approaches, and drops in pitch (a longer wavelength) as it recedes. In 1868, the Doppler effect took on a particular importance for astronomy when spectral analysis began to be used to study the stars. It had long been known that when the light of the sun passed through a slit and then a prism, the resulting spectrum included hundreds of dark lines that always occupied the same place. It came to be understood that the dark lines were caused by the absorption of particular wavelengths of light and that this absorption was the result of the presence of specific chemical elements. This made it possible to determine the chemical composition of a far-off body through a spectral analysis of the light it sent to the Earth.

In 1868, the British astronomer William Huggins (1824–1910) noted that the dark lines in the spectra of some stars were shifted slightly to the red or blue ends of the spectrum when compared with spectral lines produced by the sun. He correctly inferred that this was the result of the Doppler effect, and that some of these stars were moving closer to the Earth while others were moving away from it. During the second decade of the 20th century, Vesto Melvin Slipher (1875–1965), an American astronomer, observed that the spectral lines of many nebulas also exhibited a red or blue shift that indicated motion away from or toward the Earth, respectively. The Andromeda nebula, for example, was found to be moving toward the Earth at a velocity of about 1,000 km (620 mi) per second. Subsequent observations revealed that most galaxies were moving away from the Earth, most of them at even greater speeds.

After conducting more observations of a number of galaxies, Hubble in 1929 found that the shift toward the red end of the spectrum increased in proportion to a galaxy's distance from the Earth. This led to the formulation of what came to be known as Hubble's law: the greater a galaxy's distance from the Earth, the greater the velocity. Hubble's law is an important component of cosmology, but like some other scientific laws, its empirical undergirding was weak at first. Because they were largely based on galaxies that were relatively close, Hubble's observations did not show a clear relationship between distance and velocity. As sometimes happens in the history of science, an important principle was based on initially inconclusive data.

As observations improved in quantity and quality over the next few years, Hubble's initial formulation was borne out. In 1931, Hubble was able to state that for every million light-year distance, velocities increase by 170 kilometers per second; this came to be known as Hubble's constant. In 1952, Walter Baade's (1893–1960) research into the period-luminosity relationship in the cepheids indicated that Hubble had used distances that were about 1/10 what they should have been; consequently, Hubble's constant is now given as 15 kilometers per second per million light-years.

By using Hubble's constant in conjunction with the known distance of remote galaxies it is possible to estimate the age of the universe. Hubble's initial calculations gave an age of about 1.8 billion years, but the revised Hubble constant gave an age of 20 billion years. Significantly, estimates based on other sources—especially rates of decay for radioactive materials and regular patterns of stellar evolution—provide about the same age for the universe.

Years of observations and calculations indicated that the universe was expanding, but they did not demonstrate that the universe would necessarily continue to expand. It is possible that the continual expansion of the universe might eventually be checked by the mutual gravitational attraction of the objects in the universe. Gravity depends on mass, so one determinant of continued expansion is the amount of matter in the universe. The other determinant is the speed at which objects are receding from one another and whether this speed is sufficient to overcome gravitational attraction, that is, the attainment of what is known as an *escape velocity*. Efforts have been made to use the magnitude of the red shift to determine if galaxies are attaining sufficient speed to continue their movement away from other galaxies. Up to now, these efforts have been thwarted by the inherent difficulty in determining the distance of remote galaxies. As was done decades ago, the distance of these galaxies is ascertained by comparing their apparent luminosity with their absolute luminosity, but since the light of the latter originated thousands of millions of years ago and is reaching the Earth only now, there is no certainty that their luminosity is the same today as it was when the light we receive today was first radiated. If they were brighter in the past, their true distance of these galaxies will be underestimated. Uncertainties such as this prevent the formulation any definitive answer to the fate of the universe. It may go on expanding (the "open universe" theory), or it may someday collapse back on itself (the "closed universe" theory). It is also possible that the universe may go through continual cycles of expansion and contraction.

While a great deal of uncertainty surrounds the future of the universe, there is a stronger consensus regarding its origin. Most (but not all) cosmologists now subscribe to the Big Bang origin of the universe, the theory that the universe began in an immensely compressed state that in a brief interval of time "exploded" into expanding space and matter. There is some irony in the fact that a theory that assumes that the universe began as a tiny bit of compressed space and matter began with a determination of the vastness of the present universe and the discovery that this immense collection of space and matter is still expanding.

See also BIG BANG THEORY; DOPPLER EFFECT; LIGHT, SPEED OF; SPECTROSCOPY; TELESCOPE, REFLECTING

Further Reading: Steven Weinberg, *The First Three Minutes: A Modern View of the Origin of the Universe*, 2d ed., 1993.

Experiment

An experiment is an experience or activity that is executed by design and with a purpose in mind. The goal of an experiment is knowledge. Thus, playing first base on a softball team is not an experiment, even though it is an activity, unless it was undertaken in order to learn something about one's ability, endurance, competitiveness, etc. The purpose of an experiment is usually either to gain factual information or to prove a point or demonstrate an hypothesis.

If the purpose of an experiment is simply to gain factual information, then the design of the experiment is simply to arrange suitable experiences in which the matter can be directly observed and learned. Perhaps a scientist wants to know the melting point of a chemical substance such as aspirin. We know that aspirin melts; we just do not know at what temperature. The determination is done by experiment. In particular, the scientist needs to design an experiment in which she or he can observe aspirin melting and measure the temperature simultaneously. A simple apparatus for this purpose uses a small capillary tube containing a sample of purified aspirin, attached to a mercury-in-glass thermometer and immersed in an oil bath that can be heated slowly. When the aspirin melts, as observed by the first wetting of the capillary walls, the temperature of the thermometer is recorded. The scientist can report the melting point of aspirin as the temperature observed in this way. Presumably we now know something that we did not know before.

The design of an experiment is extremely important; indeed, the result of an experiment is only as good as the design that produced it. Putting it another way, what we think we know as the result of an experiment is only as good or useful as the experiment's design and execution. If, in the example above, the sample of aspirin was tapped into a capillary tube that had been used in a previous experiment with sucrose (sugar), then the contamination of the aspirin by residual sucrose would lead to very different results—probably, in fact, a rather wide range of melting temperatures. Purity is essential; thus, clean equipment is essential. The thermometer must be well calibrated and accurate. Also, the oil bath must be heated slowly to minimize any differences in temperature between the thermometer and the capillary tube. When a scientist reports a property of something, like the melting point of aspirin, it is always the result of one or more experiments that are being reported. We do not know properties directly; we know only what we observe in well-designed experiments. Thus, our knowledge of properties is always mediated by the appropriateness of experimental designs and execution.

A more complicated type of experiment is based on hypothetical or theoretical expectations. A scientist has suggested an hypothesis such that, when certain conditions are met, a specific event or class of events should take place. In this situation, an experiment is designed to test the hypothesis by arranging to meet the hypothetical conditions and preparing to observe or measure what happens. Suppose that someone has made the hypothesis that children can learn a second language faster if they hear all of their lessons through only the left ear, stimulating the right side of the brain, where the first (native) language is apparently not stored. An experiment is designed to test this hypothesis by arranging to have a group of children attempt to learn a new language under the prescribed conditions, that is, with their hearing blocked in their right ears. If the children demonstrate accelerated learning, the scientist can report that the hypothesis has been confirmed experimentally in at least this one situation.

Experimental design is even more important when we are attempting to demonstrate hypotheses. In the example above, we might not know how fast a group of children should learn the new language; furthermore, we have to account for individual differences. The experiment design will include a control group, that is, a group of children who attempt to learn the language under suitably similar conditions except with both ears, as is normal. Since for any random group of children we would expect them to learn at different rates, we will prepare to measure our experimental results by comparing the relative positions of the mean learners of the two groups. If the mean monophonic learners demonstrate a higher rate of learning that is statistically significant than the mean biphonic learners, then the hypothesis is supported by the experiment.

Knowledge is acquired in many ways. A great deal of knowledge arises spontaneously and simply by memory of experiences that we have had. To experiment, though, is to seek knowledge directly, and this requires us to exercise control over experience. To achieve this, we have to design the experience and execute it in accordance with the design. Knowledge of this character can always be questioned by appropriate criticism of the design and execution of experiments.

It should be noted that this discussion has been directed toward empirical knowledge and confirmation of hypotheses about things we can experience. The experiments we have discussed are activities in the experienced world. Students of science will occasionally find reference to *gedanken* (from the German word for "thought") experiments or pencil-and-paper experiments. These are activities within the mental world of well-constructed scientific theories. *Gedanken* experiments are usually situations in which one set of attributes of a theory are tested against another set of attributes of the same theory as a test of the theory's self-consistency. The experiment must be conducted in a mental world because it addresses aspects of the theory itself and not the theory's relationship with the experienced world.—T.B.

See also CONTROL GROUP; EMPIRICISM AND SCIENTIFIC METHOD; HYPOTHESIS; INFERENCE, STATISTICAL; SCIENTIFIC METHODOLOGY

Explosives

Few inventions have had as much historical significance as black powder, commonly known as *gunpowder*. Yet while black powder transformed military operations; it had a number of defects, such as obscuring the battlefield with thick smoke and gumming up the mechanisms of the weapons it fired. The first explosive to take the place of black powder began as an effort to produce a synthetic material. In 1846, Christian Schoenbein endeavored to make an artificial fabric fiber by treating cellulose with nitric acid. The results disappointed Schoenbein; not only was the fiber not suitable for its intended use, it was also highly flammable. In Italy, Ascanio Sobrero took inspiration from Schoenbein's experiments in nitrating. By treating glycerin (also known as *glycerol*) with a mixture of nitric and sulfuric acid he was able to produce trinitroglycerine, or nitroglycerin for short.

Nitroglycerin was highly explosive, but it was also sensitive to any kind of shock that it could not be safely used for military or civilian purposes. The subsequent "taming" of nitroglycerin was to a very substantial extent the result of the efforts of Alfred Nobel (1833–1896) and the firm he founded in his native Sweden. In 1865, Nobel invented a blasting cap based on fulminate of mercury that allowed the reliable long-distance detonation of nitroglycerin. Nitroglycerin was still a dangerously unstable explosive, but in 1866–1867, a major improvement was effected when Nobel absorbed nitroglycerin in a porous substance known as *kieselguhr*. Nobel named the resulting product *dynamite*. Although its explosive power had been reduced by 25 percent when compared to pure nitroglycerin, the diminution in explosive force was more than offset by dynamite's stability. This property saved many lives, but military uses of dynamite claimed many more. Nobel hoped to counteract the effects of his invention by instituting a series of prizes to honor people whose accomplishments brought substantial benefits to humankind.

In 1863, another Swede, J. Wilbrand, used the nitration of toluene to create another well-known explosive, trinitrotoluene, commonly known as TNT. Meanwhile, Nobel continued to develop new explosives, including gelatin-based explosives that were patented in 1876. These subsequently gave rise to the plastic explosives often favored by terrorists.

In 1907, Nobel's firm enhanced the safety of dynamite by devising a way to maintain its effectiveness in cold weather. Dynamite had been useless below 0°C (32°F), and lives had been lost as a result of ill-considered attempts to bring the dynamite up to operating temperature. Several other improved formulations followed, and 1925 saw the introduction of an explosive based on ethylene glycol (ethylene glycol dinitrate).

This soon supplanted nitroglycerin, although this substance is still used for the medical treatment of angina.

Twenty years earlier, in 1887, Nobel invented a nitrocellulose- and nitroglycerin-based explosive he called *ballistite*. This came to be known as *cordite* after its adoption by the British military in 1891. This explosive became the standard propellant for artillery and small-arms projectiles, but some problems accompanied it. The use of cordite-powered bullets in the standard Lee-Metford rifle resulted in the premature wear of the barrel, necessitating the development of a new form of rifling. This was not the last time that a particular kind of gunpowder caused problems with infantry weapons. In the 1960s, the U.S. Army was equipped with a new infantry rifle known as the M-16. This weapon was designed to use a nitrocellulose-based propellant. The army insisted on the use of another kind of powder to increase the muzzle velocity of the rifle, even though there was no convincing reason for doing so. Critics of army policy believe that this change was motivated by the army's hostility toward the M-16 because it was initially developed outside the usual ordnance network. In any event, the change in powder led to a tendency of the M-16 to jam, often in the midst of a firefight. Many American soldiers lost their lives in Vietnam when their weapons were rendered inoperative; some were found dead next to a jammed M-16 that they had torn down in an attempt to repair it.

Although a great amount of technical sophistication has gone into the invention of new kinds of explosives, it is an unfortunate fact that deadly explosives also can be made from everyday items like fuel oil and nitrogen-based chemical fertilizers. The destruction in 1995 of a federal building in Oklahoma City and the resulting loss of life bear tragic witness to this fact.

See also GUNPOWDER; NOBEL PRIZE; NOT-INVENTED-HERE SYNDROME

Extraterrestrial Life

Extraterrestrial life refers to life that might exist beyond the Earth's biosphere, on planets in our own solar system or around other stars. The idea is very old, and appears prominently with the birth of Western science in ancient Greece. For most of this history, "life" has meant intelligence, but as biology became a separate discipline in the 19th century, increasingly the extraterrestrial life debate came to encompass microbial life. The debate is intimately entwined with the issue of humanity's place in the universe; in the 20th century it involves the sciences of astronomy, biology, and biochemistry, and is also a prominent theme in popular culture in the form of science fiction literature and UFOs. Because it is a controversial and highly visible issue almost always carried out at the limits of scientific capability, the debate tells us much about the nature of science.

Three eras may be distinguished in the history of the extraterrestrial life debate. In the "cosmological era," the idea originated and was sustained by the principal scientific world views. Thus the ancient atomists Leucippus, Democritus (c. 470–c. 380 B.C.E.), and Epicurus (341–270 B.C.E.) believed in an infinite number of inhabited world systems (*aperoi kosmoi*), a belief tied to their principles of atomism. In this they were opposed by Aristotle (384–322 B.C.E.), and the discussion of the "plurality of worlds" (*plures mundi*) became one of the chief cosmological questions of the Middle Ages. It was the heliocentric theory of Copernicus, however, that gave the debate its modern underpinning. By making the Earth and the planets potential Earths, the Copernican cosmology launched a long tradition, aided by increasingly sophisticated instruments, to explore the physical nature of the planets and their suitability for life. The debate was carried beyond the solar system by the 17th-century vortex cosmology of René Descartes (1596–1650) and his followers, who believed every star to be surrounded by planets. This view reached its popular height in Bernard le Bovier de Fontenelle's *Conversations on the Plurality of Worlds* (1686). The Newtonian cosmology quickly superseded Cartesian vortices, but with the help of natural theology retained the idea of other planetary systems.

The second era in the debate may be called the "philosophical era." During this time, from the middle of the 18th through the 19th centuries, the philosophical ramifications of extraterrestrial life were examined. In particular, the implications for Christian doctrines such as the Incarnation were explored. Some, including Thomas Paine, believed that Christianity and plurality of worlds could not coexist. Others offered ingenious arguments showing the two could be reconciled. Secular philosophies also grappled with the ramifications of the idea. The anthropocentrism of many 19th-century German philosophers, including G. W. F. Hegel (1770–1831), led many of them to oppose other worlds. Writers ranging from Milton to Pope and Tennyson vilified or applauded extraterrestrials.

The "modern scientific era" of the debate had its underpinnings in the 19th-century developments of spectroscopy and evolution. Spectroscopy showed that the universe was composed of the same elements that existed on Earth. Thus, in a very general sense there was no reason that life similar to ours should not arise under similar conditions. Darwinian evolution was gradually expanded to the idea of cosmic evolution, which held that the entire universe was evolving, from the Big Bang to intelligent life.

These ideas, along with increasingly sophisticated telescopes and instrumentation, were used in 20th-century attempts to expound on the question. The planet Mars held a special place in the debate. Percival Lowell's (1855–1916) idea that the Martian canals were proof of an intelligent civilization began an uproarious controversy. After this died down beginning about 1910, the idea of vegetation on Mars was very much alive from the 1920s to the beginning of the Space Age. Finally, the search for microbial life on Mars was one of the drivers of the Space Age, culminating with the Viking landers in 1976, which proved to the satisfaction of most that there was no extant life on Mars. However, 20 years later NASA scientists made serious claims of fossil life on Mars, based on an analysis of Martian meteorites discovered in the Antarctic. Thus, from Lowell to Martian meteorites, the debate over life in the solar system was sustained over the entire 20th century. Notwithstanding extreme difficulties of observation and inference, the debate progressed from intelligence to vegetation to microbial life, and finally focused on past Martian life. And at century's end, there were still intriguing hints of conditions favorable for life on the Jovian moon Europa and the Saturnian moon Titan. All of these developments are an intriguing case study of evidence, inference, and the dynamics of science when it attempts to function at its outermost limits on a subject of extreme public interest.

Beyond the solar system, the search for planetary systems was a Holy Grail of astronomy during much of the century, consummated only at its end with the confirmation in 1996 of some eight planetary systems. Many of these were not suitable for life, however, and the number of planets in the Milky Way galaxy that might develop intelligent life remained a favorite guessing game of scientists. The discovery of organic molecules in many locations in outer space gave much hope to the optimists, even though they realized it was a long way from organic molecules to life, much less highly developed life. The question was more than academic; serious programs were undertaken to search for artificial radio signals of extraterrestrial origin. The Search for Extraterrestrial Intelligence (SETI), undertaken at several observatories around the world, was the ultimate development in the long history of the extraterrestrial-life debate.

In connection with SETI programs, since 1961 the idea of cosmic evolution has been encapsulated in the famous Drake equation, $N = R^* f_p n_e f_l f_i f_c L$, where each symbol represents a factor affecting the possibility of communicating with other civilizations in the galaxy (N). The first three factors were astronomical: estimating, respectively, the rate of star formation, the fraction of stars with planets, and the number of planets per star with environments suitable for life. The fourth and fifth factors were biological: the fraction of suitable planets on which life developed, and the fraction of those life-bearing planets on which intelligence evolved. The last two factors were social: the fraction of cultures that were communicative over interstellar distances, and the lifetime (L) of communicative civilizations. The uncertainties, already shaky enough for the astronomical parameters, nevertheless increased as one progressed from the astronomical to the biological to the social. Depending on whether one was an optimist or a pessimist, the number of civilizations in the galaxy comes out 100 million or 1, the latter number representing our own Earth.

The significance of the Drake equation is not that it gave any definitive answer, nor even that it was something around which an uncertain discussion could focus, but that it was the very embodiment of the concept of cosmic evolution. Cosmic evolution was a concept that NASA wholeheartedly embraced as a unified research program. It was NASA that carried out the Viking project, NASA that funded the flagship SETI program until terminated in 1993 by a skeptical Congress, and NASA that made the announcement of possible microfossils in Martian meteorites. As such, with the Space Age the idea of extraterrestrial life entered the realm of science policy. Belief in extraterrestrial life remained very high among the public, but the question of how many resources to devote to the subject remained open.

In the history of science and culture, the extraterrestrial-life debate is perhaps best seen as a fundamental shift in world view from the physical world to the biological universe. While most of the history of science has attempted to demonstrate the role of physical law in the universe, proponents of extraterrestrial life attempt to show that biological law (perhaps in the form of Darwinian natural selection) reigns throughout the universe, that the end point of cosmic evolution is not merely planets, stars, and galaxies, but also life, mind, and intelligence. In the 20th century, the concept of otherworldly life has become a world view in itself comparable in status to the Copernican or Darwinian world views. It is a kind of "biophysical cosmology" that encompasses not only science; it also has profound implications for religion, philosophy, and humanity's image of its place in the universe. It is for this reason that the debate has been so passionately prosecuted and its outcome so eagerly awaited.—S.J.D.

See also BIG BANG THEORY; EVOLUTION; NATURAL SELECTION; SOLAR SYSTEM, HELIOCENTRIC; SPECTROSCOPY; TELESCOPE, REFLECTING; TELESCOPE, REFRACTING; UFO

Further Reading: Steven J. Dick, *The Biological Universe: The Twentieth Century Extraterrestrial Life Debate and the Limits of Science*, 1996. Karl S. Guthke, *The Last Frontier: Imagining Other Worlds, from the Copernican Revolution to Modern Science Fiction*, 1990.

Eyeglasses

Human beings always have had vision problems, but they were very slow in doing something about them. Lenses were used by the Greeks for starting fires around 300 B.C.E. The magnifying properties of these lenses must have been observed, and why they were not used for this purpose is hard to understand. In medieval times, Arab scientists investigated the refraction of light and knew that a spherical container of water magnified objects, but this knowledge was not used for practical optical devices. The first recorded use of a lens for magnification purposes was by the English monk Roger Bacon (1220–c. 1292), who noted that he used a lens for magnification of small objects.

It will probably never be possible to determine who made the first eyeglasses. About all that can be said with certainty about the invention of eyeglasses is that it took place in Italy around 1280. Credit is sometimes given to Salvino degli Armati (1245–1317), whose epitaph refers to him as "the inventor of spectacles." Another possibility is a neighbor of Armati, the Dominican priest Allesandro della Spina, who, if he wasn't the inventor of eyeglasses, may have been the first person to wear them. Within a century, the wear-

A 16th-century drawing of a man reading with the help of eyeglasses (courtesy National Library of Medicine).

ing of eyeglasses must have been commonplace, for they begin to appear in portraits, perhaps as a sign of the erudition of the wearer. The earliest of these is a portrait of a Dominican friar that was painted by Thomasso a Modena in 1352.

Early lenses were made of polished quartz; glass was not suitable for this purpose because it could not be made without flaws. By the 16th century, improvements in glass making achieved in Venice and Nuremberg allowed the manufacture of eyeglasses that were actually made from glass. The curved pieces of glass (or quartz) used for eyeglasses and magnifiers were called *lenses* because their shape is similar to that of a lentil (the Italian word for lentil is *lenticchie*; the word for lens is *lente*). A lentil-shaped lens is a convex lens, and the first eyeglasses used these lenses for the correction of farsightedness (hypermetropia), a common problem encountered as part of the aging process. It is also known as *presbyopia*, which literally means "old sight." The use of concave lenses for the correction of myopia (nearsightedness) was a later development; the German cardinal Nicholas Krebs, also known as Nicholas of Cusa (1401–1464), is known to have made them around the middle of the 15th century. A famous depiction of concave eyeglass lenses can be found in Raphael's 1517 portrait of Pope Leo X.

The wearing of eyeglasses at this time was stimulated by the spread of literacy, which in turn had been substantially boosted by the invention of printing with movable type. Eyeglasses and printing thus were mutually reinforcing technological innovations that together paved the way for the emergence of the modern world. The development of an eyeglass industry was also of cardinal importance to the emergence of two scientific instruments that revolutionized our view of the world: the telescope and the microscope.

From the 17th century onward, a great deal of scientific inquiry centered on the study of optics. Yet very little of the theoretical understanding that was gained was translated into improvements in the optical quality of eyeglasses. At least they became a bit easier to wear as the result of the introduction of framed lenses, which were introduced in France in the mid-18th century. The greatest innovation in eyeglass design came in 1784 when Benjamin Franklin (1706–1790) invented bifocals. Franklin had grown tired of carrying two separate pairs of eyeglasses, one for reading and the other for seeing long distance. He fashioned a pair of eyeglasses with two different lenses, one above the other. With eyeglasses such as these, Franklin noted, "I have only to move my eyes up and down as I want to see distinctly far or near."

The last common vision problem to be corrected through the use of eyeglasses was astigmatism (an irregularity in the lens of the eye that prevents bringing images into sharp focus). George Airy (1801–1892), a renowned English astronomer, corrected his own astigmatism through the use of a sphero-cylindrical lens that had been made according to his instructions in 1827. Frans Donders, a Dutch opthamologist, is also credited with creating some of the first lenses that effectively corrected for astigmatism.

During the 20th century, considerable improvements were made in the testing of eyes and the development of instrumentation for the fitting of eyeglasses exactly suited to the optical needs of the wearer. Plastic lenses and lenses that darkened on exposure to bright light added to the comfort of wearing eyeglasses. The major innovation, however, was the development of contact lenses that did away with the need to wear conventional eyeglasses.

See also CONTACT LENS; GLASS; MICROSCOPE, OPTICAL; PRINTING WITH MOVABLE TYPE; TELESCOPE, REFLECTING; TELESCOPE, REFRACTING

Factory System

To manufacture literally means to make something by hand. For most of human history, life's necessities and luxuries were made by artisans whose energy and dexterity were complemented by a few simple tools. Manufacturing workers went about their tasks as solitary craftsmen or in small groups. They developed their skills through apprenticeships, and in turn they passed on these skills to the next generation of apprentices. In many instances, artisans banded together to form guilds that, among other things, controlled the way the work was done. Guild regulations allowed few deviations from the accepted way of doing things and often limited the number of apprentices that a master craftsman could have in his employ. In sum, manufacturing was a small-scale enterprise with few prospects for significant improvements in production and productivity.

This situation began to change in 18th-century Europe. In a few industries, production by independent artisans was supplemented or even replaced by what has been called the "domestic" or "putting-out" system. In this manner of production, the workers did not own their tools and the materials with which they worked; these were supplied by a "putter-out" or "factor," who paid the workers after collecting the finished products. This was not an altogether satisfactory arrangement for the putter-out, who ran the risk of being cheated by workers who sometimes sold the materials given them, replaced them with inferior ones, and pocketed the difference. It was also hard to maintain adequate quality standards when production workers labored in their own homes, free from direct supervision.

While some nascent capitalists were developing alternatives to craft production, new power sources allowed sizable increases to be made in the scale of production. Waterwheels, long used for certain industrial operations, became more powerful and efficient as a result of systematic improvement. The steam engine, originally used for pumping water out of mines, began to supply power to a number of industries. During the early decades of the 19th century, the hydraulic turbine emerged as an important source of industrial power.

The early 19th century also saw the initial stages in the development of mass production technologies. Machine tools like lathes and milling machines made possible the production of large numbers of identical parts. Used in conjunction with specialized jigs, templates, and gauges, these tools produced components that could be assembled into final products without the need for extensive hand finishing and fitting.

All of these devices were gathered together in a new setting for productive work: the factory. Large production sites like shipyards had been in operation for centuries, but it wasn't until the late 18th and early 19th centuries that significant numbers of workers began to labor in a factory setting. To take one notable example, by the 1830s the textile mills of Lowell, Mass., were employing 6,500 workers, of whom 5,000 were women, the majority of them quite young. As the 19th century came to a close, many factories were vast enterprises; the United States had more than 400 factories with at least 1,000 workers, while three steel plants and a locomotive works each had more than 8,000 employees.

The spread of the factory system meant that few manufacturing workers earned their living as self-employed artisans. Most workers were hired employees, dependent on their employers for the tools they used, the facilities they worked in, and, of course, the wages that supported them and their families. The supervision and coordination of these workers required new organizational methods. The face-to-face relationships typical of the craftsman's shop gave way to impersonal managerial methods. Written rules, clear definitions of responsibilities, the precise scheduling of work activities, and bureaucratic procedures formed the basic elements of factory organization. During the late 19th and early 20th centuries, a movement known as Scientific Management attempted to make factory labor even more machinelike through the application of "scientifically" derived principles.

In part, the management methods used in factories reflected the demands of industrial technologies. Steam engines or water turbines and the machinery that they powered were expensive items, and they had to be kept going for long hours if they were to pay off. As a result, workers often had working schedules that fitted poorly with the needs of their families. Machine-based production did not tolerate workers who wanted to work at their own pace, nor did it allow much worker input into how things should be done. But technological imperatives are not the sole explanation for the rise of the factory. No less important was the factory's contribution to the imposition of managerial control. Within the walls of the factory, it was possible to supervise workers, control the pace of work, and prevent any deviations from standard procedures. The development of assembly-line manufacture intensified managerial control by making the factories that used them even more rigid in their work organization.

Today, assembly lines are still prominent features of many factories, but a considerable amount of the work done on the line is now performed by industrial robots. As a result, far fewer workers find employment in factories than was the case a generation or two ago. However, while many operations requiring little skill have been taken over by robots and other computer-controlled devices, the need for skilled work may well have increased. A modern factory cannot be expected to hum along indefinitely; breakdowns and other disruptions are inevitable, and dealing with them requires prompt attention, a sense of responsibility, and problem-solving abilities on the part of the workers. Even routine tasks often need intimate knowledge of the production process and all of its quirks; not everything can be anticipated by the designers of the machinery or controlled by management. Workers who have been reduced to mindless operatives are rarely capable of rising to the challenges posed by sophisticated factory operations.

In the past, factory production involved the large-scale manufacture of standardized products. But as consumers largely take for granted the satisfaction of their basic needs, a significant number now wants some distinctiveness in the goods that they purchase. Flexible-production systems allow for some customization of consumer goods, but to be effective these systems need workers capable of adapting to changing procedures, as well as a work environment that gives them the autonomy necessary for the effective development and use of their skills.

See also ASSEMBLY LINE; AUTOMATION; DESKILLING; GAUGES; LATHE; MASS PRODUCTION; MILLING MACHINE; SCIENTIFIC MANAGEMENT; STANDARDIZATION; STEAM ENGINE; TURBINE, HYDRAULIC; WATERWHEEL

False Teeth

Loss of teeth due to dental caries (cavities) and gum disease is a common occurrence among human beings. Over the centuries, many attempts have been made to develop suitable replacements for lost teeth. One of the oldest surviving examples of replacement dental work is a bridge that dates to between the 5th- and 4th-centuries B.C.E. A product of ancient Phoenicia, the bridge consists of two carved ivory teeth connected to the surrounding teeth by gold wire. A similar design was followed by the Etruscans, who sometimes used human teeth for the bridge, but more often the teeth of calves or oxen. In some surviving examples, the artificial teeth are held together by gold bands instead of wires.

Probably the first people to make full dentures were the Chinese, a practice that began as early as the 12th century. By the early 16th century, the Japanese were making complete lower and upper dentures, which—well in advance of Western practices—were kept in place by natural adhesion and air pressure. These dentures used a base carved from a single piece of wood. The shape of the base was guided by a wax impression of the patient's mouth. The inside of the mouth was then painted with ink that revealed the presence of high spots when the base was fitted. The teeth themselves were carved from animal bones or marble, although real teeth also were used.

In 18th-century Europe, dentures were commonly connected to the lower teeth when there were remaining teeth. In the absence of natural teeth, full dentures were equipped with springs that kept them properly positioned. In France, Etienne Bourdet (1722–1789) pioneered the use of gold denture bases, which received natural teeth that were pinned in place. Ivory was the standard material for artificial teeth at this time, but it became stained and malodorous with prolonged use. In the late 18th century, Alexis Duchâteau (1714–1792), a Parisian druggist, and Nicholas Dubois de Chémant, a practicing dentist, jointly developed the first porcelain dentures. When de Chémant received a royal patent for porcelain dentures, his former partner sued him. But since de Chémant had done most of the development work, his patent was upheld.

Porcelain dentures were a significant advance, but they tended to shrink and become distorted over time. These problems were rectified in the early 19th century by Giuseppangelo Fonzi (1768–1840), an Italian dentist working in Paris. Fonzi used porcelain teeth that were cast in molds and equipped with a pin that was subsequently soldered to a gold or silver denture base. This was a substantial improvement, but the time and cost of casting individual teeth restricted their use, and ivory and other materials continued to be used in many dentures. George Washington, for example, had false teeth made from hippopotamus tusk, ivory, gold, and human teeth. Contrary to popular belief, Washington never wore wooden dentures.

As with many earlier dentures, Fonzi's porcelain teeth used a gold backing to hold them in place. A major improvement came with the use of vulcanized rubber for denture bases. In 1851, Nelson Goodyear (the brother of Charles Goodyear, the inventor of vulcanized rubber) developed Vulcanite, a hard rubber substance that soon replaced gold as the favored denture base. The process of making dentures with Vulcanite bases was covered by a patent that was vigorously, even ruthlessly, defended in the face of opposition from dentists everywhere. The patent monopoly stimulated efforts to make use of other denture bases like gutta-percha (the latex of a tropical tree) and aluminum. In 1870, celluloid was introduced for denture bases, but while initial results were promising, celluloid bases soon warped out of shape and shed teeth. The situation was only remedied in 1881, when the Vulcanite patent expired. The next major improvement in denture bases came in 1919, when pink rubber was introduced. In the mid-1930s, methyl-methacrylate resins began to take over as the preferred material for denture bases.

No matter what the material from which it is made, a denture will be effective only if it is a good fit. And the closeness of the fit depends in large measure on the accuracy of the impression taken of the wearer's mouth. The making of impressions was substantially improved around 1820, when Christophe Delabarre introduced a metal tray for holding the substance that was pressed against the gums. Plaster of paris began to take the place of wax as an impression medium in the 1840s, and in 1857 Charles Stent, an English dentist, invented a compound that was softened in hot water prior to making an impression and then allowed to harden as it cooled.

The proper fitting of dentures was further improved in the 1840s by the development of *articulators* that imitated and measured the movements of the jaws. Credit for the first articulator that was altogether satisfactory goes to Alfred Gysi of Switzerland, who brought out his device in 1909. The fitting of dentures also benefited from George B. Snow's face bow, a device that allowed more accurate measurements of the spatial relationship between the upper and lower jaw. Finally, dentures gained a more pleasant appearance through the efforts of James Leon Williams, an American who practiced dentistry in London. In 1914, Williams published a study of the relationship of facial form to tooth form; this facilitated the manufacture of artificial teeth that looked more natural because they were matched to specific facial types.

See also CELLULOID; PORCELAIN; VULCANIZED RUBBER

Further Reading: Malvin E. Ring, *Dentistry: An Illustrated History*, 1985.

Farm Tractor

In 1910, American farms were home to 24 million horses and mules. These animals supplied the motive power for essential farm tasks: plowing, seeding, cultivating, harvesting, and general transportation. On today's farms, beasts of burden are rare, their place taken by tractors. This process began in the late 19th century when tractors began to be commercially marketed in significant numbers; 3,000 of them were sold in 1890, and 10 years later the number had risen to 5,000. These early devices were powered by steam, and many of them bore a strong resemblance to railroad locomotives. Most were massive machines, weighing more than 20,000 kg (44,100 lb) and generating 120 horsepower. On the open road they topped out at 2 or 3 miles per hour. Although by 1910, steam engines were supplying 3,600,000 horsepower to American farms, their vast size restricted their use. Their most appropriate locales were the expansive wheat fields of Kansas, Nebraska, and the Dakotas, where one of their most important uses was to supply power for threshing. Here they brought substantial productivity improvements, but they still required a great amount of labor: typically, two people were required to run the engine, two to haul coal and water, and two to operate the thresher, along with several people to haul sheaves to the thresher and to load the grain into waiting wagons.

Tractors began to develop into practical multipurpose implements when they were equipped with internal combustion engines. The first tractor of this sort was built in 1892, when John Froelich combined a single-cylinder gasoline engine with running gear from a steam tractor. Commercial expansion came swiftly; by the end of the century about 100 firms were producing tractors powered by internal combustion engines. The first successful gasoline-powered tractor was built in Iowa by the Hart-Parr Company in 1902–1903. The first machine in actual use to be called a *tractor*, Hart-Parrs constituted one-third of the 600 gasoline-powered tractors in use on American farms in 1907.

Tractors multiplied rapidly on American farms in the years after WWI, their utility enhanced by improvements like the power takeoff introduced by International Harvester in 1918, which harnessed the tractor's power for machines such as sprayers, binders, balers, and corn pickers. In 1925, 600,000 tractors were in operation, 2 million in 1935, and 2.4 million in 1945. Many manufacturers got into the tractor business, their

A 1924 McCormick-Deering tractor (courtesy State Historical Society of Wisconsin;. negative number WH: [x3] 26155).

number peaking at 186 in 1921, although much of the market was supplied by a few large firms such as International Harvester and John Deere.

In terms of market share, the biggest player was a subsidiary of the Ford Motor Company, whose kerosene-powered Fordson tractor duplicated the firm's success with the Model-T automobile. In 1925, Ford produced nearly a half-million Fordson tractors, and 750,000 in 1927, at that time accounting for half of total tractor production in the United States. The Fordson tractor's influence was not limited to the United States; 25,000 of them were shipped to the Soviet Union between 1920 and 1927. Seemingly a harbinger of a new age in a backward land, this tractor was held in such high regard that newly formed collective farms and even the children born on them sometimes were given the name Fordson.

In the United States, the Fordson and other tractors were making significant contributions to increasing agricultural productivity, but at a considerable cost. The rapid spread of tractors coincided with a chronic decline in the prices of agricultural commodities. In fact, the increased crop yields made possible by tractors and their associated machines played a significant role in depressing prices and intensifying the farmer's economic plight. Burdened with payments for tractors and other machinery at the same time that farm income was stagnating or even declining, farmers were caught in a squeeze from which many never recovered. Over the long term, the result was a massive exodus from the countryside and the consolidation of small farms into larger ones.

Surviving farmers were able to make use of increasingly versatile, durable, and safe tractors. One indication of the future evolution of tractors came in 1924 with the introduction of the International Harvester Farmall, which has been characterized as the first truly successful all-purpose tractor. It had good ground clearance, closely spaced front wheels that allowed it to do row-crop work, and was equipped with a hitch, allowing it to be used for plowing, cultivation, and many other tasks. Another significant improvement came in 1931, when Caterpillar marketed the first diesel-engined tractor. Caterpillar had also led the way with the design and manufacture of tractors equipped with tracks instead of wheels. The idea went back to the middle of the 19th century, and in 1904 Benjamin Holt successfully demonstrated a track-equipped tractor that was the direct ancestor of subsequent Caterpillar tractors. Holt's tractor was turned through the use of clutches that engaged and disengaged the tracks on each side, a system still in use to day. Although tracked tractors were not widely employed on American farms, they served as the inspiration for the armored vehicles first used in World War I. For the substantial majority of farmers who continued to use wheeled tractors, significant benefits came with the introduction of pneumatic tires in the 1930s, making riding easier, dampening vibration, diminishing wear on components (and the operator), and allowing higher speeds on the roads and in the fields.

Today, there are about 4.6 million tractors in operation in the United States, approximately 5.5 tractors for every 1,000 acres under cultivation. The operator of a modern tractor may enjoy power steering, comfortable seats, and even an air-conditioned cab. Due to federal safety standards, tractors are also less likely to tip over than in the past, although this advantage has been partially negated by some farmers using their machines on steeper hillsides. As with many examples of rural technological modernization, the tractor increased production and lightened the burden of physical labor, but only for those who have been able to hold on to their land and earn their living through farming.

See also ENGINE, DIESEL; ENGINE, FOUR-STROKE; REAPER; TANKS; TIRE, PNEUMATIC

Further Reading: U.S. Department of Agriculture, *Power to Produce*, 1960, pp. 25–45.

Fax Machine

A facsimile machine (usually abbreviated as "fax") sends an image electrically over a distance. The message is scanned, transmitted, and then reproduced at the receiving machine. These operations are the same in 1997 as they were in 1843 when the first patent was issued for a fax machine. What has changed are the technologies employed and the social and economic environments in which they operate.

The history of the fax machine can be divided into

its electromechanical, electrical, and electronic stages. The first phase used physical contact to scan and reproduce an image. The second used photoelectric cells, and the third employs computer processing and a range of solid-state scanning devices.

In 1843, British clockmaker Alexander Bain (1818–1903) patented the first facsimile machine as an outgrowth of his inventions of an electric telegraph and a synchronized electric clock. To scan, a metal stylus moved line by line over a message that literally stood out from the medium it stood on. In Bain's machine, the message was engraved on a metal block, which was then washed with acid so that the message was higher than the rest of the block. The message was only a solid line; no gradations were possible. The sending telegraph transmitted the entire motion of the stylus, including the times when it hit part of the message. At the receiver, electric impulses recreated the message line by line on chemically treated paper.

In 1865, Giovanni Caselli introduced the first commercial facsimile system. Connecting Paris and Lyons, Caselli's system dispensed with Bain's time-consuming engraving process, and instead used nonconducting ink on tinfoil. Although it was moderately successful, this early service encountered a basic problem, synchronizing and phasing the message so that the receiving machine would accurately reproduce the message transmitted by the sending machine. Several 19th-century inventors, Thomas Edison among them, employed pendulums and elaborate gearing in an attempt to meet this challenge. Their solutions were mechanically complex and needed constant attention to ensure precise alignment. In the 20th century, tuning forks, radio signals, and, eventually, signals embedded in the message were used for synchronization.

The fax machine received a significant boost in the 1870s as a result of the development of photoelectric cells. A photocell scanned faster than a mechanical stylus and also transmitted gradations of gray. This capability created the first real market for fax: the transmission of newspaper photographs. Germany's Arthur Korn (1870–1945) demonstrated the first successful photographic transmission in 1902. By 1911, a wirephoto circuit connected Berlin, London, and Paris. A decade later, European postal administrations established picture telegraphy services using German Siemens Karolus or French Belin equipment. Similar services in Japan and the United States used indigenously developed equipment. Some systems employed photographic paper and negatives to record images; others used paper sensitized to light or electricity. Transmission was by telegraph or, increasingly, by the much faster telephone line or by radio. Newspapers were the main customer for these services because they could receive pictures in less than an hour instead of the days or weeks needed to physically receive a negative.

Fax usage grew slowly through the creation of specialized applications like weather maps or military needs. One limiting factor was the refusal of telephone companies to allow fax machines to use regular telephone lines. Fax users had to rent special lines, considerably adding to the cost of faxing.

In the mid-1960s, Magnavox and Xerox began marketing general-purpose analog equipment. These machines took 4 or 6 minutes to transmit and reproduce an image of medium quality. Fax offered speed, accuracy, and low labor costs because employees could use the machine with only a little training. As a result, some firms employed facsimile in order to replace costly, skilled teletype and telex operators with less-skilled workers. This illustrates a common paradox of technological development: New technologies sometimes eliminate jobs, but they usually enable larger numbers of less-skilled people to find employment.

By the mid-1970s, some machines could transmit a page in 2 to 3 minutes and dial automatically. A number of firms offered a wide range of equipment, and American leases and sales of fax machines grew from 25,000 in 1970 to 250,000 by 1980. By the mid-1990s, an average of 88,000 faxes were transmitted in the United States each minute. The deregulation of the telecommunications market assisted fax by reducing long-distance rates and lifting restrictions on the attachment of nonsystem equipment to the telephone system. The increasing internationalization of commerce and the growing need for immediate information also pushed the use of fax.

In the 1970s, the expansion of faxing had been hindered by a deliberate incompatibility of equipment and communications protocols. The purpose of this lack of standardization was to lock the customer into a particular manufacturer's equipment, but it also meant a smaller network of users, which limited the incentives to use fax machines.

This situation changed dramatically in 1980 with the creation of international standards and the development of digital fax machines. The International Telegraph and Telephone Consultative Committee of the United Nations International Telecommunication Union had approved recommendations for analog fax transmission in 1968 and 1976, but these group a and 2 standards were established after standardization had already occurred.

The new group 3 standard marked the first standard that was approved prior to significant investment in new equipment. This recommended standard proved to be very adaptable and allowed substantial technological innovation within the framework of the standards. Consequently, the latest fax machines offer far more options than earlier G3 machines, but still communicate with them. The result has been a continuing evolution of compatible, yet competing machines.

Components also have changed significantly. Digital scanning and signals have reduced transmission times to three pages a minute. Inexpensive thermal printing has replaced expensive electrolytic paper. Computer chips have made automated operations possible, such as broadcasting to multiple receivers. Finally, good design has packaged these complex technologies into equipment so easy to use that no special training is required to operate it.

Despite the popularity of electronic messaging (which includes e-mail and EDI), faxing has maintained its popularity due to its integration into computer-based messaging. The ongoing technical evolution of computer fax modems and sophisticated software has blurred the distinction between different types of communications.—J.C.

See also DESKILLING; ELECTRONIC MAIL; MICROPROCESSOR; TELETYPEWRITER; UNEMPLOYMENT, TECHNOLOGICAL

Featherbedding

Featherbedding encompasses work rules or procedures that require unreasonably large numbers of workers to perform a task. The practices that fall under the term *featherbedding* include limiting the amount of work that may be done by an employee, requiring unnecessary work, prohibiting the use of labor-saving devices, and requiring the presence of more workers than are necessary to get the job done.

The origin of the word *featherbedding* is obscure. According to one account, soldiers with cushy jobs at headquarters were known as "featherbed soldiers." Another version traces the term to train crews who complained about having to sleep on mattresses stuffed with corncobs; the response of their supervisor was, "What do you want—featherbeds?"

Work rules and procedures labeled as featherbedding are usually a response to possible wage reductions or unemployment due to changes in the workplace. Since technological change is often associated with the displacement of workers, featherbedding often represents an effort to block or mitigate the consequences of a new productive technology. To take one oft-cited example, when the diesel-electric locomotive replaced the steam locomotive, the locomotive fireman was left with little to do. Even so, union rules required the presence of a fireman in the cab. This served to preserve jobs, but it was also responsible for high labor costs in an industry that was facing competitive pressures from trucks and other modes of transport.

Although the railroads have been the scene of the greatest amount of featherbedding, other industries also have exhibited the practice. In the construction industry, workers were able to block the use of many prefabricated and modular assemblies. In the newspaper business, printers routinely broke up and reset type that had been set by advertisers or other newspapers. From 1959 to 1962, some airlines had to employ airline cockpit crews that included a redundant pilot. There was even a time when double crews of studio musicians had to be hired when a radio program was being broadcast on both an AM and an FM station.

Featherbedding has been less evident in recent decades. The Taft-Hartley amendments to the National Labor Relations Act, which was passed in 1947, contains a section that labels as an unfair labor practice any attempt to make an employer pay for services not actually performed. In the years that followed the passage of Taft-Hartley, a series of rulings by federal courts took most of the teeth out of this section. More important over the long run were the aggressive efforts of employers to eliminate or reduce featherbedding. At the same time, the diminished influence of labor unions left them less able to maintain restrictive work practices. It should be noted, however, that most labor unions have never espoused featherbedding as a legitimate practice.

Although featherbedding is not a significant source of productive inefficiency today, the conditions that engendered the practice have not disappeared. Many workers still face substantial job insecurity, some of it the result of technological advance. The complete elimination of featherbedding and other make-work practices will be possible only when workers no longer fear job losses or pay reductions resulting from technological change.

See also LOCOMOTIVE, DIESEL-ELECTRIC; LOCOMOTIVE, STEAM; UNEMPLOYMENT, TECHNOLOGICAL

Further Reading: Robert D. Leiter, *Featherbedding and Job Security*, 1964.

Federal Communications Commission

Since the late 19th century, a portion of the electromagnetic spectrum has been of great commercial value, for it is the basis of radio and television transmission and reception. When radio and television stations have to broadcast within a specific range of frequencies, this portion of the spectrum is a valuable resource that exists in finite supply. In general, the distribution of resources such as these can be done in one of two ways. The resource can be bought and sold in a free market with its price set by supply and demand. Alternatively, the procedure used for decades in the United States can be employed: A government agency grants legal rights to broadcast through the issuance of a licenses.

During the formative years of radio in the United States, no allocation process was used. Individuals and firms that wanted to broadcast were free to do so, although they had to adhere to a specific frequency. A license from the Department of Commerce was required, but this license was issued to everyone who applied. This liberal policy caused few problems when radio was primarily used for wireless telegraphy. In the early 1920s, the situation began to change dramatically as commercial broadcasting began to take off. Although stations were supposed to confine their broadcasts to a specific frequency, the transmission equipment of the time did not always allow such accuracy. In similar fashion, radio receivers did not always distinguish between signals with closely adjacent frequencies. Making matters even worse, stations with increasingly powerful transmitters were sending their signals considerable distances, where they interfered with other stations' broadcasts. The situation was often described as chaotic, an aural cacophony of interfering signals.

In 1926, matters came to a head when a federal court overthrew the Commerce Department's authority to assign frequencies to individual radio stations. In response, Congress passed the Radio Act of 1927. This created the Federal Radio Commission (FRC), which was given broad powers to issue licenses and assign frequencies. When the Radio Act was passed, the FRC was intended to be a temporary organization, but in 1934 government oversight of the communications industry was made permanent with the creation of the Federal Communications Commission (FCC), an independent regulatory commission comprised of seven members (today, the number is five) appointed to 7-year terms by the president.

The legislation that created the FCC gave the federal government a firm legal basis to control many aspects of radio and television broadcasting. Since its creation, the FCC's most important weapon has been its ability to issue and terminate broadcasting licenses. Radio and television broadcasting is a free-enterprise activity, and individual stations can be bought and sold, but the FCC has the power to revoke a license if a station's programming is deemed not to serve "the public interest, convenience, and necessity."

This, however, has rarely happened. For decades the radio and television industry has enjoyed a symbiotic relationship with the FCC. License holders have a secure market niche, while the FCC has what all bureaucracies seek: its own survival. As a result, the relationship has been a non-antagonistic one for the most part. The FCC maintains an orderly environment for broadcasting, while the industry provides most of the commission's political support. FCC commissioners and staff members often maintain close personal ties with industry representatives, and moving on to a job within the industry is not unusual. As often happens in regulated industries, the regulators tend to view the regulated industries as clients to be served, sometimes to the detriment of the public interest.

One aspect of the willingness to maintain a comfortable status quo has been the FCC's technological conservatism. In the past, the FCC did not enthusiastically lend its support to innovations like FM broadcasting, UHF broadcasting, and pay and cable TV. But technological advance continues, posing many challenges for the FCC. In addition to conventional radio and television broadcasting, the FCC must now deal with a growing number of communications technologies: cellular telephones, pagers, fax machines, cable television and radio, high-definition television, and computer networks. At the same time, the political support for federal regulation has been waning; some have even called for its dissolution. In 1994, the FCC departed from its decade-old practice of awarding licenses and conducted an auction, raising $9 billion for the government in the process. Other changes in policies and procedures may be expected.

See also CABLE AND SATELLITE TELEVISION; FREQUENCY MODULATION (FM); RADIO; TELEVISION

Further Reading: Erwin G. Krasnow, Lawrence D. Longley, and Herbert A. Terry, *The Politics of Broadcast Regulation*, 3d ed., 1982.

Fermentation

Fermentation is a natural process that converts carbohydrates (sugars and starches) into carbon dioxide and ethyl alcohol, also known as ethanol. For millennia, people throughout the world used fermentation to produce a great variety of alcoholic beverages and other food products, even though they had no understanding of the underlying biochemical processes. Fermentation is an outstanding example of a technology that was used to good effect long before science was able to explain what was actually going on.

Some insight into the process was provided in 1815 when the French chemist Joseph Louis Gay-Lussac (1778–1850) depicted the conversion of sugar into alcohol as a chemical equation:

$$C_6H_{12}O_6 \rightarrow 2CO_2 + 2C_2H_5OH$$

The actual mechanism through which this occurred was still uncertain, however. Many chemists, such as Justus von Liebig (1803–1873) firmly believed that fermentation was a purely chemical phenomenon that had nothing to do with living things.

This view began to be challenged in the 1835, when observations conducted by Cagniard de la Tour in France and Kützing, Müller, and Schwann in Germany demonstrated that yeast was a living organism, and that it was the cause of fermentation. But this fundamental insight lay dormant for more than 2 decades, until in 1854 a French industrialist asked the renowned French biochemist Louis Pasteur (1822–1895) to help him solve an annoying and expensive problem, the occasional spoilage of the alcohol he was producing through the fermentation of beet juice. One of the substances Pasteur found in the fermenting juice was amyl alcohol, a substance he believed could only be produced through organic processes. Pasteur went on to study milk that had soured and came to the conclusion that this too was the result of a biological process, in this case one that converted glucose into lactic acid. In 1857, he wrote a brief paper entitled *Memoire sur la Fermentation Appelee Lactique*, a key document in the history of microbiology. Pasteur went on to study the souring of wine, which he concluded was caused by a bacterium, *Mycoderma aceti*, that converted alcohol to acetic acid. He also studied beer microscopically and came to the same general conclusion about the cause of spoilage. One of the motivations for his work with beer stemmed from a desire to improve the quality of French beer, a quixotic effort to beat the Germans at their own game after being defeated by them in the Franco-Prussian War (1870–1871).

Pasteur also was able to demonstrate that harmful microorganisms entered fermenting liquids from the air. Moreover, he showed that they did their work after the fermentation process had begun. The remedy, therefore, lay in heating milk, wine, or beer to a temperature of 55°C (131°F) in order to kill the harmful organisms, a process that soon came to be known as *pasteurization*. More than a useful production technique, pasteurization also complemented another one of Pasteur's scientific accomplishments, the refutation of the theory of spontaneous generation.

Pasteur's research had convinced him that fermentation was a biological process that allowed certain organisms to live without oxygen. Still, the exact mechanism through which this occurred was not understood at the time. It was later discovered that fermentation is caused by the action of particular enzymes in the yeast. Some enzymes convert starches into sugars; other enzymes convert the sugars into different sugars; and still other enzymes do the actual fermentation of sugar into alcohol.

Different sources of carbohydrates produce different amounts of alcohol because they differ in their sugar content or their capacity to be converted into sugar. Wine is about 12 percent alcohol, whereas beer usually has less than half this concentration. This is because grapes used to make wine have more sugar than the barley malt used to make beer. It is possible to increase the alcoholic content of a fermented beverage by adding sugar before the fermentation begins, but when the alcohol concentration gets beyond a certain point it kills the yeast, putting an end to the fermentation process. Beyond this level, alcohol content can be increased only by subjecting the beverage to distillation or by adding alcohol, as is done to produce "fortified" wines like sherry.

Fermentation remains an important industrial process, not just for beer and wine but for foods and condiments like vinegar, cheese, yogurt, and soy sauce. When produced industrially, these products are made in large batches under carefully controlled conditions that maintain key variables such as temperature, amount of dissolved oxygen, and acidity or alkalinity. After fermentation, the product is recovered by separating it from the detritus through centrifuging or filtration.

See also ALCOHOL, ETHYL; BREWING; CARBON DIOXIDE; CHEESE; DISTILLATION; ENZYME; PASTEURIZATION; SCIENCE-BASED TECHNOLOGIES; SPONTANEOUS GENERATION

Fertilizer, Chemical

Repeated attempts to grow crops on the same acreage results in an exhausted soil and falling crop yields. The soil can regain its fertility if it is taken out of cultivation (left fallow), but this is not an option when large numbers of people are dependent on a limited amount of cropland. Generations of farmers have been able to till productively the same plot by practicing crop rotation and by adding nutrients to the soil. For centuries these were supplied by human and animal wastes, bones, mud scraped from the bottoms of ponds, and any other organic substance capable of replenishing the soil. In the 19th century, farming began to change in a fundamental way when chemicals unrelated to organic processes began to be added to the soil. Although the consequences were revolutionary, the application of chemical fertilizers took many decades to be established.

The importance of chemical processes for plant growth was first noted by Johannes Baptista van Helmont (1579–1644), a Dutch chemist/alchemist. In 1609, he planted a tub containing 100 kg (220.5 lb) of soil with the sapling of a willow tree that weighed 2.5 kg (5.51 lb). After 5 years, the tree had gained 82 kg (180.8 lb) in weight, yet less than 100 grams (3.5 oz) of soil had disappeared from the tub. This was the beginning of the realization that plant growth requires air, water, and chemical reactions of some sort.

After much experimentation, in the following centuries it was learned that plants need a number of trace elements, but the crucial requirement is the availability of three chemical elements: potassium, phosphorous, and nitrogen. This basic principle was elucidated by the German chemist Justus von Liebig (1803–1873), who noted in 1840 that the growth of a plant is limited by the element in the lowest concentration; this has come to be known as Liebig's law. Today, the concentrations of these three key chemicals are identified on most packages of fertilizer by an N-P-K code. This code indicates the percentages by weight of compounds containing nitrogen, phosphorous, and potassium, respectively. For example, 8-16-8 indicates that the fertilizer is 8 percent nitrogen, 16 percent phosphorous, and 8 percent potassium.

In the early 17th century, Johan Rudolf Glauber (1604–1668) hypothesized that saltpeter (potassium nitrate, KNO_3) was crucial to the growth of plants. At that time, the primary source for potassium was potash (potassium carbonate, K_2CO_3), a substance made by leaching wood ashes, boiling down the resulting solution, and then heating it until it was fused into a solid. It was a major product of North America (the first patent granted by the U.S. government was awarded in 1790 to Samuel Hopkins for an improved method of making potash). In the 1860s, deposits of naturally occurring potassium salts in Germany became the main source of supply, remaining so until the requirements of World War II stimulated the Unite States to rely on local sources in New Mexico, California, and Utah.

The importance of phosphorous, the second key element, was discovered by Liebig. Phosphorous was available in phosphate-bearing rocks, but it was not readily absorbed into the soil. Absorption was improved by converting the rock into superphosphate. In 1842, what might be described as the first chemical fertilizer factory was built in England by John Bennett Lawes for the purpose of producing superphosphates by reacting phosphate rocks with sulfuric acid. This technique became common from 1870 onwards.

Although a plant's need for nitrogen was not recognized until 1857, for nearly 3 decades before that time, nitrogen-rich South American guano (bird droppings) had been used to improve the farm soils of North America and Europe. In the 1860s, nitrogen began to be obtained from ammonium sulfate, a byproduct of the coking ovens that produced illuminating gas in many cities. The key development in the production of nitrogen compounds came in 1908–09 when Fritz Haber (1868–1934) invented the process that bears his name. The Haber process consisted of reacting hydrogen and nitrogen at high temperature and pressure in the presence of iron or some other catalyst. First commercially produced in Germany in 1913, ammonia and other nitrogen compounds are still made through this process, usually with methane providing the nitrogen, although the cryogenic separation of air is sometimes used to supply the nitrogen.

The widespread use of chemical fertilizers is largely a post–World War II phenomenon, but once this occurred the results were dramatic; more than half the increase in American productivity per acre from 1940 to 1955 (a time of rapid production increases) is attributed to the use of chemical fertilizers. Yet, as with most technological advances, the extensive use of chemical fertilizer has not been an unmixed blessing. Chemical fertilizers require massive amounts of energy for their production, distribution, and application. Agriculture of this sort has been accurately described as converting fuel into food. Second, chemical fertilizer runoff can get into water supplies, polluting groundwater and causing the eutrophication of rivers and lakes. The latter occurs when chemicals in the water stimulate the excessive growth of algae, and cause a consequent deterioration of water quality and loss of other forms of aquatic life. Chemical fertilizers have helped make possible the great expansion in agricultural production that has sustained a constantly expanding world population, but care must be taken so that these gains are not negated by extensive environmental destruction.

See also CROP ROTATION; GREEN REVOLUTION; HABER PROCESS

Fertilizer, Organic

In general, a fertilizer is any material that is added to the soil to compensate for the lack of nutrients essential to crop growth. Deficiencies in nutrients may be inherent to a particular soil, or they may be the result of prolonged crop cultivation. Early agriculturists overcame the loss of soil nutrients by growing a crop and then moving on to more fertile lands. Today, the lack of virgin land prevents the employment of this strategy in most parts of the world. Crop rotation on permanent plots of land can make up for the loss of some soil nutrients, but it is not a complete answer. The cultivation of legumes returns atmospheric nitrogen to the soil, but not in quantities that are completely adequate. In most places, the regular application of fertilizer is an essential agricultural practice.

The most important constituents of fertilizer are the chemical elements nitrogen, phosphorous, and potassium. Also needed for plant growth, but in smaller quantities, are three "secondary elements": sulfur, calcium, and magnesium. Other elements, known collectively as *micronutrients*, are required in small or trace quantities but are nonetheless essential for healthy plant growth. This group includes boron, cobalt, copper, iron, manganese, molybdenum, and zinc. Some plants also require additional elements; rice, for example, needs small amounts of silicon.

A great variety of organic substances have been used to enhance the fertility of the soil: algae, animal manures, "green manure" (alfalfa, clover, and other plants), bone meal, blood meal, human excrement (sometimes euphemistically known as "nightsoil"), cottonseed cakes, and kitchen scraps. In many cases, plant and animal wastes are composted so that they can be more safely and easily stored, handled, and applied. Different kinds of organic fertilizers do not contain the same proportion of nutrients. For example, blood meal has a relatively high nitrogen content and a low potassium content, whereas bone meal has a high potassium content and a low nitrogen content. For this reason, farmers and gardeners need to use organic fertilizers that meet the particular needs of their soil.

Organic fertilizers can also be differentiated in terms of the availability of the nutrients in them. Fertilizers like animal manures have a high nutrient availability index; i.e., they are rapidly decomposed by microorganisms in the soil and can be readily absorbed by growing plants. Conversely, other fertilizers—composted materials, for example—decompose and add their nutrients to the soil at a slow rate. It is often beneficial to combine the two types to ensure that nutrients are released throughout a crop's growing cycle.

Although many organic fertilizers have a high nutrient index, they still release their nutrients into the soil at a lower rate than chemical fertilizers do. This, in fact, is one of the primary advantages of organic fertilizers. Not only does the slow release of nutrients make them available to crops for a long period of time, it also reduces nutrient losses through runoff and leaching. Also, unlike chemical fertilizers, organic fertilizers improve the physical structure of the soil, increasing aeration and improving water retention.

A primary disadvantage of organic fertilizers compared to chemical fertilizers is that they contain less nutrient material relative to their bulk. Moreover, a large proportion of nutrients are lost in the course of collecting and applying organic fertilizers. As a result, massive quantities of organic fertilizers may have to be used in order to provide adequate amounts of soil nutrients. Also, some organic fertilizers can create a health hazard by facilitating the spread of infectious diseases. Agricultural societies that use human excrement for fertilizer are particularly vulnerable in this regard. On the other hand, the proper composting of wastes can diminish this hazard substantially.

Although chemical fertilizers are an essential component of modern farm production, organic fertilizers still play a vital role. About half the nitrogen added to farmland in the United States comes from organic sources. Crop residues provide about 13 percent of the nitrogen, while 6 percent comes from animal manures, and 31 percent is derived from legumes and green manure crops.

There is no denying that the application of chemical fertilizers has made an essential contribution to the substantial increase in agricultural productivity that has occurred over the last 50 years. At the same time, however, the production of chemical fertilizers is an energy-intensive process. Consequently, an increased reliance on organic fertilizers may become necessary in the event of significant increases in energy costs.

See also COMPOSTING; CROP ROTATION; FERTILIZERS, CHEMICAL; LEGUMES; NEOLITHIC AGRICULTURAL REVOLUTION

Fetal Monitor

Fetal heart monitors were developed in the 1950s in order to keep track of the conditions of a fetus during the mother's labor and subsequent delivery. In particular, these devices are used to determine if the heartbeat of the fetus has diminished to such an extent that a danger of neurological damage exists. Fetal monitors can be either internal or external. An internal monitor measures the heartbeat of a fetus by means of an electronic catheter that is passed through the vagina and cervix and then attached to the scalp of the fetus. It can also be used to take a blood sample from the scalp to determine if the fetus is suffering from oxygen deprivation. Another type of internal monitor measures uterine contractions by means of a catheter located between the wall of the uterus and the fetus. This type of monitor is used largely as research tool, not for normal medical practice. An external monitor uses two straps placed around the abdomen of the mother. One is attached to a pressure gauge that records contractions, the other is connected to a sensor that measures the fetal heart rate. Both of these devices indicate any changes occurring in the course of labor. In some cases, changes are deliberately induced through the intravenous administration of the hormone oxytocin to the mother; the need for immediate delivery is indicated when the resultant contractions cause a slowing of the baby's heart rate.

Fetal monitoring is not completely risk-free. The most common problem is injury to the baby's scalp. More serious is the compression of a large vein known as the vena cava and the restriction of uterine blood flow. This may have adverse effects on contractions and may slow down labor. And, according to one study, electronic monitoring may increase the incidence of cerebral palsy. Fetal heart monitoring also has been implicated as a cause of a rise in the number of caesarean sections that have been performed in recent years. These are performed because many obstetricians are reluctant to wait for a natural birth when the monitor indicates the presence of an abnormality, even though nothing is really wrong. In fact, a study of fetal monitors conducted in the mid-1990s at Brigham and Women's Hospital in Boston found that 99.8 percent of the indications of possible neurological damage were false alarms.

Since fetal heart monitors do not appear to diminish the risk of perinatal mortality or neurological damage—even for infants who have been born prematurely—most medical professionals now agree that fetal heart monitors, while useful, should only be used in the case of high-risk births (for example, when the mother has diabetes, or Rh-factor incompatibility) or when complications develop during labor. For normal births, the heartbeat of a fetus can be adequately monitored through the use of a stethoscope especially designed for this purpose.

See also BLOOD GROUPS; STETHOSCOPE

Fiberglass

Fiberglass (known as *glass fibre* to writers of British English) is one of the earliest examples of a composite material. Consisting of spun glass fibers imbedded in a polyester resin, it is light and strong, and does not corrode. The fibers can be loose, or they can be woven into a mat before the resin is added. Glass fibers were first produced in 1893 by Edward Libbey, but they were not combined with resin until the 1930s, when the Owens-Illinois Glass Company began to manufacture a product they called Fiberglas (the spelling with one *s* is the trade name used by Owens-Illinois, but it is commonly used as a generic term as well).

Unlike many other synthetic materials that require elaborate molds and expensive equipment, fiberglass is easily molded. All that is required is a wooden or plaster mold, or an item to be copied, which also can serve as a mold. The resin does not give off toxic fumes, so the material can be laid on by hand. A more rapid means of production that uses a special gun to spray a mixture of fibers and resin was developed in the late 1950s. Another way of making objects from fiberglass uses a technique known as "filament winding." Used for cylindrical objects like large pipes, it starts with a long strand of glass fiber coated with resin. The filament is then wound around a form until the proper thickness is reached, and is then fused solid and cured by the application of heat.

Many production processes using fiberglass do not require the application of heat; after being molded, the fiberglass is simply allowed to cure at normal temperatures and pressures. Alternatively, pressure can be added by surrounding the molded item with a flexible bladder and then applying compressed air or a vacuum. This procedure allows a higher ratio of glass to resin, resulting in a stronger product. Pressure also can be applied through the use of metal dies.

Like a number of other plastic materials, fiberglass was given a boost during World War II, when it was used in large quantities for enclosing airborne radar antennae. In the early 1950s, it took on a glamorous aspect when General Motors brought out a fiberglass-bodied sports car, the 1953 Chevrolet Corvette. A more technically impressive vehicle was Britain's Lotus Elite. Introduced in 1957, it used fiberglass for a unibody that combined chassis and body in one structure. Fiberglass bodies continue to be used for the Corvette, as well as for a variety of replica automobiles. On a more prosaic level, fiberglass has found an important niche in the furniture industry. The most famous piece of fiberglass furniture is the shell chair designed by Charles and Ray Eames, produced from 1950 to the present. The commercial success of these chairs helped to increase the visibility of fiberglass and inspire other uses for it. Today, fiberglass is used for a diverse number of products, including surfboards, fishing rods, safety helmets, and boats.

See also COMPOSITE MATERIALS; RADAR

Further Reading: Jeffrey I. Meikle, *American Plastic: A Cultural History*, 1995.

Fiber-Optics Communications

Ever since the invention of the telegraph, messages have been sent by transporting electrical signals through copper wires. In the 1970s, a new means of transmitting messages emerged, one that used pulses of light traveling through cables made from glass fibers. In a fiber-optics communication system, each pulse of light encodes a digital signal that carries part of the information being transmitted. The information can be voices, pictures, or data—anything that can be digitized.

Fiber-optics cables are superior to conventional wires in a number of ways. They are lighter, more compact, more flexible, and easier to repair. Of particular importance, fiber-optics communications provide a larger bandwidth for communications. Bandwidth is a measure of the range of frequencies that can be sent along a transmission line. The range of frequencies, in turn, determines the information-carrying capacity of a line. To put it slightly differently, the greater bandwidth provided by fiber-optics cables allows them to carry more circuits, which in turn lowers the cost of long-range communications.

The source of the light sent through a fiber-optics cable can be a light-emitting diode or a laser. Lasers are used for telephone systems because they produce narrowly focused ("coherent") beams. Although LEDs are cheaper to produce, the light they generate is unfocused and hence susceptible to scattering. As a result, LEDs are unsuitable for long-distance communications, but they are adequate for local computer networks, and they are frequently used for this purpose.

The information to be sent by means of fiber optics is first converted to digital form through a process known as *sampling*. The digital code is then converted into pulses of light that are sent through the fiber at the rate of millions of pulses per second. Each pulse of light enters the fiber like a droplet of water, sliding along the transparent core of the fiber at the speed of light and separated from the light pulses in front and behind by a tiny dark space. However, if the distance traversed by a fiber is too long, the pulses may blur and eventually overlap one other. To prevent this from happening, the signals are intercepted by a photodetector in a device known as a *repeater/regenerator*. There they are converted to electrical signals, reshaped, and pumped back into the next fiber link by a laser similar to the one at the transmitting end. After arriving at its destination, the light is converted back to an electrical signal by a demodulator.

Each optical fiber is formulated to transmit specific wavelengths of light without distorting the individual pulses. This capability is instilled during the fiber-manufacturing process. The fiber starts out as a glass rod about .9 m (3 ft) long and less than 2.5 cm (1 in.) thick. The rod is formed in a special furnace, where vapors from melted chemicals are sequentially deposited as the rod rotates. This creates a cross section similar to the rings of a tree. The rod is then heated and drawn, leaving the concentric layers in the fiber the same as they were in the rod before it was drawn. As a final step, the fiber is encased in coatings that protect it during storage, shipping, installation, and maintenance. Individual fibers are bundled together with reinforcing materials to form a cable.

The idea of using light to transmit information goes back to Alexander Graham Bell's unsuccessful "photophone" of 1880. Glass fibers also have a long history, having been made since the 19th century. From the late 1930s onwards, their main use was for reinforcing plastic resins. In 1955, an Indian-born American researcher, Narinder Kapany, hit upon the idea of using glass fibers for medical exploration. The resulting instrument, known as an *endoscope*, allows the examination of body cavities that cannot be directly observed. It consists of two bundles of fibers, one for transmitting light along a nonlinear path, the other for transmitting an image. The optical fibers used for long-distance communications are similar, but they have to be free of contaminants that would absorb or scatter the light. Fibers of this quality were first produced by the Corning Glass company in the early 1970s.

By this time the potential advantages of using fiber optics for communications had become evident. The first fiber-optic communications system was installed in 1976 in Atlanta, Ga., where it traversed a distance of 10.9 km (6.8 mi). When the first commercial fiber optic system went into operation in Atlanta during that same year, it transmitted at a rate of 45 million bits per second. In 1977, a fiber-optics communication system installed in Chicago was carrying a full range of voice, video, and data signals.

By the early 1990s, the transmission capacity of telephone fiber systems had reached 565 million bits per second, equal to 16,000 voice circuits. However, in transoceanic submarine cable applications, digital-circuit multiplication techniques have increased the capacity to 80,000 simultaneous conversations. By the mid-1990s, transmission capacity was up to 5 billion bits per second in high-traffic transoceanic routes.

Although light pulses transmitted through fiber-optics cables can carry immense amounts of information very rapidly, fiber-optics communications systems encounter bottlenecks when the pulses have to be converted to and from electrical signals. Every conversion of a signal from digital to analog and back entails some degradation of the information being sent. Consequently, today's communications systems try to minimize the use of analog systems in the telephone trunk lines between switching offices or across the long-distance networks. Even so, telephone companies have to change information signals from digital to analog at most central offices, where the local telephone lines to homes, small businesses, and institutions usually consist of analog copper wires.

These wires, called the *loop plant*, will remain in place until the telephone companies have completely written off their investment in the wires, or until changes in regulatory laws allow telephone companies to enter the video cable business, in which case they would need the extra capacity offered by fiber cables. Conversely, some cable television companies, anticipating possible entry into the telephone business, already have equipped their distribution systems with fiber optics.

In the 1960s, the future of communications seemed to lie with orbiting satellites and microwave ground stations. In less than 20 years, however, fiber-optics cables had emerged as the preferred means of transmitting information over long distances. Compared to geostationary satellite systems, terrestrial fiber optics provides long-range communications without the annoying time lag that occurs when messages have to be relayed to and from satellites orbiting tens of thousands of kilometers above the Earth. In addition to being annoying, these time lags disrupt computerized error-checking methods, even in one-way transmissions, which require return pulses that confirm the arrival of the signal.

See also CABLE AND SATELLITE TELEVISION; COMMUNICATIONS SATELLITES; COMPUTER NETWORK; FIBERGLASS; LASER; LIGHT-EMITTING DIODE; MICROWAVE COMMUNICATIONS; SUBMARINE CABLE; TELEGRAPHY

Film, Roll

During photography's early decades, the glass plate that received the photographic image was a major obstacle to easy picture taking. The invention of dry plates in the 1870s helped somewhat, but each photograph still required the loading and unloading of a separate glass plate and the exercise of considerable care to prevent its breakage. The idea of putting separate frames on a single roll of film was realized as early as 1855, and a roll film holder for 100 exposures was designed by Leon Warneke (1837–1910) in 1875, but these were not commercially successful ventures. Roll films began to come into their own in 1884 when George Eastman (1854–1932) used them as the basis for a complete system of photography. Roll film was used in conjunction with Eastman's Kodak camera, a relatively inexpensive ($25) device that was aimed specifically at amateurs. With dimensions of 3 $\frac{1}{4} \times$ 3 and $\frac{3}{4} \times$ 6 $\frac{1}{2}$ inches (7.9 × 9.5 × 16.5 cm) and weighing 1 lb 10 oz (.46 kg), it was easily manipulated. The lens had a 27-mm focal length and an aperture of f/9, which allowed good depth of field, so the subject was always in focus except when it was very close.

The camera came loaded with a roll of film good for 100 exposures. The film itself was made from paper coated with gelatin and a top layer of light-sensitive gelatin emulsion. After being processed, the negative was stripped from the paper and mounted onto a glass plate to allow prints to be made. In 1889, paper was replaced by a transparent film based on nitrocellulose. Again, this was not a completely new invention, for roll film using celluloid strips had been produced by Rev. Hannibal Goodwin in 1887. A patent suit was

A contemporary advertisement for a 19th-century Kodak camera. Note that the roll contained 100 frames. The film had to be developed by Kodak, so the creation of a photograph was certainly not "instantaneous."

settled in Goodwin's favor, but not until 8 years after his death in 1890, and 17 more years elapsed until the final monetary judgment was rendered.

Important as the film technology was, the real key to Eastman's success was the simplification of taking pictures, as expressed in the Kodak slogan, "You press the button, we do the rest." After taking 100 shots, the user returned the camera to the factory, where the pictures were developed and the camera reloaded with film before being sent back to the owner. This fit perfectly with Eastman's strategy, which was to make money not by selling cameras but by selling film, a strategy that continued with later generations of Kodak cameras, such as the ubiquitous Instamatic, that were designed to use special Kodak films. The ultimate expression of this philosophy is the disposable camera, described by Kodak as "the film that is also a camera."

Early Kodak cameras can also be seen as leaders of the movement to make devices "user friendly" to the point that they can be successfully manipulated by people who are completely ignorant of their inner workings. As we have seen with successive generations of automobiles, radios, and computers, many consumer items have required less and less effort and skill on the part of the operator, but often at the cost of increasing technological complexity for the device itself.

See also CELLULOID; PHOTOGRAPHY, EARLY

Fireproof/Fire-Resistive Buildings

In the 19th century, and indeed until as recently as the 1960s, the term "fireproof building" was a category of fire-resistive construction recognized by fire protection experts and was used in building codes. While the term is no longer used, buildings today that are highly fire-resistive once would have been called "fireproof," and are descendants of the fireproof buildings of the 19th century. The virtual end of urban conflagration—an all-too-common event in the 19th century—has been one of the important results of the spread of fire-resistive construction.

Fire-resistive buildings are designed so that their structures will not collapse in a blaze, and their interiors are subdivided to check the spread of flame. These goals are accomplished by protecting the vulnerable load-bearing parts of the frame with noncombustible, insulating materials, and by making the walls, internal partitions, and floors into barriers that can hold back a fire. Various building materials can be used for these purposes, which are rated according to the length of time they can withstand a fire. Today, with respect to fire resistance, buildings are rated according to the length of time they should remain intact in a fire, whereas in the past, buildings were classified according to the materials and methods of construction.

In its earliest meaning, a fireproof building was simply one in which no wood was used for its construction. Buildings of this sort were introduced during the last decade of the 18th century. The first fireproof buildings were solid masonry construction), where the walls, the floors, and often the roof were made of brick or stone. No serious attention was given to separating horizontal from vertical spaces in these massive structures. And while they certainly were fire resistive, they were also costly and impractical.

In the 1840s, a new fireproof system was introduced in the United States, one that used iron for interior beams and posts, and brick arches for floors. Iron was considered fireproof because it was noncombustible. But actual fires showed that while iron could not burn, it could still succumb in a blaze.

In the 1870s, two basic principles of fire safety in buildings were established. The first was the protection of structural materials—even noncombustible ones—in order to prevent collapse. Structural iron was no longer left exposed but was covered ("fireproofed") with insulating materials such as clay blocks and concrete. The second principle was "compartmentation," the subdivision of a building's interior so that a fire could be confined to the area where it started. The idea of using barriers to check the spread of fire between buildings was an old one; it was the reason that early building codes required thick exterior walls. What was added to this traditional idea was the thorough subdivision of the interior of a building, with the floor areas separated from stairways and elevator shafts by fire- and water-resisting floor/ceiling assemblies and partitions.

After the terrible fires in Chicago and Boston in the early 1870s, many new kinds of noncombustible building materials were introduced. Most notable were hollow clay blocks, which replaced bricks for making floors and partitions. Later, in the 1890s, structural steel began to replace iron in the frames of fireproof buildings. Also in this decade and first 2 decades of the 20th century, structural concrete, both plain and reinforced, began to be used in the United States on a commercial basis in fireproof buildings.

One drawback to these new materials and construction methods was that they increased construction costs, and therefore many owners were reluctant to adopt them. The spread of fire-resisting buildings, consequently, followed the enactment and enforcement of building codes. Beginning in the 1880s, a number of cities began to require that certain kinds of buildings—specifically theaters and tall buildings—be fireproof. In time, other kinds of buildings were added to the list.

The focus of fire protection of buildings in the last century was the preservation of individual properties and prevention of disastrous conflagrations; the principles governing emergency egress developed later. Today, preventing injury and loss of life is a major element in designing buildings for fire safety. Most fires start in the contents of a building rather than in the structure itself, and the smoke from burning contents is the main threat to life. Therefore, even fire-resistive buildings must have exit facilities of sufficient capacity to allow the occupants to escape quickly.

Fire suppression also is necessary to fire safety. Except when there is so little fuel that a fire will burn itself out without causing structural damage, fires have to be extinguished. Buildings can be equipped with automatic sprinklers, which can control fires and extinguish small ones. Vertical pipes and hose connections at each floor also are important for fighting fires in large buildings.

The term *fireproof* eventually was discarded because it was so widely misunderstood by the public. Modern methods for categorizing buildings involve grading them according to the fire resistance of the construction materials. The fire resistance of a building's components—the exterior walls, structural frame, and floors—is established by subjecting a sample of the material to a standard fire test, which follows a uniform time-temperature curve. The materials or assemblies are rated in hours, according to their performance in the test. The level of performance required of particular kinds of buildings depends on their size, location, and occupancy, and is specified in local building codes. The fire-safe building today combines insulation of vulnerable structure materials, compartmentation, and adequate, protected egress.—S.E.W.

See also BRICKS; BUILDING CONSTRUCTION, STEEL-FRAME; CONCRETE; CONCRETE, PRESTRESSED; MASONRY

Flight Data Recorder

A flight data recorder (FDR) is used to provide information that can be of great value in determining the cause of an airplane crash. These devices are sometimes known as *black boxes* even though their outside surface is painted bright orange for maximum visibility. Recovery of an FDR is aided by a locator beacon on the box's outer surface; after being activated by contact with water, the box produces a beep that can be detected by microphones on board search vessels. FDRs are usually located in the tail sections of airplanes, where they are somewhat less susceptible to crash damage.

Flight recorders are almost as old as powered flight. The Wright brothers used a simple device to record distance flown, time in the air, and engine revolutions, while Charles Lindbergh carried a simple barograph (an instrument for recording barometric pressure, and hence the altitude of an airplane) on his epic solo flight across the Atlantic. The post–World War II expansion of commercial aviation made safety a primary concern, spurring the invention of devices capable of tracing what happened to an airplane prior to an accident.

The first commercially successful flight recorder was invented by James J. Ryan and manufactured by General Mills, a firm better known for its grain products. Known as the Ryan VGA flight recorder, it measured a plane's airspeed (or velocity, hence the *V*), its vertical acceleration (G forces), and altitude. The recordings were made by three styluses, each one hooked up to a particular instrument (airspeed indicator, accelerometer, and altimeter). Each stylus traced a path across a slowly moving sheet of aluminum foil as the instrument readings changed, while a fourth stylus indicated the passing of time. The works were encased in a sealed box that was resistant to crash and fire damage, was unaffected by vibration and extreme temperatures, and could endure long periods of immersion, even in seawater. Beginning in 1953, the works were packaged in a sphere, returning to a rectangular configuration in 1969.

The use of flight recorders surged after 1957 when the U.S. Civil Aeronautics Board stipulated that they had to be standard equipment on all passenger-carrying aircraft weighing more than 12,500 lb (5,670 kg) and operating above 25,000 ft (7,620 m). In 1960, the requirement was extended to cover operations at all altitudes. FDRs were required to record five parameters: airspeed, altitude, compass heading, vertical acceleration, and time.

Moving-stylus recorders (known colloquially as "scratch recorders") proved to be quite reliable; from 1959 to 1973, only 8 percent of recorders involved in accidents were unreadable due to damage or equipment malfunction. But they were inherently limited in the number of parameters that they could record. The next generation of flight recorders could accumulate much more varied data by digitally recording on magnetic tape. Increased data acquisition allowed the recording of new parameters like pitch and roll attitudes, engine thrust, and flap positions. Today's FDRs may routinely record hundreds of flight parameters. Analyses of these parameters have allowed more thorough investigations of airplane crashes. The study of flight recordings has even led to the identification of a hitherto unrecognized flying hazard, wind shear.

Interpreting the information provided by FDRs makes extensive use of computers. Third-generation FDRs are particularly well suited to the computerized analysis of flight data; in the place of tape, these devices record their data electronically through the use of solid-state technology. The data can be immediately downloaded into a computer, which then generates graphs and tables. In addition, computer graphics can be used for a video simulation of an airplane's last moments. Most aircraft currently in service are not equipped with FDRs that allow this level of sophistication, and retrofitting can cost as much as $70,000 per airplane. Still, this is a small price to pay when the long-term benefits of having more extensive flight data are taken into consideration.

See also AIRPLANE; WIND SHEAR

Further Reading: Vanda Sendzimir, "Black Box," *American Heritage of Invention and Technology*, vol. 12, no. 2 (Fall 1966): 26–36.

Fluidized-Bed Combustion

The burning of coal provides a significant amount of the world's energy, but at the same time it makes a major contribution to air pollution, especially through the production of oxides of sulfur (SO_X). This problem can be substantially alleviated through the use of fluidized-bed combustion, a technology that confers a number of benefits in addition to cleaner emissions. Fluidized-bed combustion was originally developed for the gasification of coal, and it is widely used for the incineration of wastes, the roasting of metal ores, and petroleum refining. A fluidized-bed combustor can operate at atmospheric pressure, or for greater efficiency it can be pressurized to 6 to 16 times atmospheric pressure by an exhaust gas-powered turbine. Fluidized-bed combustion is now beginning to be used to supply heat to the

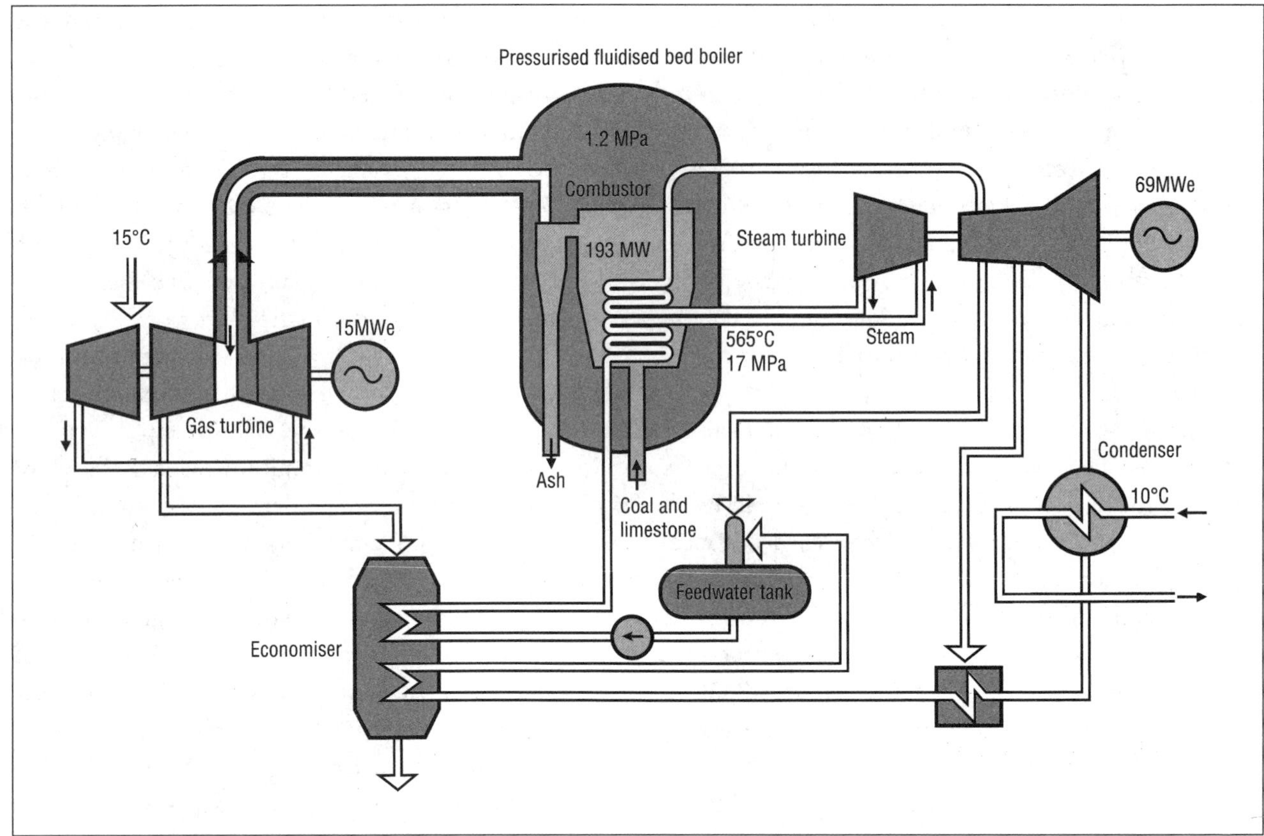

Schematic illustration of an electrical power plant employing fluidized-bed combustion (courtesy World Coal Institute).

steam generators of electric power plants, and it is likely that its use will grow rapidly in the years to come.

The basic technology of fluidized-bed combustion is simple: A rapidly moving stream of air moves upward through a perforated or slotted floor on which rest particles of coal pulverized to less than .16 cm (.06 in.). When the air travels at the proper velocity, the coal particles go into suspension with it. The mixture of coal and air takes place with great rapidity, giving the bed the appearance of a boiling liquid. The turbulent mixing of the coal and air promotes the rapid transfer of heat and the constant replenishment of oxygen. The continual addition of oxygen helps to react the coal at lower temperatures (730°–1,000°C, 1,350°–1,850°F) than those of conventional furnaces, which range from 1,510° to 1,815°C (2,750°–3,300°F). Lower combustion temperatures are advantageous because they diminish the formation of oxides of nitrogen, another atmospheric pollutant.

The lower combustion temperature is also crucial to the control of sulfur oxides. Limestone (calcium carbonate, $CaCO_3$) or dolomite [$CaMg(CO_3)_2$], each of which decomposes at higher temperatures, can be added to the bed, where they trap sulfur emissions in the form of calcium sulfate when limestone is used and magnesium sulfate when dolomite is used. This is a significant advantage, for even though the calcium sulfate or magnesium sulfate has to be removed from the furnace, this is a less-expensive operation than adding scrubbers to catch exhaust emissions. Exhaust scrubbers can account for as much as a third of the total capital cost of a new coal-fired power plant.

Fluidized-bed combustion has several other advantages. A variety of fuels can be used in addition to coal: heavy oil, coal wastes, wood, and even fuel derived from garbage. A fluidized-bed combustor is flexible in its operation, for the rate of combustion can be altered by varying the coal feed. The combustor can even be shut down for a few hours and then restarted, provided that the temperature is never allowed to drop below 590°C (1,100°F). Finally, a fluidized-bed combustor with a convection pass produces much less ash, thereby reducing maintenance costs.

A few power plants using fluidized-combustion technology are now in operation. However, at present the application of the fluidized-bed principle is restricted by the relatively small size (boiler capacities of less than 1 megawatt) of these installations. A 20-megawatt pilot plant has been built and even larger ones are in the planning stage. In the years to come, growing power demands will require extensive use of the world's vast coal deposits; fluidized-bed combustion will play an important role in converting coal to energy cleanly and efficiently.

See also COAL; COAL GASIFICATION

Further Reading: David Merrick, *Coal Combustion and Conversion Technology*, 1984.

Fluorescent Lights—see Lights, Fluorescent

Fluoridation

One of the most common bacterial diseases is tooth decay. Although the actual mechanism producing decay is still not completely understood, its source has long been known: a number of different streptococcus bacteria, of which *Streptococcus mutans* is the most important. In all likelihood, these bacteria use sugar present in the mouth to produce acids that take minerals out of the teeth, resulting in the formation of cavities. Dental cavities (properly known as *caries*) have afflicted the majority of the population, bringing with them toothaches and dental bills. Fortunately, in recent years there has been a considerable reduction in the prevalence of caries. Although in the mid-1980s more than half the children in the United States had at least one cavity, this represented a considerable improvement over the 72 percent that were so afflicted in the early 1970s. One of the primary reasons for this change was the widespread use of fluoride as a tooth decay preventative.

The therapeutic value of fluoride was discovered in an indirect manner. Early in the 20th century, a dentist practicing in Colorado Springs, Colo., named Frederick McKay was puzzled that many of his patients' teeth had a mottled appearance. But while their teeth were unattractive, they were less prone to decay than those of people in other communities. The locals were of the opinion that the mottling was being produced by "something in the water," but nothing had been detected through standard testing procedures. A similar pattern was observed in a small town in Idaho, and McKay noted that it began to disappear when the town shifted to another source of water. The breakthrough came in 1930 when, at McKay's request, the water from the town of Bauxite, Ark., was tested by H. V. Churchill, a chemist in the employ of the Aluminum Company of America. Churchill and his company had reason to be concerned about the mottling phenomenon. Mines near the town supplied aluminum ore (also known as *bauxite*), and Churchill's company wanted to squelch a rumor that aluminum pots and pans were poisonous. As Churchill's analysis revealed, the cause of the mottling was not aluminum but another metallic element, fluoride.

Subsequent tests demonstrated that other areas with high levels of teeth mottling also had water supplies with significant quantities of fluoride. Laboratory tests and field studies implicated fluoride in the discoloration of teeth, but at the same time they showed that fluoride significantly contributed to the reduction of dental caries. By the 1940s, it had been learned that only 1 part per million (1 ppm) of fluoride reduced cavities but produced no tooth discoloration. Although the effectiveness of fluoride in preventing or retarding tooth decay is now well proved, the actual process through which this occurs is imperfectly understood. Fluoridation is another example of a technology that works, even though scientific understanding lags.

In 1945, the Michigan town of Grand Rapids became the first to have fluoride deliberately added to its water supply. Fluoridation then spread throughout the United States, until two-thirds of the U.S. population are now provided with fluoridated water. Yet it was precisely "mass medication" of this sort that agitated fluoridation's opponents. As they saw it, fluoridation was part of a larger plot to undermine American society. As one fringe group asserted, "The MAIN Communist plots by the internal traitors of the United States are as follows—fluoridation, disarmament of American citizens, federal aid to education."

Although some communities were the scene of widely publicized political conflicts over fluoridation, the extravagant charges made by opponents of fluoridation were overcome by the obvious benefits of adding fluoride to the water supply. It is rather curious that opposition to fluoridation was largely located on the extreme political right, whereas opposition to other technologies, most notably nuclear power, often has been identified with the political left.

Fly-by-Wire Technology

It should be no surprise that the field of aviation, like so many other facets of modern society, has felt the hot breath of the computer revolution. Not long after their invention in the 1940s, electronic digital computers were employed to assist with the difficult calculations involved in aerodynamic design, structural analysis, and reduction of test data. Still, the computer was used in a prosaic way, doing the same work that its ancestor the mechanical calculator had done, only faster. As computer technology advanced, especially after the introduction of transistors in place of vacuum tubes in computer circuits, more dramatic applications emerged. Among the latter was the use of the computer to assist the human pilot in actually flying the craft. Thus was born the notion of "fly-by-wire" technology, loosely defined as the use of digital computers that combine information about an airplane's status with a pilot's actions, and whose output is a set of commands that actually move an aircraft's control surfaces and throttle.

The Wright brothers' key breakthrough was their understanding of the need not only to get a machine into the air but also to control it once aloft. But controlled flight, especially in the machines the Wrights designed, was not easy. Other airplane pioneers configured the positions of the wings, center of gravity, and tail to give more inherent stability so that there was less need for vigilant attention from the pilot. Such a design was less agile, but most felt that the trade-off was worthwhile. By the 1930s, a basic design pattern had emerged, of a craft with a good deal of inherent stability, augmented by a mechanical, gyroscope-based autopilot. This combination produced aircraft with safe, controllable characteristics.

There were a few cases where this pattern proved inadequate. One of the first was the Northrop XB-35 "Flying Wing," which first flew in 1946, followed by a jet-propelled version, the YB-49, which took to the air in 1947. With a design that consisted of almost all lifting surfaces, these aircraft had good performance, but their lack of a tail meant that stability and control were difficult to achieve. Northrop developed a sophisticated, analog flight control system to assist the pilot, but stability problems (among others) led to the project's cancellation in 1949. Forty years later, the flying-wing concept would return as the B-2 Stealth bomber, this time with a digital control system.

The flying machine that was the first to use digital fly-by-wire successfully did so because there was no alternative. That machine was the Lunar Module, the spidery bug that took men from orbit to the surface of the moon, beginning with the historic flight of Apollo 11 in 1969. Landing a rocket on its own exhaust plume required a deftness and control that no human being, no matter how quick his reflexes, could master. And the distance at which this was to occur made it impossible to control things from the Earth (as most of the Apollo mission was controlled). A 32-kg (70-lb) Apollo guidance computer, designed by the MIT Instrumentation Laboratory and built by Raytheon, handled the central task of keeping the rocket's thrust vector aligned with the LM's center of gravity, while the pilot maneuvered by using small thrusters located at the craft's edges. It was not a coincidence that this computer was one of the first anywhere to use integrated circuits, which had just been invented.

The Lunar Module's requirements were "out-of-this-world," but they were not unique. New-generation military aircraft included design features that also severely taxed the reflexes of even the best pilots. At General Dynamics, engineers designing a lightweight and agile fighter for the Air Force recognized that by relaxing the inherent stability demanded by "textbook" design configuration, and augmenting that with a fly-by-wire system, the result would be a much more agile, yet still a safe, airplane. The result was the very successful F-16, which had relaxed stability and an analog fly-by-wire system designed in from the start. A similar analog system was developed for the F-117 Stealth fighter, although the existence of this craft was not made public until years later. The F-117 wings and fuselage were designed first and foremost to scatter radar beams; its aerodynamic capabilities, including stability, had to take second place. The airplane is not known for its stability; without fly-by-wire, it probably could not be flown at all.

Meanwhile NASA had taken a surplus Apollo guidance computer, left over after a later mission to the moon was canceled, and installed it in an old F-8 they were using as a test bed. Beginning in 1972, these tests showed that digital systems could be as effective as analog on aircraft. The higher speeds offered by integrated circuits finally gave digital computers the ability to operate in "real-time," i.e., as quickly as data are coming in, which fly-by-wire of course requires. The Apollo systems consisted of a single digital computer with a crude, analog backup. As adapted for aircraft, this evolved into a system of three or more identical digital computers, which "vote" on every command. Like having multiple engines on a plane, this redundancy means that the craft can continue flying even in the unlikely event of two computer failures. Digital fly-by-wire was successfully adopted for the Space Shuttle, the experimental, forward-swept-wing X-29 (1984), later models of the F-16, and the B-2 Stealth bomber (c. 1989) previously mentioned. For the B-2, the single Apollo guidance computer, with less than 100 kilobytes of memory, had evolved into an onboard local-area network of computers with a power equal to that of the fastest ground-based supercomputers.

By the early 1980s, the military had embraced the fly-by-wire concept, but commercial aviation was less convinced. Commercial jets did not need the agility required by a fighter, nor did they have to worry about designing for stealth. But fly-by-wire did offer lower fuel costs and smoother flights through bad weather. Boeing chose conventional control systems for its 757 and 767 aircraft, but the European consortium Airbus Industrie introduced digital fly-by-wire in its A320 transport. Boeing followed suit with its 777, which entered commercial service in 1995. Digital controls, a produce of the race to put a man on the moon, have now become an accepted part of aviation design.—P.C.

See also APOLLO PROJECT; COMPUTER, MAINFRAME; INSTRUMENT FLYING; INTEGRATED CIRCUIT; SPACE SHUTTLE; STEALTH TECHNOLOGY; TRANSISTOR

Flying Buttress

The gothic cathedrals of medieval Europe are among the world's most impressive structures. An expression of an age of faith, they also were based on the application of solid principles of structural engineering. One of the most important elements of cathedral design was the flying buttress. Although some 19th-century architectural historians believed that the buttresses were purely ornamental, 20th-century research has made it clear that they were essential to the structural integrity of the cathedrals. Flying buttresses allowed cathedral vaults to soar to unprecedented heights, while at the same time making possible the insertion of large stained-glass windows in the walls.

A flying buttress is an arch-shaped structural member that helps to support a building by absorbing some of the lateral forces acting on the building's walls. These forces are generated by the weight of the walls, but no less important are the forces caused by high winds, a problem that increased as cathedrals soared to greater heights, thereby encountering winds that blew more strongly than they did at lower levels. Exacerbating the problem is the fact that the force of a wind is proportional to the square of its velocity. Because the flyer of a buttress swept downward, the forces that fed into the flyer compressed it, and the upright buttress it rested on. This was precisely what the builders wanted to happen. The stone and mortar from which the buttresses were made are weak when subjected to tensile (pulling) forces, but they are very strong when subjected to compressive forces. In many cases, the upright buttress was topped with a pinnacle, not just because it added a decorative touch but because the pinnacle prestressed it, and in so doing prevented the buildup of tensile forces.

Flying buttresses were probably first used around 1180 with the building of the Cathedral of Notre Dame in Paris. As more cathedrals were constructed, builders gained knowledge that could be applied to the new buildings. In regard to flying buttresses, they discovered

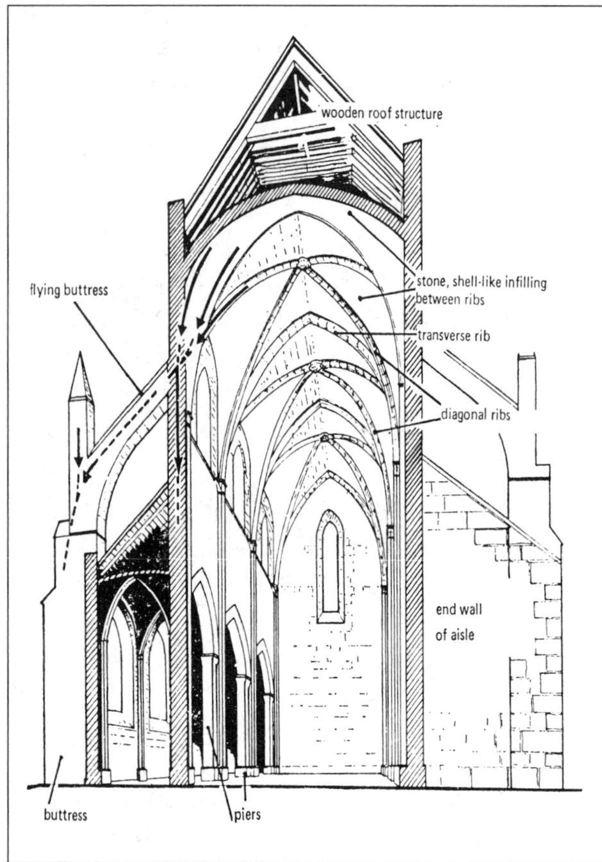

Diagram showing how forces from a structure's walls, ceiling, and roof are partially carried by flying buttresses (from A. Pacey, *The Maze of Ingenuity,* 1992, reprinted with permission).

that considerable savings in stone and mortar could be effected by increasing the angle of the flyer and reducing the height of the upright buttress accordingly. In general, the development of flying buttresses and other structural innovations emerged in the course of actual construction. The people who designed and built the cathedrals went about their work in an empirical spirit, for there was little in the way of codified information to guide the design and construction of these buildings. Moreover, the master masons that designed the cathedrals and supervised their construction had none of the scientifically based knowledge that can now be taken for granted; even multiplication tables were unknown to them. Many useful skills were learned in the course of apprenticeships, but aspirations often outran existing knowledge, and even skilled builders often had to proceed in a tentative manner. Cathedrals were built one bay at a time so that any problems that emerged could be addressed by modifying the design. In this way, for example, it was discovered that a flying buttress could be strengthened by adding a pinnacle to the pier that supported its outer end.

See also ARCHES AND VAULTS; MASONRY; STAINED GLASS

Further Reading: Robert Mark, *Experiments in Gothic Structure*, 1982.

Flywheel

A flywheel is a rotating mass that is used to maintain the speed of a mechanism. For example, the power impulses provided by a four-stroke internal combustion engine are not delivered evenly; in the absence of a flywheel this would make for very rough operation. A flywheel bolted to one end of the crankshaft smoothes out the engine's operation by serving as an energy storage device. It gains kinetic energy as its speed increases, and it loses this energy, which is transferred to the crankshaft, as it slows down. The amount of energy interchange between a flywheel and its power source is a function of the flywheel's mass, its diameter, and the extent to which its velocity changes. In addition to taking these factors into consideration, the designer of a flywheel for a particular application has to take into account the stresses developed while the flywheel is rotating. For example, a cast-steel flywheel can safely withstand speeds of 3,000 m/min (9,840 ft/min), but greater speeds can turn it into shrapnel.

The principle of the flywheel was first embodied in grindstones, with which the stones themselves acted as flywheels. The flywheel became a separate component with the invention in medieval times of the spinning wheel, which takes its name from the large wheel that maintains the rotation of a spindle. By the 16th century, the flywheel had been adapted to one of the key inventions of the Renaissance era, the rotating lathe.

While the flywheel has a long historical pedigree, it may take on new importance in the future. Concern over automobile-borne emissions has led to a search alternatives to the internal combustion engine. One possible alternative is the electric car, but the deficiencies of the batteries used as sources of power may be impossible to overcome. A better alternative may be a flywheel-powered automobile. This would use two flywheels spinning in opposite directions to counteract gyroscopic forces. After being set in motion by an integral motor, the spinning flywheels would function as alternators for the production of an electric current. This current could then be used to run a motor that powered the wheels of the car.

The development of a practical flywheel-powered car will require the solution of a number of technical problems. The flywheels would have to operate at high rotational speeds, on the order of 20,000 to 30,000 revolutions per minute. Conventional materials are not capable of withstanding these speeds, so special materials like carbon-fiber composites would have to be used. Composites also have the advantage of turning into harmless balls of fluff should excessive speeds cause them to disintegrate. It would also be necessary to enclose the flywheels in a vacuum chamber to eliminate air resistance, and the flywheels would have to be supported by bearings that were as friction-free as possible.

These will be difficult obstacles to overcome, although several functional automobiles were built by Harold A. Rosen and Benjamin M. Rosen. Their propulsion system used a 45-kW (60-hp) gas turbine engine in conjunction with a 150-kW (200-hp) flywheel that engaged when more power was needed. Lack of interest by car manufacturers caused this project to be terminated in 1997.

See also ALTERNATORS AND GENERATORS; AUTOMOBILES, ELECTRIC; BATTERY; BEARINGS; COMPOSITE MATERIALS; EMISSIONS CONTROLS, AUTOMOTIVE; ENGINE, FOUR-STROKE; GYROSCOPE; LATHE; SPINNING WHEEL

Food and Drug Administration, U.S.

The U.S. Food and Drug Administration, located within the Department of Health and Human Services, is the federal government agency authorized by Congress to inspect, test, approve, and set safety standards for marketed foods and food additives, drugs, cosmetics, and medical devices. The present agency was preceded by the Bureau of Chemistry, established within the U.S. Department of Agriculture by President Abraham Lincoln in 1862. In 1927, a separate law enforcement agency, the Food, Drug, and Insecticide Administration was created within the Department of Agriculture (USDA), and this agency became the Food and Drug Administration (FDA) in 1930. In 1940, FDA was transferred from the USDA to the Federal Security Agency. which became the Department of Health, Education and Welfare (HEW) in 1953. HEW, in turn, became the Department of Health and Human Services in 1988. FDA currently operates as one of eight health agencies of the Public Health Service.

Following the aggressive political campaigning of the head of the Bureau of Chemistry, Harvey W. Wiley, in the late 1890s, the original Food and Drugs Act was passed by Congress and signed by President Theodore Roosevelt in 1906. It was amended repeatedly over the next 30 years before being completely revised by USDA Secretary Rexford Tugwell and passed by Congress as the Federal Food, Drug, and Cosmetic Act of 1938. That act too has been subject to repeated amendments, most significantly the 1954 Pesticides Amendment, which outlines procedures for setting safety limits for pesticide residues on agricultural products, the 1958 Food Additives Amendment, which requires manufacturers to demonstrate the safety of food additives, the Delany Amendment, which bans any food additive shown to cause cancer in human beings or animals, and the 1962 Kefauver-Harris Drug Amendments, which require that manufacturers of drugs establish the effectiveness as well as the safety of newly marketed drug products. The FDA is also responsible for administering a wide range of more recently authorized legislation: The Fair Packaging and Labeling Act (1966), the Radiation Control for Health and Safety Act (which in 1971 transferred the Bureau of Radiological Health to FDA), the Public Health Service Act, the Prescription Drug Marketing Act (1987), the Nutrition Labeling and Education Act (1990), the Safe Medical Devices Act (1990), the Prescription Drug User Fee Act (1992), the Mammography Quality Standards Act (1992), and the Dietary Supplement Health and Education Act (1994).

The FDA is headed by a commissioner appointed by the President and approved by the U.S. Senate. The Commissioner is assisted by four deputy commissioners who head the Office of Policy, the Office of Operations, the Office of Management and Systems, and the

Office of External Affairs. In 1996, the agency had a budget approaching $1 billion and employees numbering about 9,200. Of those employees, about 1,100 are inspectors and investigators working in 151 district and local offices, and 2,100 are scientists who work in 40 FDA-run laboratories, including the National Center for Toxicological Research.

As a regulatory agency of long standing, the FDA has considerable experience balancing the goals of fostering the development of new products that can improve health and welfare, and at the same time protecting the public from unsafe and ineffective goods and devices. Drug industry leaders sometimes complain that FDA testing and approval procedures unjustifiably delay the introduction of new products that could relieve patients' suffering from chronic diseases and at the same time bolster industry revenues that could be invested in future research. These critics argue that FDA procedures encourage pharmaceutical companies and patients to go abroad where regulatory restrictions are less onerous. Public-health advocates argue that FDA procedures are not rigorous enough and that industry leadership unfairly influences agency decisions. These critics argue that recent efforts to streamline and abbreviate approval mechanisms for new biotechnology products are evidence of the agency's recent willingness to put corporate profits ahead of public health. The FDA has generally remained above the political fray on these and other contentious issues and therefore enjoys a level of public trust that is envied by many newer and less well-established regulatory agencies.—T.I.

See also BIOTECHNOLOGY; DELANY CLAUSE; DEPARTMENT OF AGRICULTURE, U.S.; MAMMOGRAPHY; PESTICIDES

Food Chain

Most biological processes require the acquisition and conversion of energy. Analyzing energy flows is essential for understanding the functioning of individual organisms and entire ecosystems. One can understand the flow of energy through an ecosystem in terms of three sets of interlinked food chains, each of which begins with the primary production of carbohydrates out of sunlight, carbon dioxide and water by some photosynthesizer, such as a plant or photosynthetic bacterium. Photosynthesizers are said to constitute the first, or highest, trophic level. All organisms that eat the primary producers constitute the next trophic level. These in turn divide into three types, each of which initiates one of three primary types of food chain. First are the herbivores, an extremely varied group that includes deer, cattle, grasshoppers, termites, water fleas, and anchovies, which eat the living photosynthesizers. Herbivores, in turn, are eaten by primary carnivores, such as wolves, spiders, warblers, and trout. These constitute the next trophic level, and they in turn are eaten by secondary carnivores such as falcons, tuna, and killer whales, which constitute yet another trophic level. The set of linkages from primary producer through herbivores, primary carnivores, and a number of secondary carnivores constitutes the "grazing" food chain. Biologists now believe that less than 10 percent and often less than 5 percent of plant production is ever grazed.

In many ecosystems, especially in the sea and in the rhizosphere (root-soil zone), a substantial part of primary production is in a liquid form that is extracted or exuded from living cells. This dissolved organic matter, or DOM, may be eaten by nectar-feeding insects or birds, such as bees and hummingbirds, which then join the grazing chain when they become the prey of carnivores. But a much greater and more important segment is consumed by nitrogen-feeding microorganisms. In the microbe-based food chains that develop (the DOM-microbial food chains), the DOM is eaten by bacteria and fungi, which are in turn consumed by protozoa, microarthropods, nematodes, etc. These in turn become food for small animals that become food for larger ones. In some ocean ecosystems, up to 50 percent of the total energy flow may be contained in DOM-microbial food chains.

In most nonocean ecosystems, the bulk of primary production energy is transferred through the third type of food chain, that associated with dead plant and animal tissue, or detritus. Except in very hot, moist environments, detritivores work relatively slowly, allowing a storage and buffering biomass to build up. For the most part, bacteria and fungi act as primary detritivores, although many worms, millipedes, and even vultures can act as detritivores. In many cases, detritus consumption depends on cooperation between different organisms. In the cases of termites and ruminants (deer, cattle, antelope), the host animal breaks up cellulose and other dead matter into small bits and provides in its intestinal tract an environment where symbiotic microbes then convert the cellulose and wood bits into sugars that the host can use as food.

It should be fairly clear from the above that none of the three main types of food chain is completely independent of the others. Some organisms, such as human beings, opossums, and crabs, function as omnivores; these may enter any of the chains at almost any trophic level. Carnivores may eat members of the DOM or detritus chains at almost any level other than that of primary production, and members of a grazing chain may die to become the food of detritivores, thus producing a complex "food web" in any ecosystem.

The complex interrelationships found in a food web mean that a change in one small part may have substantial repercussions throughout the system. For example, human beings once overhunted sea otters, leading to a rise in the population of sea urchins, their customary prey. This in turn caused a substantial increase in the algae that previously had been controlled by the urchins. In similar fashion, the food chain concept helps us to understand the environmental consequences of our actions, particularly our dietary habits. To take an often-cited example, the production of a single kilogram of beef requires 16 kilograms of grain; accordingly, "eating low on the food chain" can substantially lower the energy costs absorbed in food production.—R.O.

See also ECOSYSTEM; PHOTOSYNTHESIS

Food Preservatives

Many techniques have been used to prevent or arrest the spoilage of food, including smoking, salting, pickling, drying, canning, pasteurization, freezing, and radiation. Other processes, most notably brewing and the making of cheese, were partially motivated by the need to convert a nutritious substance into something more long lasting.

The food industry underwent massive changes in the 20th century as producers catered to ever-expanding markets while distribution links became longer. The needs of consumers also changed as a faster pace of life and the growth in the number of women working outside the home made daily shopping difficult. Both producers and consumers had an interest in the development of long-lasting food products; their needs stimulated research that brought forth many new chemical preservatives.

A quick reading of the labels of most commercially prepared foods will usually reveal the presence of at least one and often several chemical preservatives. Each type of preservative performs a particular function. Some preservatives, most notably sodium benzoate, sodium nitrite, and sorbic acid, are intended to kill potentially harmful microorganisms. Others, such as sulfur dioxide, serve to inhibit the action of naturally occurring enzymes that cause spoilage. The type of food will also determine the kind of preservative used. Foods containing fat often have added to them the antioxidants butylated hydroxyanisole (BHA) and butylated hydroxytoluene (BHT), which prevent or delay rancidity. Soft drinks contain benzoic acid or sodium benzoate, preservatives that work especially well in foods with high acidity. Baked goods make use of sodium propionate or calcium propionate along with sorbic acid to retard the growth of mold. Sorbic acid and its potassium or sodium salts are used to retard the growth of molds and yeasts in a variety of products ranging from pie fillings to pickles. In addition to preservatives, many prepared foods contain emulsifiers, thickeners, stabilizers, coloring agents, foaming agents, and artificial flavors and sweeteners in some combination.

Preservatives and other additives have long been a source of concern for many consumers. This anxiety sometimes takes the form of a generalized fear of "chemicals," presumably unnatural substances that have no business being in the food we eat. In reality, all foods, even the most "natural" ones, are collections of chemicals. Some of these naturally occurring chemicals are toxic when consumed in large quantities. For example, in India a poor person's diet that largely consists of chick pea meal has been found to cause lathyrism, a spastic paralysis of the legs that sometimes ends in death. At the same time, the *absence* of preservatives in some foods can result in the production of dangerous chemicals by bacteria. One of the most deadly of these is the toxin produced by the botulinus bacteria that causes the frequently fatal form of food poisoning known as botulism.

In the United States and other industrial countries, the use of chemical preservatives and other additives is closely regulated by the government. In the United States, efforts to ensure the safety of food began with the passage of the Pure Food and Drug Act in 1906. This act, along with the 1938 Food, Drug, and Cosmetic Act, stipulated that substances added to food had to be safe for human consumption and that they had to serve a useful purpose. However, for many years, government enforcement was reactive: The key enforcement agency, the Food and Drug Administration (FDA), had to prove that a substance was harmful before it could be taken off the market. In 1960, the Color Additive Amendment turned this around by assuming that a substance was harmful unless proved otherwise. This means that a preservative has to undergo extensive testing before it can be released for consumption. In the United States, any substance that is found to cause cancer, no matter how large the dosage employed, cannot be marketed. FDA procedures may also result in the removal of well-established preservatives and additives. When the 1938 Act was passed, it contained a "grandfather clause" whereby any of the 687 existing substances "generally recognized as safe" could still be used. However, after laboratory tests seemed to indicate that cyclamate might cause cancer, this sugar substitute was removed from the market in 1970.

The banning of cyclamate was based on laboratory tests with questionable relevance to real-world circumstances. Some widely publicized apprehensions about food preservatives and additives have had even less basis in fact. One example is the public concern regarding nitrosamines that surfaced in the 1980s. Some commonly consumed foods, most notably hotdogs and bologna, have nitrites added to them to retard spoilage. These chemicals are converted into nitrosamines during the digestion process. Nitrosamines have been shown to be mutagenic; that is, they are capable of causing mutations in DNA, the substance that encodes genetic information and transmits it from generation to generation. Because they act as mutagens, nitrosamines have the potential to act as carcinogens, agents that cause cancer, hence the concern about foods containing nitrites. But these foods are far from being a unique source of nitrites; bacteria in the mouth continually convert nitrates, which are found in many vegetables, into nitrites. In a typical diet, the contribution of prepared meats to the total intake of nitrites is quite small.

The use of preservatives may create health risks for some people, for it cannot be said with certainty that every additive and preservative is safe for every person. No two people are exactly alike, and individuals may vary in their reactions to a substance. Still, the dangers posed by preservatives are slight, and in any event, a proper risk assessment has to take into account the much larger risks presented by the consumption of foods with harmful bacteria and other contaminants. Moreover, by preventing or diminishing spoilage, preservatives also increase the amount of food available in a world where adequate supplies cannot be taken for granted.

See also ANIMAL RESEARCH AND TESTING; ASPARTAME; BREWING; CANCER; CANNED FOOD AND BEVERAGES; CHEESE; CYCLAMATE; DELANY CLAUSE; DNA; FOOD AND DRUG ADMINISTRATION, U.S.; FROZEN FOOD; IRRADIATED FOOD; PASTEURIZATION; RISK EVALUATION; SACCHARIN; SALT

Further Reading: Melvin A. Benarde, *The Chemicals We Eat*, 1971.

Football Helmet

In 1888, a rule change revolutionized the traditional British sport of rugby and gave rise to the American game of football. The rule change, making it legal to tackle below the waist, transformed the agile game of rugby into the collision sport revered by U.S. fans, thus creating the need for player protection. Tackling below the waist put heads in contact with knees, a combination that caused—and still causes—many severe injuries. This new rule also made running around the line of scrimmage or running along the sidelines less productive. Without the forward pass, which had not yet been introduced, football became predominantly a game of crashing up the middle.

Protective headgear, adopted as a result of the new style of play, was originally produced and designed in the home. Players tied padded bands around their heads to protect their ears if they had a previous injury, or added pads under their team cap. In the 1890s, head protection moved from being nothing more than long hair and decorative caps to padded helmets and leather harnesses. Leather harnesses consisted of three thick leather straps that fit tightly around the head. To this harness, designed by George Berkeley of Lafayette College in 1896, players began adding ear flaps, and then flaps with ear holes.

During the early years of the 20th century, helmets and other protective devices for football began to appear in sporting equipment catalogues. Throughout the 1920s, manufacturers like Rawlings experimented with helmet designs to improve ventilation and comfort. Until 1917, commercially manufactured helmets did not differ appreciably from homemade harnesses and wool protective hats. With the introduction of the felt suspension system in that year, a new type of helmet and a new protection mechanism came into use. Helmet manufacturers offered a variety of styles and combinations of suspension and padding.

Half a century after the transformation of rugby into football, a technical development in the helmet began to revolutionize the game. The John Riddell Co. patented and produced the first plastic (tenite) football helmet in 1939. This plastic helmet offered a suspension system riveted to the shell, something impossible to achieve with leather helmets. Also, the new helmet would not mildew when wet, as leather helmets did, and it was lighter by 226 g (8 oz), yet more durable than leather helmets. Finally, plastic helmets eliminated the need to repaint logos after each game since the color and design could be baked into the plastic. Players in the 1939 College All-Star game in Chicago were the first to wear the new helmets in a game.

Mass production of the plastic helmet was delayed until the end of World War II. Riddell sold its patent to the military to produce plastic liners for steel battle helmets. After the war, plastic was more readily available and Riddell began to mass-produce plastic football helmets. However, in 1948 Riddell suffered a setback when the National Football League (NFL) banned the use of the plastic shell. Fred Naumetz of the Los Angeles Rams had split nine helmets in one season, thus prompting the NFL to deem them unsafe. But the ban was short-lived, and in 1949 the plastic helmet was restored in the NFL due to the efforts of George Halas of the Chicago Bears, who actively campaigned for its reinstatement. By the late 1950s, helmetmakers were no longer offering leather helmets.

Although head protection had been mandatory in the college game since 1939 and in professional football since 1942, until 1979 no regulations existed about the type of helmet that had to be worn. From the perspective of safety and protection, the plastic helmet has been a mixed blessing. Since the plastic helmet dominated the gridiron from the late 1950s onwards, head injuries have decreased, but severe cervical spine injuries have risen dramatically. Although the plastic football helmet offered protection from injuries to the skull, jaw, and face, it also appeared to contribute to catastrophic neck injuries. And a more brutal style of football accompanied the introduction of the lighter plastic helmet that did not need logos repainted after each game and would not mildew when wet.—J.N.G.

Forging

Forging is the shaping of ferrous and nonferrous metals while in a plastic (heated) state by means of hammers and presses. Iron that is worked this way is a relatively pure and malleable form called *wrought iron*; locally applied force partly displaces it and allows it to be shaped as desired. Forging is one of the most venerable of metalworking processes, for it is at a forge where a blacksmith works. The blacksmith begins by heating iron bars in a furnace kept hot with bellows. The smith then uses hammer and tongs to fashion objects like horseshoes against an anvil. Working something like a horseshoe by flattening a bar along its axis is called *drawing*. When a workpiece is thickened by hammering it on the end against its length, this is called *upsetting*; for example, a rivet is secured by upsetting it. Forging is also a basic technique for welding two pieces of metal together. Indeed, before the development of electrical and gas welding in the late 19th century, the only way to bond two pieces of metal was to hammer them together at a suitably high temperature in the presence of a flux (which

An 18th-century forge in operation (courtesy Arthur C. Bining Collection, Pennsylvania State Archives).

aided the flow of the molten and semimolten metal).

What the fabled village blacksmith did throughout the ages was draw, upset, and weld. Muscle power had its limits, however, and the progress of forging required finding ways to use greater amounts of energy to the process—that is, turning forging into an industrial process. Waterpower could be applied to trip hammers, and towns like Valley Forge, Pa., and Clifton Forge, Va., were sited because of their proximity to the essential natural resources. Purely mechanical hammers such as board hammers were lifted high through arrangements of pulleys and dropped by gravity. Another type of hammer, the helve hammer, was actuated by rotating cams. Hammers of this sort exerted relatively little impact, but the blows could be landed in rapid succession. Steam drop hammers, an invention of the British engineer James Nasmyth (1808–1890), appeared later in the 19th century. These were followed by hammers powered by compressed air. Both types were capable of making a tremendous impact.

The techniques of forging may be employed to reduce a metallic ingot to the approximate dimensions of the intended product or to change its granular alignment to coincide with the contours of a product after it has been machined to its finished dimensions. This is essential when the conditions under which a component is intended to operate are characterized by high stress, as in crankshafts for engines of all kinds; indeed, among the very largest forgings ever made are the crankshafts intended for enormous marine steam engines.

As with many similar activities, forging now covers a range of processes. In *drop forging*, a ram is impelled downward against an anvil. When this operation is performed in conjunction with hardened dies placed on both sides of a workpiece is to be shaped, the process is known as *die forging*. Sometimes workpieces are successively forged in a series of dies. *Machine forging* is an upsetting technique used to make small parts and involves squeezing rather than pounding. The most powerful machine-forging presses, which are hydraulically actuated, are capable of thousands of tons of force and indeed are rated by that force; hence a 2,000-ton press is capable of exerting that much pressure. Machine presses can be used for many related manufacturing processes such as swaging (a process that uses a series of rollers to give a set of dies a reciprocating motion as they press down on the workpiece), bending, punching, and shearing, and in fact some of them are called punch, press, and shear machines.

The machines that forge the countless fasteners that hold our industrial civilization together are essential components of an industrial economy. Yet they are hardly any more important to our present age than was the blacksmith in times past. The Greeks worshipped Hephaestus (Vulcan to the Romans) as the divine smith and patron of craftsmen, and the ancient smiths of Africa, Asia, and Europe were always indispensable members of any community because they produced and repaired the implements essential to the work of artisans and farmers. Even today, smiths still perform essential repair work and occasionally flourish where horseback riding is in vogue, practicing an art to which the Bible pays tribute in the person of the smith Tubalcain.—R.C.P.

See also HORSESHOES; WELDING, ELECTRIC; WELDING, GAS

Forklifts and Industrial Trucks

Forklift trucks are widely used to move materials in a great variety of industrial settings. The genesis of the modern forklift truck goes back to the late 19th century, when relatively small electric motors began to be used to power cranes used for the dockside loading and unloading of bulky ship cargoes. The evident advantages of these motors motivated a few companies such as the Elwell-Parker Co. of Cleveland to initiate the production of electric vehicles, primarily automobiles and a few trucks. A request to Elwell-Parker from the Pennsylvania Railroad in 1905 to produce an electric-powered baggage mover, however, quickly refocused the company's direction. The company soon expanded its production of what were then called power trucks to wider industrial applications, beginning with a line of custom-built trucks for a number of production and materials-handling duties. Their first attempt was nothing more than an electric flatbed truck that had to be loaded and unloaded by hand. Around 1914, the company added a projecting platform, and later a set of forks, that could be run under a skid containing a load and then raised and transported to a different location. These were the ancestors of the forklift trucks that initiated a revolution in materials handling. The outbreak of World War I, with its attendant increase in materials-handling demands, coupled with labor shortages, spurred the development of the industry, and other producers of industrial and lift trucks quickly entered the field.

Most early industrial trucks were electric because they offered the advantage of little noise and no exhaust fumes when utilized in indoor applications. They ran at about 10 kph (6 mph), which, while appropriate for inside work, limited their usefulness for longer outside runs with heavier loads. Accordingly, industrial trucks with gasoline engines for such applications were soon added to product lines. A wide variety of specialty application attachments and vehicles

A modern forklift truck (courtesy Toyota Industrial Equipment).

also quickly emerged, including cranes, scoops, and rams for picking up coils, as well as rotating cradles that could handle irregularly shaped materials such as paper rolls and turn them for vertical or horizontal storage. Today, huge lift versions of such industrial trucks can position themselves over 12-m (40-ft) containers and move them about storage and shipping facilities at will.

Industrial trucks offered a link between the fixed elements of production and storage facilities: conveyor belts, overhead cranes, storerooms, and loading platforms. Starting in the 1920s, new factories began to incorporate the availability of industrial trucks into their designs through single-story construction, the substitution of straight-line production for the grouping of machines by departmental type, and the mechanization of all materials handling. Older factories were retrofitted by smoothing out floors and adding ramps between irregular floor heights to accommodate the new industrial trucks. Collectively, these changes increased efficiency tremendously, and although at first they did not reduce overall labor levels due to the rapid expansion of industry after the war, they reduced labor requirements over the long run. By 1936, an Elwell-Parker advertisement for a coil strip lift truck could proclaim: "These pictures of efficient coil-handling in a modern steel plant, fairly shout at you: Materials-handling costs here have been cut to the last cent!" The development of multiple-stage, hydraulic forklift trucks, in conjunction with the inexpensive wooden pallet, allowed floor-to-ceiling stacking, which previously had been labor intensive and dangerous to workers, if not physically impossible in many cases, and thereby turned the waste of one-level, square-foot storage into modern multilevel, cubic-foot warehousing.

The military materials handling needs of World War II, again combined with the wartime shortage of labor, led to another rapid expansion of industrial trucks and forklifts of all varieties. Rapid industrial expansion with its attendant materials-handling needs after the war, rising labor costs, and the wide variety of industrial truck applications meant that many new companies could find a market niche for themselves in a rapidly expanding industry. By the mid-1950s, there were almost 250 U.S. producers of industrial trucks. Today the lift-truck industry is global in nature, with such companies as Clark and Yale in the United States; Boss in the Great Britain, with subsidiaries in Germany and France; and Toyota, Nissan, and Mitsubishi in Japan producing ever-wider product lines ranging from pedestrian or "walkie" pallet loaders to specialized side-loaders and container lifts. The ubiquitous forklift has evolved to include automatic, guided material-handling robots and storage-retrieval systems. Overall, the impact of industrial trucks and forklifts on materials handling and storage, factory design, and labor, both in terms of skill and strength requirements and in terms of job reduction, has been tremendous.—S.H.C.

See also AUTOMATION; AUTOMOBILES, ELECTRIC; ENGINE, FOUR-STROKE; MOTOR, ELECTRIC; TECHNOLOGICAL UNEMPLOYMENT; TRANSPORTATION, INTERMODAL

Fossils

Fossils are the remains or evidence of organisms that lived in the distant past and have been preserved in rock. Primarily, fossils are a record of the history of life on Earth. Through the study of fossils, or paleontology, the course of evolution can be followed through time. Fossils are also used to delineate the geochemical processes that formed the Earth. Changes in the chemistry of the oceans and the atmosphere are tied to changes in the fossil record. For example, fossils and mineralization changes are used to document the development of photosynthesis by plants and the subsequent changes in the oxygen content of the atmosphere. Fossils also help to determine the age of rocks, since life constantly changes. The distribution of different fossils throughout time also is important in paleogeography and paleoclimatology studies. Since environmental conditions for the development of oil and mineral deposits are known, exploring for oil and minerals frequently involves the reconstruction of ancient geographic and climatic conditions. Organisms are much more environmentally sensitive than rocks, so fossils are therefore a much better tool for understanding and mapping ancient environments.

Fossils are formed by various methods depending on the type of organism and the environment in which the organism died. Organisms consist of soft, organic compounds of mostly carbon, hydrogen, and oxygen—referred to as "soft parts" such as muscles and skin—that are easily decomposed. There may or may not be reinforcing "hard parts," generally made of calcium carbonate (shells) or calcium phosphate (bones and teeth). When an organism dies, its body becomes chemically unstable, and the soft parts of the organism decay into more stable compounds such as water, carbon dioxide, and methane, destroying all traces of the soft parts. The hard parts are relatively resistant to decay but may be destroyed by abrasion or by bioerosion. Very few organisms are preserved as fossils. Millions die every day, and nothing is generally left of them unless the remains are somehow protected from decay. Usually only the hard parts are preserved; rapid burial is the most common method of protection from decay. Fossils are therefore almost always found in sedimentary rock.

The fossil record is biased. Organisms are more likely to be preserved as fossils if they have hard parts. Soft parts are much more likely to decay, so organisms made entirely of soft parts are less likely to become fossils. Organisms that die in depositional rather than erosional environments are more likely to become fossils. The ocean is the ultimate depositional environment; land is more likely to be an area of erosion. Marine organisms are therefore more likely to become fossils than terrestrial organisms, especially the marine organisms that die in shallow areas close to land. Such areas are in the photic zone and are more likely to have abundant life. The more individual the organism, the better the chance that at least one will die in conditions that permit fossilization. Also, the closer to land, the better the chance of rapid burial from the sediment eroding off the land.

Among land organisms there is a preservation bias towards organisms that die near lakes, streams, or in lowland areas. Organisms that die in mountains and highland areas are much less likely to be fossilized. An organism's mode of life also effects the possibility of its becoming a fossil. Burrowing organisms have a better chance of being preserved as fossils than flying or swimming organisms.

There is also a bias in the fossil record that is unrelated to the process of fossilization. Fossils are more likely to be found in relatively young rock. The older the rock, the more likely it is to be either eroded or metamorphosed. Either process can destroy existing fossils. There is also a collection bias. For example, fossils are more likely to be found in the Northern Hemisphere than the Southern Hemisphere, because there are more paleontologists living and working in the Northern Hemisphere. Fossils are also less likely to be collected in areas where it is difficult to sample the rock, such as under ice caps, in tropical rain forests, or in the deep ocean basins. There are also areas where access is restricted for political or cultural reasons. Given the events necessary for the creation and collection of fossils, it is evident that the fossil record is incomplete. Only a small amount of the diversity of life on this planet over the eons has been preserved and studied as fossils.

The modes of fossil preservation may be divided into four major groupings: preservation without alteration, preservation with alteration, molds and casts, and trace fossils. Preservation without alteration is relatively rare for soft parts. Usually, this occurs when an organism is frozen or mummified. These processes rarely happen since climatic conditions stop the decay. There are few areas where both the climatic conditions have remained unchanged *and* the remains of the organism have been undisturbed. While such preservation is rare, it is important because the entire organism may be preserved. Unaltered hard parts are relatively more common, but these are restricted to very young rock where there is little mineralization or deformation.

Preservation with alteration is much more common than preservation without alteration. Preservation with alteration includes the processes of leaching, carbonization, permineralization, recrystallization, and replacement. Leaching is the preservation of hard parts when the more soluble parts of bones and shells are chemically dissolved. The remaining bone or shell is generally bleached and pitted and found only in very young rock. Carbonization most commonly occurs in the preservation of plant fossils or occasionally as the preservation of the soft parts of marine organisms. The organism is preserved as a thin carbon film pressed within the rock, revealing an outline or silhouette of all or part of the organism. An anaerobic environment is necessary for carbonization to occur; such fossils are generally found in black shales.

Permineralization occurs when the pore spaces within the hard parts of an organism are filled by minerals that precipitate from fluids. These minerals reinforce the original remains, increase the possibility of preservation, and sometimes can preserve the microstructure within the organism. The most common minerals involved are forms of calcium carbonate, silica, pyrite, and dolomite. Petrified wood is sometimes preserved through the permineralization of silica within the cell structure. Recrystallization, on the other hand, frequently destroys the microstructure of the organism. The unstable forms of mineral material that had made up the hard parts recrystallize into more stable forms. This is relatively common in the preservation of shells. The original shell secreted by the organism may be made of the calcium carbonate mineral aragonite. Over time, the shell recrystallizes from aragonite into calcite, a more stable form of calcium carbonate. The recrystallization preserves the overall external shape of the shell, but the fine structure is destroyed by the growth of the calcite crystals. Recrystallization changes only the mineral composition of the fossil; the chemical content is unchanged.

Replacement is the final type of preservation with alteration. It is also most common in the preservation of hard parts, although, on rare occasions, soft parts of organisms have been preserved this way. The original material is dissolved, then new material is deposited in its place, the most common way that wood is petrified; the cellulose of the wood is replaced by silica. Replacement can occur on a microscopic scale, and the microstructure of the organism may be preserved. Silica, especially in the form chalcedony, is the most common replacement mineral. Sometimes the organism is not replaced as it is dissolved. The rock around the organism preserves the external shape of the organism, creating a void, or mold, of the organism. Nothing remains of the organism itself. If the mold of the organism is later filled by new material, a cast of the organism is created. The cast is a replica of the shape of the dissolved organism; the mold is a negative impression of its surface.

Trace fossils are the final type of fossils. Trace fossils are not a preservation of the actual organism; rather, they are evidence that the organism existed. Common examples of trace fossils are footprints, burrows, borings, and coprolites. Footprints, burrows, and borings give evidence on how the animal moved and its living conditions. Coprolites are fossilized feces that provide information on the organism's diet, size, and habits. Whether the organism itself is preserved, or only evidence of its existence is preserved, fossils are the way that the history of life has been recorded on Earth.—D.A.B.

See also EROSION; EVOLUTION; OIL AND GAS EXPLORATION; PHOTOSYNTHESIS

Further Reading: Colin W. Stern and Robert L. Carroll, *Paleontology: The Record of Life*, 1989.

Fractal

Lines, planes, spheres, cylinders, cones: These are objects from classical Euclidean geometry. It is with these shapes that we can describe the designs of our engineers, the products of our technology. But as Benoit Mandelbrot has pointed out in his groundbreaking work *The Fractal Geometry of Nature*, "Clouds are not spheres, mountains are not cones, and lightning does not travel in straight lines." To describe the patterns of nature we need a new geometry, a new morphology. Mandelbrot coined the word *fractal* in 1975 to describe the fragmented and irregular objects of this new geometry.

It is not easy to define a fractal. Indeed, Mandlebrot himself has altered the definition at least once, and a concise and complete characterization of fractals is still missing. Nevertheless, it can be said that fractals generally have two signature features: They often (but not always) have fractional Hausdorff dimension (as is explained below), and they exhibit in some way a similarity between their parts and their whole.

We are accustomed to the usual notion of dimension, living as we do in a three-dimensional space. If we enlarge our consideration to the space-time of Einstein, then the dimension increases to four. A surface has two dimensions, and a line but one. In each case the dimension is equal to the number of coordinates that are needed to locate a point in the space. Thus, two coordinates, latitude and longitude, are needed to locate a ship at sea, while three coordinates, latitude, longitude, and height are needed to describe the location of an airplane. Equivalently, the dimension is the number of independent "degrees of freedom" that a creature living in that space would experience in movement.

But topologists have devised other notions of di-

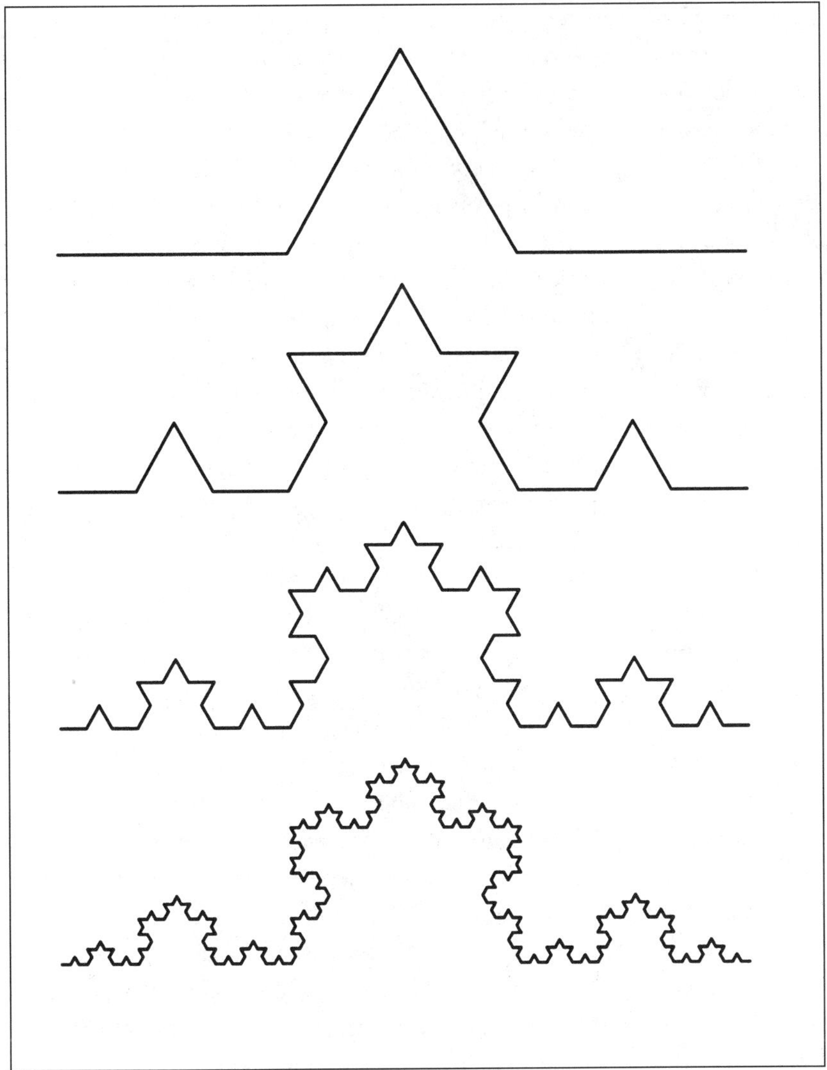

The principle of the Koch curve.

mension as well. The various definitions coincide for the smooth objects of classical geometry, but such is not the case for fractals. To see how this is possible, we consider a simple fractal, the Koch curve, first described by Helge von Koch in 1904.

The figure shows four subsets of the plane, each made up of line segments. The first has 4 segments, the next 16, and so on. Each set is built from the previous one by replacing each segment with a scaled-down version of the original set. None of these sets is the Koch curve. Instead, the Koch curve is the limit of this sequence of sets, the set we would get if we repeated this construction forever!

Notice that if a small piece of the Koch curve is cut from itself, it can be magnified (or demagnified) by a power of 3 so as to coincide with a larger (or smaller) piece of the whole. This kind of self-similarity, or invariance with respect to scaling, is one of the characteristic hallmarks of a fractal.

For a smooth curve, we may approximate its length by walking a pair of dividers along the curve. Multiplying the number of steps by the length of each step gives a number close to the length. Moreover, this approximation will converge to the true length as the dividers are made ever smaller. But this is not so for the Koch curve! As we use smaller and smaller steps, we walk in and out of more and more wiggles, and the total length approaches infinity! Remarkably though, if we plot the logarithm of the total length versus the logarithm of the stepsize, we obtain a line. In the case of the Koch curve, the line has slope $1 - \log 4/\log 3$, and the number $\log 4/\log 3 = 1.26185$ is known as the Hausdorff dimension of the curve. In general, the Hausdorff dimension of any fractal exceeds its usual integral dimension and is often a fraction.

While fractals are interesting shapes whose self-similarity makes them useful as models for natural patterns, they also arise naturally in the study of dynamic systems. Consider a set together with a function from the set to itself. If we pick any point of the set and then follow that point as it is moved by repeated application of the function, a number of interesting things can occur. It may be that the point is actually held fixed, or converges to a fixed point, or perhaps after a finite number of iterations it returns to its original position, exhibiting a periodic behavior. However, it might happen that some points never return to their starting point and never fall into any kind of stable pattern. Instead they appear to bounce around the set in an apparently random, unpredictable way. There may even be pairs of points that are arbitrarily close together and yet whose trajectories differ wildly. This sensitive dependence on initial conditions can make it a practical impossibility to predict the future course of a point, even though the system may be completely deterministic in theory. These kinds of phenomena occur in what have become known as *chaotic systems*. In general, the regions in which the chaotic behavior occurs is usually a fractal. In this way, fractals emerge in a variety of fields that are commonly modeled with nonlinear differential equations, such as fluid dynamics, or population biology.—J.H.

See also CHAOS THEORY; GEOMETRY; RELATIVITY, SPECIAL THEORY OF; TOPOLOGY

Further Reading: Benoit Mandelbrot, *The Fractal Geometry of Nature*, 1982.

Fraud in Science

In May 1981, colleagues of a highly successful and prolific research physician at the Harvard Medical School caught him fabricating raw data for an experiment that was to be the basis of a published article. Before long it became clear that he had forged data for experiments that had led to almost 100 publications in a 2-year period. The case was particularly disturbing because after he confessed to forging his data, and even though Harvard officials revoked his title, he was allowed to continue working in the laboratories of the school's physician-in-chief on a project funded by the National Institutes of Health. No outsiders were warned that their work might be based on results produced by faked data, and only after he was caught by NIH officials for faking additional results did the whole episode become public knowledge.

According to one point of view, in this case the scientific community eventually caught the perpetrator and corrected potentially misleading results. But there is no getting around the fact that he successfully published nearly 100 papers based on faked data, that senior scientists colluded in keeping his actions hidden, and that only his continuing fraudulent activity led to public exposure. In sum, the case poses a severe challenge to assertions that science is inherently self-correcting and that we can be more confident of the veracity of the published claims of scientists than of claims made by members of other groups.

In the idealized version of science that most of us are taught in schools, these events should not happen, because scientists are supposed to be motivated solely by a love of truth, and because the institutions of modern science are supposed to be structured so that attempts to mislead will be caught and automatically corrected. In regard to the latter, when a paper is submitted for publication, it is sent to experts in the field for evaluation. These referees are supposed to check that the work is new, that it properly acknowledges previous work, that the methods used in producing the results are appropriate and explicated in such a way that other researchers can replicate the experiments, and that the arguments used to draw conclusions from the data are legitimate. Moreover, it is assumed that once they are in the open literature, the results will be verified or challenged by other researchers.

Unfortunately, two factors mitigate strongly against this second assumption. In the first place, scientists are rewarded with publication and recognition almost exclusively for new knowledge; thus, there is almost no incentive for anybody to repeat a previous experiments unless it happens to constitute a stage in some new experiment. In the second place, due to the high cost of scientific equipment, sometimes experiments utilize instrumentation that is unavailable for the replication of results. Though this is particularly true in high-energy physics and the space sciences, the high cost of doing experimental work in any field militates against using scarce resources for replication of work already done, increasing the likelihood that fraud will be undetected.

Since the ideology and institutional structure of science promote honesty, however imperfectly, it is likely that intentional deceit is less prevalent in science than in some other professions. Still, the presumption that scientists are motivated solely by a love of truth is highly suspect. Like human beings involved in any other activities, scientists are motivated by many factors, some of which lead to questionable practices not limited to the deliberate faking of data. The desire for recognition and fame is an important feature of scientific life. So is the pressure to publish in order to become tenured, which is a major concern of academic scientists. Scientists in industrial settings are under pressure to produce commercially viable products. These situations have led scientists to appropriate the ideas of others without acknowledgment, and to underemphasize or to deny the unintended negative side effects associated with the products of scientific investigations.

Such practices as the intentional failure to disclose all the information discovered may be understood as fraudulent by some and not by others. These practices shade off into a number of practices that do not necessarily involve the intentional deceit of others but might be construed as fraudulent. In some cases—such as that of Samuel George Morton (1799–1851), a 19th-century physical anthropologist who seems to have compacted the mustard seed used to measure the cranial capacity of Caucasians more than the seed used to measure the capacities of non-Caucasian skulls (thereby increasing the apparent cranial capacity of Caucasian skulls)—there is reason to believe that some form of self-deceit led to an unconscious difference in the way he went about his work. While the results were certainly misleading, it is not certain that they should be considered fraudulent, since there is no evidence of an intent to deceive. In a slightly more complex case, the Nobel Laureate physicist Robert Millikan (1868–1953) ignored some observations in determining the charge on the electron because they violated his expectations. As far as he was concerned, the spurious results were caused by electrons attaching themselves to contaminating dust particles rather than to the oil drops that were the focus of the investigation. Once again, it is fairly clear that Millikan was not intentionally deceiving people; he was making a judgment based on his familiarity with his apparatus. Still, he was undeniably withholding information.

Scientists in other times and places on occasion have engaged in practices that would be considered fraudulent today but were widely accepted practices within the scientific community then. For example, in his optical work, the Hellenistic Alexandrian Greek astronomer and optical writer Claudius Ptolemy (c. 100–c. 170) described a set of experiments in which he measured the bending of light at an air-water interface. He gives what we now call the angles of incidence and their corresponding angles of refraction both for light coming from a source in air and being bent into the water, and for a source in water and being bent into the air. We now know that for light passing from an optically more dense medium into an optically less dense medium, there is an angle of incidence called the critical angle. Beyond this angle, total internal reflection occurs, and no refracted ray can be found. Yet Ptolemy's table of results indicates angles of refraction corresponding to angles beyond the critical angle. It is relatively certain that Ptolemy did, in fact, do some of the experiments described, and that he then extrapolated from his experimental results conditions under which only a "thought experiment" was performed. Such a practice could not be condoned today, but there is no reason to think that Ptolemy was intentionally misleading his readers. Indeed, this kind of extrapolation process was still common as late as the time of Galileo (1564–1642), who, in describing the data that led him to discover the isochronism (i.e., the constant period) of pendulum motions (which hold only for small angles of displacement from the vertical), included displacements up to 90 degrees. Moreover, in describing the experiments that led him to the discovery that the period of a pendulum is proportional to the square root of its length, Galileo recorded the periods of pendulums whose length was too long to have been suspended down the deepest mine shaft in Europe. Once again, there is no reason to believe that Galileo was intending to deceive anyone, but he was extrapolating results in a way that the would be considered fraudulent today.—R.O.

See also ELECTRON; NATIONAL INSTITUTES OF HEALTH; PEER REVIEW; SCIENTIFIC PAPER

Further Reading: William Broad and Nicholas Wade, *Betrayers of the Truth: Fraud and Deceit in the Halls of Science*, 1982.

Freeway

The term *freeway* suggests certain engineering features, such as a median strip to divide vehicles traveling in opposite directions, grade separation by bridges rather than intersections with crossing streets, and access by acceleration ramps. The word itself, however, originated in the field of city planning as a legal concept to define public spaces dedicated solely to vehicular transportation. The legal definition has a distinct point of origin. The engineering features, more diffuse in their origins, developed between the 1860s and the 1930s.

Edward M. Bassett, a lawyer and the president of the (U.S.) National Conference of City Planning, introduced the term in 1930. "A freeway," he wrote, "is a strip of public land, dedicated to movement, over which the abutting owners have no right of light, air or access." His explanation for the coinage revealed the justification for this new type of facility, as well as something of Bassett's cultural milieu:

> This word is short and good and Anglo-Saxon. It connotes freedom from grade intersections and from private entrance ways, stores, and factories. It will have no sidewalks and will be free from pedestrians. In general, it will allow a free flow of vehicular traffic. It can be adapted to the intensive parts of great cities for the uninterrupted passage of vast numbers of vehicles.

Bassett distinguished the freeway from the highway, which had no limitation to access, and the parkway, which had access restrictions but was less an artery than a linear park, devoted to recreation.

Authority over access to public lands resided in state legislatures. In the 15 years after Bassett introduced the term, every state enacted legislation enabling transportation agencies to designate selected rights-of-way as freeways. The Bureau of Public Roads, a federal agency, facilitated the process by drafting model language for states to adopt. The distinction between freeways and parkways was important in the conception of this new legal authority because many of the design features first appeared on park roads. The earliest grade separation between roads is generally credited to Frederick Law Olmsted's plans for New York's Central Park in the 1860s. Divider strips for carriages traveling at different speeds, and to separate carriageways from walkways, also appeared in park plans of that period, such as the promenade along Boston's Charles River. The earliest known acceleration and deceleration ramps connecting separate rights-of-way were also built along the Charles. In the 20th century, the roads between Manhattan and Jones Beach, and between George Washington's home and the nation's capital, incorporated these features in integrated designs for recreational motoring.

The application of the legal and technical features that achieved limited access was not simply adapted whole from parkways to freeways. The justification for controlling access to parkways was to preserve the scenic qualities of the land through which these roads passed. A different goal, the accommodation of escalating automotive traffic, had occupied city planners and engineers from the early years of the 20th century. City transportation engineers typically had trained in railroad work, in which practices such as dedicated rights-of-way and grade separation provided a technical background not connected to parkways. Systematic attempts to apply these concepts might well have drawn from the parkway idea, as in Daniel Burnham's celebrated plan for Chicago (1909), but economic and political opposition generally thwarted such ambitious schemes. City engineers won approval for tunnels and grade-separated intersections at isolated locations rather than as parts of fully realized high-volume arteries. Freeways, as codified by Bassett, resulted from the joining of these urban transportation practices with the limited-access authority first applied to parkways. Nor were the German *autobahnen* a direct predecessor to urban freeways in the United States, as the German roads bypassed population centers rather than slicing through them.

The first dedicated rights-of-way in cities were the 4-mile long Henry Hudson in New York (1936), the Arroyo Seco in Los Angeles (1940), and the Davison in Detroit (1942). Though called *parkways*, they were something very different. They did not fulfill any recreational purpose; rather, they had the utilitarian goal of moving the maximum amount of traffic in the shortest possible time.

Other state and local initiatives produced several hundred miles of urban freeways in the 1940s and early 1950s, but the freeway era in the United States dates from the enactment in 1956 of massive federal aid for highway construction. By 1960, with the administrative mechanisms in place to regulate the flow of federal dollars to the states, freeway construction began in earnest. The tens of thousands of freeway miles constructed over the next 15 years rank among the largest public-works projects in history. However, increasing concerns over the social and environmental impacts of freeways in urban areas, as well as inflation that eroded the value of appropriations, brought a virtual end to freeway construction by the mid-1970s.—M.W.R.

See also INTERSTATE HIGHWAY SYSTEM

Further Reading: Bruce E. Seely, *Building the American Highway System: Engineers as Policy Makers*, 1987.

Freeze Drying

For thousands of years, people have preserved perishable substances by dehydrating them. In most cases, this entailed nothing more complex than leaving meat, fish, or fruit out in the sun so it would slowly dry out. In the early 20th century, a new technique of drying was invented. *Lyophilization*, or *freeze drying* to use its popular name, entails the freezing of a substance, followed by the removal of the frozen solvent (typically water) by adding just enough heat to cause sublimation (sublimation is similar to evaporation, except that it involves the vaporization of a solid rather than a liquid). Freeze drying is usually conducted in a partial vacuum, which accelerates the process and allows it to be more easily controlled. Freeze-dried materials undergo little shrinkage because they dry while frozen solid. The freeze-dried substance is stored until it needs to be reconstituted through the addition of water or the particular solvent that had been removed.

A kind of freeze drying was used by the Incas in Peru 1,000 years ago. The modern method was developed in France in 1906 by Jacques Arsène d'Arsonval (1851–1940) and George Bordas as a means of storing biological material for a long time. The first food product to be freeze dried was coffee; freeze-dried instant coffee was first produced in Switzerland in 1934. Freeze drying was introduced to the United States in 1940, where it was first used to preserve blood plasma. A year later, freeze dried orange juice was being supplied to the U.S. Army.

Due to relatively high production costs, freeze drying for the preservation of food products has not attained the importance of canning or conventional freezing, although it is often employed when weight reduction is important, as with food taken on camping trips. Freeze drying is extensively employed for the purpose for which it was first devised, the preservation of medical and biological materials such as blood plasma, vaccines, tissues, and bacterial cultures.

See also CANNED FOOD AND BEVERAGES; COFFEE, INSTANT; FROZEN FOOD

Frequency Modulation (FM)

In the early years of the 20th century, radio was demonstrating its potential as a practical means of communication and entertainment. Technical improvements such as the regenerative circuit and the superheterodyne receiver, coupled with the development of vacuum tubes, increased radio's ability to reach a large and growing audience. Yet despite all these advances, radio reception was still plagued by a major problem: "static," an unpleasant background noise caused by electrical discharges in the atmosphere. It was the hope of industry leaders like David Sarnoff (1891–1971), the president of the Radio Corporation of America (RCA), that a clever inventor would someday devise a "black box" to eliminate this nuisance.

It fell to a friend of Sarnoff's, Edwin Herbert Armstrong (1890–1954), to come up with a solution. Financially independent through the sale of patent licenses for his radio inventions and extensive stock holdings in RCA, Armstrong was free to experiment while holding a chair at Columbia University's department of electrical engineering. In late 1933, in the Hartley Laboratory of that institution, Armstrong demonstrated to Sarnoff not a black box but a static-free radio system based on a novel principle: frequency modulation. In existing radios, the audio signal was carried on a radio wave that was controlled by variations in the wave's amplitude.

This universal system of radio transmission and reception was known as AM, for amplitude modulation. AM had the disadvantage that unwanted noise strongly affected wave amplitude. This was not the case for radio transmissions that employed frequency modulation (FM).

An FM system would be much less subject to static, but it required nothing less than completely new equipment for transmitting and receiving radio signals. Moreover, the signals would have to be put into the very-high-frequency band, where the advantages of FM were more evident and where they would not conflict with the existing AM band of 500 to 1,600 kilocycles. More was involved than the invention of new equipment. Armstrong had to develop his system in the face of a widespread belief that FM was technically impossible. A few years before the successful demonstration of FM, one eminent mathematician had "proved" that FM couldn't be done, concluding that "static, like the poor, will always be with us."

Armstrong dispelled these notions first in the laboratory and then through long-distance broadcasts. These demonstrated that not only was FM feasible but also that it conferred many advantages in addition to being almost completely static-free. FM's signal conveyed a wider range of audio frequencies, thereby increasing the fidelity of the transmitted sound. It also manifested a "capture effect" that allowed a receiver to pick up only the strongest of several signals of the same wavelength; this allowed more broadcasting stations to send their programming without having it interfere with other broadcasts.

FM was technically superior, but it posed a problem for Sarnoff and RCA. The firm was the largest manufacturer of radios as well as the parent company of the National Broadcasting Company. Conversion to FM would necessitate considerable expenses for the firm and its customers. Also, RCA under Sarnoff's leadership was beginning to go beyond radio. It was strongly interested in an infant technology then being developed in its laboratories: television. As a result, RCA's support of FM was half-hearted. After sponsoring some tests, it pulled out, leaving Armstrong to promote FM on his own. On July 18, 1939, his station began broadcasting from New Jersey, its programs received by General Electric radios licensed by Armstrong.

RCA's lack of interest was not the only problem. Many obstacles to the commercialization of FM were deliberately planted by entrenched radio interests. In the 1930s, the Federal Communication Commission (FCC), which had close ties to the industry, often denigrated FM. In 1944, the FCC moved FM's frequency band from 42 to 50 megacycles to 92 to 106 megacycles, thereby rendering obsolete existing transmission and receiving equipment. The old FM frequency band was assigned to the audio portion of television broadcasts. The FCC also lowered the maximum permissible power of FM broadcasting stations to 1.2 kilowatts, drastically reducing their range.

Still, FM had showed signs of becoming a commercial success. Recognizing FM's potential, in 1940 RCA tried to purchase nonexclusive rights to Armstrong's patents for $1 million, only to be rebuffed by the inventor. World War II halted further commercial

Edwin H. Armstrong (courtesy Institute of Electrical Engineers).

work with FM, but in 1946 RCA began to produce its own FM radio sets, claiming that the technology it employed did not infringe on Armstrong's patents. In 1948, the inventor filed suit against RCA and some other firms that were producing FM radios. Drawing only on his own financial resources, Armstrong was seriously handicapped in his battle against RCA. He could have achieved a reasonable out-of-court settlement, but Armstrong, who previously had engaged in a lengthy and unsuccessful court battle over the patent rights to the regenerative circuit, wanted a complete victory. The drawn-out court case put him under great financial and psychological stress. Seemingly defeated, on Jan. 31, 1954, Edwin Armstrong jumped to his death from the 13th floor of his apartment building. His wife, Marion Armstrong, continued the struggle to a successful conclusion. An arbitrated settlement concluded later in the year awarded $1,050,000 to Armstrong's estate. Legal proceedings against other radio firms dragged on. The last one was not settled until 1967, when the Supreme Court ruled that Motorola had infringed on Armstrong's patents. Today, about 5,000 FM stations broadcast their programming in the United States, all of them operating on principles first developed by Edwin H. Armstrong.

See also AMPLITUDE MODULATION (AM); FEDERAL COMMUNICATIONS COMMISSION; RADIO; REGENERATIVE CIRCUIT; SUPERHETERODYNE CIRCUIT; THERMOIONIC (VACUUM) TUBE

Further Reading: Tom Lewis, *Empire of the Air: The Men Who Made Radio*, 1991.

Front-Wheel Drive

For many decades, the power trains of automobiles followed a basic pattern: a front-mounted engine drove the rear wheels via a long drive shaft. In contrast, a small number of cars employed front-wheel drive (fwd). In this configuration, the front-mounted engine powers the front wheels, which also have the usual function of steering the car. Front-wheel drive entails some mechanical complications, but it also has a number of advantages. The engine's weight is over the driving wheels, where it provides the car with superior traction on slippery surfaces. The nose-heavy weight distribution causes the car to understeer (tend to run wide on turns and scrub off speed), making it somewhat safer in the hands of unskilled drivers. Most importantly, front-wheel drive cars have superior packaging when compared to rear-drive equivalents. Because there is no drive shaft, there is no obtrusive driveshaft tunnel. The absence of differential gears at the rear also frees up space at the back of the car. Space saving became especially important in the United States from the 1970s onwards, when rising gasoline prices necessitated the production of smaller cars; front-wheel drive allowed the retention of generous passenger space for cars that had been "downsized."

Although a substantial portion of the world's cars did not have front-wheel drive until the mid-1960s, the principle is as old as the automobile itself. The first self-propelled vehicle, Nicholas Cugnot's steam-powered three-wheeler of 1769, employed frontwheel drive, two vertical cylinders were connected to a single front wheel through rods and levers. The first fwd automobile powered by a gasoline engine was the Austrian Gräf & Stif, produced from 1895 to 1897. A few other front-wheel drive cars were made in the early days of automobiling, most notably a series of racing and experimental cars designed by John W. Christie, an American naval engineer. By 1911, Christie had built seven cars (including two with engines displacing more than 20 liters) before turning his attention to tracked military vehicles. In the years between the world wars, the most prolific builder of fwd automobiles was the German firm DKW. Like many fwd cars that came later, DKWs had their engines mounted transversely (i.e., with the crankshaft at right angles to the car's centerline). Another noteworthy fwd car was France's Citroen *Traction Avant*, which was introduced in 1934 and remained in production until 1955, at which time it was replaced by an even more radical fwd design. In 1959, the British Motor Corporation introduced one of the most significant automobiles of all time, the Mini. Front-wheel drive and a transverse engine were essential elements of a design that provided seating for four in a car that was all of 3 m (10 ft) in length.

The Mini transmitted power to the front wheels via universal joints that followed a design of Donald Bastow and were produced by GKN-Birfield. Adequate universal joints were key components of the

Mini and other fwd automobiles, for they allowed the front wheels to both steer and receive power. Early fwd cars used conventional cardan joints, but these were not completely adequate, for turning the wheels in order to steer the car produced excessively sharp angles between the shafts, as well as cyclical acceleration and deceleration of the shafts. The solution to this problem is the constant-velocity joint, of which the Bastow-Birfield joint is one example.

Early fwd cars suffered to some extent from torque steer, the tendency of the steering to tighten up when power is applied. This problem has been largely eliminated through the improvement of universal joints and careful attention to steering geometry. Although sporting drivers usually dislike the understeer inherent in front-wheel drive cars, the packaging advantages of fwd are undeniable. Today, the majority of cars with engine displacements under 2 liters (121 in.^{3}), as well as many others with larger engines, have front-wheel drive.

See also AUTOMOBILE; DIFFERENTIAL; CONSTANT-VELOCITY JOINT; UNIVERSAL JOINT

Further Reading: Jan P. Norbye, "Pulling to the Front: Front-Wheel Drive," *Automotive Quarterly*, vol. 35, no. 2 (May 1996): 24–39.

Frozen Food

For most of human history, people subsisted on locally grown foods. Variety was provided by seasonal fruits, vegetables, and meats, but when these were gone, there was nothing left until they reappeared the following year. Some foods were preserved through smoking, salting, or drying, but such foods bore only a vague resemblance to their fresh form. Preserving food through freezing was possible in cold climates; one recorded effort is that of Francis Bacon (1561–1626), who used snow to preserve a chicken. Unfortunately, in the course of doing so, he caught a chill that led to his death a month later. By the 19th century, commercial freezing using an ice-and-salt mixture was being employed to preserve fish, and in the early 20th century, freezing was used to preserve fruit destined for the commercial production of pies, ice cream, and jam.

These techniques evened out seasonal fluctuations in the availability of certain foods, but they suffered from a fundamental shortcoming. Slow freezing damages food by extracting water from the colloids of individual cells, leading to the collapse of their walls, the concentration of salts, and the precipitation of proteins. Making matters worse, slow freezing results in the formation of large ice crystals that rupture cell membranes and break up tissues, thereby harming the food's flavor and texture.

These problems can be avoided by freezing the product quickly at temperatures below –4°C (25°F). The most commercially successful of these was the one pioneered by Clarence Birdseye (1886–1956), who gained insight into the problem while living in Labrador from 1910 to 1917. During his residence in that frigid region, he observed that fish subjected to very cold temperatures retained all of their freshness when they were defrosted months later. His freezing process, first employed in 1924, entailed prepacking the food in a small carton and then holding it between two metal belts while they were chilled by a calcium carbide solution, the lower one by a spray, and the upper one by a constant flow of the solution. An alternative method, which came to be the more widely employed, relied on two hollow metal plates that clamped over the package at a pressure of 150 pounds per square inch (150 psi), while vaporized ammonia within the plates brought temperatures down to –32°C (–25°F).

Although the basic process was conceptually simple, its implementation required many auxiliary devices and procedures. By the time it had been put into commercial operation, the Birdseye process was covered by 168 patents. A complete system of quick freezing required slicing and filling machines, sealing machines, and special papers and packages. Furthermore, in 1930, scientists at the U.S. Bureau of Plant Industry discovered that destructive enzymatic action could be forestalled if vegetables were briefly scalded prior to freezing. This process, illogically dubbed "blanching," became an essential process in the quick freezing of fruits and vegetables.

Once the operational details had been worked out, the use of the Birdseye process required a considerable amount of capital investment. This came from the Postum Company (subsequently renamed General Foods), which was eager to expand its product line. In 1928, amidst some questionable financial dealings, the company paid $11,650,000 to acquire Birdseye's fledgling operation. At the time of the acquisition, the consuming public had a low opinion of frozen foods, for those that had been available were of poor quality. Another barrier to commercial success was the fact that refrigerators capable of storing frozen food were by no means universal. In the late 1920s, only half of American households had refrigerators of any sort, the majority of them old-fashioned iceboxes.

Fortunately for General Foods, refrigerators were becoming more common just as the firm was beginning to market frozen foods under the Birdseye label. In 1930, for the first time sales of mechanical refrigerators exceeded those of iceboxes, and within 15 years, 70 percent of American homes had one in their kitchen. The spread of refrigerators was due in part to technical improvements, but no less important was the aggressive promotion of these appliances by their manufacturers and by utilities companies eager to expand the market for electric power.

Complementing the spread of domestic refrigerators were freezing cabinets for the storage and display of frozen foods in grocery stores. In the hard economic times of the 1930s, most retailers were reluctant to pay as much as $2,000 for such a fixture. To increase market coverage, General Foods had a new design for the cabinet, and then rented them out at low cost. The transit of frozen foods was another problem, one not solved until the early 1950s when railroads began to use fleets of mechanically cooled refrigerator cars.

The widespread availability of adequate home and retail storage facilities was a necessary condition for the spread of frozen foods, but it was by no means sufficient. Quick-frozen foods had many advantages—freshness, elimination of waste, easier preparation, and year-round availability; still, the consuming public had to be sold on their virtues. Test marketing and substantial advertising campaigns were mounted in order to convince people of the benefits of frozen food. Progress was slow in the 1930s, but the industry was given a substantial boost by America's entry into World War II, due to the restriction of the sales of canned goods as a result of metal shortages, an obstacle not faced by frozen foods. Even more important, the requirements of feeding military forces created a vast new market. The armed forces were also powerful exponents of standardization, so a hitch in the army or navy did much to prepare a generation of Americans for a postwar society built on a foundation of standardized products. Frozen foods were the embodiment of standardization, but, paradoxically, standardization was paralleled by increased variety, as a vast number of new products appeared in the frozen-food cabinets. Complete ready-to-heat meals, exemplified by the "TV dinner" that first appeared in 1954, became staple food items in millions of households.

The rapid expansion of the frozen-food industry that began in the 1950s was due in part to the consumer boom of that era and the expansion of refrigerator and freezer sales. But no less important were cultural shifts. Work and domestic life were changing, altering conventional beliefs of what constituted a proper meal and who had the responsibility for preparing it. Increasing numbers of women began to find work outside their homes, commuting distances were lengthening, and families no longer necessarily took their meals together. Frozen foods of increasing variety and sophistication were an integral part of lifestyle changes encompassing far more than altered eating habits.

See also FROZEN PREPARED MEALS; REFRIGERATOR; REFRIGERATOR CAR; STANDARDIZATION

Further Reading: O. E. Anderson, *Refrigeration in America: A History of a New Technology and Its Impact*, 1953.

Frozen Prepared Meals

The development of the frozen-food industry in the 1930s and 40s brought many benefits to consumers. One important class of frozen foods, however, has not always met with unqualified praise. Beginning with the TV dinners of the 1950s, frozen prepared meals have been criticized for their nutritional and gustatory deficiencies. They have also raised the ire of those who view the preparation of a "proper" meal as an essential household task. Still, there is no denying that to many busy people frozen dinners provide an adequate meal with a minimum of preparation time.

The idea of offering a complete frozen meal goes back to the early days of the frozen-food industry, but at that time it was a assumed that sales opportunities would be limited. Instead of being marketed to households, the first frozen meals were produced for military and civilian airplane passengers. In 1945, Maxson Food Systems introduced frozen meals on a tray that it called Strato-Plates. These were produced for only 2 years and failed to win a large market. This was in part due to problems with the onboard reheating equipment, but the product also suffered from bad timing. In the immediate postwar years, the size of the military was shrinking, and airline travel had not yet entered its period of rapid growth.

Despite this early failure, development of frozen meals continued, entailing the solution of a number of technical problems. Food technicians knew that covering a meat entree with gravy would keep it from drying out when it was reheated, but then the gravy would curdle; this necessitated the creation of new thickeners. Potatoes were another source of difficulty, as most prepared potato dishes did not reheat well. As a result, early prepared meals often featured other kinds of potato dishes: potato croquettes, scalloped potatoes, and potato puffs. Getting all of the dishes properly heated at the same time was also a problem; since meat cooked more slowly than vegetables, it was necessary to only scald the vegetables prior to freezing so that the actual cooking took place in the oven.

These efforts culminated with the successful introduction in 1955 of the Swanson "TV Dinner." Part of the product's appeal lay in a marketing plan that capitalized on the rapid spread of television sets into American homes. The TV dinner could be eaten in front of the TV, and no viewing time was lost while the cook prepared a meal in the kitchen. The package was even made to look like a TV set, reinforcing the relationship between television and the TV dinner. In 1955, in the year it was introduced, 70 million TV dinners were sold, making them a major marketing triumph. Yet these sales pale in comparison with the 2 billion prepared dinners and entrees that were sold annually in the United States in the first half of the 1990s.

The "classic" TV dinner in its aluminum tray is no longer produced; since 1986 frozen prepared meals have been packed in microwavable plastic trays. Also, the complete dinner has been overshadowed by the single entree. In 1990, in their never-ending quest to gain market share, producers of frozen foods introduced 651 entrees, but only 55 complete meals. But however it is prepared and packaged, the frozen meal continues to be emblematic of a changing society with a looser definition of what constitutes a family and a proper meal.

See also FROZEN FOODS; MICROWAVE OVEN

Further Reading: Frederic D. Schwartz, "The Epic of the TV Dinner," *American Heritage of Invention and Technology*, vol. 9, no. 4 (Spring 1994).

Fuel Cell

Energy is usually extracted from a fuel by burning it. An internal combustion engine takes a fuel such as gasoline or methanol, mixes it with oxygen, and ignites the mixture in the combustion chamber, thereby producing pressure to drive a piston. The energy so produced can be used to turn a generator in order to produce electricity. It is possible, however, to convert a fuel directly into electrical energy. The device that accomplishes this is a fuel cell. Invented in 1839 by William Grove (1811–1896), the fuel cell may be an important source of energy in the 21st century.

In its simplest form, hydrogen is introduced into

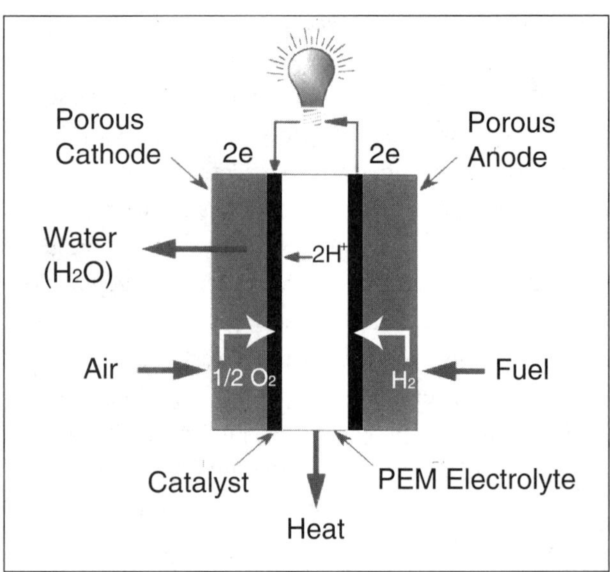

Simplified diagram of a fuel cell (courtesy Ballard Power Systems).

one of the fuel cell's electrodes (the anode). Here the hydrogen molecules lose their electrons, which migrate to the other electrode (the cathode), and in so doing produce an electrical current that can be used to power a motor. At the cathode, the electrons then combine with oxygen molecules, ionizing them. Meanwhile, the ionized hydrogen molecules travel through an electrolyte located between the electrodes and react with the ionized oxygen molecules to form water. These reactions are greatly accelerated by the presence of a catalyst surrounding the electrolyte. The catalyst, which is made from platinum, silver, palladium, or other suitable substance, is particularly important if the fuel cell is to function at relatively low temperatures, on the order of 95°C (203°F). A complete fuel cell apparatus consists of many of these individuals units, all wired together in series.

When made sufficiently large, a fuel cell can produce a fair amount of power; 15-kilowatt (20-hp) fuel cells are currently in operation, and larger ones are practical with today's technology. Fuel cells have two key advantages over conventional heat engines: they are more efficient and they produce less pollution. As far as efficiency is concerned, a fuel cell running on hydrogen is three times more efficient than an internal combustion engine similarly fueled. A fuel cell can also use other substances such as methanol or natural gas as a source of hydrogen. This is done by reacting them with steam in the presence of a catalyst, a process known as *reforming*. A fuel cell using methanol will produce energy $2\frac{1}{2}$ times as efficiently as an internal combustion engine running on this fuel.

An operating fuel cell produces small amounts of carbon monoxide, along with some oxides of nitrogen when methanol or natural gas are used as the source of hydrogen. Greenhouse gases such as carbon dioxide also are produced, but in much smaller quantities than in the case of conventional internal combustion engines. Concerns about global warming brought on by carbon dioxide emissions make the fuel cell an especially attractive alternative to the internal combustion engine.

With so many apparent advantages, the fuel cell would seem destined to replace conventional means of converting a fuel into useful energy. But more is involved than theoretical superiority. In the first place, a fuel cell is more expensive than a conventional heat engine, due to the need for exotic catalyst metals and sophisticated polymers for the electrodes. Whereas an ordinary automobile engine costs in the neighborhood of $3,000, a fuel cell of equivalent power would cost 10 times that amount. Size is another problem; a fuel cell used as an automobile power plant would be the size of a refrigerator. Further research and the application of mass production techniques will likely resolve these problems, but an automotive fuel cell probably will not be a practical proposition until the second decade of the 21st century. Also, the use of hydrogen fuel will require special techniques for production, distribution, and storage, while the extraction of hydrogen from methanol or natural gas will require some expensive new technologies. Barring a substantial long-term increase in the price of gasoline and other conventional fuels, hydrogen sources may prove too expensive, even when the superior energy-conversion features of the fuel cell are taken into account. Finally, today's energy system is the result of more than a century of growth and development. Conventional automobile engines and power stations draw on a vast array of manufacturers, refineries, transportation and distribution networks, and repair facilities. A fuel cell–based energy system would need a similar infrastructure, and this would be the work of many years, if not decades.

See also CLIMATE CHANGE

Fuel Injection

A fuel-injection system supplies an engine with atomized fuel (i.e., fuel that has been divided into very small droplets) that is then combined with air, compressed, and burned in the engine's combustion chamber. For most of the history of the automobile, this function was performed by the engine's carburetor. Although the carburetor had its deficiencies, it was inexpensive, simple, and easy to maintain. Fuel injection has been around since the early days of the internal combustion engine, but it only began to be widely employed in the late 1970s. The widespread adoption of fuel injection at that time is a clear example of technological change being propelled by government regulations. As fuel efficiency and exhaust emission standards tightened, the carburetor's shortcomings were all too evident. Getting more mileage out of a tank of gasoline and burning it as cleanly as possible required more precise fuel metering than the carburetor was able to provide. Atomized gasoline had to be delivered at exactly the right moment, and it had to be mixed with air in a precise ratio. Only fuel injection could do this.

As with much of the history of technology, it is difficult to determine who built the first fuel injection system, but credit is often assigned to Robert Bosch in Germany, who produced a commercial fuel injection system in 1912. For many years, the only road vehicles to use fuel injection were those powered by diesel engines, the design of which eliminated the use of carburetors. Fuel injection became increasingly attractive in aircraft engines in the 1930s, for airplanes flew at a wide range of altitudes and were put through severe maneuvers. As has often been the case, developments in aviation eventually spilled over into automotive technology. In the early 1950s, Daimler-Benz, a major producer of aero engines during World War II, equipped its highly successful Mercedes 300SLR sports racing car with fuel injection and soon began to use fuel injection in some of its passenger cars. Other manufacturers fitted fuel injection to limited numbers of high-performance cars, but, as noted above, it was not until the 1970s that fuel injection came to be widely adopted.

There are two basic types of fuel injection systems: mechanical and electronic. The basic elements of a mechanical system are a fuel pump, regulating valve, distributor, and injectors for each of the engine's cylinders. The pump supplies fuel at a pressure of about 100 psi (7 bar) for gasoline engines, and much higher for diesels. Fuel pressure is kept constant by the regulating valve. The distributor supplies fuel sequentially to the injectors, where it is atomized as it leaves the injector nozzle's very small aperture (the diameter of a human hair or less). In diesel engines, the atomized fuel usually is delivered directly to the combustion chamber, while in gasoline engines the fuel is delivered to an intake manifold, entering the combustion chamber when the intake valve begins to open.

In an electronic system, fuel moves at a pressure of about 30 psi (2 bar) along a pipe known as the *fuel rail*, which delivers it to the individual injectors. The timing of the injection and the amount of fuel passing through the injector nozzle is controlled by electromagnetic (solenoid) valves. These valves are opened and closed by an electronic control box that receives signals from sensors measuring engine speed, throttle position, and manifold vacuum. In this way, the injectors deliver fuel at just the right moment and for the exact duration required. The electronic controller also receives information on the amount of oxygen in the exhaust gas. All of the signals supplied by the various sensors allow the controller to maintain the optimal 14.7:1 ratio of air to fuel (by weight) in the engine's combustion chambers, optimizing performance and keeping emissions at as low a level as practically possible.

See also CARBURETOR; ENGINE, DIESEL; ENGINE, FOUR-STROKE

Fuels, Fossil

Fossil fuels appear as gases, liquids, and solids. Coal can be converted to a liquid or gas, and natural gas can be liquefied for ease in shipping. Fossil fuels were formed when organic materials were subjected to heat and pressure for many millions of years. Since they were made from organisms that were once alive, fossil fuels are in effect repositories of stored-up solar energy, for organisms cannot exist without the energy of the sun. Like most other organically based fuels, fossil fuels are hydrocarbons: molecules composed of atoms of hydrogen, carbon, and often other elements. Fossil fuels vary substantially in their ratio of carbon to hydrogen. Anthracite coal is close to being pure carbon, whereas an atom of methane (CH_4), the main constituent of natural gas, has four hydrogen atoms for each carbon atom.

Excluding biomass fuels (which supply less than 5 percent of the world's energy), 88 percent of the world's commercially produced energy comes from natural gas, petroleum, and coal (hydroelectricity and atomic fission supply the remaining 12 percent). Fossil fuels also serve as important feedstocks for many industrial products such as plastics, dyes, and chemical fertilizers.

A common measurement of large amounts of energy is the quad: one quadrillion [10^{15}] British thermal units, or 1.055×10^{15} kilojoules. At present, natural gas provides approximately 65 quads, or 20 percent of the world's energy consumption. For petroleum the figures are 121 quads and 38 percent; coal supplies 96 quads and 30 percent of global energy use. Since modern society is heavily dependent on the use of vast amounts of energy, there is an understandable concern that the eventual exhaustion of fossil fuels will cause the collapse of industrial civilization. But determining when (or even if) this will happen is a difficult endeavor. For one thing, much of the world's land and sea remains unexplored as far as fossil fuels are concerned. Even when deposits are known to exist, they are of value only when the cost of extracting the fuel is lower than its selling price. In turn, the price of a fuel is strongly affected by the extent of the demand for it, as well as by the availability of substitutes. Finally, the level of technology has to be factored in, for it determines the feasibility of extracting gas, petroleum, or coal from a deposit. In the past, large deposits of fossil fuels may as well not have existed, for there was no way of extracting them; today, they can be counted as recoverable reserves. It is likely that technological advances will continue to increase our ability to locate and exploit new sources of gas, petroleum, and coal. In similar fashion, technological advances can generate more economical ways of using fuel and a consequent stretching out of existing supplies.

However, even if the supply of fossil fuels can be ensured for many decades to come, other problems must be confronted. The combustion of fossil fuels is always accompanied by emissions that are the primary source of air pollution. To mention only two, the burning of coal produces oxides of sulfur and nitrogen that cause acid rain, while gasoline-powered engines are primary contributors to the formation of smog. An even greater long-term problem of fossil fuel use may be the production of large quantities of carbon dioxide (CO_2) and a consequent increase in global temperatures through the "greenhouse effect." The burning of any carbon-based fuel is accompanied by the production of carbon dioxide; burning 1 kilogram of carbon in a fossil fuel generates 3.5 kilograms of carbon dioxide. Put another way, the typical car produces in 1 year approximately its own weight in carbon dioxide. Some fossil fuels are better than others in this regard. For an equivalent output of carbon dioxide, the combustion of natural gas provides 70 percent more energy than the combustion of coal. Still, there is no getting around the fact that all fossil fuels generate CO_2 when they are burned. Should global warming become a severe problem, it may become necessary to replace fossil fuels with other sources of energy, even though abundant resources still remain in the ground.

See also ACID RAIN; BIOMASS FUELS; CLIMATE CHANGE; COAL; COAL GASIFICATION; COAL LIQUEFACTION; DYES, ANILINE; ENERGY, MEASURES OF; ENERGY EFFICIENCY; ENERGY INTENSITY; FERTILIZER, CHEMICAL; GASOLINE; NATURAL GAS; OIL REFINING; POLYMERS; SMOG, PHOTOCHEMICAL

Furnace, Reverberatory

A reverberatory furnace supplies a great deal of heat while keeping the material to be heated separated from the source of that heat. In a reverberatory furnace, the burning of coal, coke, or some other fuel produces heat, but the smoke and fumes of the fire do not come in contact with the material being heated. Simple reverberatory furnaces were described in a widely read handbook of Renaissance-era technology, Vanoccio Biringuccio's *De la Pirotechnia* (1540). Furnaces of the sort described by Biringuccio had dome-shaped roofs lined with refractory clay, a material capable of withstanding high temperatures for prolonged periods of time. Heat from the burning fuel reflected off the inside surface of the dome, while the material to be heated was located off to one side, partitioned off from the burning fuel. The heat of combustion was intensified by drawing air into the furnace through underground pipes.

One of the major uses of early reverberatory furnaces was for the manufacture of glass, a process that began in England in the early 17th century. In the late 18th century, reverberatory furnaces took on a new importance as they began to be used in the English iron industry. Until then, wrought iron had been made by laboriously hammering heated pig iron. In 1784, Henry Cort (1740–1800), an English ironmaker, began to use the "puddling process." A coal- or coke-burning reverberatory furnace heated cast iron until it reached a white-hot, molten state. The liquefied iron was then stirred through apertures in the furnace, causing the impurities to clump together. These impurities were then removed from the metal by manual hammering or by passing the metal between rollers.

It has been estimated that the puddling process increased the productivity of an ironworker by 20 to 30 times. The process also changed the economic geography of ironmaking. With coal or coke now being used to fuel reverberatory furnaces, it was no longer necessary to site ironworks in forested areas where they could be near sources of wood. Instead, the iron industry came to be closely associated with the coal industry, and places situated close to iron ore and coal (or having easy access to it by rail or water) became the major centers of iron production.

See also IRON

Fusion Energy

When the nuclei of two atoms fuse together, a new element is produced and a considerable amount of energy is released—a million times more energy than the combustion of the same quantity of a chemical fuel. Fusion occurs continuously within the interior of the sun and other stars, where massive amounts of energy are produced as hydrogen atoms fuse to form helium atoms. Fusion reactions began to be investigated in the 1920s, and culminated in the most destructive weapon of all time, the hydrogen bomb. Efforts also have been mounted to produce energy by means of controlled fusion reactions since the 1950s.

Because the nucleus of an atom carries a positive charge, fusing it with another positively charged nucleus requires overcoming the strong repulsive forces of the two nuclei. All fusion research centers on molecules with very low atomic numbers because their nuclei carry a smaller charge than the nuclei of heavier elements. In particular, fusion research makes use of hydrogen, the element with the smallest atomic number, and its two isotopes, deuterium and tritium. Deuterium is fairly abundant; water (H_2O) contains 1 atom of deuterium for every 6,500 atoms of hydrogen. Tritium, however, is very rare; 5 atoms of tritium, most of them the result of thermonuclear bomb tests, occur for every 10^{16} atoms of hydrogen. Due to its scarcity, tritium is artificially produced by alloying an isotope of lithium with magnesium or aluminum, and bombarding it with neutrons produced by a nuclear reactor.

Instigating a fusion reaction begins with converting the hydrogen, deuterium, and tritium gas into a state of matter known as a *plasma*, a gas that has been heated to a very high temperature. This causes the molecules to ionize, that is, to lose electrons, so the plasma consists of positively charged nuclei and negatively charged electrons. Because all of the particles carry a charge, the plasma can be controlled by magnetic fields. When a plasma is subjected to a magnetic field, magnetic lines of force cause the individually charged particles to move in circles around these lines, keeping the plasma from moving towards the walls of the reactor, where it would rapidly cool.

Preventing the cooling of the plasma is critical, for self-sustaining fusion requires very high temperatures: about 100 million degrees centigrade, and perhaps twice that amount in the core of the reactor. It is also necessary to keep the plasma confined by the magnetic field for enough time for fusion reactions to occur. In contemporary fusion research, these times are measured in seconds or fractions of seconds. Using fusion to produce energy will require much longer confinement times.

The most commonly employed fusion reactor is known as a *tokamak*. Invented in the 1950s in the Soviet Union, tokamak is a Russian acronym for "toroidal chamber with magnetic coil." The term *toroidal* refers to the fact that the basic shape is that of a torus, or to use everyday language, the shape of a doughnut. A tokamak uses external electromagnets that encircle the toroidal chamber to create a magnetic field. At the same time, an electric current is run through the plasma, also generating a magnetic field. Together, they produce a helical (twisted) magnetic field that confines the plasma.

In the 1970s, researchers began to pursue a new fusion technology known as *inertial confinement fusion*. This entails the use of laser beams directed at the molecules of deuterium and tritium to be fused. The resulting implosion compresses and heats these molecules to the point where fusion occurs and energy is released. An alternative approach to inertial confinement fusion uses the energy supplied by a laser to heat up an intermediate medium. This causes the emission of X rays that act on a capsule containing the material to be fused and their subsequent fusion.

Supporters of fusion energy hope that tokamak fusion, inertial confinement fusion, or some other means of fusion will eventually be used for the commercial production of energy. Should it become feasible, fusion energy would be superior in many respects to current methods of providing energy. The hydrogen isotopes that serve as fusion's raw materials are readily obtainable, and their production causes few environmental problems. Unlike fission energy, the use of fusion does not entail the use of highly radioactive materials and the accumulation of large quantities of nuclear wastes. Tritium is radioactive, but its half-life of 12 years compares very favorably with uranium-235's half-life of 713 million years, and plutonium's half-life of 24,360 years. The main products of fusion reactions are helium (an inert gas) and free neutrons. Some radioactivity does result from the bombardment of a fusion reactor's walls by neutrons, necessitating the replacement of the walls every 5 years or so. Finally, unlike the burning of fossil fuels, the generation of energy through fusion would not result in the accumulation of carbon dioxide and possible global warming.

Unfortunately, all of these advantages are only hypothetical, as fusion continues to be a laboratory exercise and not an actual source of power. Just experimenting with fusion has been a very expensive proposition; the cost of constructing the Tokamak Fusion Test Reactor at Princeton University came to about $1 billion, and the costs of a single experiment may be in the neighborhood of $20,000. Most importantly, fusion research still has not reached "breakeven," the point where the energy produced by a fusion reaction equals the energy that caused the fusion to occur. Even if breakeven occurs, many years will pass before laboratory success can be translated into the actual generation of power.

In 1989, the high-energy, high-cost approach to fusion was challenged by the remarkable assertion of two chemists that they had created fusion reactions in a small container at low temperature. But when other researchers were unable to replicate these results, "cold fusion" was shown to be a false hope. Fusion holds out the promise of clean, virtually limitless energy, but its realization is not ensured. In the meantime, national governments will have to expend huge sums on fusion research. In an era of stagnant or shrinking governmental budgets, even this cannot be taken for granted.

See also ATOMIC NUMBER; CARBON DIOXIDE; CLIMATE CHANGE; COLD FUSION; FUELS, FOSSIL; HYDROGEN BOMB; ISOTOPE; LASER; NUCLEAR WASTE DISPOSAL; PLASMA; RADIOACTIVITY AND RADIATION; X RAYS

Further Reading: Robin Herman, *Fusion: The Search for Endless Energy*, 1990.

Gaia

Gaia is the name of an early Greek goddess of the Earth. In the late 1970s, Gaia began to be used as an alternative name for the Earth. In this usage, the Earth is not a simple mix of the living (plants and trees) and the nonliving (rocks, bodies of water, and the atmosphere); instead, it is in a certain sense a living entity. This view of the Earth is not completely novel. In the late 18th century, James Hutton (1726–1797), sometimes called "the father of geology," considered the Earth to be a kind of "superorganism"; he believed that physiology was a good model for understanding how the Earth functions. A revitalized view of the world as a kind of organisms arose following the 1979 publication *Gaia: A New Look at Life on Earth* by James Lovelock (1919–), an English biologist, atmospheric science, and inventor. Lynn Margulis, an American microbiologist, also played an important role in formulating the concepts underlying the Gaia hypothesis.

Not all adherents of the Gaia hypothesis believe that the Earth is literally alive in the way that dogs, whales, sycamores, and ferns are alive. For them, the Gaia hypothesis is best treated as an analogy; the Earth is not literally alive but the way that it operates bears many similarities to a living organism. Of course, whether the living-Earth hypothesis is treated literally or analogically depends on one's definition of *life*. In this endeavor there is considerable latitude, for the distinction between living and nonliving is not always a distinct one. For Lovelock, there is only a "hierarchy of intensity" that extends from rocks and the atmosphere to cells and organisms.

One of the basic concepts of the physiology of living beings is homeostasis. Organisms are homeostatic in that their physiological processes are automatically regulated in order to maintain the life of the organism. For example, the body temperature of a "warm-blooded" animal is kept at the same level despite considerable variations in the ambient temperature. In similar fashion, the Earth adjusts to changes brought on by geological and biological activity. It has long been understood that animal and plant life have undergone alterations as a result of physical changes on the Earth. According to the Gaia hypothesis, the physical features of the Earth are changed by the activity of organisms. In sum, the evolution of living things and the evolution of their physical environment are inextricably conjoined.

To supporters of the Gaia hypothesis, the most important example of life processes altering the physical environment is the creation and regulation of atmospheric oxygen by an emergent organism. A new phylum, the cyanobacteria, 2.5 billion years ago began to provide energy for itself through photosynthesis. Through this process, cyanobacteria were able to convert water and carbon dioxide into carbohydrates, leaving oxygen as a residue. The subsequent evolution of animal life was powerfully conditioned by the availability of oxygen for physiological processes.

Another important, and more recent, example of the Earth's being affected by biological activity is the accumulation of carbon dioxide in the atmosphere. Some of this is the result of the burning of fossil fuels, which may bring on global warming through CO_2 accumulation. Lovelock, however, has placed a sizable amount of blame on the cutting down of forests to provide pasturage for cattle. Deforestation and cattle raising combine to generate important ecological changes. CO_2-absorbing trees are lost, and the pastured cattle produce significant quantities of methane, a gas that has been implicated as a leading cause of global warming. In addition, cattle compact the Earth, causing the runoff of rainwater and loss of topsoil.

In some quarters, the Gaia hypothesis has taken on religious overtones, with Gaia serving as a kind of Earth deity. The hypothesis also has been attractive to some feminists, who ascribe a feminine nature to the planet named for a goddess. These developments have not been welcomed by all adherents of the Gaia hypothesis, who feel that they have detracted from the scientific foundations of the hypothesis.

When considered solely in terms of its underlying logic, the Gaia hypothesis can be criticized for being tautological. In the simplest version of the hypothesis, it is asserted that life creates the conditions necessary for its survival. But, as critics point out, these conditions are defined as necessary because life as we know it needs them. This appears to be circular reasoning. Even if a more sophisticated version of the hypothesis succeeds in being nontautological, it will still require comprehensive empirical research to determine its validity. A considerable amount of this has already been done, and if nothing else, Gaia can be considered a success for the role it has played in stimulating research into the interactions of life and nonlife.

Although it has been associated with a particular environmental perspective, in fact the Gaia hypothesis can be used to support diametrically opposed views. On the one hand, the Gaia vision of the interconnectedness of everything on the planet can be employed to demonstrate that human activities have the potential to do vast damage to many components of the environment. On the other hand, the Gaia hypothesis can be used to stress the self-healing aspects of the environment, thereby obviating the need to take corrective action against problems like the depletion of the Earth's ozone layer.

See also CARBON DIOXIDE; CLIMATE CHANGE; DEFORESTATION; FUELS, FOSSIL; OZONE LAYER; PHOTOSYNTHESIS

Further Reading: Lawrence E. Joseph, *Gaia: The Growth of an Idea*, 1990.

Game Theory

Modern game theory was initiated in 1928 by John von Neuman (1903–1957), whose "Zur Theorie der Gesellshaftspiele" delivered at Gottingen, Germany, discussed the development of a best possible strategy for the game of matching pennies. Von Neuman showed that for certain classes of games, such as backgammon and special forms of chess and card games, an optimal strategy for minimizing long-term losses always exists for each player. In such cases, the game is said to have a "solution." The classic work in the field, von Neuman and Oskar Morgenstern's (1902–1977) 1947 *Theory of Games and Economic Behavior*, showed that most problems of economic behavior can be represented mathematically as games of strategy and dealt with by game theory. Beginning with World War II, game theory has been extensively applied to military strategy problems as well. In general, game theory is the most appropriate mathematical theory to use to model social situations in which there are multiple "players" or "opponents" such that only some of the salient variables can be controlled by any single player, or in which there may be only one player facing a situation in which information about some of the salient variables is unavailable.

One commonly encountered term that has come from game theory is "zero-sum game." In a zero-sum game, each player's winnings come as a result of another player's (or players') losses, so the sum of wins and losses is zero. For example, when you win $50 (+$50) while playing cards with me, I lose $50 (-$50); the sum of the two is zero. Many of life's situations are necessarily zero-sum games, but many others are not. For individuals as well as nations, the ability to distinguish a zero-sum game from one that is not may determine if one's behavior is wise or foolish.

Perhaps the most famous early discussion of several key features of game theory appears in Blaise Pascal's wager regarding the existence of a traditional Judeo-Christian God in fragment 233 of his 1662 posthumously published *Pensées*. According to Pascal, there is no way to know whether God exists or not. Yet one must wager one way or the other. If one wagers that he does not exist and he does, then the wagerer is doomed to eternal damnation, but if one wagers that he does and he does not, there is not great loss. Consequently, in the absence of adequate "information" on the subject, the only rational choice—i.e., the choice that can minimize long-term loss—is to believe in God. However, as in the case of virtually all applications of modern game theory, the assumptions brought to this problem have a substantial bearing on the conclusion to be drawn.—R.O.

Gamma Rays

The discovery of radiation in the late 19th century presented many new puzzles for science. One of the first steps in understanding the nature of radiation was taken by Ernest Rutherford (1871–1937). In 1899, he determined that radioactive substances gave off two types of emanations that he called *alpha rays* and *beta rays*. In less than a decade, Rutherford and other investigators determined that the rays were in fact particles; alpha particles were helium nuclei (two protons and two neutrons), and beta particles were high-speed electrons. In 1900, while studying uranium, the French physicist Paul Ulrich Villard (1860–1934) discovered a third form of radiation, which, unlike alpha and beta particles, produced rays that were not bent by a magnetic field. These came to be called *gamma rays*, after the third letter in the Greek alphabet (alpha and beta being the first two).

Gamma rays had great penetrating power—a rather thick piece of lead was needed to stop them—but little else was known about them. Many scientists doubted that they were rays at all, that they instead were some kind of particle. That they were rays was demonstrated in 1910 by the British physicist William Henry Bragg (1862–1942) when his experiments revealed that gamma rays ionized gas in much the same way that X rays did. In 1914, Rutherford and E. N. Andrade used the recently devised technique of crystal diffraction to determine the wavelength of gamma rays. This was done by directing a beam of gamma rays through a crystal; the wavelength could then be calculated after measuring the angle of diffraction as the beam passed through the crystal. The resulting wavelengths proved to be similar to those of X rays of very short wavelengths.

It is now known that gamma rays are produced along with alpha or beta particles, or both, as a result of nuclear decay, most likely due to changes in energy levels within the nucleus. With their extremely short wavelengths of 10^{-11} to 10^{-14} meters, gamma rays comprise one end of the electromagnetic spectrum first described in the 1860s by James Clerk Maxwell (1831–1879). Like all constituents of the electromagnetic spectrum, gamma rays have the characteristics of both particles and waves. The belief of early researchers that gamma rays were particles has thus been vindicated, although for rather different reasons.

Gamma rays are often used as tracers while conducting scientific and medical research. For example, the rate at which blood flows in an organism can be determined by injecting a small amount of a gamma ray–emitting material into the bloodstream and tracking the radioactivity. Astronomers have been intensively studying gamma rays of extraterrestrial origin as a way to learn more about the history and composition of the stars and other constituents of the universe. Because most gamma radiation is absorbed by the sun, the monitoring of gamma rays is done by orbiting satellites.

See also ALPHA PARTICLES; BETA PARTICLES; MAXWELL'S EQUATIONS; QUANTUM THEORY AND QUANTUM MECHANICS; RADIOACTIVITY AND RADIATION; X RAYS

Garbage Disposer

"Taking out the garbage" is a necessary if unpleasant domestic task. The storage of garbage also creates an ideal environment for the breeding and accumulation of flies and rats, as well as the generation of nauseating odors. For municipal authorities, dealing with household wastes entails considerable costs for collection and disposal. In the 1930s, a new technology promised the eventual elimination of these problems by grinding up household garbage and flushing it into a community's sewer system.

The first grinders were installed in the 1920s and 1930s for use by municipal sewage facilities. By the mid-1930s, the availability of small electric motors and the spread of domestic electricity made household units feasible. In 1935, General Electric introduced the first domestic garbage disposer, the "Disposall," an appliance whose name is now generic. By 1941, an estimated 175,000 of them had been installed in American homes. World War II caused a temporary hiatus in the installation of garbage disposers, but sales took off in the years that followed. In 1948, 17 manufacturers produced garbage disposals; 200,000 had been installed by the end of that year.

A number of forces were responsible for the widespread adoption of garbage disposers. Postwar economic prosperity created an expanding market for household appliances. Suburbanization and a housing construction boom also contributed. No less important, an expansion of the number of women workers provided many families with financial resources for the purchase of appliances like garbage disposers, as well as strong motivation to obtain anything that promised to lift the housekeeping burdens of working women.

For a few years immediately after World War II, some municipalities in the Northeast banned disposers because they feared the overloading of sewer systems. But other town and city governments took the opposite tack. The installation of garbage disposers was strongly encouraged by some municipal governments, who hoped that their widespread use would lower garbage collection costs and promote a healthier environment. The one that took the lead was Jasper, Ind., a town of 6,800; in 1950 it sought to make garbage disposals universal within its borders. The town did not specifically mandate the use of disposers, but it achieved the same effect by discontinuing garbage collection and prohibiting the storage of garbage in receptacles left outdoors. Other towns and cities followed suit; while not mandating a complete changeover to disposers, Denver, Detroit, Columbus, and many cities in California required their installation in all new homes. Most local governments refrained from issuing requirements of this sort; in most places consumer demand was a sufficient stimulus to the installation of garbage disposers. Despite some strains on household budgets caused by increased water and electricity costs, and the need for occasional repairs, in millions of households the garbage disposal had become an essential component of domestic modernity.

See also MOTOR, ELECTRIC; WASTE DISPOSAL; WATER SUPPLY AND TREATMENT

Further Reading: Suellen Hoy, "The Garbage Disposer, the Public Health, and the Good Life," *Technology and Culture*, vol. 26, no. 4 (Oct. 1985).

Gas Chamber

Poison gas came into prominence early in the 20th century when it was used as a World War I battlefield weapon. Although the consequences of being gassed with chlorine or mustard gas could be horrific, other lethal gases, notably hydrogen cyanide, produce little or no pain. Very small concentrations of hydrogen cyanide bring on death by blocking the uptake of oxygen in the blood: 100 to 200 mg (0.0035–0.007 oz) is considered a lethal dose, and even smaller quantities may have fatal consequences. In contrast, electrocution and hanging, the most common methods of execution used in the early 20th century, did not always produce quick, painless death. Therefore, in an effort to make execution more humane, in 1924 Nevada became the first state to use lethal gas to execute condemned prisoners.

In the years that followed, more states used the gas chamber for executions. Although gas was not employed by the military in World War II, it was used for even more ghastly purposes, the extermination of millions of Jews and other "undesirables" by the Nazis. The evil associations of poison gas did not prevent its use by the American penal system; by the early 1950s, 11 states were executing condemned prisoners in gas chambers.

In 1972, the U.S. Supreme Court by a 5 to 4 margin ruled that capital punishment as it had been administered violated the 8th and 14th Amendments to the Constitution. No executions were performed until 1976; in that year the Court gave individual states the right to resume executions provided that certain legal procedures were followed. By this time, the gas chamber had fallen out of favor. From 1976 to 1995, only 9 people in the United States were executed in the gas chamber; 156 were executed through lethal injection, 119 were electrocuted, 2 were hanged, and 1 died before a firing squad.

Although the gas chamber originally was introduced as a humane means of terminating life, its continued use is in doubt. In early 1996, a federal appeals court ruled that executions conducted in the gas chamber of San Quentin penitentiary violated the 8th amendment to the U.S. Constitution because it constituted a "cruel and unusual punishment." Today, death by lethal injection is the preferred means of execution in states that still employ capital punishment.

See also ELECTRIC CHAIR; EXECUTION BY LETHAL INJECTION; GAS, POISON

Gas Lighting—see Lighting, Gas

Gas, Poison

World War I was the scene of many horrors brought on by new technologies. Not the least of these was poison gas. Poison gas deeply influenced tactical and strategic thinking for many years after the war, yet its use came about rather indirectly. Like all parties to the conflict, Germany in 1914 thought that the war would be a brief affair. But after several months of furious fighting that resulted only in stalemate, the German Army began to suffer from serious shortages of explosives for artillery shells. Attempts were made to replace TNT with various chemical irritants, but they were battlefield failures. Fritz Haber (1868–1934), a chemist who played a key role in the commercial production of nitrogen compounds, suggested the use of chlorine. On Apr. 22, 1915, 5,700 gas cylinders spewed 168 tons of chlorine in the direction of the British, French, and Canadian troops holding their positions near the Belgian town of Ypres. By the time the Second Battle of Ypres ended in late May, 20,000 soldiers had been gassed, a quarter of whom died a horrible death as the gas seared their lungs.

The Allies responded in kind, aided by a prevailing wind that usually blew in the direction of German lines. Bombs and shells containing poison gas were swiftly developed, allowing more precise targeting. More lethal compounds, notably mustard gas $(ClCH_2CH_2)_2S$, were used. But despite being used by both sides, poison gas had limited effects, causing less than 6 percent of all casualties and one-third of 1 percent of fatalities. Even so, poison gas strongly affected military thinking in the years following World War I. Advocates of air power saw poison gas as a terror weapon that could be dropped from high-flying aircraft to kill, incapacitate, and demoralize a civilian population, thereby forcing a rapid surrender. Statesmen who abhorred this prospect tried to hold the use of poison gas in check through international treaties banning its use. All but one of the signers of the 1922 Washington Conference on the Limitation of Armaments agreed to refrain from the use of poison gas in times of war. A similar resolution was included in the Geneva protocol of 1925 and was ultimately ratified by 42 nations, although 19 of them reserved the right to use gas if they were attacked by an enemy that used it first.

During the early months of World War II, gas attacks were widely anticipated, and in Britain every man, woman, and child was issued a gas mask that they were supposed to keep with them at all times. Yet the attacks never came. World War II gave the world increasingly sophisticated killing technologies, but poison gas was not one of them. International treaties were not the main reason, for many of the provisions contained in them were widely violated. Part of the reluctance to use poison gas was due to the logistical problems its use presented. Troops had to be protected from the gas they used, hindering their operational effectiveness. Also, a conquered area might require time-consuming detoxification. But military factors alone do not explain why gas was not used. Any radically different weapon requires adjustments if it is to be used successfully, and ways could have been found to make gas an effective weapon. Military and civilian leaders were unwilling to do so in large measure because they knew that the use of gas would be followed by retaliation in kind. The nonuse of gas during World War II shows that deterrence can be effective; a combatant may refrain from using a weapon if he is sure that his opponent will also use it. Combatants assumed that the deployment of poison gas would have subjected them to a great deal of suffering for only a transitory advantage.

Still, it is possible that military leaders would have taken the risks inherent in the use of gas if they had felt a greater affinity for it. But poison gas seemed unlike all other weapons of war: silent, insidious, and even dishonorable. As one World War I German general put it, "I must confess that . . . poisoning the enemy just as one poisons rats struck me as it must any straightforward soldier; it was repulsive to me." At the same time, the widespread use of poison gas would have elevated the military status of chemists, technicians, and other nonmartial types, a situation that traditional military leaders did not want to encourage.

Although poison gas was eschewed in World War II, it has been used in more recent conflicts, most notably the war between Iran and Iraq. Other forms of chemical warfare have been pursued, although not with lethal consequences, at least initially. During the Cold War, the Soviet Union and the United States produced and stockpiled large amounts of binary nerve gas. Although they were never used in anger, there is no guarantee that poison gas will never be deployed by a country that sees an advantage in doing so.

See also AGENT ORANGE; BOMBING, STRATEGIC; FERTILIZERS, CHEMICAL

Gasoline

Gasoline (*petrol* in British English) is not a single substance but a mixture of more than 100 liquid hydrocarbons, all of which have 4 to 12 carbon atoms per molecule. These hydrocarbons are refined from crude petroleum and fall into one of three basic groups: paraffins, olefins, and aromatics. To meet changing driving needs, refiners alter the exact mixture of particular hydrocarbons over the course of a year. In winter, more volatile mixtures are used in order to improve cold-weather starting, while less volatile mixtures are produced for warm-weather periods in order to forestall vapor lock. In the 1980s, efforts to produce gasoline with higher octane ratings led to fuels with increased volatility. In 1995, refiners reduced volatility in order to decrease emissions of gasoline vapor, a major source of photochemical smog. Modern gasoline fuels also have a number of additives, such as antivarnishing agents, detergents, and oxidation and rust inhibitors. Finally, oxygenating agents are added to the gasoline consumed during cool-weather months in parts of the country with serious air pollution problems.

Until the 1970s, tetraethyl lead was commonly added to gasoline in order to reduce engine knock. But in addition to being a toxic substance, lead has a deleterious effect on catalytic converters. Consequently, the widespread adoption of these key antipollution devices necessitated the removal of lead from gasoline. Cars meant to use unleaded fuel are equipped with narrow filler necks to prevent the insertion of the larger pump nozzles used for leaded fuel. In 1996, federal regulations banned leaded fuel altogether. Since lead deposits help to cushion the impact of valves when they seat, older cars that were designed to run on leaded gasoline may have a problem known as valve recession. But this problem occurs only when engines are run at continuous high speeds or with high loads; in normal operation the use of unleaded gasoline results in little or no damage to valve seats.

Concerns about poor air quality also have motivated efforts to reduce tailpipe emissions through the use of substitutes for conventional gasoline. One such example is gasohol, a blend of gasoline and either ethanol (ethyl alcohol) or methanol (methyl alcohol), usually in a ratio of 85 percent gasoline to 15 percent alcohol. The use of gasohol reduces hydrocarbon emissions but generates a new pollutant, formaldehyde (HCHO), a suspected carcinogen. Moreover, the energy costs incurred in the production of ethanol and methanol, the ease with which they combine with water, and other drawbacks make gasohol of limited value as an automotive fuel.

A much better proposition is reformulated gasoline (RFG), which is produced by removing volatile olefins and aromatics, and replacing them with high-octane alkylates. This results in a clean-burning fuel that brings substantial improvements in air quality. In 1995, the U.S. Environmental Protection Agency mandated the use of low-benzene RFG in nine smog-prone regions, and a year later the California Air Resources Board required the use of even cleaner gasoline. Federally mandated RFG has reduced daily hydrocarbon and nitrogen oxide emissions by 85 million tons, while the state requirement reduced them by 215 million tons. Together, the state and federal RFG mandates resulted in a 15 percent drop in smog-causing automotive emissions. The negative consequence have been modest: an increase in the price of gasoline by about 7 cents per gallon, and scattered instances of fuel system failures in older cars.

See also ALCOHOL, ETHYL; ALCOHOL, METHYL; CATALYTIC CONVERTER; CATALYTIC CRACKING; ENVIRONMENTAL PROTECTION AGENCY; FUELS, FOSSIL; LEAD POISONING; OCTANE NUMBER; OIL REFINING; SMOG, PHOTOCHEMICAL; THERMAL CRACKING

Gauges, Industrial

Industrial gauges play an essential role in the manufacture of products made from separate components. Products can be successfully manufactured only if their parts fit together properly. The amount of allowable space between components that fit together is known as *tolerance*. A gauge (or gage, as it is sometimes spelled) measures the size of a part in order to determine if that part falls within an acceptable range of tolerances; that is, within specified limits it must be neither too large nor too small. Some gauges give the numerical size of a part; others simply indicate if the part falls within the accepted range of tolerances. Today, a component is usually discarded if a gauge indicates that its size is incorrect. In the past, parts were often made slightly oversize, and gauges were used to check on them as they were filed to the proper size.

Much of the improvement in manufacturing has come from gaining the ability to make components that fall within closer tolerances. When Matthew Bolton and James Watt began to produce steam engines in the late 18th century, they were satisfied if the clearance between the piston and the cylinder wall was equivalent to the thickness of a sixpence. This situation began to change in the 19th century, spurred by advances in the manufacture of firearms in the United States. Although many firearms were made from parts that were not completely interchangeable, there was nonetheless enough uniformity that only a small amount of hand finishing was required. Gauges made that uniformity possible by determining the size of a part to 1/1,000 of an inch (.0025 cm). The accuracy of these gauges was periodically checked by comparing them with a master gauge, or by using them on model components that were known to be of the proper size. If this were not done, the gauges eventually would lose their accuracy due to wear.

Few 19th-century industrial enterprises approached the degree of accuracy found in the most advanced armories. During the first decade of the 20th century, however, the rapid development of the automobile as an item of mass consumption required substantial changes in manufacturing processes. The introduction of the assembly line usually receives most of the attention, but the widespread use of gauges was critically important for ensuring that all the parts coming down the line would fit together properly. Henry Ford was so convinced of the importance of gauges that in 1923 he purchased the renowned gauge-making company founded by C. E. Johansson, and moved it to his headquarters in Dearborn, Mich.

In addition to making for greater accuracy, the widespread use of gauges changed the nature of work for many production workers. In the absence of the need to make interchangeable parts, an artisan only had to be concerned with the overall functioning of the final product. As interchangeable parts took on greater significance, much of the worker's skill came to reside in determining whether or not a part was in compliance with the gauge. Over time, even these skills were downgraded through the use of "limit" or "go-no-go" gauges that largely eliminated the need for judgment on the part of the worker. For example, the dimensional accuracy of a rod could be determined through the use of a gauge with two holes; the rod was deemed suitable if it could fit into the larger hole but not the smaller one.

While gauges made many production efficiencies possible, they sometimes acted as a conservative force in manufacturing, since a redesigned component might necessitate the expense of supplying workers with a new set of gauges. This problem could be overcome by providing adjustable gauges, such as micrometer calipers, that could be easily changed to allow the gauging of redesigned components.

See also ASSEMBLY LINE; MASS PRODUCTION; STEAM ENGINE

Gears

Gears transmit power from one part of a mechanism to another. Most gearing systems use two or more gears to transmit rotary motion from one shaft to another. It is also possible to use gears to convert rotary motion to linear motion, or vice versa. A common example of this is the rack-and-pinion gearing used in the steering mechanisms of many automobiles; this form of gearing translates movements of the steering wheel to the road wheels through the use of a rotating pinion that is engaged with a horizontal rack. In addition to transmitting power, gears can be used to change the direction in which a force moves. For example, the ring-and-pinion gearing of a car's differential turns the rotary motion of the driveshaft 90 degrees in order to power the rear axles.

Gears also perform the very important function of changing the speeds at which various parts move. When a gear with 20 teeth (also known as cogs) drives a gear with 10 teeth, the latter will rotate at twice the speed of the former. The difference in the speed of two gears, as well as the relation between the number of teeth they have, is given by the *gear ratio*. This is the number of revolutions (or the number of teeth) of the driving gear divided by the number of revolutions (or the number of teeth) of the driven gear. In the above example, 20 is divided by 10, so the gear ratio is 2:1. Many transmissions use a gear train containing more than two gears. In this case, the gear ratio refers to the relative speeds of the gear that initiates the motion and the gear that completes it.

This ability to turn gears at different speeds allows the gears to be used as torque multipliers. Torque is the product of the force applied and the distance through which it is applied. Consequently, a relatively large, slowly rotating gear will multiply the torque supplied by a relatively small, rapidly rotating gear. Torque multiplication is important for the functioning of many mechanisms. For example, an automobile engine produces little torque at low rotational speeds. It therefore needs a low gear that allows the engine to turn at a high speed even though the car's wheels are moving slowly at first. As the car picks up speed, the transmission is shifted in succession to higher gears that keep the engine within its proper range of engine speeds.

As with many fundamental innovations, the origins of gearing are obscure. Gears were undoubtedly constructed before they were written about, making it difficult to trace their origins. Making matters worse, many texts that had something to say about gears have been lost. One of the earliest known applications of gearing were water-lifting devices that date back to Egypt in the 3d- or 2d-centuries B.C.E. It is possible that at about the same time Archimedes (287–212 B.C.E.) used worm gears (gears with spiral ridges, much like the threads of a screw) to actuate the mechanisms of a planetarium. In the 1st-century C.E., Hero (or Heron) of Alexandria (c. 20 C.E.–?) described a weight-pulling device that used a train of gears to produce a theoretical advantage of 200:1. This may have been only a theoretical exercise, and the actual device may not have been built. The first known use of gearing for the transmission of power is provided by the Roman architect Vitruvius (c. 70 B.C.E.–?), who described the gearing that was used for waterwheels that turned on a horizontal axis.

The invention of the mechanical clock gave a great stimulus to the development of mechanisms that used metal gears. At the same time, wooden gears continued to be used for the transmission of wind and water power, much as they had been in the time of Vitruvius. The gears that were used were often made of different kinds of wood, with the teeth being made of an especially hard wood like chestnut. Gears made from wood may not have been as durable as those made from cast iron, but they ran more quietly and were easier to repair. Many textile mills and other industrial operations used wooden gears well into the 19th century, while smaller wooden gears were used for clocks that were sold in large numbers.

As increasing numbers of gears came to made from metal, improved methods of gear cutting became necessary. Small gears could be cast as single units, but larger gears were made by cutting teeth in a solid disk. Some clockmakers were using gear-cutting machines of their own invention by the middle of the 16th century, and by the end of the 17th century they were in common use. Machines for cutting large gears took longer to develop, but by the middle of the 19th century a variety of gear-cutting machines were available.

Along with improved methods of making gears came improvements in their design. In particular, it was important to design gears that produced little sliding motion when they meshed, for this was a cause of power losses and excessive wear. In the 16th and 17th centuries, mathematicians decided that the best shape for the face of a gear tooth was some type of cycloid. A cycloid is the path of a point on the circumference of a circle when the circle is rolled in a straight line. If a circle rolls around the outside of a larger circle, a point on it will describe a curve known as an *epicycloid*. If the rolling takes place along the inside of a circle, the result is a *hypocycloid*. One mathematician in 1660 designed machinery with gears with epicycloidal teeth, and by 1750 the basic geometrical principles of gear tooth design were well established, although it took many decades for them to be put into practice. An involute curve (one traced by the end of a taut cord as it is unwrapped from a circle) was later found to be an even better shape for the face of a gear tooth.

During the 1830s and 40s, the British engineer-mechanic Joseph Whitworth (1803–1887) built machines that cut gear teeth that had involute curves, and by the middle of the 19th century these machines also had self-acting feed mechanisms. Gear-cutting machines of this sort and the ones that succeeded them have not enjoyed the historical notoriety of steam engines and blast furnaces. But without these machines, industrialization would have proceeded at a slower pace, for gears are to be found in the great majority of mechanical devices, as well as the machinery that is used to make them.

See also CAST IRON; CLOCKS AND WATCHES; DIFFERENTIAL; PLANETARIUM; WATERWHEEL; WINDMILLS

Further Reading: Robert S. Woodbury, *History of the Gear-Cutting Machine: A Historical Study in Geometry and Machines*, 1958.

Geiger Counter

The first decade of the 20th century was a time of intensive efforts to understand the nature of atomic radiation and to discern the structure of the atom. One of the most significant advances in this area was Ernest Rutherford's discovery in 1899 of two of the key products of radioactive decay: alpha particles and beta particles. By 1903, he was able to demonstrate that the two types of particle were deflected in opposite directions by magnetic and electrical fields. Their charge-to-mass ratio had been determined, and it was known that alpha particles were much heavier than beta particles. Gaining further knowledge about these particles, most importantly their mass and charge, required a way to count with some precision the number of alpha particles given off by a known quantity of a radioactive substance.

This was accomplished in 1908 through a collaborative effort of Rutherford and Hans Geiger (1882–1945). Geiger, who had received a doctoral degree in physics at the University of Erlangen in Germany, was a research fellow at the University of Manchester. Shortly after Rutherford's arrival in Manchester, he set Geiger to the task of devising an instrument for counting alpha particles. Rutherford and other experimenters already knew that radioactive materials ionized the air and other gases, and that the resultant ions could be collected in a space between two electrified plates. By determining the extent of ionization produced by a radioactive material, Rutherford and Geiger hoped to be able to count alpha particles.

The instrument that Geiger and Rutherford constructed was quite simple, consisting of a long, slender brass tube, stoppered at both ends, with a thin wire that passed from one end to the other through the center of the tube. At one end, the stopper had a small opening that was covered with a thin plate of mica. Through this window, a very small number of alpha particles from a distant radium source were able to enter the tube. The instrument was then hooked up to a 1,400-volt battery in such a way that the tube was negative and the wire was positive.

When alpha particles entered the tube, they ionized some of the gas molecules that remained in the partially evacuated tube, liberating electrons that gravitated towards the wire. In the course of moving in this direction, the electrons collided with other gas molecules. This often gave them enough energy to ionize another atom, which set free another electron, resulting in more ionization. Through this process what is often described as an "avalanche" of electrons settled on the central wire; the resulting electrical charge was then read by an electroscope. This allowed the number of alpha particles to be inferred. It was not possible to directly ascertain the mass of the tiny radium sample that had been the source of the alpha particles. Rather, its mass was determined by comparing its gamma ray emissions with the emissions of a sample of radium with a known mass.

At the time that the counter was devised, it was hypothesized that beta particles were either ionized hydrogen atoms or ionized helium atoms. Using the counter, Rutherford and Geiger were able to calculate that a gram of radium would emit 3.4×10^{10} alpha particles per second. After measuring the amount of electrical-current gain produced by the particles, it was a simple matter to determine the charge of a single alpha particle by dividing the current gain by the total number of particles. The number came fairly close to being double the already known charge of an electron. Since a hydrogen ion had the same amount of charge (albeit an opposite one) that an electron had, it could not be an alpha particle. An alpha particle, it was surmised, had to be a helium atom that had lost its electrons, i.e., a helium nucleus.

Within a few years, Geiger had returned to Germany, where in 1912 he devised a counter that was able to detect beta particles as well as alpha particles. The instruments used today (properly called Geiger-Mueller counters, in recognition of the improvements made by Erwin Mueller in 1928) operate on exactly the same principles. The main difference is that the electrical impulses are now fed into an amplifying circuit and are then counted by a register or used to produce audible clicks.

Geiger-Mueller counters are widely used in industry, medicine, and research. In industry, they are used to measure the wear of bearings that have been treated with trace radioactive materials. They can also be used to determine the thickness of a moving sheet of metal that has a source of beta particles placed on the opposite side; changes in the particle count can then be related to changes in the thickness of the sheet. Geiger-Mueller counters are used in medicine and scientific research to locate the position or distribution of radioactive tracers that have been administered to a patient or subject. And of course they are used by prospectors in search of radioactive minerals.

See also ALPHA PARTICLES; BETA PARTICLE; GAMMA RAYS; RADIOACTIVITY AND RADIATION

Gemini Project

The Mercury program successfully put Americans into orbit, but several new capabilities had to be developed for the space program to meet its ambitious goal of putting a man on the moon by the end of the 1960s. Specifically, flights of longer duration had to be accomplished, for a flight to the moon would be far longer than the under-35-hour duration of the longest Mercury voyage. Also, a landing on the moon would be incomplete if astronauts could not walk on its surface, so it was necessary to develop the capability to move outside the capsule and engage in what was called an "extravehicular activity," or EVA. Finally, after much debate, the National Aeronautic and Space Administration (NASA) had decided on a plan that would put a command module into orbit around the moon, and then launch a lunar lander from it. After its stay on the moon, a portion of the lander would rejoin the orbiting command module. This meant that it was necessary to learn how to rendezvous and couple together two vehicles in space.

The spacecraft used in the Gemini project were larger and more complex than those used by Project Mercury. Each carried two crewmen and weighed about 3,600 kg (8,000 lb). Instead of returning to Earth by falling like a rock before its parachute opened, the Gemini capsule was designed to generate a slight amount of aerodynamic lift, allowing it to be steered towards its recovery point.

A larger spacecraft required a more powerful rocket to put it into space. This was provided by a second-generation intercontinental ballistic missile, the Titan II. Its first stage put out nearly half a million pounds of thrust; the second stage supplied 100,000 pounds of thrust. To develop rendezvous and docking capability, a target vehicle was separately launched by a two-stage Atlas-Agena rocket. The Titan used a special hypergolic fuel, a blend of hydrazine with unsymmetrical dimethyl hydrazine that was oxidized by nitrogen tetroxide. This combination had the advantage of being self-igniting, obviating the need for a potentially troublesome ignition system. It was also safer than the fuel-oxidant combination (kerosene and liquid oxygen) used in Project Mercury's Atlas launchers, for these reacted with much greater suddenness and violence.

After being put into orbit, the capsule could be steered with an array of rocket thrusters, a capability that was essential to the craft's ability to rendezvous and dock. On board, Gemini capsules beginning with Gemini 5 were powered by fuel cells, devices that converted a fuel directly into electricity. The fuel cell was also supposed to supply drinking water as a by-product, but several missions passed before this actually happened.

Gemini 4 (June 3–7, 1965) produced the first EVA by an American, although the world's first spacewalk had been done by a Soviet cosmonaut 3 months earlier. After earlier Gemini missions worked with varying success on space rendezvous, Gemini 8 was the first to dock with its Agena target, which was done on Mar. 16, 1966.

Altogether, Project Gemini consisted of 10 flights. The longest was a mission of 14 days by Gemini 7, crewed by Frank Borman and James Lovell from Dec. 4–18, 1965. Although overshadowed by Project Mercury, which put Americans into space for the first time, and Project Apollo, which put the first men on the moon, Project Gemini was an essential transition between the two. It allowed the solution of new problems, added to the experience and confidence of space and ground crews, and helped to convince Congress that the space program still merited substantial appropriations of funds.

See also APOLLO PROJECT; FUEL CELL; MERCURY PROJECT; MISSILE, INTERCONTINENTAL BALLISTIC (ICBM)

Gene

A gene is the basic unit of heredity passed from one generation to the next. This concept was first proposed in 1865 by a German monk, Gregor Mendel (1822–1884). Based on careful plant-breeding experiments, Mendel postulated rules governing the inheritance of "factors" (later named *genes*) that control the appearance of an organism. Mendel postulated that there are a pair of genes for each trait, one inherited from each parent, and that there is a single copy of each gene in the egg and sperm. Moreover, the genes for different characteristics assort randomly in the egg and sperm. Mendel's laws contradicted the prevailing theory that the egg and sperm each contain samples from different parts of parents that are "blended" in offspring.

Mendel was the first to describe the visible appearance of an organism (its *phenotype*) in terms of its genetic composition (its *genotype*). He determined that there can be different versions of a gene, known as *alleles*, in a gene pair. When different alleles are present in a gene pair, the organism is said to be "heterozygous." Often, one of the two alleles (the dominant allele) is preferentially expressed. The other allele, the "recessive" allele, is only expressed in the "homozygous" state, when there are two copies present.

Mendel was able to show that inheritance is predictable, that it can be analyzed mathematically. For example, if the gene responsible for eye color is represented as B (brown, the dominant allele) and b (blue, the recessive allele), then individuals whose genotype is BB or Bb will have brown eyes, and blue eyes will only be present in those whose genotype is bb. From heterozygous Bb parents, both egg and sperm can carry either the B or the b allele. Possible combinations, after an egg is fertilized by a sperm, can be represented by a *Punnett square* diagram. In a Punnett square, possible male gametes are represented in one direction and female gametes in the other, with pairings represented in the center of the diagram.

CROSS BETWEEN B/b
heterozygotes

		male gametes	
		B	b
	B	BB	Bb
female gametes	b	bB	bb

Possible combinations with a "B" egg are BB and bB; with a "b" egg, bB and bb. Of the four possible combinations, there is a likelihood of 1 in 4 of having a child with a BB genotype (homozygous dominant), 1 in 2 that the genotype will be either bB or Bb (heterozygous), and 14 of bb (homozygous recessive). Since the B (brown) allele is dominant to b, there is a $\frac{1}{4} + \frac{1}{2}$, or 3 in 4 chance that a child will have brown eyes, and 1 in 4 for a child with blue eyes. Mendelian genetics thus explains how a blue-eyed child can have both parents with brown eyes.

Although Mendel provided the theoretical basis for inheritance, it was not until the early 20th century that work by Walter Sutton, Nettie Stevens, and Edmund Wilson identified chromosomes as the cellular structures containing genes. Later, Thomas Hunt Morgan, Alfred Sturdevant, Calvin Bridges, and Hermann Muller, in their classic experiments with the fruit fly *Drosophila melanogaster*, provided a wealth of new information that located genes on chromosomes. They were the first to show that the inheritance of genes that lie in close proximity to each other on the same chromosome is often linked and that they are inherited together. A revolutionary finding was that pairs of chromosomes can become closely aligned and exchange material, or "cross-over." In 1931, Barbara McClintock (1902–1992) and Harriet Creighton (1909–) were able to actually visualize in the microscope this phenomenon of crossing over. Crossing over, or recombination, creates a variety of new genetic combinations in the egg and sperm.

What is the substance in a gene, and how do genes work? The first insights into these questions were provided in 1928, when Frederick Griffith discovered that an extract from a pathogenic bacterium could transform a different, avirulent bacterial strain into one that would cause pneumonia and death in mice. Griffith's observations were pursued by Oswald Avery, C. M. MacLeod, and M. McCarty, who in 1944 showed that the "transforming principle" was DNA. Also in the 1940s, landmark experiments by George Beadle and Edward Tatum led to their one gene–one enzyme hypothesis: that genes control specific enzymes, the protein catalysts in biological reactions. Moreover, inherited characteristics are frequently the result of several enzymatic steps. A mutant phenotype can result from a defective enzyme at any of the steps.

In 1957, James Watson and Francis Crick published their model for the structure of DNA. Shortly thereafter, experiments in the laboratories of Crick and Sidney Brenner led to the cracking of the genetic code. Crick proposed his "central dogma," relating DNA, the reservoir for genes, to the function of those genes. According to the central dogma, DNA is first transcribed into another nucleic acid, RNA, and RNA is "translated" into protein. In 1961, Marshall Nirenberg and Heinrich Mathaei demonstrated that the genetic code is a "triplet code": each amino acid is specified by three successive nucleotides in DNA. Moreover, specific triplets indicate the location in a DNA sequence where a protein will start and end.

In the 1970s, Walter Gilbert, Alan Maxam, and Frederick Sanger developed means to experimentally determine the nucleotide sequence in DNA. DNA sequencing has revealed an unexpected mode of organization for many genes in higher organisms. Unlike bacteria and viruses, in higher organisms, genes can be "split." DNA sequences coding for proteins may be interrupted by noncoding regions. When the gene is expressed, the entire DNA sequence is first transcribed into an RNA from which the noncoding sequence is eliminated prior to translation. In this process, segments of noncoding regions called *introns* are cut ("spliced") out, and the coding sequences (*exons*) are fused together.

Gene expression is highly regulated. Some genes specify essential proteins that are required all of the time and that have a high turnover rate. These genes, called *housekeeping* genes, are expressed all of the time. Other genes may be expressed only occasionally, such as during development or in response to environmental conditions. A wide variety of regulatory mechanisms exist: Genes may be regulated at the level of DNA, through genomic rearrangements or changes in gene copy number. More often, gene expression is induced or repressed when appropriate signals (such as hormones) bind to DNA, at sites near target genes that then are turned on or off. Regulation is frequently at the level of transcription, although any step between DNA and protein could be a target for regulatory activity, and current research is unraveling many complex patterns of gene regulation.—C.I.

See also ALLELE; CHROMOSOME; DNA; ENZYMES

Gene Therapy

A novel approach for treating diseases caused by defective genes, gene therapy is directed not toward relieving disease symptoms but instead to the replacement of defective genes with normal genes. Gene therapy is grounded in experiments using organisms such as the bacterium *Echerichia coli* and baker's yeast, *Saccharomyces cerevisiae*. Recombinant DNA technology provides the means to isolate specific genes. In yeast, cloned copies of normal genes can "heal" a genetic defect due to a mutant gene, because the copies replace the defective gene at precisely the same chromosomal location.

Animal cells do not incorporate cloned genes with the same precision. In animal cells, including those of human beings, cloned genes are inserted into the chromosomes by an apparently random process. Cells that have incorporated the cloned gene still carry the mutant gene in addition to the normal copy. Moreover, there is no assurance that the normal gene will not destroy another gene already at that location or that its expression will be controlled in the same way.

The kind of cell targeted in gene therapy also is of profound significance in terms of its effect on someone with a genetic disease. If the cell is a somatic cell (all cells except egg and sperm are somatic cells), the effect of gene therapy is limited to the lifetime of that particular cell in the patient's body. If reproductive cells (known as *germ cells*) are targets of gene therapy, then the effect is more significant, as the new gene may be transmitted to future generations. To date, all clinical trials have been limited to somatic calls, and it is unlikely that any gene therapy experiments with germ cells will be attempted in the near future.

At least 4,000 inherited diseases have been known to afflict human beings. The massive effort called the Human Genome Project has identified many of the genes responsible for these diseases, and a large number of them have been cloned, making them candidates for use in gene therapy. Gene therapy could be a possible treatment for inherited diseases. For example, gene therapy could become a way to target drugs to cancer-specific genes in cancer cells. A third, highly controversial, possibility is to cure genetic defects with germ-line gene therapy, so that future generations do not inherit a defective gene.

There are certain technical requirements for gene therapy. First, the causative gene for a genetic disease must be identified. The second step is to clone the region of human DNA that codes for that gene. The cloned gene is then inserted into another DNA, called a *vector*, which can infect the target cell and deliver the cloned DNA to human chromosomes. Vectors include modified or "disabled" versions of viruses known to infect human cells. Genes have also been delivered in fatty particles called *liposomes*.

Currently, human gene therapy is in the experimental stage. The first inherited disease chosen for gene therapy was severe combined immune deficiency, a recessive condition that is due to a deficiency in adenosine deaminase (ADA). Adenosine deaminase is a critical enzyme in the immune system, and children born with ADA deficiency have a poor prognosis because they are unable to fight infectious disease. During the 1970s, there was a well-publicized case of a boy with this rare disease. Identified as the "bubble baby," he spent most of his life in the germ-free environment of a sterilized chamber before eventually succumbing to an infection. Gene therapy for ADA deficiency was first performed on a patient in 1990. The protocol requires that white blood cells called T lymphocytes be removed from the patient, infected with a virus carrying the ADA gene, and then returned to the patient. The transplanted ADA gene corrects the deficiency for a period of 3 to 5 months, until the white cells die. Improved delivery systems use *stem cells* from the bone marrow or umbilical cord that have a longer life span.

A different approach to gene therapy has been taken with some heritable diseases affecting the lungs. The cloned gene has been presented in aerosol form to treat a deficiency of the enzyme alpha-1-antitrypsin (AAT) that leads to a form of emphysema, and cystic fibrosis, a relatively common inherited disease caused by a defective protein called cystic fibrosis transmembrane conductance regulator (CFTR).

Clinical trials involving a few patients are also underway to treat familial hypercholesterolemia (inherited high blood cholesterol), AIDS, and cancers including melanoma, neuroblastoma, brain tumors, and breast cancer. Gene therapy for cancer is directed toward enhancing the immune system's rejection of cancer cells or making cancer cells specific targets for drugs.

For all its potential benefits, gene therapy remains a controversial medical technology. Concerns are focused on such issues as: the type of cells treated, possible effects on other cells (particularly germ cells), the duration of the therapy's effectiveness, possible risks from overexpression of the gene, and whether the cell altered by the new gene will be recognized as foreign by the immune system. Critics also warn of the risk of gene therapy being used for cosmetic or eugenic purposes in order to enhance "desirable" human traits.

All gene therapy protocols are carefully screened before they are used. A gene therapy experiment must be approved by at least two review boards at the requesting institution, and then by the Recombinant DNA Advisory Committee (RAC) of the National Institutes of Health. In addition, they must be approved by the Food and Drug Administration.—C.I.

See also AIDS; CANCER; CHROMOSOME; CLONING; DNA; *E. Coli*; ENZYMES; FOOD AND DRUG ADMINIS-

TRATION, U.S.; HUMAN GENOME PROJECT; NATIONAL INSTITUTES OF HEALTH

Generator, Electric

A small electrical charge can be produced by rubbing certain materials such as amber or hard rubber. This fact was known in the ancient world, but there was no further development for hundreds of years. A variation on this old theme was the simple electric generator built in 1663 by Otto von Guericke (1602–1686). It consisted of a ball of sulfur that rotated on an axle; a momentary touching of the ball caused it to be electrically excited. Isaac Newton (1642–1727) constructed a similar apparatus using a glass globe in 1709, as did Francis Hauksbee (c. 1666–1713) at about the same time. Hauksbee's device was for many years the standard means of producing an electrical charge. As such, it was useful for electrical experimentation, but it had no practical application.

The same can be said of the device that is often described as the first electric generator. It was constructed by Michael Faraday (1791–1867) in October 1831, a decade after he had built a type of electric motor. It is a bit curious that so long a time elapsed between the two, for the basic principles are the same: A motor produces motion when electricity is fed to it, whereas in a generator, motion is used to produce the electricity. The operation of both is a consequence of the production of a magnetic field by an electric current, a principle discovered by Hans Christian Oersted (1777–1851) in 1820. Faraday's first generator was nothing more than an magnetized iron rod and a surrounding coil; when the magnet was moved back and forth it induced an electric current in the coil. A few days later, Faraday built another generator that consisted of a copper disk that rotated between the poles of a magnet. An electric current was manifested when one wire was attached to the disk's spindle and another was connected to a brush in contact with the disk itself. A year later, a Frenchman with the unlikely name of Hippolyte Pixii (1808–1835) constructed a device with two wire-wound bobbins that were rotated through the magnetic field produced by a horseshoe magnet. A second machine improved the generator's output by collecting the current through a commutator, an arrangement that allowed the production of direct (albeit fluctuating) current. By the middle of the 19th century, generators producing direct current were being employed for electroplating, the first commercial use of electricity. Within a few years, generators were also supplying electricity for the next major commercial use of electricity, the powering of arc lights.

Most early generators used permanent magnets. As early as 1838, however, experimenters such as Moigno and Raillard in France and Wheatstone, Varley, and Wilde in England were using battery-powered electromagnets to produce the necessary magnetic field. The greatest early success in the production and sale of generators of this sort was achieved by William Siemens (1823–1893), a German living in England.

Early generators produced a fluctuating current, which limited their usefulness for many applications. In

Michael Faraday (courtesy Institute of Electrical and Electronic Engineers).

1860, this problem was essentially solved by a 20-year-old Italian, Antonio Pacinotti (1841–1912), when he invented the ring-wound armature, although it was forgotten until the Belgian Zenobe Theophile Gramme (1826–1894) reinvented it 10 years later. Gramme's generator went on to a significant commercial success as the power source for carbon arc lighting systems. Another major drawback of early generators, especially when they began to put out a fair amount of power, was that the brushes that rubbed against the commutator wore rapidly and produced a great amount of sparking. By the end of the 1880s, this problem was solved through the adoption of carbon brushes.

Thomas Edison (1847–1931), who had developed a practical incandescent lamp that eventually made the carbon arc obsolete, built a direct-current generator in 1878 to power the first commercial electrical system. Edison's first generator was rather inefficient, but after a considerable amount of empirical work his second generator achieved an efficiency of 82 percent. Generator design also profited from the more theoretical researches of England's John Hopkinson (1849–1898). By the end of the 19th century, generators were operating at 95 percent efficiency.

In the early 20th century, hundreds of generator sets were being sold to electric utility companies. By this time the basic configuration of the generator had been determined: rotating electromagnets surrounded by a fixed armature (the part that contains the coils). As is often the case with a developing technology, alternators grew in size and output. In 1900, the largest generator was a two-1,500-kW set installed in Germany. Four years later, the generating capacity at one installation in northern England was 6,000 kW. In the 1930s, the United States had a 260-MW station, and in 1963 the first 1-million-kW generating facility was installed. Powered by steam turbines and hydroturbines, or by the power of the wind and transmitted through high-voltage transmission lines, electric generators produced the power essential to a modern society. At the same time, and on a much smaller scale, engine-driven generators are essential to the operation of the electrical systems of cars, trucks, and other vehicles.

See also ALTERNATING CURRENT; ALTERNATORS AND GENERATORS; DIRECT CURRENT; MOTOR, ELECTRIC; TURBINE, HYDRAULIC; TURBINE, STEAM; WINDMILLS

Further Reading: Percy Dunsheath, *A History of Electrical Power Engineering*, 1962.

Genetic Screening

In recent years, genetic screening increasingly has referred to DNA testing, but since the beginning of the 20th century, the term has encompassed an enormous number of means for evaluating populations, individuals, body fluids, cells, chromosomes, and genes for actual and alleged inherited disorders.

Charles Darwin's cousin Francis Galton believed that statistical analysis could be applied to what he called "anthropometric research"—i.e., quantitative studies of human populations to reveal frequencies for any "heritable" trait from physical stature, to intelligence, to moral behavior. Among other aspects of "biometric" thinking, he copioneered fingerprinting as a means of personal identification. In 1883, Galton also coined the word *eugenics*, referring to practices or policies that applied selection to the "improvement" of human beings. He lived to see both Mendel's inheritance ratios rediscovered in 1900 and Archibald Garrod's description of alkaptonuria (1902) as arising from an inherited failure to produce an essential enzyme, or, as he later wrote, from an "inborn error of metabolism." Eugenics gained support on both sides of the Atlantic since it also held a scientific rationale for overcoming social problems. In 1907, for example, Indiana passed a eugenics act "to prevent the procreation of criminals, idiots, imbeciles, and rapists"—the first sterilization law in the United States.

Together, genetics and eugenics found a home together in Cold Spring Harbor, N.Y., where Charles B. Davenport directed the Carnegie Institution's Station for the Experimental Study of Evolution from 1904 on, and established the Eugenics Record Office in 1910. In effect, the ERO carried out the first genetic screening operations in the United States, sending its workers to gather data on such things as polydactyly (having an excess of fingers), albinism, "feeblemindedness," and tendencies toward personal violence among the "races" of immigrants arriving at Ellis Island. Eventually, the ERO discredited itself by association with the Nazi race hygienists. It closed in 1939. And after the Nazi atrocities were known, anything that suggested eugenics, such as genetic screening, stood under a pall.

In 1959, researchers in France reported on the trisomy of chromosome 21 in individuals afflicted with what was then known as "mongolism" (Down's syndrome); this discovery reignited interest in prenatal genetic diagnostics. Two years later, American pediatrician Robert Gurthrie introduced a simple assay for measuring serum phenylalanine in newborns to test for the hereditary disease phenylketonuria (PKU). Because infants who tested positive could be placed on a phenylalanine-restricted diet until their brain development reduced the risk of mental retardation, screening offered a real health benefit. Massachusetts passed the first mandatory PKU screening law in 1963, but other states soon followed. As similar tests for other metabolic defects were developed, state legislators added appropriate amendments to the PKU screening. New York, for example, by 1977 had made neonatal testing mandatory for PKU, congenital hypothyroidism, maple syrup urine disease, glalactosemia, histidenemia, homocystinuria, and adenosine deaminase deficiency.

Some states also mandated screening for sickle-cell anemia, not only for newborns but also for marriage license applicants (as already existed for syphilis) and for school children and prisoners (as already existed for tuberculosis). Since as many as 1 in 10 African Americans carried the sickle-cell trait (with an actual disease incidence of 1 in 400–600), testing mainly aimed at identifying "carriers." But, in the racially tense atmosphere that followed Dr. Martin Luther King's assassination, mandatory screening for sickle cell was alleged to have racist motivation. In response, the 1972 National Sickle Cell Anemia Control Act restricted Federal funding to voluntary testing programs.

Prospects for prenatal screening arose when British physician Douglas Bevis developed amniocentesis in 1952. By inserting a slender needle into the amniotic sac and extracting cells the fetus produced, it became possible to detect chromosome abnormalities first, and then enzyme deficiencies. By the 1980s, tests of amniotic fluid existed for nearly 200 diseases. Ultrasonography, fetal skin sampling, chorionic villus sampling, and percutaneous umbilical blood sampling expanded prenatal diagnostics. The Supreme Court's striking down of Texas's criminal abortion laws in Roe v. Wade (January 1973) increased demands for tests and involved genetic screening in the highly charged debate over abortion.

Opponents in both controversies—mandatory sickle-cell testing and abortion—could agree, at least, that genetic screening needed an infusion of training and public education. Consequently, when the National Sickle Cell Anemia, Cooley's Anemia, Tay Sachs, and Genetic Diseases Act became law in 1976, federal funding went into counseling and community information as well as to basic and applied research. A number of schools then followed the lead of Sarah Lawrence College, which in 1969 offered the first master's degree in genetic counseling, and several universities added genetic counseling courses in their M.D., Ph.D., and postdoctoral programs. Enough trained counselors existed in 1981 for the American Board of Medical Genetics to begin certification.

The Omnibus Budget Reconciliation Act of 1981 reversed the trend in federal support for genetic screening. Block grants for maternal and child health did not specify genetic testing or counseling, leaving the states to choose between spending on nutrition, vaccination, and other needs. Funding for screening diminished further, right up to the eve of one the biggest science projects ever, the Human Genome Project. In 1990, the Human Genome Organization began an international project to locate the estimated 100,000 sequences of DNA that comprise the full gene complement of the human being. Congress authorized the National Laboratories of the Department of Energy and the National Institutes of Health (National Center for Human Genome Research) to conduct the effort in the United States, anticipating 15 years work and $10 billion in expense. As researchers identified genes, they also found mutations that result in disease, some of which were defined clearly, such as those for cystic fibrosis and muscular dystrophy, others that were implicated in multiple diseases, and some that appeared to be linked to conditions that are nearly as murky now as they were in the age of eugenics: alcoholism and criminal behavior, for example. Forty years ago, about 400 genetic diseases were known, with only a handful linked to actual chromosome locations; today, more than 4,000 genetic disorders are known, with nearly 1,000 specifically mapped. Moreover, more than 100 gene therapy clinical trials are now in process, holding some prospect for treating unregulated, single genetic disorders, and for giving cells new traits, such as susceptibility to replication-inhibiting drugs in tumor cells, or chemotherapy resistance in hemopoietic bone marrow cells.

Most researchers suppose that the ability to find genes responsible for disease will exceed the possibility of treating those disorders for many years. Further, as is the case with Huntington's disease, early detection can predict a near-certain development of disability later in life, raising those individuals' concerns over employment and insurability status. The age of genetic screening has dawned only recently, and it looks like a cloudy day.—G.T.S.

See also AMNIOCENTESIS; BIOLOGICAL VARIATION, HUMAN; CHEMOTHERAPY; DNA; EUGENICS; HUMAN GENOME PROJECT; INVOLUNTARY STERILIZATION; GENETICS, MENDELIAN; SICKLE-CELL ANEMIA; ULTRASOUND

Genetics, Mendelian

People had long observed that offspring resemble their parents, sometimes in a very marked fashion. On a practical level, farmers were able to use their knowledge of heredity in order to improve their domesticated plants and animals. The mechanisms of heredity were poorly understood, however, and in many ways the depiction of hereditary processes was altogether wrong. This situation began to change when the invention of the microscope made the close examination of cells possible. By the end of the 18th century, the biological basis of the sexual reproduction of plants was understood, and systematic studies of plant hybridization were underway.

In the century that followed, studies of cells led to the identification of the parts responsible for the transmission of traits from generation to generation. But the key discoveries regarding the processes of hereditary inheritance were not concerned with what went on at a cellular level. In fact, they involved nothing more complicated than breeding the common pea plant, *Pisum sativum*, and observing how traits were handed down from generation to generation. The researcher was Gregor Mendel (1822–1884), an Augustinian monk who initially had learned about the plant world while working in the orchards of an Austrian nobleman.

When Mendel began his inquiries in 1857, inheritance was assumed to entail the blending of the characteristics that originated with each parent, much as the combination of two different metals produces an alloy. Mendel discovered that this belief was erroneous; traits inherited from each parent were retained even after being incorporated into the biological makeup of the offspring. Mendel came to this conclusion by careful experimentation, but with nothing in the way of sophisticated apparatus. He pollinated his pea plants by hand and then wrapped them to prevent accidental pollination by insects. After they reached maturity, he collected the seeds produced by each new generation of pea plant, planted them, and then noted the characteristics of the resulting plants. The results were strangely inconsistent. Dwarf pea plants produced more dwarf pea plants; in the language of horticulture, they "bred true." Tall pea plants were another story; about a third of them bred true for generation after generation, while the remainder gave rise to both tall and dwarf plants in a ratio of three tall plants to one dwarf.

Mendel then crossbred some true-breeding tall plants with dwarf plants. All subsequent generations of plants were tall. He then bred these tall plants with one another. This time the results were mixed. A quarter of the next generation were tall plants that subsequently bred true. Another quarter of this generation were true-breeding dwarf plants, and the remaining half were tall plants that did not breed true. The genetic constitution of the last group apparently included both tall and dwarf characteristics. Tallness was dominant, but dwarfism had not been extinguished. It was a "recessive," so dwarfness would reappear after its bearer was bred with another plant bearing the same recessive trait.

Mendel also looked into the inheritance of other physical features. The same basic pattern presented itself: Characteristics like the skin texture of individual peas (smooth or wrinkled) did not blend into an intermediary state. Rather, pairs of characteristics transmitted from two parents remained separate, so the offspring manifested one or the other of these features in their original form. Mendel accordingly postulated that there must be two factors for a given trait, each parent contributing one factor.

We now know that inherited characteristics are borne by chromosonal materials known as *genes*, but this term was not coined until 1909. In modern terminology, paired genes—one from each parent—in an organism are known as *alleles*. If the two genes in an organism are alike, the organism is said to be a *homozygote*. If the genes are dissimilar, the organism is said to be a *heterozygote*. Mendel found that when a homozygous plant with smooth-skinned peas was bred with a plant bearing either smooth or wrinkled skinned peas, the result always was the same: plants with smooth-skinned peas. In modern parlance, the gene responsible for smooth skins was "dominant" over the "recessive" trait for wrinkled peas.

Mendel also noted that the material responsible for heredity (what we now know to be genes) in each pair of alleles separated from each other during the formation of reproductive cells (sperms and ova), so each reproductive cell contained only one member of an allele pair. Mendel called this the "law of segregation," because when reproductive cells are formed, the alleles are always separated from each other, each ending up in a different ovum or sperm. Furthermore, Mendel discovered that the separation of each pair of alleles occurred independently of the separation of other allele pairs; he termed this "the law of independent assortment." Due to the operation of this law, each reproductive cells ends up with a random collection of the genetic materials provided by each parent.

Mendel codified his observations into some rather simple statistical rules that governed the acquisition of inherited traits. Subsequent research uncovered the biological mechanisms that underlay the processes described by Mendel's statistics. Accordingly, Mendel's research provided the foundation for the scientific study of inheritance. No less important, Mendel's discoveries undercut a major objection to Charles Darwin's theory of evolution by natural selection. According to Darwinian theory, variations occur randomly in groups of organisms. The few variations that help the organism to survive and reproduce will be handed down to succeeding generations of organisms, resulting over time in the evolution of these organisms into something different. Not everyone was convinced; opponents of the theory objected that useful variations would not persist, for they would be diluted as subsequent generations interbred with individuals lacking the variation. Mendel showed that blending in fact did not occur, and that genetic variations could be transmitted intact to succeeding generations.

The significance of Mendel's research for evolutionary theory was not immediately appreciated. Although he had owned and even had annotated a copy of *The Origin of Species*, Mendel made no mention of Darwin or evolution when he published the results of his research. In fact, Mendel's discoveries lay unappreciated for many years. His research had met with complete indifference when he presented it to a local natural history society, and when he sent a paper describing his experiments to Karl Wilhelm von Nägeli (1817–1891), a prominent botanist, the latter returned the paper with comments that were anything but encouraging. He also sent a copy of the paper to Darwin, but there is no evidence that he ever read it. Mendel went on to publish his results in a rather obscure scientific journal in 1865, and followed it with another published paper in 1869.

Mendel's work lay dormant until 1900, when the Dutch botanist Hugo de Vries (1848–1935) performed experiments that led him to the same conclusions as those of Mendel. Before publishing his results, he reviewed the existing literature, and found, much to his surprise, that an obscure monk had anticipated his research by several decades. At about the same time, Karl Erich Correns in Germany and Erich von Tschermak in Austria were rediscovering Mendel in the course of their own research. To their credit, all three gave full acknowledgment to Mendel's prior discoveries. Within a few years, the significance of chromosomes for heredity had been grasped, and the cellular mechanisms responsible for the patterns observed by Mendel began to be understood.

See also ALLELE; CELL; CHROMOSOME; EVOLUTION; GENE; MICROSCOPE, OPTICAL; NATURAL SELECTION

Further Reading: Robert C. Olby, *Origins of Mendelism*, 1966.

Geodesic Dome—see Dome, Geodesic

Geological Epochs and Periods

A geological epoch or period is a unit in a system used by geologists to divide up approximately 4.6 billion years of the Earth's history. The basic nomenclature of geological epochs and periods is simple, but the names sometimes differ from country to country. Europeans use the original form, while Americans have modified it to correspond to the rocks present in the United States. There have been various efforts to reform the naming of periods, including one by the Intentional Commission on Stratigraphic Terminology that began in 1952 and continues to this day.

Geologists have complicated the naming process by using different terms to describe the different ways of dividing up time. If you divide up time on the basis of the different fossils assemblages found in the rocks, that is called a *chrono-stratigraphical* or *geological* time unit. Other terms used for the same principle are *time-rock* units and *biostratigraphic* units. Periods may also be delineated by *litho-stratigraphic* units, which reflect the kinds of rocks found in sedimentary layers. Also used in geological texts is the term *zone*. This term corresponds to the International Commission's fifth-order name of *substage*, as indicated below.

Looking at the overall picture of time, the International Commission recommended that the following terms be used by geologists:

Rank	Geological time	Chrono-stratigraphical unit
1st order	Era (e.g., Palaeozoic)	Erathem (rarely used)
2d order	Period (e.g., Cambrian)	System
3d order	Epoch (e.g., Pliocene)	Series
4th order	Age (e.g., Oxfordian)	Stage
5th order	Time (rarely used)	Substage

Accordingly, an era divides into periods, which divide into epochs, which divide into ages. The correct order of the periods is something that many people have heard but find hard to remember. Here is a simple mnemonic to help:

Pretty	Pre-Cambrian
China	Cambrian
Owls	Ordovician
Seldom	Silurian
Devour	Devonian
Clay	Carboniferous
Pigeons	Permian
They	Triassic
Just	Jurassic
Chase	Cretaceous
Them	Tertiary
Quickly	Quaternary

The names of the geological periods/systems came about as the result of research carried out in the 18th and 19th centuries by scientists in Great Britain and Europe. In 1756, Johann Lehmann put forward a geological timescale based on the type of rocks thought to have been formed at the time:

Primitive:	all crystalline rocks (e.g., gneiss, granite, basalt)
Secondary:	consolidated sedimentary rocks that contain fossils
Alluvial:	soils and gravels

This timescale was further developed in 1760 by Giovanni Arduina, an Italian mining geologist:

Primitive:	crystalline rocks in the cores of mountains
Secondary:	sedimentary rocks
Tertiary:	unconsolidated sediments
Volcanics:	extrusive igneous rocks

In 1788, Georges de Buffon (1707–1788), a French naturalist, proposed a historical timescale in which there were six epochs with a precise chronology that conformed to biblical eras.

The timescale with which geologists are familiar today was developed in the 19th century. William Smith put forward the idea that a bed or group of beds were arranged in a regular order, bed succeeding bed in a definite sequence. He also said that each bed contained fossils that were more or less peculiar to it. Because of the fossils that a bed or group of beds contained, these beds could be traced from place to place.

When Charles Lyell (1797–1875) began his geological studies in the early 19th century, he used the following timescale, in order of increasing age:

Era/group	System/period
Quaternary	Recent or Holocene
	Pleistocene
Cainozoic (or Cenozoic)	Pliocene
or Tertiary	Miocene
	Oligocene
	Eocene
Mesozoic or Secondary	Cretaceous (including chalk)
	Jurassic (including lias)
	Rhaetic
	Trias
Palaeozoic	Permian
	Carboniferous (including coal measures, millstone grit and carboniferous limestones)
	Devonian (including old red sandstone)
	Silurian
	Ordovician
	Cambrian
Eozoic or Precambrian	

The era names translate from the Greek as:

Eozoic:	dawn life
Palaeozoic:	ancient life
Mesozoic:	middle life
Cainozoic (or Cenozoic):	recent life

These era names refer to the gradual increase in the number of fossil-sized animals and plants that resemble or were related to those still living. In 1831, Adam Sedgwick (1785–1873) started a detailed study that led to the establishment of the periods of the Palaeozoic era. In north Wales, he studied a sequence of deformed volcanic and sedimentary rocks that he proposed to be called the Cambrian system after Cambria, the Roman name for Wales.

At about the same time, Roderick Murchison (1792–1871) published a study of rocks in southern Wales, which he called the Silurian system after a Welsh tribe, the Silures. Further study showed that the upper Cambrian system and the lower Silurian system overlapped. What turned into a bitter conflict between Sedgwick and Murchison was resolved by Charles Lapworth in 1879. Lapworth recognized that the intermediate rocks were in fact distinctive and that these rocks should be put in a new system, the Ordovician, after the Welsh Ordovices tribe.

With the resolution of their feud, Sedgwick and Murchison studied a thick marine sequence together in Devon, southwest England. Initially they thought that the sequence was either Ordovician or Cambrian in age. Further study of the fossils by the palaeontologist William Lonsdale led to his conclusion that the sequence came between the Silurian system and the Carboniferous system. This was accepted by Sedgwick and Murchison, and the sequence became known as the Devonian system.

The Permian system also was defined by Murchison. He traveled at the invitation of the czar of Russia to an area west of the Ural mountains, where he found a thick sequence containing fossils older that the Triassic system but younger than those in the Carboniferous system. It was called the Permian system after the Russian province of Perm.

In 1822, two English geologists, William Coneybeare and William Phillips, named the Carboniferous system. This name came about because of the huge quantities of coal found in rock sequences of this age. The systems of the Mesozoic were named from sites in Western Europe. In 1834, Van Alberti named the Triassic system after the tripartite nature of the sequence that he studied. The German geologist Friedrich von Humbolt (1767–1859) named the Jurassic system after the Jura mountains in 1799. The Cretaceous system was named in 1822 by the Belgian geologist d'Halloy. The system name was derived from the Latin for chalk—*creta*. The system names for the Cainozoic (or Cenozoic) hark back to the original timescale put forward by Lehmann and Arduina. The deposits of the Tertiary and Quaternary are mostly unconsolidated sediments. These sediments contain many modern-looking fossils, although the life that they represent is now extinct.—J.G.

Further Reading: Carl K. Seyfert and Leslie A. Sirkin, *Earth History and Plate Tectonics*, 1973.

Geometry: Euclidean and Non-Euclidean

Geometry is prized both for its wide range of applications and for the way it proceeds by logical deduction from obvious assumptions to what seem to be absolutely true statements. For applications, the key results are the Pythagorean theorem (that the sum of the squares on the shorter sides of a right triangle equals the square on the longest side or hypotenuse), and the theory of similar figures (if two triangles have their respective angles equal, they have the same shape, and the corresponding sides are proportional). The Pythagorean theorem allows us to find exact distances, even of inaccessible objects. The theory of similar figures allows the measurement of distances between nearby, small objects to be used to determine the distances or sizes of faraway objects, including astronomical bodies. For geometry considered as a model of "truth," what counts is the logical structure of Greek geometrical texts, notably the *Elements of Geometry* (written \c. 300 B.C.E.) of Euclid of Alexandria.

Though the ancient Greeks are often credited with being the discoverers of geometry, both the Pythagorean theorem and the theory of similar triangles were independently discovered and were used for surveying and geometrical problem solving in several ancient societies, including China and India. Furthermore, Euclidean plane geometry is acknowledged by the Greeks to have begun in ancient Egypt because of the need to resurvey fields after the Nile's floods had subsided, as the word *geometry*—"earth-measurement"—indicates. The Egyptians soon went beyond mere surveying to develop an understanding of symmetries and of similar figures, and to calculate areas and volumes. The pyramids at Gizeh (c. 2800 B.C.E.) are the best-known symbol of Egyptian geometric knowledge, although it has been said that "the greatest Egyptian pyramid" was the one in the mind of the anonymous Egyptian writer who in about 1900 B.C.E. first correctly found the volume of the frustum of a pyramid.

Greek writers often said that they learned geometry in Egypt, although recent scholarship has revealed that the Babylonians, with whom the Greeks also had contact, had considerable geometric knowledge. In particular, the Babylonians by 1700 B.C.E. not only had a very accurate approximation to the square root of 2 but also had an extensive table of whole numbers that, according to the Pythagorean theorem, can be the sides of a right triangle (such as 3, 4, and 5; 5, 12, and 13; 65, 72, and 97; etc.). Thus the Greeks were heir to two sometimes conflicting mathematical traditions, Babylonian and Egyptian, and may have wanted a method to establish which results were true. Early Greek geometers like Thales (624–546 B.C.E.) and Hippocrates of Chios (fl. 5th-century B.C.E.) established that some results could be derived from earlier, simpler ones. Greek philosophy and politics involved logical discourse and arguments by contradiction. So the Greeks worked out the method of deducing some truths from earlier, better-known ones. Ultimately they tried to prove *all* geometric truths logically, on the basis of the smallest possible number of self-evident truths or *postulates* of geometry, and on clear definitions of the terms used. Euclid's *Elements of Geometry* did this successfully and has been the model of such reasoning ever since.

Euclid gave five postulates. The first three establish the straightedge-and-compass constructions familiar to geometry students: Between two points a straight line can be drawn; a straight line can be produced to any desired length; a circle can be drawn with any center and any size radius. The fourth postulate states that all right angles are equal. Postulate 5, the so-called parallel postulate, is not as obvious:

> If a straight line falling on two straight lines makes the interior angles on the same side less than two right angles, the two straight lines, if produced indefinitely, meet on that side on which are found the angles less than two right angles.

Though Euclid needed postulate 5 and its consequences to prove most theorems about parallels, it seemed less obvious than the other postulates. Many geometers—including the Greek Proclus (410–485); the Muslims ibn al-Haytham (965–1039), 'Umar al-Khayyāmī (1048–1131), and Nasīr al-Dīn al-Tūsī (1201–1274); and the 18th-century Europeans Girolamo Saccheri (who knew al-Tūsī's work) and Johann Heinrich Lambert (1728–1777)—thought it ought to be a theorem and tried to prove it, a point to which we shall return.

On the basis of his definitions and postulates, Euclid proved a wide range of results in elementary geometry, including the Pythagorean theorem (*Elements*, Book I, Proposition 47), theorems about areas, similar triangles, circles, and so on. Drawing on the theory of ratios of irrationals developed by Eudoxus, Euclid also proved results that anticipate aspects of the 19th-century theory of real numbers, and proved that the areas of circles are proportional to the squares of their diameters. Moving to solid geometry, Euclid showed how to construct the five regular solids (cube, tetrahedron, octahedron, icosahedron, and dodecahedron) and proved that there can be only five.

Both the content and the method of the *Elements* had great influence. Archimedes of Syracuse (c. 287–c. 212 B.C.E.) extended the methods of areas and volumes to the conic sections and to solids of revolution, also finding centers of gravity, thus anticipating (and influencing the discovery of) the integral calculus. Apollonius (c. 260–c. 190 B.C.E.) studied the conic sections in great depth and detail. Later geometers extended the methods of Euclid, Archimedes, and Apollonius. This was especially important in the medieval Islamic world, where men like al-Haytham and al-Khayyāmī studied the conic sections and other curves and their applications, from optics to astronomy. Trigonometry, which in the Islamic world fused Indian and Greek methods, also advanced greatly. In particular, methods of spherical trigonometry were used by men like Muhammad al-Bīrūnī (973–1055) to determine the direction of prayer, that is, the correct great-circle direction to Mecca from every point on the globe. Especially important was the Arabic synthesis of the Greek tradition of geometry and proof with the computational traditions of China and India, which can be viewed clearly in the famous 9th-century book *al-Jabr* (the origin of the term *algebra*) by Muhammad ibn Mūsā al-Khawārizmī (c. 780–850).

Arabic-language mathematical works (and Arabic versions of some Greek ones) were translated into Latin in the 12th century as part of the revival of learning in Europe. In the Renaissance, printing allowed the widespread publication of the Greek geometrical texts. Together with the algebraic work inherited from Eastern sources and the development of general symbolic notation in algebra by the 16th-century French mathematician François Viète (1540–1603), the geometric revival led René Descartes (1596–1650) and Pierre de Fermat (1601–1665) to apply algebraic methods to solve Greek geometric problems and, in the process, to invent analytic geometry.

Meanwhile, the logical structure of geometry decisively influenced philosophy. Plato's theory of knowledge (4th-century B.C.E.) is modeled on geometry in assuming that certainty can be achieved only when reasoning about objects—which, like ideal triangles and circles—do not change. Followers of Plato thought that science could be exact only if it had a geometric form, and in the 17th century Galileo and Kepler successfully developed important physical and astronomical laws on this basis. Aristotle believed that all scientific reasoning should have the same logical structure as geometry, and works from the 17th-century theology of Baruch Spinoza to Isaac Newton's *Mathematical Principles of Natural Philosophy* have followed this model.

The success of Euclidean geometry caused some mathematicians, especially in the 18th century, to feel that the postulate 5 was a "blemish" on the beauty of Euclid's structure and to try anew to prove it from other postulates, using Euclid's definition of parallel lines as lines that, however far extended, never meet. The method chosen was indirect proof: Assume that postulate 5 was false, and try to deduce a contradiction. But instead of a contradiction, mathematicians produced a set of odd, counter-intuitive statements such as: The sum of the angles of a triangle is less than two right angles; on the same surface, more than one parallel can be drawn to a given line through a point outside the line; two lines parallel to a third line can intersect; parallel lines are not everywhere equidistant. Saccheri thought he had found a contradiction and thus had succeeded in proving postulate 5, whereas Lambert, understanding the logical situation better, thought that postulate 5 must forever remain a postulate.

In the early 19th century, though, several mathematicians—notably Carl Friedrich Gauss (1777–1855), Janos Bolyai (1802–1860), and Nikolai Ivanovich Lobachevsky (1793–1856)—found a different and exciting perspective on the same logical situation. Instead of absurdities, these conclusions of the negation of postulate 5 were perfectly valid theorems in an entirely different—a *non-Euclidean*—geometry. On a non-Euclidean surface, through a given outside point, more than one parallel *can* be drawn to a given line, the sum of the angles of a triangle *is* less than two right angles, two lines parallel to a third *can* meet, parallel lines are *not* everywhere equidistant, and so on. Another way in which a surface can be non-Euclidean is for there to be no parallels at all. A familiar example of this second kind of surface is the surface of a sphere such as the Earth. If one defines *line segment* as the shortest distance between two points, on the sphere line segments are arcs of great circles—circles on the sphere whose radius equals the radius of the sphere. The equator and the meridians of longitude are great circles, while the "parallels" of latitude are not. It is easy to see that all great circles on a sphere intersect. Now consider the sum of the angles of the triangle formed by two meridians of longitude and the equator. Each meridian of longitude is perpendicular to the equator, so the two base angles add up to two right angles, and when one includes the angle between the meridians at the pole, the sum of the angles of this triangle exceeds two right angles. Geometries of the spherical type were studied in the mid-19th century by Bernhard Riemann (1826–1866). More important, Riemann began to look at non-Euclidean geometries in a more general way, treating them analytically, studying them for *n*-dimensional spaces, and observing that the "distance" between two points in a non-Euclidean space was no longer given by the Pythagorean distance formula but by a more general function called a *metric*.

Since Euclidean geometry is so useful, one might think that non-Euclidean geometry is nothing but a clever logical game. But it turned out otherwise. When Albert Einstein was studying gravitation as part of his general theory of relativity, he found that the way to describe physical space-time was by a geometry of the Riemannian type, where space was positively curved and where bodies with large amounts of mass "distort" the space around them so that "shortest distances"—for instance, the paths of light beams—are curved. Black holes are a well-known example of a phenomenon requiring general relativity to explain it. So it appears that Euclidean geometry is not the best model for physical space after all. Still, Euclidean geometry remains supremely useful for a vast number of applications, from surveying to perspective drawing, and remains the foundation of the classical physical sciences and engineering.—J.V.G.

See also BLACK HOLES; CALCULUS, DIFFERENTIAL AND INTEGRAL; MECHANICS, NEWTONIAN; PRINTING WITH MOVABLE TYPE; PYRAMIDS; RELATIVITY, GENERAL THEORY OF; THEODOLITE

Further Reading: George Gamow, *One, Two, Three, Infinity: Facts and Speculations of Science*, 1961. Jeremy Gray, *Ideas of Space: Euclidean, Non-Euclidean, and Relativistic*, 1989. T. L. Heath, ed., *The Thirteen Books of Euclid's Elements*, 1956.

Geophysics

Geophysics is the study of the Earth through the application of physics. Through geophysics, it is possible to learn about the Earth's composition and structure, and how it has changed over time. Examples of the physical properties measured and mapped include the velocity of energy moving through the Earth, the density of Earth materials, their magnetic susceptibility and electrical conductivity, and radioactivity.

For example, energy from earthquakes or explosions moving through the Earth is mapped as seismic waves. These waves may reflect or refract at compositional or structural boundaries inside the Earth. By mapping changes in seismic waves and their movement through the Earth, geophysicists discovered that the Earth's interior consists of a core, a mantle, and an outermost crust. Variations in the internal structuring of the Earth are inferred by analyzing data obtained through the measurement of changes in seismic waves that have traveled through the Earth. Since the technology to actually sample deep within the Earth does not exist, geophysical methods provide the only way to obtain knowledge of Earth's interior.

On a smaller scale, consider prospecting for iron ore. Magnetic susceptibility, or how magnetic a rock is, is related to the amount of iron-based minerals in the rock. Geophysicists can measure changes in the magnetic susceptibility of an area from an airplane and map the location of possible iron deposits. Geologists can then map and test for iron in those areas. In general, geophysics is used as a reconnaissance method to delineate possible areas for study. Environmentally, there is less disruption within the region, while financially, money is spent with less risk.

Geophysics cannot answer all questions about the Earth. Geophysics is only helpful when the question asked or problem to be solved is related to changes in the physical properties of the rock. Under many circumstances, this is not true, and understanding is best obtained by geological techniques.—D.A.B.

See also EARTHQUAKES; RADIOACTIVITY AND RADIATION

Further Reading: Martin Harold Phillips Bott, *The Interior of the Earth*, 1971.

Geothermal Energy

Geothermal energy is a resource found throughout the world but concentrated in six belts that coincide with active junctures of the Earth's crustal plates. These are the "fire belt" around the Pacific Ocean, the mid-Atlantic Ridge, the Alpine-Himalayan mountain chain, eastern Africa and the western Arabian peninsula, central Asia, and some archipelagos in the central and south Pacific. Hot springs are plentiful in these belts, and people always have used them for bathing. They became immensely popular in Japan and in ancient Rome, where water also was used to heat bathhouses. In the Tuscany region of Italy, geothermal steam heated houses and provided industrial heat for boric-acid factories during the 1800s, but it was not until the 20th century that geothermal energy was exploited widely.

The first large-scale geothermal district-heating systems were introduced in Iceland during the 1930s, and subsequently spread to Europe, New Zealand, and the United States. Today, direct use for space heating and in horticulture, aquaculture, and industrial processing accounts for approximately one-fourth of overall geothermal energy use. However, electric-power production, although it represents only about 0.5 percent of total world electrical generating capacity, is the most flexible and important use of geothermal energy.

So-called "volcanic steam" was first tapped to generate electricity in the well-known geothermal field at Larderello, Italy, in 1913. It turned low-pressure steam turbines that eventually produced electricity for a number of towns in the Tuscany region. This achievement inspired the drilling of experimental wells at the Geysers in California in 1923, but in a time of flush oil and gas supplies and abundant hydroelectricity, sufficient capital could not be raised to make them a commercial success. During the 1950s, however, rising demand and costs for energy encouraged geothermal development.

After World War II, Italy led the way, rebuilding and enlarging its Larderello plants. Japan and Zaire followed, building experimental plants, and Mexico constructed a 3.5 megawatt (mW) plant. Not surprisingly, the Geysers caught the attention of American geothermal entrepreneurs. They persuaded Pacific Gas and Electric Company (PG&E) to open a 12-mW plant there in 1960, which stimulated further geothermal power development. Expanding its energy interests, Union Oil Company drilled wells at the Geysers, and PG&E constructed additional generating plants with capacities ranging from 28 to 110 mW. By 1980, drilling companies provided steam for several power plants that together provided 2 percent of California's electric power. Geysers power production soon reached over 2,000 mW, making it the largest power-producing geothermal field in the world.

Meanwhile, the energy-crisis decade of the 1970s spurred additional geothermal development throughout the world, power plants appearing in the Philippines, Japan, Mexico, El Salvador, Kenya, Turkey, China, the Soviet Union, and other countries. In the United States, the Public Utility Regulatory Policy Act (1978), which required regulated public utilities to purchase electricity produced by small power producers at a price equivalent to the utilities' production costs, inspired drilling in fields beyond the Geysers. In 1980, two 10-mW plants were dedicated in Southern California's Imperial Valley. By 1990, 15 Imperial Valley power plants drew on five geothermal fields to generate 379 mW of electricity. Significantly, the San Diego Gas and Electric Company built the first commercial-scale binary-cycle (organic Rankine cycle) plant to generate power from low-to-moderate temperature geothermal fluid. With about 80 percent of U.S. geothermal resources in the low-to-moderate range, the 45-megawatt project opened the way to unlock this vast resource.

Several additional geothermal fields also were

tapped in California. At the Coso field near China Lake in the Mojave Desert, nine power plants provide 240 mW of electricity, and smaller plants in northeast California also utilize geothermal energy. In Southern California, the City of San Bernardino, which received funding from the California Energy Commission to convert its wastewater treatment plant to geothermal heat in 1983, expanded its initial project into the largest geothermal district-heating systems in the United States, providing space- and water-heating to 27 major public and commercial buildings.

Geothermal energy has been one of the brightest spots in the world's renewable energy picture, both for generating power and for agribusiness and district heating, but it is not without its drawbacks. Exploitation of the geothermal resources, for example, has an aesthetic impact on the environment, and geothermal plants release into the atmosphere quantities of mercury and arsenic plus gaseous pollutants such as hydrogen sulfide, radon, and carbon dioxide. To halt air pollution, various abatement techniques at the Geysers have converted the greatest pollutant, hydrogen sulfide, into sulfur, while reinjection of geothermal gas back into the reservoir has been adopted at the Coso geothermal field. Compared to conventional fossil fuel and nuclear energy production, however, geothermal energy is much more environmentally benign.

Another difficulty with geothermal energy rests in the long-term sustainability of steam pressure in geothermal fields. In 1987, PG&E noticed a decline in steam production at the Geysers and delayed and then canceled expansion plans because of insufficient steam. Steam production at Geysers has declined precipitously, pressures dropping from 3,546.6 kPa (514 psi) in the 1960s, to 2,070 kPa (300 psi) by 1990. Some new and deeper wells seemed only to bring steam with increasing amounts of corrosive hydrogen chloride and noncondensible gases, and instead of producing power at the field's gross capacity of 2,093 mW, 1992 output reached only 1,226 mW. Although the California Energy Commission and others are studying the steam pressure problem, the long-term prospective for geothermal energy at the Geysers remains unclear.

Nevertheless, geothermal energy remains an important energy resource. Because as many as 40 countries stand to benefit substantially from its development and because conventional power production will continue to increase in cost, geothermal energy development is likely to experience continued expansion during into the 21st century.—J.W.

See also FUELS, FOSSIL; NUCLEAR REACTOR; PLATE TECTONICS; RANKINE CYCLE; TURBINE, STEAM

Further Reading: Robert Bowen, *Geothermal Resources*, 2d ed., 1989.

Germ Theory of Disease

The creation and spread of the modern germ theory of fermentation and disease was largely the work of three mid-19th-century scientists: the chemically oriented Louis Pasteur (1822–1895), the surgeon Joseph Lister (1827–1912), and the physician Robert Koch (1843–1910). Each speculated that many diseases might be spread by tiny self-replicating seeds, or "seminaria," that were passed on by contact with infected individuals, by contact with materials they had touched, or by passage through the air. These ideas had been anticipated in the 16th century by Girolomo Fracastoro, who had presented his theories in a book entitled *De Contagione*. Fracastoro's views had been updated, with the additional insistence that the sources of contagia were living organisms, by the German Pathologist Jakob Henle in 1840, but Henle was unable to offer any compelling evidence for his views.

This situation began to change dramatically in 1857 with the publication of Louis Pasteur's *Memoire sur la fermentation appellée lactique*. Two years earlier, Pasteur had been asked to help solve the problems of an industrialist who produced alcohol from sugar beet juice. Some of his vats had gone "sick" and were producing unmarketable lactic acid instead. Pasteur was able to demonstrate that under normal circumstances alcohol was produced as a by-product by living globules of yeast which fed on the beet juice. In the "sick" vats, rod-shaped microorganisms killed off the yeast and converted the beet sugar into lactic acid instead. In this paper, both microbiology and the germ theory of disease emerged for the first time, for Pasteur pointed out that human diseases might well be caused by microorganisms that might be airborne and which came into contact with and grew in patients, producing toxic effects. Many scientists were skeptical, however, and argued that Pasteur's theory seemed to require a vast concentration of germs in the air.

In the early 1860s, Pasteur did an additional set of experiments to undermine the claims that living organisms might be produced by spontaneous generation. He was able to show that if a fermentable solution from which all living materials were removed was allowed to stand indefinitely without communicating with the air, no fermentation would take place. As soon as the liquid came into contact with the air, however, fermentation began. Subsequent experiments demonstrated that it was airborne microorganisms, rather than the air itself, that were the active agents of fermentation.

One of the immediate consequences of the demonstration that microbes cannot appear spontaneously was the development of the process of pasteurization. In this process, liquids such as wine, beer, and milk, that often "spoil" with unpleasant or dangerous consequences, are exposed to heat long enough to destroy the undesirable microorganisms. Equally importantly, Joseph Lister transformed surgical practice when, following Pasteur's lead, he postulated that microorganisms caused wound suppurations and other infections, just as they produced fermentation and putrefaction of dead organic matter. This was a radical break with contemporary medical practice, which actually welcomed the appearance of "laudable pus" as an indication that a wound was healing. In contrast, Lister urged extreme cleanliness, phenol (carbolic acid, C_6H_5OH) sprays of instruments and wounds, and the use of antiseptic surgical dressings as detailed in his book *On the Antiseptic Principle in the Practice of Surgery*.

The final step in establishing the germ theory was taken by Robert Koch. One of his major contributions was the development in the 1870s of techniques for staining bacteria to make them more visible. For this he used aniline dyes, which were widely employed in the textile industry. In 1881, Koch introduced the use of jellylike culture media that allowed the growth of only a single species of bacteria; this greatly facilitated the study of bacteria and the means to control them. Taken together, Koch's technical expertise in staining, microscopy, microphotography, and culture growth and purification allowed him to isolate and identify the particular microorganisms that produced most of the common diseases that had killed surgical victims before the time of Lister, as well as the anthrax bacillus and the tuberculosis bacillus. By 1877, less than 20 years after Pasteur's first paper, the germ theory had become almost universally accepted. Much still remained to be discovered, such as the existence of viruses , but the germ theory had demonstrated that specific microorganisms were the source of many diseases.—R.O.

See also BACTERIA; DYES, ANILINE; EPIDEMICS IN HISTORY; PENICILLIN; SPONTANEOUS GENERATION; SURGERY, ANTISEPTIC; VIRUS

Glass

Glass is an amorphous (i.e., of undefined molecular structure), supercooled liquid made by heating complex silicates of soda with a basic oxide (such as lead oxide) and fusing them. Its primary characteristics are its transparency or translucency, and its resistance to weathering and to nearly all chemicals, the primary exception being hydrofluoric acid. Hard and rigid at ordinary temperatures, glass can be readily molded or shaped at elevated temperatures into all conceivable forms. It welds when red hot and can be cut with a knife at lower heat, but when cooled it is extremely brittle. When glass was first produced is not known, nor is the place, though the oldest known specimens are Egyptian. It was extensively used during the time of the Roman Empire, and glassmaking continued to flourish in Constantinople during the early Middle Ages and in Venice during the later Middle Ages and the Renaissance; Venetian goblets and mirrors were ultimately exported worldwide. Despite Venetian efforts to monopolize the manufacture of fine glass, Venetian glassworkers came to Paris in the mid-17th century, and they were ultimately responsible for the famous Hall of Mirrors at Versailles. Magnifying lenses were known to Roger Bacon in the 13th century and to Al Hazan even earlier. Glassmaking in the United States dates from the early 17th century at Jamestown and Salem, with place glass first being produced in Pittsburgh in the mid-19th century.

Glass is an extraordinarily diverse material, there being hundreds of different compositions and perhaps as many as 50,000 formulas. The oldest, cheapest, and easiest to work are the soda glasses, which account for 90 percent of all glass production in the United States. Soda glass is the stuff of tumblers, windows, and ordinary bottles—the greenish hue of "bottle glass" is due to iron impurities. Lead or lead-alkali glasses, also inexpensive, are resistive to electricity, but compared to other glasses they are not so resistive to acids. They are used, among other things, in optical components and neon tubing. Borosilicate glasses are resistant to heat, thermal shock, and chemicals, and have many applications as optical and laboratory devices. Of the six types, the best known is a low-expansion variety called Pyrex. Fused silica glass refers to a material that is 100 percent silicon dioxide; in its natural form it is called quartz glass. It too has many special applications in laboratory optical systems and instruments. Flint glass is a highly transparent variety of soda-lime-quartz glass. Industrial glass is a generic name for glass that is molded; optical glass is usually a flint glass of special composition, and plate glass is any glass that has been cast or rolled as a sheet, then polished.

Like the varieties of glass, the methods of manufacture are extremely diverse. The manual art of glassblowing still provides a form of entertainment if nothing else, but at one time almost all glass objects were hand-blown, including windows and bottles. Flat pieces, such as window panes, were first blown as cylinders and then flattened while still in a semimolten state. Today, sheet window glass is drawn directly from a bath of molten glass, while heavy plate glass for storefronts is either rolled or cast. Tumblers and dishes are often pressed, sometimes in imitation of "cut" glass, which is actually shaped by grinding. Like glassblowing, glasscutting is a highly skilled occupation. Glass can also be etched with hydrofluoric acid, one chemical to which it is not impervious, and it can be spun into a textile material.

Safety glass, sometimes also referred to as shatterproof glass, is made by laminating a sheet of transparent resinoid between two sheets of plate glass and molding everything together under heat and pressure. Presently the primary use for safety glass is for automobile windows. Bulletproof glass is similar to automotive safety glass, but extremely thick, some types being as much as 6 inches overall with five or more plies, although a thickness of 5 cm (2 in.) is sufficient to stop a 30/30 bullet. A variation of safety glass is wire glass with embedded woven wire netting, which is often used in schools and other public buildings where fireproofing is critical.—R.C.P.

See also EYEGLASSES; FIBERGLASS; GLASS, HIGH-TEMPERATURE; GLASS, SAFETY

Glass, High-Temperature

Among the significant achievements of modern glassmaking are the high-temperature glasses introduced in Germany and the United States during the late 19th and 20th centuries. Before then, glassmakers had primarily depended on two families of glasses, soda-lime-silica and potash-lead-silica batches, that melted within a wide temperature range. Although these glasses are suitable for windowpanes, bottles, and many consumer, bar, chemical, and apothecary products, they lack the physical properties essential for many modern technical and consumer applications. The demands of various late-19th-century "high-technology" industries for glasses that were temperature tolerant and chemically stable provided glassmakers with the impetus for creating new families of glasses whose high-silica and low-alkali contents imparted resistance to pressure, thermal shock, and corrosion.

The earliest and most prominent group of high-temperature glasses were the borosilicates, a family of soda-boron-silica compositions that melt at temperatures from 1,550° to 1,575°C (2,820°–2,870°F). In the early 19th century, British chemist Michael Faraday first used borax and boric oxide as fluxes in batch formulas in his quest to produce better glasses for scientific instruments, but these experiments reached a dead end as English optical glass manufacturers yielded their market to Bavarian competitors. The first commercially successful borosilicate glasses were created by Otto Schott, whose Thuringian glassworks added the world-renowned Jena laboratory glass to its innovative product portfolio, which included optical, thermometer, and gauge glasses in the 1890s.

By the early 20th century, managers at Corning Glass Works developed nonexpansion glasses initially in response to the railroads' needs for shatterproof globes used in signalmen's lanterns. In collaboration with scientists at leading public and private research laboratories, Corning researchers developed their own borosilicate glasses, introducing Nonex-brand signal globes and Pyrex-brand household, chemical, and electrical wares by the time of World War I. During the war, scientists affiliated with Corning and several other American glassworks collaborated with the Bureau of Ordnance and the National Bureau of Standards to adapt borosilicate formulas for the production of optical glass intended for field glasses, fire-control instruments, and other military devices. Glass manufacturers have continually refined borosilicate formulas and manufacturing processes, devising methods for introducing more and more silica into batches to enhance the stability and the performance of the final product.

In the interwar years, research-oriented manufacturers perfected numerous high-temperature glasses. Many new glass families were developed to fit specific customers' technical specifications, but others had general-purpose applications. Aluminosilicate glasses were super-borosilicates with lower coefficients of expansion than their predecessors, making the earliest glasses of this family ideal for stove-top cooking ware. Developed in the 1930s, fiberglass was spun into pads of insulation for buildings and refrigerators, woven into fabrics for curtains and clothing, pressed into acoustical board, tile, and other sound-reducing construction materials, shaped into fishing rods, and molded with plastic into components for appliances and vehicles. Another product of the Depression years, Corning's Vycor-brand glass, consisting of 96 percent silica, could withstand temperatures slightly greater than 3,000°C (5,430°F) without deformation. This glass could be subjected to red heat and immediately plunged into cold water without fracturing; it was the ideal material for furnace crucibles, furnace windows, and see-through panels on the headgear of the fiberglass suits used by firefighters.

Today, many commonplace technical and consumer products are fabricated from high-temperature glasses or from a successor group of post–World War II materials, glass ceramics. High-temperature glasses are appropriate for a wide variety of manufacturing and engineering applications that require strong, stable, and transparent containers or components. In the petroleum and chemical industries, these glasses are used to construct large pipelines that transport hot and cold solutions and to build towers for distillation, filtration, and absorption processes. The mechanical strength and electrical resistance of high-temperature glasses make them, like porcelain, suitable for insulators used in high-voltage power transmission. Some high-silica glasses have applications as mirrors in missile, satellite, and airborne telescope systems, and others are appropriate for windshields in military aircraft. Electrical and electronic components large and small, from color television tubes to sealed-beam automobile headlights, are also made from high-temperature glasses. Indeed, high-temperature glasses are ubiquitous materials that permeate all aspects of daily life in the late 20th century.—R.L.B.

See also ENGINEERING CERAMICS; FIBERGLASS; GLASS; OIL REFINING; PORCELAIN

Glass, Safety

Although ordinary glass has many uses, it is brittle and prone to shatter. For many applications, this is not much of a drawback, but for others it is very serious. A broken water tumbler is an annoyance; a shattered windshield is a potential cause of death. Shatterproof glass antedates the automobile; one of its first applications was to protect machinists from flying chips. Even so, the rapid spread of the automobile in the early decades of the 20th century provided the major incentive to make glass that did not shatter, or *safety glass* as it came to be known.

The origin of safety glass can be traced to 1903, when Eduard Benedictus, a French chemist, accidentally dropped a flask on the floor of his laboratory. To his surprise, the flask fractured, but retained its shape. This happened because the flask had contained a solution of cellulose nitrate, commonly known as *collodion*. The solution had evaporated, leaving behind a thin, flexible film that held the flask together after the glass had broken. At first, Benedictus took little notice of what had occurred, but after reading about several automobile accidents where flying glass had caused serious injuries, he realized the practical value of his discovery. In a short space of time, Benedictus produced examples of "triplex" glass, so named because it consisted of a layer of cellulose nitrate sandwiched between two layers of glass. The three layers were bonded together by the application of heat and pressure.

As is often the case, accommodating a novel material like laminated glass to mass production took some time, so it did not appear in commercial quantities until 1910. At first, its main application was not for automobiles but for the lenses of gas masks that were to be used in World War I. Some automobile manufacturers began to equip their cars with safety glass in the 1920s, but others balked, convinced that "safety doesn't sell." This attitude was eventually overcome, due in part to the adoption of safety glass as a standard by the U.S. Society of Automotive Engineers in 1930.

Laminated glass was more slowly adopted in Europe and Japan. There, a cheaper version of safety glass known as "toughened" or "tempered" glass was commonly used. This kind of glass is heat treated so that on impact it breaks into small fragments with no sharp edges. However, tempered glass has the disadvantage of becoming opaque when it breaks, unlike laminated glass, which typically produces a few cracks radiating out from the point of impact.

In the early 1930s, manufacturers of laminated glass began to use cellulose acetate in place of cellulose nitrate because of the latter's tendency to yellow with age. Cellulose acetate was not a perfect laminate, however, for it tends to open and admit moisture in cold weather, resulting in a hazy appearance. This problem was largely overcome in 1939, when a synthetic polymer known as *polyvinylbutyral* began to be used as the laminate. In recent years, safety glass has been further improved by adding a plastic layer to the inside of a windshield; this protects the car's occupants in the event of an accident by preventing lacerations caused by contact with the edges of broken glass.

See also CELLULOID; GAS, POISON; GLASS; PHOTOGRAPHY, EARLY

Gliders, Sailplanes, and Hang Gliders

The modern, high-performance sailplane represents the acme of performance for unpowered, heavier-than-air aircraft. It is to the glider as the high-performance sports car is to the family automobile. Similarly, the hang glider may be viewed as being to the glider as the unicycle is to the ordinary bicycle, i.e., a machine requiring some daring to ride. All of these variations on the theme of unpowered flight require, firstly, a means to gain (relative) altitude and then means to maintain that altitude or even gain more altitude. At the least, the machine's descent should occur slowly.

The history of unpowered flight has mythological origins going back to the Greek legend of Daedalus and Icarus. More factual episodes date to as early as the 9th century, when the Spanish Moorish savant Abbas bin Firnas covered himself in feathers and attempted a flight around 875 C.E. About 150 years later, the English Benedictine monk Eilmer was reported to have equipped himself with wings and glided about 200 m (660 ft) after jumping off the tower of Malmesbury Abbey. Having broken both legs in his adventure, Eilmer abandoned further such attempts at flying. Unless the tower was at least 30 m (100 ft) high, the reported distance covered by Eilmer would seem to have been exaggerated.

The modern history of human flight may be taken to start with the not-completely-successful effort of J. J. Montgomery, who made a single flight in 1883 in a glider having rudimentary controls operated by the pilot from a seat suspended beneath the machine. Eight years later, the German engineer Otto Lilienthal (1848–1896) made his first flight in a glider he had designed and built after more than 20 years of aeronautical experimentation. Lilienthal held on to the framework of his glider and exerted control over the flights by means of shifting his weight. His first glider, built in 1891, weighed little more than 18 kg (40 lb).

Lilienthal's successful flights were widely reported and closely followed on both sides of the Atlantic, most notably by the prominent American civil engineer Octave Chanute. Chanute had long been interested in the possibility of human flight and by 1896 had designed and built a structurally sound hang glider. Chanute, by then in his 60s, delegated flying the machine to assistants who made many successful flights. Comparable efforts were made by the Englishman Percy Pilcher. Perhaps the greatest achievement of all these pioneers was to inspire the Wright brothers, Wilbur (1867–1912) and Orville (1871–1948), to take up the efforts that culminated in the first successful powered flight in 1903. Early on, they had decided that the key problem to be solved was the matter of control. In the course of solving the problems of control, they began with gliders, and then went on to achieve powered flight only $4\frac{1}{2}$ years after their initial efforts of 1899.

The advent of powered flight led to a loss of interest in gliders, so that by the start of World War I in 1914 they virtually had vanished from the consciousness of aeronautical enthusiasts. It is interesting to note, however, that the Wright brothers themselves maintained an interest in unpowered flight to the extent that in 1911 Orville Wright set an endurance record for gliders of 9 minutes 45 seconds, which stood until broken by a German pilot in 1921 with a flight lasting for 15 minutes 40 seconds. Orville's flight was the first recorded soaring flight, i.e., a flight making use of rising air currents to maintain or gain altitude. Gliding regained popularity in Germany after 1919 because the Versailles Treaty, which ended World War I, forbade powered flying in Germany but said nothing about gliding.

The pioneers had launched their gliders by jumping off natural or artificial hills or dunes. Clearly such launches limited their flights, so that when gliding resumed in Germany after World War I other means of launching gliders were devised. The most successful such means was to tow the glider behind an airplane, and this still remains the most common method for bringing a glider to flight altitude. The usual release altitude now is about 600 m (2,000 ft). The glider pilot controls the release mechanism. A less common but very elegant way for a glider or sailplane to gain altitude is through the use of an engine that is shut off and, possibly, retracted when the machine reaches the desired altitude. Remarkably, the best of these engine-assisted craft perform as well as their unpowered, somewhat lighter, and more streamlined counterparts, e.g., the power-assisted Finnish Eiri PIK-20E vs. the unpowered PIK-20B. Both of these machines have glide ratios exceeding 40 and minimum sinking speeds of about 0.6 m/sec (2 ft/sec). If such a machine glides from 600 m (2,000 ft) without any assistance from rising air currents, it will stay aloft for more than 16 minutes while covering a distance of about 24 km (15 mi).

Sailplanes, the term applied to high-performance gliders, soar on rising columns of heated air called *thermals*, which frequently terminate in cumulus clouds. Rising air currents also may be found along slopes and mountain ridges where the interaction of the wind and the topographical features produces the currents. One frequently sees large birds, most commonly seagulls, soaring on rising air currents. Such a sight gives one some appreciation of the elegance of such flight.

The modern sailplane with its long, slender wings and sleek configuration looks deceptively simple to the

Otto Lilienthal's pioneering glider (from R. E. Bilstein, *Flight in America*, 1984).

casual observer. In fact, such a machine incorporates a variety of exceedingly sophisticated technologies in its design and construction. These include, among others, the application of the latest knowledge of low-speed aerodynamics, as well as structural design incorporating the most advanced composite materials. In addition, the modern sailplane is equipped with sophisticated communications equipment, as well as a limited number of high-precision instruments. Among the latter, and unique to high performance sailplanes, is the *variometer* or *vario*, which indicates the sailplane's rate of climb (*lift*) or rate of descent (*sink*) very precisely. Without the instrument, when those rates are small, the pilot has difficulty in making good use of the rising air currents around the sailplane. Sailplane pilots cannot fly "by the seat of their pants." Record flights of hundreds of kilometers at altitudes of more than 10 kilometers and lasting many hours require great skill on the part of highly trained pilots and superb machines equipped with the finest gadgetry. Due to its remarkable performance in remaining airborne, the modern sailplane sometimes needs assistance when the pilot, for whatever reason, wishes to increase the rate of descent. To do this, sailplanes are equipped with spoilers or air brakes much like those found on jet-powered airliners.

In spite of what has just been said about sailplanes, a person of average ability may join a local glider club and soon learn to safely achieve flights of as much as an hour after a course of training lasting a few hours. As in many sports, gliding and soaring may be enjoyed by those of varying levels of skill.

Gliders have been used for more than sport and recreation. During World War II, both Germany and the Allied Powers used large, specially designed gliders as troop and cargo carriers. A large transport or bomber airplane was used to tow as many as three loaded gliders, each almost as large as the towing aircraft. Since such glider trains provided excellent targets for enemy fighter aircraft, they could be employed successfully only when friendly airpower controlled the skies through which they flew. The Germans used gliders with great success in the 1941 invasion and capture of the Greek island of Crete, while the Allied Powers made successful and important use of gliders in the invasions of Sicily, Normandy, and southern France in 1943 and 1944. It seems unlikely that gliders will ever again play a significant role in warfare.

Hang gliding, after all but disappearing by the start of WWI, reappeared in the 1960s. This followed by about 15 years the invention by Francis Rogallo, then working for the National Advisory Committee for Aeronautics (the NACA was NASA's predecessor), of a kitelike aircraft with minimal structure and a flexible, fabric lifting surface. The original intention was to use the easily collapsed, stored, and transported Rogallo Wing as equipment for frontline troops. This use never occurred. A Californian, Dick Miller, devised a bamboo and clear-sheet-plastic version of the Rogallo Wing, named the Bamboo Butterfly, which he first flew successfully in January 1966 while literally hanging on to its fragile framework. Within a short time, Rogallo Wing–type hang gliders became a common feature of the California beach scene. All of these hang gliders depended on shifting body weight for control as had the pioneers three-quarters of a century earlier. Because this new generation of hang gliders were no safer than their predecessors, some of the more sophisticated hang-gliding enthusiasts started to design and build hang gliders incorporating conventional control systems. These machines frequently were not unlike improved versions of the gliders built by Octave Chanute and Percy Pilcher in the 1890s. In time, these newer hang gliders became increasingly sophisticated and made use of instruments including variometers.

Modern hang gliders achieved a level of performance that would have amazed Otto Lilienthal and his fellow pioneers. By the mid-1970s, they had flown as far as 160 km (100 mi), reached altitudes of 6,000 m (20,000 ft), and stayed aloft for as long as 15 hours. Yet by 1980, they had largely vanished from the scene, because in the late 1970s a few hang-gliding enthusiasts had mounted small engines on their machines and had introduced the ultralight aircraft to a new generation. Something akin to the hang glider survives in the parawing, a type of parachute, which is frequently seen at airshows where teams perform what might be called aerial ballets.

See also AIRCRAFT, ULTRALIGHT; COMPOSITE MATERIALS; NATIONAL AERONAUTICS AND SPACE ADMINISTRATION; PARACHUTE

Further Reading: Charles Coombs, *Soaring*, 1988.

Global Positioning System

A global positioning system (GPS) uses orbiting satellites and ground receivers to give precise determinations of one's position on the surface of the Earth. First envisaged by the U.S. Department of Defense, GPS now encompasses a large and growing variety of civilian applications. Currently, the basis for GPS is provided by 25 satellites (3 of which serve as spares in the event that one or more of the other 22 suffer a breakdown) orbiting at a height of about 20,000 km (12,500 mi) above the Earth. The theory underlying GPS is grounded in fairly simple geometry. Each satellite orbits at a fixed distance from the Earth, and in so doing it forms an imaginary sphere with a radius equal to its distance from a point on the ground. In similar fashion, the other satellites are at the center of their own imaginary spheres. The surfaces of these spheres will intersect at only two places. One of these will be deep in space or below the surface of the Earth, and hence can be disregarded. The other place will be the point on the ground where the receiver is situated. By ascertaining the exact distances to each satellite, the position of this point can be determined with a high degree of accuracy.

Finding the distance from each satellite is done by measuring the exact moment that a radio pulse arrives at the receiver. Since the pulses travel at a known speed, it is an easy matter to determine the distance from the satellite to the receiver if the precise time when the pulse was sent is known. The satellites use atomic clocks that are accurate to 1 second every 30,000 years to generate the time signals, but cost and complexity rules out their use for ground receivers. Instead, the receivers use simple electronic clocks that allow the calculation of approximate "pseudo ranges" that correspond to imaginary spheres surrounding each satellite. Unlike spheres whose distances have been determined with complete accuracy, the surfaces of these spheres do not intersect at a single point. But with appropriate computations, it is possible to find a value for the amount of clock error that prevents the spheres from meeting at a single point, and this value is then used to determine the correct distance of each satellite from the receiver on the ground.

A further complication arises from the fact that the Department of Defense has been willing to allow nonmilitary personnel and organizations to access GPS signals, while at the same time reserving the highest degree of accuracy for military users. This has been done by deliberately altering the signal sent by the orbiting atomic clocks. In theory, this allows the fixing of one's ground position at an accuracy that is not off by more than 100 meters. In fact, nonmilitary personnel have been able to correct for the deliberate error in the signal, making it possible to determine ground positions that are accurate to as little as 1 centimeter. The GPS corrections necessary to do this are readily available to the general public. Not only are they sold by private firms, even the U.S. Coast Guard and the Federal Aviation Administration provide them. A study conducted by the National Academy of Sciences concluded that the government's efforts to restrict precise GPS data were ineffective and should be terminated, but the Department of Defense has been reluctant to do so.

This slight handicap has not prevented the rapid diffusion of GPS technology. In addition to guiding ships and aircraft, GPS is used by geologists to measure tiny movements in the Earth's crust caused by the movement of tectonic plates. Meteorologists measure delays in GPS signals induced by the atmosphere in order to produce more accurate weather forecasts. Some farmers have used GPS to survey small portions of their fields so that fertilizer can be applied in the most effective manner. For the average person, the greatest direct benefit of GPS is likely to take the form of a receiver carried in one's car that provides exact driving instructions, thereby eliminating the need to stop and ask for directions that often end up to be misleading or worse.

See also CLOCK, ATOMIC; DEPARTMENT OF DEFENSE, U.S.; PLATE TECTONICS; WEATHER FORECASTING

Further Reading: Thomas A. Herring, "The Global Positioning System," *Scientific American*, vol. 274, no. 2 (Feb. 1996): 44–50.

Global Warming—see Climate Change

Golf Balls

Early golfers knocked around balls that had been carved from boxwood. In the 14th century, a different kind of ball known as a *featherie* made its appearance. The featherie was a hand-sewn cowhide shell stuffed with boiled feathers. A skilled golf ball maker could complete only four or five in a day, and due to lung trouble caused by inhaling feathers, few skilled ball-makers enjoyed long careers. Moreover, the featheries would split from weather and wear. In 1843, a radically different kind of ball appeared on the scene. According to legend, Dr. Robert Paterson, an avid golfer and professor at St. Andrews University, received a statue from Singapore that had been carefully packaged in shavings of a dried vegetable gum known as *gutta-percha*. He heated these shavings and formed them into a ball.

At first the ball was a failure, but by 1848, manufacturing refinements produced a better product, and the gutta-percha ball began to dominate golf courses throughout Scotland. Manufacturing changes yielded a cheaper and sturdier ball, but it was still an aerodynamic failure, for the smooth ball would not fly very far and would bounce unpredictably. Golf aficionados began to notice that nicked and scratched used balls traveled farther and truer than smooth new balls. Inspired by this apparent anomaly, manufacturers began to hammer grooves into new balls. This laborious process gave way to molds that mass produced grooved or dimpled shells.

In 1899, manufacturers redesigned the core of gutta-percha balls. The Haskell balls continued to use a rubber center, but rubber bands were wound around this core, making the ball bouncier. This three-piece method of ball construction has changed very little, although gutta-percha gave way to *balata* as the cover rubber. Premium balls used well into the 1970s consisted of a rubber core wrapped with rubber bands and covered with a balata shell.

The balls used by Sunday duffers were less expensive and more durable. The core was solid rubber and the rubber thread wrapping the core was thicker, making it more like rubber tape. This tape took less time to wind and broke less frequently than did the thread. The covering on the ball was a synthetic material, *surlyn*, which cut less frequently than balata. In the 1970s, Spalding initiated a change in the golf ball manufacturing industry. This new ball was only a core and a cover, having no rubber thread or tape that required winding. In the 1990s, a golfer can purchase either a two- or three-piece ball, and one with either a surlyn or balata cover.

Despite these changes in the manufacturing process, golf balls from different eras look about the same. Since the 1930s, typical balls have had 336 dimples about 0.33 mm (0.13 in.) in diameter and 0.030 to 0.033 mm (0.012–.013 in.) in depth. The traditional pattern had identical hemispherical-shaped dimples arranged symmetrically so it could be easily manufactured. In the 1970s, ball manufacturers began to experiment with different kinds of dimples. Manufacturers offered balls with fewer dimples or larger dimples, as well as cone-shaped and hexagonal dimples, all promising longer or straighter drives. The trend in golf ball manufacturing has been to patent new and different dimple designs to give the user an advantage over opponents who are still using less scientifically designed, manufactured, and tested balls.—J.N.G.

Governor

A governor, a kind of feedback mechanism, is a device that regulates the speed of a prime mover, keeping it constant even as external factors change. There are many types, but the classic form is accessory to a stationary steam engine and consists of a pair of spherical weights called *flyballs*, which are suspended by rods from a spindle linked to the engine by a chain or belt: The faster the speed of the engine, the faster the spindle rotates and the further towards horizontal the weights are displaced by centrifugal force. The rods attached to the weights are linked to a valve that controls the admission of live steam to the engine. Ideally the engine attains an equilibrium; if the load increases and the speed drops, the balls swing inward and the valve admits more steam, while if the load decreases and the speed increases, the balls swing outward and the valve admits less steam, thereby "throttling" the engine.

Historically, the centrifugal flyball governor is attributed to Matthew Boulton, an English millwright, who developed the idea in concert with his partner, the Scotsman James Watt. In the late 18th century, the firm of Boulton and Watt became the foremost manufacturer of the steam engines that powered the Industrial Revolution. The Watt governor was first put into operation in 1788 on an engine that still survives in the collections of the Science Museum in London. Within a short time such devices became a standard component of every steam engine. There was an inherent inefficiency, however, a problem addressed in the mid-19th century by American inventors such as George Corliss. Instead of reducing the steam pressure, Corliss's governors permitted an engine to operate at constant full pressure but varied the point at which the admittance of steam to the cylinder was cut off. Engines equipped with automatic cutoff control were as much as 50 percent more efficient than those in which the governor simply throttled the steamflow. Further improvements to the design of the flyball mechanism itself followed, and by the turn of the century more than 1,000 U.S. patents had been granted for such mechanical speed-governing devices, not even counting so-called shaft governors, which are based on the action of inertia as well as centrifugal force and respond not only to changes in speed but also to the rate of change.

Besides steam engines, governors have been designed to regulate the speed of all other sources of power, from internal combustion (as with airplane propellers) to springs (as with wind-up record players). Variations include the automotive type, which allows the operator full control of the throttle up to a predetermined point but not beyond; the overspeed trip governor, which cuts off the fuel supply to an engine above a certain number of revolutions per minute, thereby preventing the possibility of a runaway should other governing systems fail; and the take-up/let-off mechanism for maintaining constant tension on the warp as it is drawn through a power loom. Not all governors are mechanical; some are hydraulically actuated, others use friction. Whatever their particular mechanism, for 2 centuries governors have been among the most ubiquitous technological devices in everyday industrial life.—R.C.P.

See also CYBERNETICS; PROPELLER, AIRCRAFT; STEAM ENGINE

Gravity

Although people have always been aware of the existence of gravity, many centuries passed before its nature and properties began to be understood. The ancient Greek philosophers attributed the lightness or heaviness of objects to their relative mix of four basic elements: earth, air, fire, and water. Earth was the heaviest element, and fire was the lightest, while air and water occupied an intermediate position. Earth had a "natural motion" downward, whereas fire's natural motion was upward. Consequently, a rock fell to the ground because it was largely earth, whereas smoke rose because fire was its chief constituent. One logical conclusion of this schema was that objects fell (i.e., were attracted to the Earth) at different rates of speed, depending on their relative mix of earth and other elements.

It is commonly believed that this notion was decisively refuted by Galileo (1564–1642) when he dropped from the tower of Pisa two objects with different weights in order to demonstrate that both reached the ground at the same time. Galileo may have performed such an experiment, but he wasn't the first to do so. Among those preceding Galileo was the Flemish engineer, mathematician, and scientist Simon Stevin (1548–1620), who in 1586 used two lead spheres, one 10 times heavier than the other, in order to show that they both hit the ground at the same time when dropped from a height of 9 m (30 ft).

Although he was not the first to show that the weight of an object did not affect the speed at which it fell (provided, of course, that the effects of air resistance are disregarded), Galileo did effect a substantial advance in experimental physics by determining the rate at which a falling body accelerates. Galileo's experimental apparatus consisted of nothing more elaborate than an inclined plane, some balls to roll down it, and a water clock to time their progress. By measuring the distances traveled over given time intervals, Galileo was able to determine that the distances covered by a falling object increase in accordance with their traveling time raised to the second power. Thus, after 2 seconds a ball will have covered 4 times the distance it did in the first second. After 3 seconds, it will have covered 9 times the distance, while after 4 seconds it will have covered 16 times the distance, and so on.

For Galileo and his contemporaries, gravity was a force solely residing in the Earth. An important step forward was Johannes Kepler's (1571–1630) conceptualization of gravity as a force that inhered in *all* bodies and in amounts relative to the size of these bodies. As he put it, "Gravity is the mutual bodily tendency between cognate bodies towards unity or contact . . . so that the Earth draws a stone much more than the stone draws the Earth."

Building on the foundation laid down by Galileo and Kepler, Isaac Newton (1642–1727) was able to show that the orbit of a planet resulted from the interaction of two forces: the planet's momentum along a path perpendicular to the radius of its orbit, and the gravitational attraction between that body and the sun. In the course of developing this idea, Newton formulated the law of universal gravitation in 1687: Two objects will attract each other with a force that is proportional to the product of their masses, and inversely proportional to the square of the distance between them. It can be written algebraically as

$$F = G\frac{M_1M_2}{D^2}$$

where F is the force of gravity, M_1 and M_2 are the masses of the two bodies, and D is the distance between them. G is a constant that is believed to be the same throughout the universe.

The value of G, as well as an empirical demonstration of universal gravitation, was supplied by Henry Cavendish (1731–1810). In 1798, Cavendish devised a highly sensitive torsion balance that consisted of a light rod that hung from a thin wire fastened to its center. Small lead balls were attached to each end of the rod, and the whole assembly was put in a glass cylinder that shielded it from air currents. When two large lead balls were put in close proximity to the two smaller ones, gravitational attraction caused the wire to twist. Cavendish had previously determined the forces required to produce particular amounts of twist, and from this he was able to infer the force of gravitational attraction. Having obtained this, and after putting in the measured values of M_1 and M_2, and D, Cavendish was able to determine the value of G, which is 6.66×10^{-8} when the lengths and masses are measured in centimeters and grams, respectively.

Newton's theories of celestial mechanics are among the greatest scientific achievements of all time. Still, there was something troubling about gravity as it was represented in these theories. To Newton's critics, gravity was being treated as an "occult property" that had the ability to act as a force while lacking any mechanical connection to the objects being acted upon. Newton himself was at a loss to explain what the actual cause of gravitational attraction might be. All that could be said was that it was a useful explanatory concept, even though its true nature remained a mystery: In Newton's own words, ". . . to us it is enough that gravity does really exist, and acts according to the laws which we have explained, and abundantly serves to account for all the motions of the celestial bodies, and of our sea."

It was not until the 20th century that a fresh insight into the nature of gravity emerged. But to accept it required nothing less than a complete reconceptualization of the physical universe. In 1915, Albert Einstein (1879–1955) presented his General Theory of Relativity, which postulated the existence of four-dimensional space-time. In this formulation, gravity was not really a force but a manifestation of the movement of bodies through a curved space-time continuum. Among other things, this theory predicted that light from the stars would be bent as it passed by the sun; this prediction was confirmed in the course of an observation of a solar eclipse in 1919. It also has been hypothesized that some collapsed stars are so dense that light cannot escape from their gravitational fields; for this reason they have been given the name *black holes*.

Einstein's explanation of gravity is not easily grasped. At the same time, it is not necessary to have a deep understanding of the nature of gravity in order to make effective use of technologies that rely on its physical properties. One such technology entails the use of instruments known as *gravimeters*. These instruments take advantage of the fact that different kinds of rock produce slightly different amounts of gravitational attraction. Detection and measurement of these differences allows inferences to be made about rock patterns beneath the surface of the Earth, which in turn can provide important clues regarding the presence of oil deposits.

See also BLACK HOLE; MECHANICS, NEWTONIAN; OIL

AND GAS EXPLORATION; RELATIVITY, GENERAL THEORY OF; RELATIVITY, SPECIAL THEORY OF

Further Reading: George Gamow, *Gravity*, 1962.

Greek Fire

One of the most terrifying weapons of medieval times was Greek fire. It was a petroleum product that discharged from tubes mounted on ships. Making it all the more fearsome, Greek fire burned on water and was thus very hard to extinguish. One of the most dramatic instances of the use of Greek fire occurred in 678, when the defenders of the Byzantine capital of Constantinople successfully used it to repel an attack by an Arab fleet. It had been brought to Constantinople by a Greek named Kallinikos, hence the reference to Greek. Although the exact composition of Greek fire remains a mystery to this day, it is likely that its main ingredient was naphtha (a distillate of petroleum), with the addition of quicklime to give it the ability to burn on water. The exact composition of Greek fire was less significant than the manner in which it was deployed. Directing Greek fire at an enemy required a sophisticated collection of pumps and siphons, and it is likely that the mixture was heated and pressurized before being discharged. It then passed through a nozzle, where it was ignited by a lamp of some sort.

Greek fire was thus more than a substance; it was a complex system that required many specific skills and techniques. At the same time, very few people had a comprehensive understanding of the system as a whole; in all likelihood only the emperor of Byzantium and a few close officials knew everything there was to know about it. This limited the danger that Greek fire might be obtained by the enemies of the Byzantine Empire, but it ultimately led to the loss of the technology. The history of the empire was one of endless coups, murders, and violent successions to the throne. Under these circumstances, it was virtually impossible to preserve the complex and multifaceted knowledge necessary for the use of Greek fire, and it is possible that it was lost as early as the 10th century. Military accounts continued to make mention of incendiary weapons identified as Greek fire, but these were pale imitations of the real thing.

Even when the emperor still had the capability of using Greek fire, he often refrained from doing so. In part, this was due to the nature of naval combat in those times. Many naval battles were fought by oar-driven galleys that tried to sink their opponents by ramming. Failing this, attackers might attempt to board the enemy vessel and engage in combat with the opposing crew. Under these circumstances, Greek fire was of limited offensive value, for an enemy ship might be able to out-maneuver a fire ship and attack it from behind. Alternatively, it could simply stay out of firing range. For these reasons, Greek fire was much more useful as a defensive weapon than as an offensive one.

The Arabs made some use of Greek fire (or something resembling it), but it never was a major weapon in their arsenal. The limited use of Greek fire may have reflected the inability of the Arabs to learn how to use the technology originally developed by the Byzantines. Arabs and others were presented with few opportunities to gain a first-hand knowledge of Greek fire, for the Byzantines rarely used it in battle. This limited use of a valuable weapon seems to be the result of a desire to keep it secret. Every use of Greek fire in combat presented the danger that it would fall into enemy hands, a potentially greater setback than the loss of a single battle. Under these circumstances maintaining monopoly control over a "secret weapon" was of greater importance than gaining a tactical advantage. As sometimes happens with advanced military technologies, the desire to keep Greek fire secret often prevented its deployment, thus negating any military benefits it might have conferred.

Further Reading: Alex Roland, "Secrecy, Technology, and War: Greek Fire and the Defense of Byzantium, 678–1204," *Technology and Culture,* vol. 33, no. 4 (Oct. 1992): 655–79.

Green Revolution

Rapid population growth, one of the main trends of recent history, has put a great deal of pressure on the world's agricultural systems. In the 1950s and 1960s, there were numerous prophecies that famine would soon be stalking the Earth. Famines did occur in many parts of the world, but they were largely the result of political and military disturbances. Hunger has by no means been eradicated—malnourished people can be found even in the wealthiest nations—but the apocalyptic vision of mass starvation never came to pass. In most parts of the world, agriculture has been able to keep up with and even surpass population growth due to the changes in agricultural practices that comprise the Green Revolution.

Like the agricultural revolution of the preceding century, the Green Revolution substantially increased agricultural productivity through technological change. At the heart of the Green Revolution was the development of high-yielding strains of wheat and rice. Through the planting of these new varieties it became possible to double and even triple grain harvests. Just as the planting of hybrid corn brought massive increases in yields in the United States, the cultivation of improved strains of wheat and rice resulted in abundant harvests for farmers in Asia, South and Central America, and some parts of Africa.

The development of high-yielding varieties of wheat, for which Norman Borlaug (1914–) received the 1970 Nobel Peace Prize, was the culmination of breeding programs that originated in Japan early in the 20th century. One of the key characteristics of the new varieties of wheat was the shortness of their stalks. Traditional varieties of wheat were limited in the number of grains that their heads contained, for if the head got too heavy the stalks would fall down and the grain would be lost before it ripened, a condition known as *lodging*. In contrast, short-stalked varieties could grow large heads and still remain upright. In 1953, Borlaug began to work with varieties of dwarf wheat that were being cultivated in Japan and the Pacific Northwest. Working in Mexico with the financial support of the Rockefeller Foundation, Borlaug and his fellow researchers bred the short-stalked wheat with many other wheat strains. By 1963, they had produced varieties of spring wheat (i.e., wheat that is planted in the spring and harvested in the summer) that matured quickly, were resistant to rust (a common fungal disease), and were suited to warm climates. In addition, these strains could be planted in the fall and harvested in the spring, making possible the cultivation of two crops a year. Adoption of the wheat quickly ensued; within 10 years, one-third of India's wheat acreage and one-half of Pakistan's acreage were planted with the new strains.

At about the time that Borlaug and his associates were working on wheat, researchers at the International Rice Research Institute (IRRI) in the Philippines were creating high-yield strains of rice. Beginning with strains first created in Indonesia and Taiwan, the researchers had developed several new varieties by the early 1970s. These strains of rice were quick to mature and, like the improved wheat varieties, they had short stalks that resisted lodging.

Development of the new varieties was an important beginning, but simply planting the new strains of wheat and rice did not necessarily result in increased yields. For one thing, the potential gains offered by these "miracle" strains could be fully realized only when they received adequate amounts of water and fertilizer. Indeed, one of the advantages that the improved varieties had over traditional strains was that they were more responsive to the application of water and fertilizer. Rice in particular had to be irrigated at specific times in its growing cycle; otherwise, much of the potential improvement of the new strains would not be realized. Yet water is a resource that cannot be taken for granted in many parts of the world, and meeting the demands of new strains has required substantial improvements in water control, storage, and distribution.

In similar fashion, planting fields with improved strains of wheat and rice produces only marginal gains unless the process is accompanied by substantial increases in the application of fertilizer. Some of this fertilizer may come from traditional organic sources, but these are necessarily limited. The main source of plant nutrients will have to be chemical fertilizer, the production of which entails substantial costs in money and energy.

Finally, growing more wheat and rice may not significantly improve food supplies if pests destroy a large portion of the crop before it reaches consumers. Consequently, increased use of pesticides is an essential part of the Green Revolution. At the same time, however, the production and use of pesticides has a number of unfortunate economic and environmental consequences that have to be taken into account when the benefits of the Green Revolution are assessed.

In sum, the Green Revolution is not simply a matter of using improved varieties of rice and wheat. Rather, it requires a number of interrelated changes in agricultural practices. In the absence of these changes, the use of improved varieties will produce some gains, but their full potential will not be realized. For example, the early use of high-yield rice on some farms in the Philippines resulted in harvests that were only half again as large as harvests obtained with traditional varieties, whereas experimental plots at IRRI had produced yields that were 3.5 to 4 times larger. The shortfall was due to inadequate supplies of water, insufficient amounts fertilizer, and the spread of plant diseases and insect infestations that were not checked by the application of pesticides.

The greater need of high-yield varieties for water, fertilizer, and pesticides has puts these improved strains of rice or wheat out of the reach of many farmers. Moreover, the introduction of new crops always entails some risk, and farmers who are close to the margin of survival are understandably reluctant to take chances. In contrast, farmers who are reasonably well-off can afford to be innovative. For these reasons, in many places it has been the richer farmers who have been the prime beneficiaries of the Green Revolution. As a result, economic inequalities, already substantial in many rural areas, have sometimes deepened.

In extreme cases, the adoption of high-yield varieties has resulted in poor farmers being pushed off the land they farm. This has occurred because increased production has made agricultural land more valuable, so it is bought or otherwise acquired by profit-minded rural entrepreneurs. Where sharecropping arrangements have prevailed, increased yields may not be equally shared. Rather than receiving their traditional portion of the crop—typically half the harvest—sharecroppers may get the same amount of grain they had received in the past. The increased output made possible by improved crop varieties is appropriated by the landowners.

To be sure, the use of high-yielding varieties of wheat and rice has not always been accompanied by the exploitative situations just described. In fact, no easy generalizations can be made about the social impact of the Green Revolution, for a great deal depends on the social structures and relationships that existed before the Green Revolution came to a particular place. In any event, there is no inherent reason that the Green Revolution cannot serve the needs of small farmers. Although there are some economies of scale in many types of farming, a small family-run farm can be highly productive. Improved strains of wheat or rice can make a major contribution to productivity increases, but only if farmers have access to adequate supplies of water, fertilizer, pesticides, and agricultural implements. Small farmers are able to acquire these essential inputs only if they have adequate supplies of capital. This means that government and private banks have to be willing and able to supply funds to farmers at reasonable rates of interest. Effective use of new crop varieties also requires the provision of adequate training and extension services. In the absence of these essential services, the promise of the Green Revolution will not be completely fulfilled.

See also AGRICULTURAL REVOLUTIONS IN INDUSTRIAL AMERICA; CORN, HYBRID; FERTILIZER, CHEMICAL; INSECTICIDES; NOBEL PRIZE

Further Reading: Maarten J. Crispeels and David Sadava, *Plants, Food, and People*, 1977.

Guillotine

A guillotine is an execution device that kills by decapitation. Its basic design is simple, consisting of a weighted blade attached to a crossbeam that slides between two grooved vertical beams. The blade is mounted at an angle, so that when it drops it severs the head cleanly. Although gruesome in its results, the guillotine has the advantage of ending life almost instantaneously. Other methods of decapitation could produce the same result, but much depended on the skills of the executioner, and many beheadings by ax or sword were bungled, resulting in needless suffering.

Devices similar to the guillotine had been used in a number of European countries, but credit for its invention is given to Joseph-Ignace Guillotin (b. 1738). Guillotin in fact did not invent the device that bears his name. Trained as a physician, Dr. Guillotin was elected to the National Assembly in 1789, the year that marks the beginning of the French Revolution. While serving in the assembly, he was instrumental in the passage of a law that required capital punishment to be done "by means of a machine." By getting this law passed, Guillotin hoped that execution would become more humane. The first instrument was actually designed by Dr. Antoine Louis of the French Surgical Academy; for this reason the guillotine was originally known as a "Louisette." The actual construction was done by a German piano maker named Tobias Schmidt.

After experiments with cadavers, the guillotine was first used for an execution on Apr. 25, 1792. Shortly after, the guillotine came to be associated with the phase of the French Revolution known as the Reign of Terror (1793–1794). During that period, several thousands of French men and women were guillotined, many of them executed for being on the wrong political side. The most illustrious victim of that period was the great chemist Antoine Lavoisier (1743–1794). The execution of Lavoisier is sometimes depicted as anti-elitist insanity—the arresting officer is said to have declared that "the revolution has no need for savants"—but in fact, Lavoisier lost his life because prior to the revolution he had been involved in the collection of taxes through a system of tax farming known as the *Ferme genérale*. Moreover, Lavoisier earlier had run afoul of Jean-Paul Marat (1743–1793) when he successfully blocked the latter's membership in the French Academy of Sciences. Marat had gone on to be one of the most important leaders of the revolution, and he used his position to convict Lavoisier on a number of absurd charges. Lavoisier was executed on May 8, 1794. As the great French mathematician and astronomer, Joseph Louis, Comte de Lagrange (1736–1813) later said, "A moment was all that was necessary to strike off his head, and probably a hundred years will not be sufficient to produce another like it."

The guillotine remained the standard instrument of execution in France during much of the 20th century, although only eight executions occurred between 1965 and 1977, the last year that the guillotine was used. In 1981, France abolished capital punishment, and the guillotine's blade fell no more.

Guitar, Electric

The electric guitar is the most recent descendent of a long line of stringed instruments. The earliest stringed instruments were lyres, which, according to archeological evidence, were first played at least 4,500 years ago. The lyre was followed by the harp, a more versatile instrument because of its greater number of strings, all of them tuned to a fixed scale. Some makers of musical instruments sought to increase the capabilities of the harp by equipping it with devices to change the tension of the strings. One example, built in 1720 by a Bavarian musician, had seven pedals; when depressed, the pedals raised by a semitone the strings to which they were attached. A more direct means of changing the pitch of a stringed instrument is a fingerboard. One early example of a fingerboard-equipped instrument is the lute. This instrument probably originated in Persia (present-day Iran) and gained considerable popularity during the Renaissance. The lute was a difficult instrument to play, however, and was largely supplanted by one of its offspring, the guitar. The guitar appeared in a variety of body shapes until the present configuration emerged in the mid-19th century.

The great popularity of the guitar in the 20th century owes much to the blues, the music of the Mississippi Delta that made its way north in the 1920s. The blues originated with Southern blacks and subsequently exerted great influence on white musicians. While the blues and its musical offshoots originated as the music of poor people, the music spread to a wider audience through two key inventions, the phonograph and the radio. Music performed on radio and captured on records exposed African-American blues and white country music to listeners in all regions and social classes, highlighting the guitar in the process. The guitar also began to take its place among the jazz bands of the 1920s and 30s, usually for rhythm accompaniment rather than as a solo instrument.

Blues and country music, either live or recorded, were also featured in clubs and roadhouses, where the prevailing level of noise could easily drown out the sounds of an acoustic guitar. At the same time, the guitar was easily lost amidst the brass instruments featured by the jazz bands of that era; the guitar needed some form of amplification in order to musically assert itself.

Efforts to amplify the guitar's sound began with three emigrants from Czechoslovakia: John, Rudy, and Ed Dopyera. In 1929, they received a patent for a guitar equipped with a metal resonator; it was called a "Dobro," short for Dopyera brothers. The brothers also fashioned a guitar with an electric pickup, as did Lloyd Loar, an employee of Gibson, an important manufacturer of guitars. These early efforts were not commercially successful. In 1931, Adolph Rickenbacker, a former associate of the Dopyera brothers, introduced an electric guitar that used a pickup that converted the string's vibrations into electromagnetic impulses that were then amplified and converted back into sounds. Due to patent problems, the "frying pan" as it came to be known, was not marketed until 1937. Another name that is closely associated with the electric guitar is Fender. In 1943, Clarence L. (Leo) Fender and his partner Doc Kauffman filed a patent application for a lap steel guitar with an electromagnetic pickup. In 1950, Fender introduced the Telecaster solid-body guitar. Distinguished by its affordability and bright tone, the "Tele" enjoyed great commercial and artistic success in subsequent years. Also important in the early history of the electric guitar was Les Paul, who, in addition to being an outstanding performer and musical innovator, built a solid-body electric guitar in 1940. Gibson's Les Paul Gold Top model, introduced in 1952, soon became a top seller.

Electric guitars do more than simply amplify the sound of vibrating strings. Electric guitars produce a distinctive sound because they are usually fitted with multiple pickups that alter the timbre of a note or chord by changing the mix of harmonics in the tone spectrum. Moreover, the sound can be electronically manipulated through the use of vibrato, reverb, and other electronic features. Electric guitars also can have a visual impact as well as an aural one; the guitar body does not contribute to the sound of the instrument, allowing it to be fashioned in a variety of expressionistic shapes.

Between the late 1930s and the 1950s, electric guitar pioneers like Les Paul, Charlie Christian, and Muddy Waters demonstrated the vast musical capabilities of the electric guitar. In the mid-1950s rock-and-roll rapidly emerged as a distinct musical form, and from Buddy Holly and Chuck Berry to Jimi Hendricks and Jeff Beck, electric guitars were the essential instruments of virtually every rock group. Adherents of nonamplified music were not favorably impressed by the electric guitar and the music it made. Bob Dylan was roundly booed when he appeared on stage with one at the Newport Folk Festival in 1965. Since that time, however, the electric guitar has taken on the character of a true folk instrument, no matter if it is played by Eric Clapton or the lead guitarist of a local garage band.

A classic 1952 Fender Telecaster (courtesy Fender Musical Instrument Corp.).

See also JUKEBOX; PHONOGRAPH; RADIO

Gunpowder

Gunpowder, an explosive mixture used to impel projectiles from firearms, has three ingredients: sulfur, charcoal, and saltpeter (sometimes called nitre). Saltpeter, the essential constituent of gunpowder, is the common term for potassium nitrate (KNO_3), which was usually made by treating potash with nitric acid. It must be purified and all three ingredients thoroughly mixed within the limits of certain proportions: 62 to

75 percent saltpeter, 10 to 20 percent sulfur, 5 to 12 percent charcoal. Ignition produces an explosion, because gunpowder contains sufficient oxygen for its own combustion. The explosion liberates gases that instantly occupy several hundred times as much space as the powder and thus will rapidly expel a projectile from the barrel of a gun.

Incendiary mixtures of sulfur, oils, pitch, turpentine, and tar had been used in warfare ever since the destructive potential of fire was first noted. Historians from antiquity such as Thucydides often refer to their use in setting fortifications ablaze as well as producing noxious smoke. Probably the most significant development prior to the advent of gunpowder was the invention in the 7th century of so-called Greek fire, which is attributed to Kallenikos from Heliopolis or Baalbec in Syria. For several centuries afterward, there was continuous experimentation with incendiary mixtures, and this ultimately led to the discovery of gunpowder. The first unequivocal mention of saltpeter or nitre appears in the writings of the Arabian Abd Allah, who was born around 1200, and the earliest known recipe containing all three ingredients of gunpowder is attributed to Marcus Graecus (a nom de plume), in a work entitled *The Book of Fires for Consuming the Enemy.* But there is no suggestion of utilizing this mixture to impel a projectile prior to the writings of Roger Bacon, an Oxford Franciscan friar. As with much extant data concerning the early development of gunpowder, there is no absolute certainty about when Bacon's *Treatise Concerning the Secrets of Science and Nature* initially appeared, though it is usually dated around 1260. Bacon was the first to note the necessity of clarifying saltpeter to render gunpowder effective for impelling projectiles; prior to this time it had been useful only in fireworks and perhaps in rockets.

Unlike the development of the proper explosive mixture, a slow and cumulative process, the military application would have represented a single momentous leap. But in an era marked by constant warfare among cities, principalities, manors, duchies, and mercenary bands, it remains unclear just when that leap took place. There is no solid evidence that firearms were known in Europe prior to the 1320s, when the first existing representation of a gun appears in a Latin illumination. Within 2 decades after that, however, the use of firearms was apparently commonplace in sieges and battles, and by the beginning of the 15th century gunpowder was also being used to mine the walls of enemy fortifications. In the years that followed, the effects of gunpowder became so significant that they produced nothing less than a revolution in military strategy and tactics.

A persistent misconception concerns the origins of gunpowder in Asia. In his 1614 treatise on artillery, Diego Ufano cited reports from Portuguese merchants in China after 1517 who referred to cannon; he assumed incorrectly that both firearms and gunpowder were of Chinese origin. During the 17th and 18th centuries, Jesuit missionaries in China perpetuated this assumption, and it lasted well into our own time. But it is now generally agreed that both technologies were of Western origin, even though the Chinese may have possessed various incendiary compounds and pyrotechnic devices at an early date, as did the Byzantines.

All gunpowder prior to the 19th century was so-called *black powder*, which releases its gases at once, emitting a great cloud of smoke in the process. The advantage of "smokeless" powder is that it releases a greater volume of gas at a more controlled rate and can thus impel a projectile further, or impel one that is heavier. While this advantage was understood by the 1840s, it took another 4 decades for smokeless powder to begin to displace black powder to any substantial degree. Chemically, smokeless powder is altogether distinct from black powder. It is composed of nitrocellulose (a nitric ester of cellulose, a primary constituent of plants), either by itself or in combination with nitroglycerin. The former is more stable and less corrosive, the latter more easily ignited and less costly. Either way, it is colloided, that is, turned into a gelatinous substance in the form of grains, flakes, or rope (the latter called *cordite*). Today, smokeless powder has almost completely superseded the classic sulfur/charcoal/saltpeter explosive mixture for use in military weaponry and sporting arms, though the latter is still employed in blasting power and fireworks.—R.C.P.

See also EXPLOSIVES; GREEK FIRE; MILITARY REVOLUTION IN EARLY MODERN EUROPE

Gyroscope

The word *gyroscope* is derived from two Greek words, *gyros* (meaning "revolution" or "turn") and *skopein* (meaning "to view"). The device itself consists of a wheel set up in such a way that it can revolve with three angular degrees of freedom—that is, in any direction—around a fixed point at its center of gravity. This is accomplished by mounting the wheel inside a gimbal ring on one axis, which is mounted inside another ring on another axis that lies at right angles to the rotation of the wheel, the latter ring likewise being free to revolve inside a support frame on an axis that lies at right angles to the axis of the inner ring.

A gyroscope embodies two fundamental characteristics. The first is called *gyroscopic inertia*, pertaining to the tendency of a rotating body to preserve its plane of rotation—in other words, not to veer off in a different direction. The second characteristic is called *precession*, which is a little more difficult to understand. When a gyroscope is subjected to force on an axis that lies at right angles to its axis of rotation, it resists that force proportionally to the velocity with which it turns around a third axis called the *axis of precession*. Assuming no inertia or friction about the precessional axis, the rate of precession will equal the applied force. No movement induced by that force can ensue until the gyroscope's plane of rotation coincides with the plane of the applied force, at which point all resistance to the applied force will cease.

Experiments with revolving spheroids mounted inside concentric rings reportedly occurred as early as the beginning of the 19th century, and in 1836 a Scotsman named Edward Lang suggested to the Royal Scottish Society of Arts that a similar device could be used to verify that the Earth rotates on its own axis. At midcentury, the French physicist Jean Bernard Léon Foucault (1819–1868) actually demonstrated this by means of a heavy bob suspended by a wire attached to a ball-and-socket joint so that it rotated with complete angular freedom. In the late 1870s, an American named G. M. Hopkins devised a means of applying electric power to such a pendulum so that it would not rapidly lose momentum. In the 20th century, so-called Foucault pendulums have become a traditional centerpiece for science and technology museums, typically with a spike on the underside of the bob that knocks down a circular arrangement of wooden pegs, one by one. So, too, simple wooden or plastic "spinning tops" operating on the principle of gyroscopic inertia have become one of the most prosaic of playthings.

Intriguing demonstrations and toys aside, the gyroscope also has practical utility. This inheres in its ability to provide a baseline that cannot be disturbed by gravity or acceleration; such a baseline may be used either as a reference point around one or more axes or as a directional reference. In the early 20th century, a prolific American inventor named Elmer Sperry (1860–1930) began to develop the commercial potentialities of the gyroscope with respect to navigation. Naval forces stood in dire need of a compass that would not be disturbed by the magnetism of a steel ship's own hull, and following the outbreak of World War I, the Sperry Gyroscope Company found a ready market for its gyrocompass. This was an instrument combining the inertia and precession inherent in a rotating iron wheel (typically 25.4 cm [10 in.] in diameter, weighing 25 kg [55 lb], and turning at 6,000 revolutions per minute by means of a self-contained induction motor) with two constants, gravity and the rotation of the Earth. By invariably aligning itself with the geographic meridian, a gyrocompass indicates true north irrespective of a vessel's pitching, rolling, and yawing. Soon, merchant vessels were being outfitted with gyrocompasses as well, and gyroscopic principles were being employed in other devices for maintaining directional control, such as torpedoes and airplanes.

Aboard ship, a master gyrocompass transmits readings electrically to the helm and other stations, while simultaneously impelling two devices, one of which records navigational headings in a continuous graphic, the other a gyropilot for steering automatically. In 1918, Sperry introduced a variant of the gyrocompass, the aircraft turn indicator. Along with another gyroscopic device called the artificial horizon, which indicates whether a plane is banking, ascending, descending, or flying level, the turn indicator enabled the amazing feat of "flying on instruments." Combined with a directional gyro and another gyro designed to indicate a plane's attitude, these are the devices that comprise an automatic pilot, which first was used to relieve pilots of the tedium of keeping to course, but now typically flies a commercial airliner at all times, even during landings and takeoffs.

While gyroscopic devices are most commonly found in aircraft, they also have many other applications in realms that range from ascertaining the proper trajectory of artillery shells to tracing the angle of oil-well casings to recording potentially dangerous deviations in the alignment of railroad tracks. Yet very few complex technological contrivances have such a simple everyday counterpart to demonstrate a basic principle of its operation as the gyroscope has in the ordinary toy top.—R.C.P.

See also COMPASS, MAGNETIC; INSTRUMENT FLYING; SERVOMECHANISM

Haber Process

Few technologies are as directly connected to the maintenance of human life as are chemical fertilizers. Although organic farming has its merits, the feeding of the world's growing population is highly dependent on synthetic nitrogen, potassium, and phosphorous. At the beginning of the 20th century, naturally occurring supplies of nitrogenous fertilizer were being used up, resulting in growing concern about the possibility of eventual famine.

The fixation of atmospheric nitrogen offered a possible solution. Henry Cavendish (1731–1810) had discovered in 1784, that running a spark through nitrogen and oxygen resulted in the formation of nitrous oxide (NO), from which nitric acid (HNO_3) could be derived. To do so on a large scale, however, required a great amount of electricity to generate the high temperatures necessary for the reaction. In the early 20th century, an electrical arc process was used in Norway, a country endowed with vast resources of hydroelectricity, but it was impractical elsewhere. In Germany, Fritz Haber (1868–1934), inspired by the theories of the French chemist Henri le Chatelier (1850–1936), discovered that small quantities of ammonia (NH_3) could be synthesized at lower temperatures by using an iron catalyst. Walter Nernst (1864–1941), another German chemist, pointed out that much larger quantities could be produced if the pressure was stepped up to 200 atmospheres.

As is often the case with chemical technologies, converting a laboratory reaction to one of commercial size was a difficult undertaking. Five years of development, overseen by Carl Bosch (1874–1940) at the Badische Analin und Soda Fabrik (BASF), was required before the process was ready for full-scale production. Among other things, the synthesis of ammonia required the selection of effective catalysts, the development of a new method of producing hydrogen (passing steam over coke), and the use of new forging techniques to produce a 65-ton pressure converter.

The Haber-Bosch process was first used on a commercial scale in 1913, just in time to provide a substantial boost to Germany's war-making capabilities through the production of large quantities of nitric acid used for explosives. In that year, 7,900 tonnes (8,700 tons) of ammonia were produced; by 1916, production was up to nearly 91,000 tonnes (100,000 tons). Haber won the Nobel Prize for Chemistry in 1918, even though his process undoubtedly prolonged the war, and he had been an important contributor to the development of poison gas. To his credit, Haber left Germany shortly before his death in protest of Nazi policies.

The Haber-Bosch process came to be widely employed outside Germany; by 1950, it was responsible for 75 percent of the world's production of fixed nitrogen. Moreover, the Haber-Bosch process was the prototype for high-pressure chemical synthesis, and it directly stimulated the development of the equipment and techniques necessary for this important mode of chemical production.

See also EXPLOSIVES; FERTILIZERS, CHEMICAL; GAS, POISON; HYDROELECTRIC POWER.

Hallucinogens

Hallucinogens are heterogeneous chemical compounds that have the ability to elicit vivid visual and auditory experiences that have no objectively verifiable basis.

Hallucinogens are broadly classified into three pharmacological categories: adrenergic, serotonergic, and cholinergic. Drugs in these three categories are chemically related to the neurotransmitters norepinephrine, serotonin, and acetylcholine, respectively.

One of the oldest of the adrenergic hallucinogens is mescaline, which was originally isolated from the peyote cactus (*Lophophora williamsii*). The above-ground portion of the cactus, known as the *crown*, is harvested, cut into disks and dried into peyote *buttons*. These buttons, when ingested, frequently cause severe nausea, cramping, and headaches. They also produce vivid visual hallucinations that are often accompanied by "out of body" experiences. Peyote has been used in religious ceremonies by Aztec, Huichal, and other Mesoamerican Indian groups since Precolumbian times. Aztec codices describe the ceremonial use of peyote as an offering to the gods. More recently, peyote has been incorporated as a sacrament into the religious practices of the Native American Church of North America. While the federal government has thus far refused to interfere with the sacramental use of peyote, the Supreme Court has recently ruled that individual states are free to do so. The peyote cactus was the first of the hallucinogenic plants studied by organic chemists. Mescaline was first isolated as the active component of peyote by Lewin and Hefter in 1896, but it was not until 23 years later that Spath characterized mescaline as 3,4,5-trimethoxyphenylethylamine.

In 1964, Dow Chemical Company synthesized a second adrenergic hallucinogen, DOM (2,5 dimethoxy-4-methyl-amphetamine). The formula for DOM was published in 1967, and it very quickly became a popular street drug known as STP. DOM is approximately 100 times as potent asmescaline and is frequently referred to as a *megahallucinogen*.

Serotonergic hallucinogens also have a long history of incorporation into the ceremonial practices of Mesoamerican peoples. In particular, the gilled mushroom *Psilocybe mexicana*, known as *teonanacatyl* ("god's flesh") by the Aztec, is a very potent hallucinogen. Representations of these mushrooms in early temple carvings document their use in religious ceremonies by at least 500 B.C.E. The active ingredient of these mushrooms was first identified by Albert Hoffman in the late 50s. In the process of chemically isolating the psychoactive agent, psilocybin, Hoffman and other volunteers ingested tiny quantities of various extracts of the mushroom. Hoffman was able to identify psilocybin as N,N-Dimethyl-4-phosphoryl tryptamine, an indole compound that is similar in structure to serotonin.

Hoffman is even better known for his work on a second class of serotonergic hallucinogens derived from the ergot fungus (*Claviceps purpurea*). This poisonous fungus that infects rye and other grains was responsible for several European plagues during the Middle Ages; 40,000 deaths were said to have occurred during one such plague in 944 C.E. Hallucinations are a common feature of ergot poisoning, along with severe vascular damage associated with prolonged vasoconstriction. As part of a research project designed to explore possible therapeutic uses of ergot derivatives, in 1938 Hoffman first synthesized lysergic acid diethylamide (LSD [the *S* is from the German word for acid, *sauer*]). In 1943, Hoffman accidentally ingested a tiny amount of LSD and graphically described that "fantastic pictures of extraordinary plasticity and intensive color seemed to surge forward to me."

In 1953, J. H. Gaddum reported that LSD inhibits serotonergic effects in the periphery and suggested that its hallucinogenic effect might result from a similar pharmacological action in the central nervous system. This speculation has since been confirmed with the demonstration that LSD is a potent blocker of two serotonin receptor subtypes, the $5HT_{1A}$ and $5HT_2$ receptors.

The final category of hallucinogens seems to act by enhancing the effects of acetylcholine neurons. Two such compounds are prominent in the pharmacological literature. The first compound, muscarine, is isolated from the fly agaric mushroom (*Amanita muscaria*), which is found in the northern European latitudes in Siberia and Scandinavia. Muscarine directly stimulates one class of acetylcholine receptors appropriately named *muscarinic* receptors. Ingestion of the agaric mushroom has been chronicled for at least 300 years. Agaric usage is accompanied by extreme physical agitation, and remarkable feats have been attributed to its users. The actions of agaric users are often described as frenzied or raving; the crazed fury of Norse "berserkers" has often been attributed to their ingestion of agaric mushrooms. Following this initial period of arousal and agitation, the user will often lapse into slumber accompanied by hallucinatory visions.

A second cholinergic hallucinogen, physostigmine, appears to inhibit the breakdown of actylcholine by blocking the enzyme acetylcholine esterase. This compound is extracted from the Calabar bean (*Physostigma venenosum*) found in the coastal areas of West Africa. In low doses, physostigmine's cholinergic properties have proved useful in treating glaucoma and myasthenia gravis.—A.J.

See also ENZYMES; NEURON

Harbors

A harbor is a body of water that provides shelter for ships and smaller vessels. Harbors can be natural or artificial. Most are natural and are located along the coastlines of oceans, bays, fjords, coves, and estuaries. Although usable without any engineering improvements, sometimes they are improved to accommodate more or larger ships or to provide greater protection. Periodic dredging usually is required to ensure that the ship channels are deep enough to provide access for all the ships using the harbor.

Artificial harbors employ structures like jetties and breakwaters to provide protection for ships. Also, they usually require more dredging than natural harbors. Artificial harbors can be traced back nearly 4,000 years. Sometime around 1900–1800 B.C.E., Cretans constructed a breakwater near the island of Pharos, close to where Alexandria, Egypt, later rose. Basically a rubble wall arising out of the water, the 2,000-m-long (6,650-ft) breakwater protected the harbor from strong currents and storms. Later, Greek engineers connected the island to the shore by erecting a 1,280-m (4,200-ft) breakwater (also called a *mole*). Drift, debris, and sediment piled up along the breakwater and enlarged it, allowing Alexandria to extend to the island.

In the period of classical Greek and Roman civilization, ships rarely traveled at night, so numerous harbors were necessary around the Eastern Mediterranean and southern and western coasts of the Black Sea. Artificial harbors often consisted of nothing more than crude breakwaters, but at important shipping points and political centers, engineers erected extraordinarily impressive and costly harbors. At Piraeus, the port of ancient Athens, Greek engineers built a harbor big enough for 400 triremes (galleys with three tiers of oars on each side). Continuing this monumental tradition, Roman engineers built a port at Ostia, at the mouth of the Tiber River, that essentially consisted of two artificial harbors. The outer one, consisting of two huge curving moles, provided covered docks for galleys numbering into the hundreds. The inner harbor served as a commercial port. Over the next centuries, however, silt rendered the harbors useless.

In terms of construction, perhaps the most impressive Roman harbor was Caesarea in ancient Israel on the eastern coast of the Mediterranean. This was King Herod's project, and he spared little expense. His engineers constructed the harbor of huge stone blocks; the foundations began some 37 m (121 ft) below sea level. The engineers also utilized an early type of concrete: aggregate bonded together with lime and *pozzalana* (volcanic ash from Italy). Two huge breakwaters extended seawards on the north and south. From the southern breakwater, another mole extended northwards, nearly enclosing the harbor. At its end was a huge lighthouse. On top of the breakwaters were statues, warehouses, and a harbor master's house.

In late medieval times, the size of ships increased tremendously. Larger ships necessitated larger harbors, as well as more drydocks. In ancient times, most ships could be repaired by purposely beaching them; once repaired, they could be maneuvered back into the water at high tide. Clearly, this was impossible with ships displacing hundreds of tons. In 1496, the English completed a permanent drydock at Portsmouth. Though built of timber and masonry, and therefore as permanent as one could expect at the time, this drydock and others that followed were somewhat misnamed. Without the benefit of steam pumps to keep water out—a development that did not occur for another 300 years—these docks were invariably wet.

Military and economic needs most commonly drive the creation or improvement of harbors. To compete with the navy of England's Henry VIII, France's Francois I decided to build a harbor at the mouth of the Seine River. Work began at Le Havre in 1516 and was not completed until the century's end. Le Havre's harbor included extensive stone jetties that were laid in cement and employed iron clamps. The engineers constructed a canal that turned the Lezarde River away from Harfleur and into Le Havre harbor. Where the canal entered the harbor, the engineers constructed gates that could be closed at high tide and opened at low tide to let the rushing water scour the bottom of the harbor.

In the 17th century, the famous French military engineer Sébastian le Prestre de Vauban (1633–1707) built the harbor at Dunkirk. It consisted of two parallel jetties that confined the flow of streams entering the ocean at Dunkirk. As at Le Havre, the stream flow helped scour the harbor and prevent siltation. However, Vauban added an interesting variation. He constructed storage reservoirs that released water into the stream to aid the scouring action when the natural flow was insufficient.

As early as 1679, Vauban had advanced plans for a harbor at Cherbourg, but another century passed before construction began. Begun in 1781, construction of the great dike or breakwater at Cherbourg severely challenged the engineers. Attempts to build the breakwater by sinking boats loaded with stone failed when the sea quickly tore apart the hulks, scattering the stone. The next attempt used 90 huge stone-filled timber cones that were intended to dissipate wave force. However, this effort also was a failure. Finally, large blocks with sloping sides were used to good effect. The laborious, accident-filled work was not completed until 1858. Although impressive, the breakwater required constant maintenance.

In response to the construction at Cherbourg, the English began enlarging and protecting Plymouth Harbor in 1812. Completed in 1847, the breakwater was one mole, whose center was 653 m (2,142 ft) in length and had arms 293 m (960 ft) in length. Inevitable comparisons arose between the two harbors. The English argued that their breakwater at Plymouth was more economical and its construction better managed. In rebuttal, some French engineers pointed to the greater stability and permanence of the Cherbourg structures (albeit, with periodic repairs). The controversy was a typical engineering (and political) issue, pitting cost against durability.

Joining Cherbourg and Plymouth among the most ambitious harbor projects of the time was Cape Henlopen on the Delaware River. This harbor was designed to improve access to the port at Philadelphia further upriver, especially during the icy winter months. Funded by the federal government, conceived by the Army Corps of Engineers, managed by the Army Quartermaster General, and built by civilian contractors, the project was loosely based on the design of Cherbourg Harbor. The cost of the breakwater far exceeded the original estimate, and construction took around 70 years rather than the few years originally projected. Still the harbor project was a failure. Shoals formed behind it, and ice still blocked the entrance to Philadelphia. Although a magnificent edifice, the breakwater did nothing to improve passage for oceangoing vessels.

Increasingly ambitious designs marked early 20th-century harbor construction. Jetties, piers, and breakwaters grew in size, and occasionally engineers built viaducts connecting shores with offshore landing stages. In moderate conditions, these viaducts also offered some protection for ships and helped reduce sediment accumulation. World War II provided many opportunities for innovation. After the Normandy Invasion in 1944, Anglo-American forces towed large concrete caissons and old ships from England to France, where they were sunk to create breakwaters and piers. Pontoon bridges connected the piers to the land.

In the second half of the 20th century, increased shipping and a larger ships have required deeper harbor depths. Some supertankers require 27-m (90-ft) depths, and large container ships also require deeper harbors. Along the coastlines of the world, where harbors have been developed into large commercial ports, authorities have enlarged and modernized facilities. They have invested in more wharves, terminals, and cranes, and in the equipment needed to transfer cargo from water to land transportation, and vice versa.

Harbors range in size from small enclosures, often primarily for recreational craft, to huge distribution and marketing centers. Whether natural or artificial, their success has directly led to the rise of cities and the shaping of inland shipping routes. The breakwaters, piers, jetties, and dikes that engineers have built to create artificial harbors or to protect natural ones illustrate slow, incremental technological progress and, almost as often, technological failure in the face of nature's forces.—M.R.

See also DREDGING; TANKERS; TRANSPORTATION, INTERMODAL

Hearing Aid

Hearing loss is a common human affliction, and one that becomes more prevalent with advancing years. Concentrating sounds by cupping a hand to the ear helps to some extent, as does the use of an ear trumpet. Over the centuries, a number of devices of this sort were used; perhaps the most elaborate was the "acoustic throne" built for a Portuguese king in 1819. Courtiers spoke into an aperture in a hollow arm, which then conducted the sound into a hearing tube. More portable devices included top hats, walking sticks, and bonnets equipped with sound receptors and hearing tubes.

A major improvement in assisted hearing came with the use of electrical amplification. Early hearing aid technology was closely linked to the development of the telephone. Significantly, Alexander Graham Bell (1847–1922) was a teacher of the deaf when he undertook to invent the device that we now know as the telephone. Telephones and hearing aids operate on similar principles: Sound is converted to electrical voltage by a microphone; the voltage is then amplified, transmitted, and turned back into sound.

Early electrical hearing aids, which date back to the early 1900s, consumed a considerable amount of power and required large batteries. This made for heavy, bulky devices; for a typical example, the 1923 Marconi Otophone weighed 7.3 kg (16 lb). In the 1930s, the use of miniaturized vacuum tubes got the weight down to about 1.8 kg (4 lb), but hearing aids were still cumbersome devices. A major breakthrough in hearing aid design came as a result of the use of transistors, which were smaller, lighter, cheaper, and more reliable than the tubes they replaced. They also drew much less power, eliminating the need for bulky battery packs. The advantages of transistorized hearing aids, which first appeared in 1950, were so evident that they were the first consumer products to use the new solid-state devices. In similar fashion, in the early 1960s hearing aids were among the first products to make use of integrated circuits, which put all the components and circuitry on a single chip.

Today's hearing aids amplify sounds that range from 200 to 6,000 Hz. In addition, they can make sounds—especially the human voice—somewhat clearer and easier to understand. Microelectronic circuits now make it possible to convert normal analog sound into a digitized signal. This signal can then be electronically processed to remove background noise and to change tonal quality so the sound can be more easily heard and understood. Although there are a number of technical advantages to digital hearing aids, in real-life situations the benefits may be modest or even nonexistent.

Hearing impairment is usually the result of abnormalities of the cochlea that come as a result of aging. This kind of hearing loss also can result from prolonged exposure to loud noises, as happens to some industrial workers or habitués of rock concerts. Conventional hearing aids may be of help in mitigating the consequences of coclear abnormalities. Some people who have been deaf from birth require a more drastic intervention, the coclear implant. This entails the surgical embedding of a receiver in the temporal bone under the scalp. The procedure usually allows a completely deaf person to hear sounds for the first time, but a coclear implant is not advised for a person with some hearing capacity, for there is a chance that the operation may destroy hearing altogether. Coclear implants also have been opposed by some deaf parents of deaf children, who prefer that their children continue to live in a deaf culture.

Hearing aids have produced great improvements in the lives of many people. The process of miniaturization has resulted in the production of hearing aids that combine the microphone, amplifier, and earphone in a single unit worn in the ear. Hearing aids are not, however, simple technological fixes that automatically compensate for hearing losses. Many people find hearing aids unsatisfactory because they cannot tolerate the amplified sound. Even a successful fitting usually entails a lengthy period of acclimation and learning before the hearing aid is fully effective. Before a person purchases a hearing aid, he or she should undergo a complete evaluation by a licensed audiologist in order to be sure that the appropriate device is selected. Many people are sold hearing aids that are unsuited to their needs, and as a result end up not being used.

See also AMPLIFIER; INTEGRATED CIRCUIT; MICROPHONE; TELEPHONE; THERMOIONIC (VACUUM) TUBE; TRANSISTOR

Heart, Artificial

Heart disease is the leading cause of death in the industrialized world. Ailments of the heart and circulatory system also dramatically lower the quality of life for many people, rendering them invalids and semi-invalids. During the 1960s, a major breakthrough occurred when it became possible to replace defective hearts with healthy ones obtained from deceased donors. Heart transplants saved the lives of many, but not everyone who needs one can obtain a donated heart. In the United States as many as 75,000 people have conditions that require a heart transplant, but only 2,000 donor hearts are available annually.

With the supply of human hearts necessarily limited, hopes came to rest on a mechanical substitute. Efforts to produce an artificial heart go back at least to the 1930s, but it wasn't until the 1970s that dreams seemed close to realization. By this time the main center of artificial heart research and development efforts was the University of Utah. With generous financial support from the Humana Corporation, a health insurance and private hospital firm, a research team produced the Jarvik-7 artificial heart, named after one of its members. In 1981, the heart was implanted in its first recipient, a Seattle dentist named Barney Clark. The deterioration in Clark's own heart had progressed to the point where death was imminent, but his life subsequent to his receipt of the Jarvik-7 hardly seems preferable. During his 112 days of life after the implant, Clark suffered from seizures, severe nosebleeds, pneumonia, kidney disease, gout, epididymitis, and an intestinal ulcer. Other patients fared little better. One lived for 620 days, but during that time he suffered four strokes and a series of infections that severely eroded his physical and mental capabilities.

These attempts at creating and using an artificial heart made it all too evident that the technology was far from being ripe. Compact, reliable, and safe implantable power supplies did not, and still do not, exit. Nonclotting surfaces for the heart and its associated plumbing have yet to be developed. The tendency of the heart to act as a culture medium for infectious bacteria still remains.

Even if the technical obstacles could be surmounted, many problems would remain. At a cost of at least $160,000 per implant, substantial financial burdens, perhaps totaling as much as $5 billion per year, would be placed on individuals, insurance companies, and the government. Under these circumstances, a technically viable artificial heart program would engender some very difficult choices about the use of limited financial resources. For example, the money spent on an artificial heart program could instead be used for an antismoking education campaign. Smoking is a major contributor to heart diseases; consequently, even a 1 percent reduction in tobacco use would save more lives than a large number of artificial hearts.

Given the inevitable financial constraints, it is possible that not everyone who could benefit from an artificial heart would get one. Some sort of rationing system might have to be set up, using criteria such as the likelihood of the patient's long-term survival. More ominously, it is possible that criteria would also include assessments of the "worth" of the patient, a

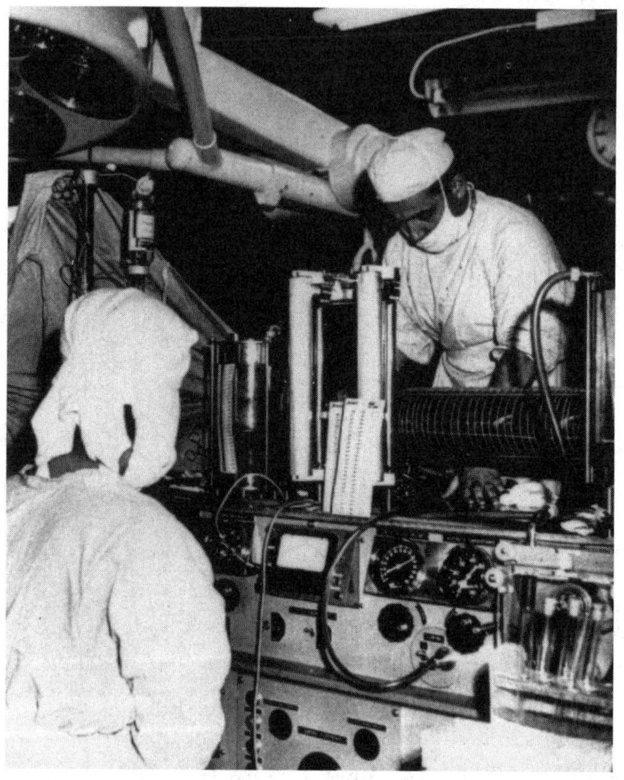

A heart-lung machine used during a surgical operation (courtesy National Library of Medicine).

situation similar to the early days of kidney dialysis. Whatever the criteria invoked, the decision to use an expensive medical technology in a world of finite resources necessitates making decisions. In many cases this can be akin to "playing God," a role that few human beings are eager to assume.

Technical, financial, and ethical considerations aside, efforts to develop an artificial heart are subject to the criticism that they embody the "technological fix" mentality, the desire to use technology as a substitute for other needed changes. Heart disease can be the result of genetic predispositions, but lifestyle choices can also play a large role. Sedentary habits, excessive consumption of salt and fats, and smoking have all been implicated as leading causes of heart disease. A healthy lifestyle does not guarantee immunity to heart disease, but it certainly stacks the odds in the person's favor.

There is nothing inherently wrong with seeking technological solutions for human problems; the danger arises when efforts are mounted to substitute technology for necessary changes. The human behaviors that can lead to heart disease produce many other physical and psychological problems. Should its shortcomings eventually be overcome, the artificial heart may become part of standard medical practice, but like all technological solutions it should not be allowed to generate false hopes and to serve as a substitute for living a healthy life.

See also KIDNEY DIALYSIS; ORGAN TRANSPLANTATION; TECHNOLOGICAL FIX

Heat, Kinetic Theory of

According to the ancient Greeks, heat was caused by the presence of fire. Fire, in turn, was thought to be an actual substance, one of the five basic elements from which everything was made. In the late 18th century, scientists worked under the assumption that heat was caused by the presence of a massless, colorless fluid known as *caloric*. Many things could be explained by this concept, such as the apparent flow of heat from a hot object to a cold one. But caloric could not be invoked to explain other phenomena, most notably the evident fact that friction produces heat. After all, how could caloric fluid suddenly appear when we vigorously rubbed our hands together? Another conundrum had to do with the expansion of gases. According to the caloric theory, when a gas expanded into an evacuated chamber, its temperature had to fall, for the concentration of caloric diminished as the gas occupied a larger space. But experimentation showed that the temperature of the gas remained the same.

A major blow to the caloric theory was struck by Benjamin Thompson, Count Rumford (1753–1814), in 1798. Rumford measured the amount of heat given off when brass cannon barrels were being bored and concluded that the caloric theory implied the removal of vast quantities of the substance, enough to melt the brass if the caloric could somehow be poured back in. Moreover, substantial amounts of heat were produced even when a drill was so dull that no metal was removed. In response, adherents of the caloric theory speculated that the particles of caloric contained in a material had to be far greater in number than the particles released in the form of heat. Although the notion of vast numbers of massless (or nearly massless) particles residing in every material stretched credulity, there were no better theories at hand.

Adherents of the caloric theory were aware that heat could do work, a phenomenon that could be explained in terms of the flow caloric (much as the flow of a stream could do work by turning a waterwheel). But if the reverse were true—if work could produce heat—this would mean that new caloric fluid was being created, something that no calorist was willing to believe. Crucial insights into this issue were provided by James Prescott Joule (1818–1889) in the 1840s. Joule lived and worked in the area around Manchester, England, a major center of industry. In this setting, Joule observed the efforts of engineers to deal with friction; as they put it, the work required to overcome friction was "annihilated power." This inspired Joule to think about how heat and work might be manifestations of a more fundamental property, what today we would identify as *energy*. In order to learn more about the connection of heat and work, he designed a classic scientific experiment. A baffled cylinder was filled with water and equipped with a thermometer. When a paddlewheel inside the cylinder churned the water, a slight rise in the water's temperature showed that the work that turned the paddlewheel had been converted into heat. Joule even was able to measure exactly how much work was required to produce a given amount of heat by using falling weights to turn the paddle; the amount of work could be calculated by determining the mass of the weights and the distance they fell.

In addition to showing that work can be converted to heat, Joule's experiments showed that heat is not conserved, which would be the case if it were a substance like caloric. But if caloric could no longer explain the presence of heat, what could? The answer lay in the modern atomic theory that began with John Dalton. By the mid-19th century, it had become evident that heat was the result of molecular motion, hence the expression "kinetic theory of heat." According to the kinetic theory, the molecules of a gas are constantly in motion. This motion is the source of heat, while temperature is a measure of the average kinetic energy of all the molecules. The kinetic theory also explained why the pressure exerted by a gas increased with its temperature. As molecules heated up, their motion increased, and their more frequent collisions with the walls of a containing vessel were manifested as the increased pressure exerted on the walls. At the other extreme, when a substance is in a solid phase (ice, for example), the molecules are not free to move about, although they do vibrate while remaining in the same place, giving off some heat in the process. If the temperature were brought down to absolute zero or 0 Kelvin (–273°C), molecular motion would cease altogether. This is purely hypothetical, for while absolute zero has been approached under laboratory conditions, it has never been achieved.

Support for the kinetic theory of heat was provided by Albert Einstein (1879–1955), who in 1905 showed that the random motion of very small particles suspended in a liquid (Brownian motion) occurs because molecules of the liquid continually slam into the particles. Because each moving molecule differs from the other molecules in the magnitude and direction of its motion, the particle being bombarded will exhibit the irregular movements characteristic of Brownian motion.

See also ATOMIC THEORY; CONSERVATION OF ENERGY; THERMOMETER

Heisenberg Uncertainty Principle

In March 1927, Werner Heisenberg (1901–1976), a German physicist who had made important contributions to quantum mechanics, published a paper "On the Intuitive Contents of Quantum-Theoretic Kinematics and Mechanics," in which he introduced the uncertainty principle, or principle of indeterminacy. One of the mathematical consequences of matrix mechanics was that there were pairs of variables, called *conjugate variables*, q and p such that $qp - pq = h/2(\text{pi})i$, where h is Planck's constant ($h = 6.6 \times 10^{-27}$ erg.seconds), and i is the square root of –1. The big question was: How could this relationship be given an interpretation that made physical sense? Heisenberg showed that if dq and dp are the uncertainties in the physical measurements of p and q, respectively, given any device to determine p and q, then the relationship between the conjugate variables was equivalent to the claim that the product of dp and dq must always be equal to or greater than Planck's constant divided by 2(pi). That is, even a theoretically perfect measuring instrument must be incapable of simultaneously providing exact measurements of conjugate variables, and any physically realizable device must do even worse. In particular, Heisenberg argued, since momentum and position are conjugate variables and since time and energy are conjugate variables, if one could discover the exact position of a particle, one could not also know its momentum. Similarly, if one could know the energy level of a particle, one could not also know precisely when it was carrying that energy. Indeed, since the product of their uncertainties was finite, as one approached absolute exactitude in the measurement of one conjugate variable, the uncertainty of the value of the other blew up toward infinity.

To suggest why this problem arose, Heisenberg considered how it is that we make measurements. Suppose we want to locate an electron, for example. We shoot a beam of photons (light particles) in the general direction of the object we want to see, then we change the aim of the photon gun until a photon bounces back to some kind of receptor. In order to accomplish this (or any) measurement, we must thus disturb the thing we are trying to get a measurement of (in this case the collision of the photon with the target electron must change the initial momentum of the electron); consequently, once we narrow down our knowledge of where the electron was, we increase the range of places that it might subsequently be found.

One important consequence of Heisenberg's considerations, explored by Heisenberg in both *The Physicist's Conception of Nature* and *Physics and Philosophy*, is that the traditional notion of an "objective" physical world, totally independent of human observers, becomes hard to defend because the behavior of a physical object depends on how it is observed. Indeed, Heisenberg pointed out, there is an important sense in which objects don't really exist at all except when they are being observed, so it becomes impossible to ask questions about what happens to an object during an interval between two observations.

A second consequence, which has received a great deal of attention from humanists, especially theologians, involves the challenge to traditional notions of determinism implicit in quantum mechanics and made explicit in discussions of the uncertainty or indeterminacy. Prior to quantum mechanics, many scientists thought, as Pierre Laplace (1749–1827) had claimed, that if they could know the present position and momentum of every particle in the universe, they could, in principle, predict the complete future history of the world. In such a completely deterministic universe, there was apparently no possibility of effective "free-will" and moral responsibility. (The behavioral psychologist, B. F. Skinner, explores the conflict between assumptions of scientific determinism and traditional notions of freedom in *Beyond Freedom and Dignity* in a particularly clear and disturbing way. The failure of psychologists to precisely predict human behaviors, he attributes to the practical problem of collecting enough initial information.) But as Heisenberg had pointed out in his 1927 article, in the statement, "If we know the present exactly, then we can calculate the future completely," the premise is demonstrably false. Furthermore, it is not just false because of practical limits to human knowledge; it is false as a consequence of the character of the quantum world. Some theologians have drawn on this idea to argue that, while quantum mechanics may not prove the existence of free will, it challenges traditional notions of causality sufficiently to remove the objections to our immediate experiences of freedom that had been grounded in classical, deterministic, physical theories.—R.O.

See also BEHAVIORISM; QUANTUM THEORY AND QUANTUM MECHANICS

Helicopter

A helicopter is a type of aircraft that is powered by a large rotor (or rotors) mounted to a vertical shaft. In addition to powering the helicopter, the rotor performs the function of an airplane's wing by providing lift. Unlike a conventional airplane, a helicopter is capable of hovering, moving in all directions, and rising and descending vertically. This capacity allows a helicopter to land and take off just about anywhere that a human being is capable of going.

Although helicopters are a recent invention, the idea of propelling a flying device with a vertically mounted rotor goes back to 4th-century China, or perhaps even earlier. At that time, the Chinese built "bamboo dragonflies," toys equipped with blades attached to a vertical shaft. The toy flew vertically in the air when the shaft was rapidly rotated by pulling a cord that had been wound around it. Much later in the West, there began a series of attempts to design aircraft based on this general principle. The earliest known example appears in one of the notebooks of Leonardo Da Vinci (1452–1519), who, around 1480, sketched a flying machine equipped with a rotating helical sail. As with many of da Vinci's inventions, the conception was brilliant, but unworkable in the absence of a compact and powerful means of propulsion. New possibilities for flight arose when such engines became available in the 20th century. In 1906, Paul Cornu (1881–1944) of France built a rotor-propelled craft that rose 6 feet into the air and remained airborne for about 20 seconds. A year later and also in France, Louis Breguet (1880–1944) built a four-rotor "helicoplane" that succeeded carrying its pilot 5 feet in the air for a duration of 2 minutes. In the years that followed, other experimenters in many countries built aircraft with rotors; many of them were able to rise into the air with their pilot, but all were plagued by strong vibration and the lack of directional stability.

Much more useful was the autogiro, which was created by Juan de la Cierva (1895–1936) and first flew in Cierva's native Spain in 1923. Like a helicopter, an autogiro has a large vertically mounted rotor, but unlike a helicopter it is not powered. The rotor acts as a rotating wing, while the craft's power is provided by a conventional engine and propeller. Although not a true helicopter, Cierva's autogiro made an important contribution to helicopter design by mounting the rotor blades on hinges that allowed the blades to move vertically and horizontally in relation to the rotor hub. This feature automatically compensated for uneven lift, thereby increasing stability to a significant degree.

The German aircraft firm of Focke-Achgelis had licensed Cierva's patents in the 1930s, and it used the hinged rotor principle for what is usually considered to be the first successful helicopter. First flown in 1937, the FA-61 featured twin outrigger-mounted rotors that rotated in opposite directions to counteract the torque they generated. In the following year, the first British helicopter, the Weir W-5, made a successful flight, while in the United States, Igor Sikorsky (1889–1972) flew his Sikorsky VS-300 the year after that. Unlike Focke's design, Sikorsky's helicopter used a single vertical rotor along with a small tail-mounted rotor to counteract the main rotor's torque. Sikorsky helicopters were delivered to the U.S. Army Air Corps early in World War II, where they saw limited service, primarily for reconnaissance and the evacuation of wounded soldiers.

The helicopter began to play a key military role during the Korean War. In addition to serving as artillery spotters, helicopters plucked downed pilots from the sea and saved many lives by rapidly moving the wounded from the front line to field hospitals. Helicopters were even more prominent in the Vietnam War, where they were widely used for the rapid deployment of combat troops. They also served in a new role as helicopter gunships, attacking ground targets with machine guns, cannon, and missiles. But battlefield successes came at a high price, as 4,200 American helicopters were destroyed during the Vietnam war. Although helicopters could be an effective attack weapon, their slow speed and dependence on a single rotor made them vulnerable to ground fire, and their mechanical complexity made them less robust than conventional aircraft. Helicopters are also inherently more difficult to fly than airplanes. Critics have argued that the extensive and at times inappropriate use of helicopters for offensive missions in Vietnam was at least partly motivated by the Army's desire to have its own

A Bell 407 helicopter used for medical evacuation (courtesy Bell Helicopter Textron).

flying combat forces. Only the U.S. Air Force and the Navy were allowed to operate fixed-wing combat craft, confining the Army's aerial combat to helicopter operations.

The Vietnam War produced large numbers of trained helicopter pilots; some of them subsequently found employment flying helicopters for civilian applications, of which there have been many. Helicopters have been extensively used in construction work for the transporting and lifting of structural components. Logging operations have used helicopters to transport timber after it has been cut, thereby eliminating the need for expensive and environment-scarring logging roads. Helicopters have been used for building high-voltage transmission lines: Surveying, lifting towers into place, stringing cables, and performing inspections. Some of the most important tasks performed by helicopters are done in the offshore oil industry, where helicopters are extensively used to bring crews and supplies to drilling platforms. This can effect a considerable savings of time, for a 30-minute helicopter flight can take the place a boat journey of several hours.

See also AIRPLANE; ENGINE, FOUR-STROKE

Herbicides

A weed can be defined as any plant that occupies space that is wanted for other purposes. Weeds are a particular problem in agriculture, where they compete with crops for water, sunlight, and nutrients. At the same time, the cultivation of crops provides an environment that helps weeds to flourish. Consequently, farmers have been at war with weeds since the beginnings of agriculture.

The earliest method of weed control is still extensively used today: digging the weeds up, either manually with a hoe or through the use of tractor-drawn implements. This is not a complete solution, however, for weeds continually reestablish themselves. Today's farmers and gardeners eliminate weeds through the application of chemicals known as herbicides, more than a hundred of which are in common use today.

One early example of a chemical herbicide appeared as an unintended consequence of a successful effort to check a fungus known as downy mildew, which was attacking vineyards in the Bordeaux region of France. Spraying the vines with a mixture of copper sulfate and lime killed the mildew, and in 1896 one farmer observed that the fungicide also attacked a weed known as yellow charlock. Other early chemical herbicides included borax, carbon disulfide, chloropicrin, and common salt. All of these substances suffered from the same shortcoming. When used for weed control they are nonselective; i.e., they kill all the vegetation standing in the places where they are applied. This property limits their value for the treatment of croplands.

A few selective herbicides, notably—iron sulfate, sulfuric acid, and various copper salts—were used to a limited extent in the early 20th century. A major advance in the development of selective herbicides came as a result of research into plant hormones during the early 1940s. At that time it was discovered that a compound known as 2,4-dichlorophenoxyacetic acid (commonly known as 2,4-D) had a chemical structure resembling auxin, a natural plant hormone. When used as a herbicide, 2,4-D causes abnormal growth in plants by affecting their hormonal control functions. As a result, the tissues of affected plants swell up and burst, leaving them vulnerable to fungi and insects. A similar compound, 2,4,5-trichlorophenoxyacetic acid (commonly known as 2,4,5-T), worked in the same way and was especially effective when used on woody plants. 2,4-D and 2,4,5-T have been extensively used on croplands because they affect only broadleaved plants, leaving grasses like corn, rice, and wheat unscathed.

Other herbicides that also can be used selectively belong to a group known as *preemergents*. These do not affect mature plants, but they are toxic to seedlings just putting down roots. As a result, they can be used in places where desired plants are already growing. Some herbicides work both ways, killing vegetation through their roots or through their foliage. These herbicides are particularly useful in areas of high rainfall where the initial application of the herbicide will quickly be washed from the leaves and into the soil.

As with other agricultural chemicals, herbicides can cause serious environmental damage if misused. Moreover, the production of herbicides can itself be hazardous. The manufacture of 2,4-D, for example, may be accompanied by the production of tetrachlorodibenzo-*p*-dixin, or 2,3,7,8 TCDD for short, a substance that has been linked to a skin disease known as chloracne, liver damage, and other human ailments. This contaminant also has been blamed for a variety of human disorders caused by the application of Agent Orange in Indochina during the Vietnam War. In addition to causing health problems, this herbicide has done extensive environmental damage; many decades will pass before the region fully recovers.

See also AGENT ORANGE; HORMONES; INSECTICIDES; NEOLITHIC AGRICULTURAL REVOLUTION

Heroin

Morphine, an opium derivative, was widely used to alleviate pain throughout much of the 19th century. But as its addictive properties became known, pharmaceutical firms began to search for a substitute. In 1898, Heinrich Dresler (1849–1929), a researcher at the Bayer pharmaceutical company in Germany, rediscovered *diacetylmorphine*, a substance obtained through the acetylation of morphine. This substance had been produced in the 1870s by treating morphine and codeine with organic acids, but it was not put into use at that time. Bayer was quite enthusiastic about the substance, and began to market it for the treatment of severe coughs. Because it worked so well and was thought to be nonaddictive, it was considered a "heroic" drug, hence the name heroin.

Within a few years it had become evident that one of heroin's main virtues, its presumed nonaddictiveness, was illusory. In fact, heroin is one of the most addictive substances known. Heroin acts on the central nervous system in the same way that morphine does, but it does so much more rapidly, heightening the temporary sense of euphoria. But euphoria is soon followed by physical and psychological discomfort. Some withdrawal symptoms—anxiety and a craving for the drug—begin to appear a few hours after injecting heroin. These are followed by other withdrawal symptoms, such as aching bones and muscles, hot and cold flashes, and nausea. These symptoms intensify 26 to 36 hours after the last dose, accompanied by vomiting and diarrhea, as the user assumes a curled-up position.

Far worse than the discomforts of withdrawal are the fatalities that result from heroin use. Every year, 2,000 to 3,000 people die from heroin overdose in the United States. Many of these fatalities are the result of injecting heroin that has been mixed with quinine or the artificial narcotic fentanyl. Heroin sold on the street is usually diluted with lactose (milk sugar), but on occasion much purer heroin will appear on the streets. Heroin users unaware of the changed concentration may die from an overdose. Heroin deaths are also the result of combining the habit with the use of alcohol and barbiturates. Finally, many heroin users share needles with other users. As a result, 40 to 50 percent of heroin users have been exposed to the virus that causes acquired immune deficiency syndrome (AIDS).

Heroin has been an illegal drug in the United States since 1924, but it goes without saying that banning heroin did not put an end to its use. Heroin use expanded in the 1960s as addicts increasingly were drawn from the ranks of white, middle-class men and women. Also, just as the American Civil War produced large numbers of morphine addicts, the Vietnam War resulted in substantial increases in heroin use. As many as 40 percent of the soldiers in Vietnam used heroin and other opiates, and 7 percent of soldiers continued to use heroin after returning home.

Much debate has centered on devising a strategy for ending or at least reducing heroin addiction. Some have argued that decriminalizing drugs like heroin will ultimately result in less drug use because the huge profits of the drug trade will be eliminated, and there will be no incentive to expand the size of the market. Decriminalization also rests on the assumption that heroin use is a type of disease rather than a moral failure. Even where heroin use continues to be illegal, many treatment programs use a therapeutic approach to the problem. Some of these programs use a chemically similar substance, methadone, to reduce the craving for heroin. Most experts in drug treatment agree that methadone may be useful in many cases, but that heroin addiction is itself a symptom of deeper problems for both the individuals and society. Under these circumstances, a combination of individual counseling and addressing the conditions that give rise to drug use may be the only long-term solution.

See also ACQUIRED IMMUNE DEFICIENCY SYNDROME (AIDS); BARBITURATES; METHADONE; MORPHINE

Hip Replacement

Many people, not all of them elderly, suffer from damaged hip joints. Much of the damage is the result of arthritis, a disease that sometimes strikes early in life. Until the 1960s, a deteriorated hip usually meant living with constant pain and severely reduced mobility. Today, the insertion of artificial hip joints has become a common procedure. More than 100,000 hip replacement operations are performed in the United States every year, and a similar number are done in Europe. These have become very common operations; they are now conducted more frequently than appendectomies.

Attempts to replace natural joints with artificial ones began in the late 19th century. Hip replacements were rarely done, however, for these joints must bear much more weight than other joints. These early efforts at joint replacement used ivory, an obvious substitute for bone, but not a perfect one. The development of plastic materials in the 1930s created new possibilities for artificial joints, but these took many years to be realized. In the United States, Marcus Smith-Petersen had some success with a highly polished metal cup that was interposed between the ground-down head of the femur and the socket within which it fitted (the acetabulum). In France, two brothers, J. Judet and R. Judet, used a plastic hemisphere to replace the head of a femur, connecting it to the remaining portion of the femur with a spike. This arrangement worked better than the metal cup, but all too often the joints loosened over time.

In 1961, Great Britain's John Charnley (1911–1982) published details of the technique that would greatly increase the success of hip replacements. It entailed replacing the head of the femur with a small (22 mm [.88 in.]) stainless steel prosthesis. This in turn fit into an acetabular cup made from Teflon. The Teflon substantially reduced friction in the joint, but its wearing properties left much to be desired. It was subsequently replaced with a high-density polyethylene that lasted up to 1,000 times longer.

None of this would have made much difference if the artificial materials could not be permanently bonded to the patient's bones. A considerable improvement in the cementation of artificial parts to bones was effected through the use of polymethyl methacrylate, an acrylic cement developed in Great Britain in the late 1960s that proved to be both strong and long-lasting. Successful artificial hip operations also require aseptic surgery of the highest level. Hip replacement operations are conducted in special, clean rooms or under tents that exclude the entry of bacteria. Even when all precautions are taken, a successful operation runs the risk of postoperative infection. This danger has been greatly reduced by the use of antibiotics, which have held the level of infection to less than 2 percent of all hip replacement surgeries.

Unlike many high-tech medical procedures, hip replacements have lowered overall medical costs. In the past, a severely malfunctioning hip might have confined an elderly person to a nursing home and resulted in the expenditure of substantial personal or governmental funds. Today, a hip replacement allows many older people to live independently. At the same time, many nonelderly recipients of artificial hips can continue with their usual work, thereby contributing to the economy and avoiding dependency on others.

See also ARTIFICIAL HEART; KIDNEY DIALYSIS; POLYETHYLENE; PROSTHESIS; TEFLON

History of Science, Internal vs. External

Most histories of science written during the 19th and early 20th centuries accepted either implicitly or explicitly some variant of the claim first made by August Comte (1798–1857) that each science follows its own internal logic of development through three stages. According to this scheme, the pursuit of knowledge begins with a stage of speculation and premature systematization, followed by a stage of experimental exploration and clarification of phenomena, and culminating in a stage of precision and quantification that allows the formulation of positive laws. These views in turn were heavily dependent on those of Francis Bacon (1561–1626), for whom the life-transforming power of science could only be won by first renouncing the premature search for "fruits" of knowledge in favor of seeking the pure "light" of knowledge. Bacon also warned that biases associated with social interests erected a barrier in front of nature that made her secrets inaccessible.

According to this Baconian and Comtian perspective, specific sciences might develop at different rates because social conditions led to a greater or lesser investment of human and material resources in them. At the same time, however, the sequence of conceptual and empirical discoveries that had to occur were presumed to be determined "internally" by the interaction between the "real," objective, world of phenomena and the universal human intellect that sought knowledge of its laws. To the extent that any group could appropriately exercise authority in guiding the search for knowledge, it was the scientific community, composed of those who had been trained to distance themselves from individual and social biases so that they might approach nature in a disinterested and objective fashion.

Even some mid-20th-century sociologists of science, including Joseph Ben David and Warren Hagstrom, continue to operate within these basic assumptions. Hagstrom, for example, argued that

> . . . the scientific community is relatively autonomous, and the group of colleagues is the most important source of social influence on research. Colleagues influence decisions to select problems and techniques, to publish results, and to accept theories.

Similarly, Ben-David clamed that

> . . . the influence of internal disciplinary traditions is permanent and ubiquitous, since these traditions determine more or less what can be done in a science at any given time, [while] the external influences are ephemeral and random.

The "internalist" tradition in the history of science that these assumptions spawned came under concerted attack initially by Marxist-oriented historians for whom all human intellectual activity must be interpreted in terms of the material interests it serves. In the words of Benjamin Farrington, science is nothing but "the system of behavior by which man acquires mastery of his environment." Science also "develops in close correspondence with the stages of man's social progress," and "The history of science can only be understood as a function of the total life of Society." The Marxist "externalist" historiographical tradition found its first major expression in *The Social and Economic Roots of Newton's Principia* (1931) by Boris Hessen and in the essays of the Polish Scholar Edgar Zilzel. These were soon followed by the writings of a series of British scientists and mathematicians-turned-historians including J. D. Bernal, Lancelot Hogben, and Joseph Needham, and by the Dutch-born American mathematician and historian Dirk Struik, whose *Yankee Science in the Making* (1948) brought the externalist tradition to the United States.

Since 1939, the internalist tradition has been joined by a more "idealistically" oriented internalist approach embodied in the writings of Alexandre Koyre, whose *Études Galileennes* (1939) argued that a neoplatonic philosophic tradition played a major role in stimulating and forming Galileo's contributions to astronomy and early modern physics. This expanded internalist tradition allows for some consideration of the metaphysical, epistemic, and sometimes even the religious assumptions that may be incorporated into scientific thought; but it continues to see these features of intellectual life as fundamentally decoupled from social and economic life.

Since the late 1960s, a whole series of new externalist approaches to the history of science involving various levels of commitment to "social constructivist" philosophies of science have begun to dominate French, British, and American historiography of science. One of the sources of these new trends grows out of the international radical student politics of the 1960s with its associated attack on all forms of political and intellectual authority. In the history of science, this attack on traditional notions of scientific authority has usually been articulated in connection with the claim that traditional internalist history of science has been guilty of a major inconsistency in the formulation of its historical explanations. The formulation of "good" or "correct" scientific theories has traditionally been seen as unproblematic and as the natural outcome of the nature of "reality," disclosed by appropriately clever human strategies. The formulation of "bad" or false science, on the other hand, is understood in terms of social accounts; i.e., the "bad" racial theories prevalent in Europe and America are to be understood as ways of preserving the economic and social status of wealthy and powerful white elites. For social constructivist thinkers, there should be a symmetry about the way in which knowledge is accounted for, and putatively "correct" knowledge should not be privileged relative to "false" knowledge. If we can only give a social account of the construction of the latter, we should seek a social account of the former as well. Perhaps the most outstanding example to date of social constructivist history of science is Steven Shapin's and Simon Schaffer's *Leviathan and the Air Pump* (1985), which seeks to understand the rise to dominance of the experimental methods associated with Robert Boyle in terms of a set of social processes associated with the growing need for trust among 17th-century gentlemen.—R.O.

See also SOCIAL CONSTRUCTION OF TECHNOLOGY

Further Reading: Warren Hagstrom, *The Scientific Community*, 1965. Joseph Ben-David, *The Scientist's Role in Society*, 1984. Benjamin Farrington, *Greek Science*, 1953.

Hologram

A hologram produces photographic images that give the appearance of being three-dimensional. Holography was invented by Dennis Gabor (1900–1979) in 1948, but its full capabilities could not be exploited prior to the invention of the laser. A hologram is produced by dividing light from a laser into two beams. One of these beams (known as a *reference beam*) hits a mirror positioned near the subject, and the reflected beam then falls onto a photographic film or plate. The other beam (the *signal beam*) illuminates the subject and is then reflected onto the film or plate, where it interacts with the reference beam to form interference patterns.

The generation and "reading" of interference patterns are the basis of holography. In general, interference patterns are produced when two beams of light intersect. Because light is propagated as a wave, the wave patterns of the two beams interfere with each other to create a new wave pattern. For example, when the crest of a wave from one beam coincides with the crest of a wave from another beam, a wave crest of greater height will appear. Conversely, when a crest of one beam meets the trough of another beam, the two cancel out, and both the crest and trough disappear. This phenomenon was first described by the English physician and physicist Thomas Young (1773–1829) in 1801, and it formed the basis for his wave theory of light.

In holography, the interference patterns provide a complete record of the amplitude and phase distributions of the light waves emanating from the subject (this is the source of the term *hologram*, which means "total recording"). This record, which appears after the plate or film is developed, looks nothing like the subject, for it appears as an apparently random assortment of swirls and blotches. The holographic image appears when a laser beam, preferably the reference beam that was used to make the plate, shines through the developed film or plate (some holographic processes now allow the use of ordinary light). This produces light waves that are identical to the waves that had been reflected from the subject. The observer sees a three-dimensional image of the subject, which seems to float behind the film or plate containing the hologram. The image replicates such an accurate three-dimensional image that the observer has to refocus his or her eyes in order to clearly see different parts of the image. This three-dimensional view is maintained even when the holographic image is viewed at an extreme angle. Even more remarkably, when the developed hologram is cut into smaller pieces, each piece will reproduce the same holographic image as the original.

Although holography is often employed solely for amusement, it also has a number of practical applications. Gabor originally invented holography to improve electron microscopes, and it presently is used for certain kinds of microscopy. Holographic microscopes have not displaced conventional electron microscopes, but they have been used for microscopic examination of moving biological specimens. Holography also is used for bar-code readers. In this application, the hologram appears on a continually rotating disk. A laser beam shines through the disk, sweeping through the disk's holographic interference patterns. This causes the beam to follow a complex scanning pattern, which in turn allows the reading of each product's distinctive bar code.

Computer engineers have extensively experimented with holography as a means of storing binary data in a computer's memory. A memory of this sort consists of an array of holograms. When a laser beam shines through an individual hologram, it generates an image consisting of bright and dark spots, each representing the presence or absence of a binary digit. These spots are then read by a detector and translated into whatever the binary digits are supposed to represent. Up to now, this technology has failed to supplant conventional memory technologies. However, some cognitive psychologists believe that holograms may provide important analogies to the way that human memories store and retrieve information.

See also BINARY DIGIT; DISK STORAGE; LASER; MICROSCOPE, ELECTRON; MEMORY; UNIVERSAL PRODUCT CODES

Further Reading: Graham Saxby, *Practical Holography*, 2d ed., 1994.

Homo erectus

Homo erectus is one species of the lineage leading to modern human beings. The oldest and most complete material comes from East Africa and dates to 1.8 million years ago (mya). One of the most complete specimens was discovered by Kimoya Kimeu, Alan Walker, and Richard Leakey in the Lake Turkana region of northern Kenya; it dates to 1.6 mya. This 80 percent complete skeleton is the fossilized remains of a young boy. As an adult, this individual would have stood approximately 1.8 m (6 ft) tall. The body reflects complete adaptation for bipedal locomotion, and is much more modern than the bodies of earlier hominids. The dentition suggests an omnivorous diet. The skull reveals continuing expansion of the brain; cranial capacities for *Homo erectus* range from of 800 to 1,100 cc (compared with a modern average of 1,400 cc). However, the shape of the skull is long and low, and prominent brow ridges form a bony shelf over the eyes.

H. erectus is also found at South African sites; more striking, however, is its expansion outside of Africa. In 1891, Dutch anatomist Eugene Dubois discovered a primitive-looking skull cap and a modern thigh bone near the Solo River of Java, Indonesia. Assuming that the two specimens belonged to a single individual called Java Man, Dubois concluded that they represented an extinct hominid and gave it the name *Pithecantropus erectus*, erect ape man. Later, Davidson Black and Franz Weidenreich discovered similar specimens in China (Peking Man); these were eventually assigned to the same species, *Homo erectus*. European material exhibits similar characteristics, but many taxonomists prefer to classify these specimens as early *Homo sapiens*.

The timing of *H. erectus*'s expansion beyond Africa is currently under reconsideration. Traditionally, the early Asian material has been dated at 1.0 to 0.5 mya. However, reevaluation of the sites, along with discovery of new material, indicate that this species may have reached Asia by 1.8 mya. Thus, their colonization of Asia may have occurred soon after their appearance in Africa. This model is still under consideration. Also in question is the taxonomic status of the species. Some scientists propose that the African and Asian materials are sufficiently different to assign them to separate species; in this case, the African specimens would be classified as *Homo ergaster*.

In addition to their anatomical differences, African and Asian *H. erectus* used distinct tools. African sites are replete with tools belonging to the Acheulean lithic culture. This culture is more advanced than the previous Oldowan, including a wider variety of tool types. Large cores were formed into hand axes and cleavers. Flakes were reshaped into straight knives, serrated knives, and scrapers for preparing hides. However, the Acheulean is unknown at Asian sites; there, stone tools resemble the chopper-chopping pebbles of the Oldowan culture. If groups of *H. erectus* did leave Africa as early as 1.8 mya, their emigration would have preceded the development of the Acheulean (at about 1.6 mya); this could account for its absence at Asian sites. Asian populations may have made more complex tools from bamboo, a material that would not have been preserved.

One cultural innovation commonly attributed to *H. erectus* is the controlled use of fire. Early evidence for this comes from the South African site of Swartkrans, dating to approximately 1.7 mya. However, considerable debate surrounds these data. More widely accepted is the hearth at Zhoukoudian Cave near Beijing, China, dating to roughly 0.5 mya. Evidently, groups of *H. erectus* inhabited this cave for many thousands of years, hunting game such as deer and gathering plant material such as hackberries. We can only imagine the change in social life and cultural transmission that accompanied the use of fire: to cook food, to harden wooden spear points, and to extend the hours of light. *H. erectus* was a highly successful, if somewhat static species, existing virtually unchanged for over 1 million years.—L.E.M.

See also AUSTRALOPITHECINES; EVOLUTION, HUMAN; *Homo habilis*; NEANDERTHALS

Homo habilis

Homo habilis is currently the earliest known member of our own genus. It was originally discovered by Louis Leakey, in the Olduvai Gorge of Tanzania. Most material dates from 2.0 to 1.8 million years ago (mya), though a recently reevaluated skull fragment may be as old as 2.4 million years. In its general bodily features, *H. habilis* resembles earlier hominids, the australopithecines, with adaptations for bipedality but retention of some apelike traits. However, in contrast with the australopithecines, *H. habilis* shows the first stages of the cranial expansion that is characteristic of the genus; cranial capacities for habiline specimens range from 630 to 750 cc. They are also distinctive in the reduction of the chewing apparatus: Unlike the robust australopithecines, *H. habilis* has relatively small molars, gracile facial bones, and no skull ridges for muscle attachment. These traits, along with tiny striations on the molars, indicate a shift to a more omnivorous diet.

H. habilis made both morphological and behavioral advancements over its predecessors. The earliest stone tools are attributed to this species. The Oldowan lithic culture is characterized by crude pebble choppers made by banging two or three flakes off of a core. A considerable assemblage of cores, flakes, and animal bones—excavated by Mary Leakey at Olduvai Gorge—perhaps represents a home base of hominid activity. Another collection of bones, described by Bunn and Kroll, suggests a site where animals were butchered. Richard Potts and Pat Shipman have recently analyzed bones from habiline sites and have noted both cut marks from stone tools and also tooth marks of large predators. They conclude that *H. habilis* procured meat through scavenging the kills of these predators, rather than through extensive, cooperative hunting.

The material now classified to *H. habilis* varies considerably in size. Some scientists attribute this variance to sexual dimorphism: Perhaps males were considerably larger than females. Others propose that two species of *Homo* were in existence, and place the larger individuals in a separate species, *Homo prometheus*.—L.E.M.

See also AUSTRALOPITHECINES; EVOLUTION, HUMAN

Homo sapiens sapiens

Homo sapiens sapiens comprises hominids of modern anatomical configuration, including all human beings in existence today. The subspecies is distinguishable from earlier *H. sapiens* by the more gracile skeleton with thinner bones and reduced musculature, the globular cranium with a high forehead and rounded occipital and parietals, a lack of prognathism, reduction in both anterior and posterior dentition (incisors and molars, respectively), and the presence of a distinct chin.

When and where this subspecies evolved remains open to debate. Some scientists propose that *H. erectus* made the transition to *H. sapiens* and then to *H. sapiens sapiens* in several places simultaneously. Others assert that anatomically modern humans arose in Africa and eventually migrated across the continents, replacing populations of more primitive hominids that they encountered. The fossil data currently available appear to support the latter model. The oldest material attributed to this subspecies comes from Africa, from sites such as Klasies River Mouth and Border Cave in South Africa, Omo in Ethiopia, and Jebel Irhoud in Morocco. Dates for these specimens are controversial, but they are probably between 80,000 and 120,000 years old.

Following the fossil trail, it seems that anatomically modern humans migrated northward and populated western Asia (the Middle East) by 90,000 years ago. Modern-looking skulls have been yielded from sites such as Skuhl and Qafzeh in Israel. Neanderthals apparently entered this region about 60,000 years ago, and the two forms coexisted for approximately 20,000 years, until the Neanderthals disappeared. The relationship between Neanderthals and modern human beings is the subject of heated debate. By assigning them to the same species (*H. sapiens*), scientists suggest that interbreeding was possible. The extent to which this happened is unknown, but some specimens found in western Asia do suggest a mixture of archaic and modern traits. The cause of the Neanderthal terminus is also unknown.

At approximately 40,000 years ago, anatomically modern human beings flooded into Europe. The form is well known from many sites across the continent, including Cro-Magnon in France (Cro-Magnon Man). Perhaps it is no coincidence that within a few thousand years of modern human migration into Europe, the Neanderthals disappeared. Again, the causes are unknown. From Europe, *H. sapiens sapiens* rapidly colonized the rest of the eastern hemisphere, reaching Australia by 30,000 years ago (Lake Mungo) and China by 18,000 years ago (Zhoukoudian Cave). Migration into North and South America had occurred by 12,000 years ago (Clovis).

H. sapiens sapiens is distinct from its predecessors in morphology, but the extent to which this correlates with cultural advances is questionable. Many scientists assert that early populations of anatomically modern human beings used tools similar to the Mousterian of the Neanderthal, and a major shift in lithic industry did not occur until about 40,000 years ago. This coincides with the disappearance of Neanderthals in western Asia and the migration of modern forms into Europe where, again, Neanderthals ceased to exist. However, evidence from selected sites in Africa, such as Katanda in Zaire, suggest a shift to the more advanced Late Stone Age as early as 90,000 years ago, which matches more closely the dates for the early modern fossil specimens. Thus, the extent to which anatomical changes, particularly neural reconfiguration, correlate with cultural progress is unclear.

The Late Stone Age industry (known as the Upper Paleolithic in European sites) is characterized by a vast expansion of tool types and raw materials. Stone tools include carefully crafted blades, such as the laurel leaf points of the Solutrean culture. Bone and antler were also used, to make barbed harpoons, fish hooks, awls, and needles with eyes. Later inventions included the spear thrower and the bow and arrow. Anatomically modern human beings were also the first hominids to make art. By 20,000 years ago, European populations were carving Venus figurines and engraving animal designs into the bone handles of tools. By 10,000 years ago, these people had adorned their caves with elegant paintings of game animals and human forms, paintings that may have had ritual significance. Ritual and respect for the dead are also reflected in burials, which often include grave goods such as tools, foods, and elaborate shell jewelry. The combination of neural reorganization with cultural and social advancements clearly accounts for the successful colonization of the globe by modern human beings.—L.E.M.

See also BOW; EVOLUTION, HUMAN; *Homo erectus;* NEANDERTHALS

Homogenization

When left alone for any length of time, whole milk will separate into cream and a fluid containing very little butterfat. Sometimes this separation is desirable; "skim milk" or low-fat milk (with 1 to 2 percent fat content) is made by running whole milk through a separator. Still, when milk contains any amount of fat, it usually goes through a process known as homogenization to prevent separation. Homogenization entails breaking up the fat globules in the milk, making them smaller—2 microns or less—so that they remain in suspension with the rest of the milk.

This process begins by first heating the milk to a temperature of at least 60°C (140°F) to liquefy the fat globules and inactivate the milk's lipase, an enzyme that causes a rancid taste. This step is usually part of the pasteurization process that is used to destroy any pathogens in the milk. The milk is then put through a high-pressure pump known, logically enough, as a *homogenizer*. A homogenizer works by forcing the milk through two sets of valves, which causes a shearing of the fat globules. The homogenization process entails "shattering," which occurs when the milk makes a high-speed impact on a flat surface in the homogenizer. The process also includes cavitation, the formation of air-filled cavities in the liquid; this results from the movement of the milk inside the homogenizer from an area of higher pressure to one of lower pressure.

In the United States, the efficacy of the homogenization process is monitored by the Food and Drug Administration (FDA) of the federal government. According to FDA regulation, a quart of milk that has been undisturbed for 48 hours cannot exhibit more than a 10 percent difference between the fat content of the upper 100 ml and the rest of the container.

See also FOOD AND DRUG ADMINISTRATION, U.S.; PASTEURIZATION

Hormone Replacement Therapy

Of the issues concerning women's health, few are as controversial and confusing as the use of hormone replacement therapy (HRT) by postmenopausal women. Menopause, the cessation of menstrual periods, is usually defined retrospectively as having occurred after 1 or 2 years of no menses. Menopause also can occur as the result of a hysterectomy. The *climacteric* is a term that refers to an approximately 15-year period, from 40–45 to 55–60, during which the production of two hormones that regulate the menstrual cycle, estrogen and progesterone, gradually declines, and periods become irregular until they stop. While the production of progesterone ends with menopause, estrogen production by the ovaries and adrenal glands continues at a much reduced level. When women speak of "going through menopause," many think of a period in life that involves not only hormonal changes but psychological and social ones as well.

Common physical signs of menopause, besides changes in the menstrual cycle, include hot flashes, sweats, sleeping disturbances, and vaginal tissue dryness. Some women also experience pain in muscles and joints, headaches, and weight gain and/or psychological changes including irritability, depression, and anxiety. Far from being simply a response to hormonal change, these conditions also may be responses to changes in social roles, the marginalization of older women in our culture, and losses that affect many midlife women. In fact, recent research has shown that many women experience relief at having no more periods, welcome a phase of life with fewer domestic responsibilities and greater freedom, and have few distressing physical changes.

As early as the 1940s, physicians began to prescribe estrogen replacement therapy (ERT). With the medicalization of menopause as an "estrogen deficiency disease" and extensive advertising by pharmaceutical companies, Premarin, the leading brand, ranked fifth in drug sales by 1975. In the mid-1970s, however, research revealed the increased risk of endometrial cancer, gallbladder disease, and possible breast cancer in women taking estrogen. Sales fell for a period, but have increased again with the advent of combined hormones (estrogen and progestin), lower dosages, and research suggesting benefits for prevention of osteoporosis and coronary heart disease, the leading cause of death in postmenopausal women.

It is widely agreed that hormone replacement therapy—either "unopposed" estrogen (estrogen taken alone) or "combined" with progestin—can provide relief from hot flashes and alleviate vaginal dryness. Controversy has surrounded the long-term effects of hormone replacement therapy, however, and this debate is ongoing. Despite being universal to women who live to middle age, menopause was the subject of amazingly little research until relatively recently, and important controlled longitudinal studies are still producing results. For example, findings are still forthcoming from the postmenopausal estrogen/progestin interventions trial (PEPI) sponsored by the National Institutes of Health. This study followed 875 healthy postmenopausal women for 3 years, testing the effects of four hormone regimens: estrogen alone daily, estrogen daily and a synthetic progestin 12 days a month, estrogen and synthetic progestin daily, and estrogen daily plus a natural progesterone 12 days a month, plus a control group taking placebos.

A number of studies have found a strong association between use of HRT and reduced levels of coronary heart disease. The Nurses' Health Study, a longitudinal study of 59,337 nurses, for instance, found that women who took combined HRT or estrogen alone, had about half the risk of a heart attack as women who took no hormones. PEPI results showed all four of the tested hormone regimens reduced heart disease risk factors in several ways, e.g., raising the level of "good" HDL cholesterol and lowering the level of "bad" LDL cholesterol without increasing blood pressure. Studies also have shown that long-term use of estrogen protects against postmenopausal bone loss and osteoporosis.

The finding that unopposed estrogen use was linked to an elevated risk of endometrial cancer led to subsequent research that tested whether estrogen combined with progestin would lessen the chances of cancer by causing shedding of the estrogen-thickened endometrium. Results from the PEPI study and other investigations have confirmed that women on estrogen only are at a significantly higher risk of hyperplasia and cancer of the endometrium than are women using a combination. For the moment, there is less consensus regarding HRT and breast cancer: Some studies have shown a significantly elevated risk for women using hormonal therapy, others have not. With statistics showing that one in nine American women will be diagnosed with breast cancer in a lifetime, many women, understandably, feel confused.

Today, replacement therapies are available as tablets or patches. Tablets are by far the most common mode of administration, available in estrogen-only form (mostly prescribed for women who have had the uterus removed) or in continuous (both hormones are taken daily) or sequential "combined" forms. In Europe, Asia, and elsewhere, estrogen and progestin are formulated in a single tablet in combined therapy; in the United States, doctors use free-combination therapy, prescribing estrogen and progestin individually.

Women's health advocacy groups urge women to view menopause as a natural process, noting that many women experience few uncomfortable symptoms, are not at particular risk for heart disease, stroke or osteoporosis, and can ease what discomfort they have with nonmedical approaches. These approaches also emphasize nonmedical efforts to promote cardiovascular health and strengthen bones. Recommendations include moderate daily exercise, a good diet with supplements—especially of vitamins A, C, and D, and calcium and magnesium—and abstinence from tobacco. Some women also are helped by herbal teas, relaxation techniques, such as meditation or massage, and reduction in the consumption of alcohol, caffeine, sugar, chocolate, prescription drugs including tranquilizers, antidepressants, and sleeping pills.

Many questions, such as the optimal length of treatment if HRT is undertaken, the effects of different dosages, and the risk posed for breast cancer, remain unanswered. The decision of whether to use HRT should be carefully made, taking into consideration many factors including a woman's family history and her own medical history. A cautious approach for women who chose ERT is to take low dosages, have frequent monitoring by expert healthcare providers who are willing to discuss women's questions and concerns, and be alert to the latest research findings. A good diet, regular exercise, and other steps to promote good health are important for all postmenopausal women.—A.H.S.

See also ANTIANXIETY (ANXIOLYTIC) DRUGS; ANTIDEPRESSANTS; CHOLESTEROL; HORMONES; NATIONAL INSTITUTES OF HEALTH; VITAMINS

Hormones

Hormones are chemical messengers produced by various endocrine glands. The glands secrete the hormones directly into the bloodstream rather than through associated ducts; for this reason they are known as *ductless glands*. Each hormone has a specific function; some regulate the growth or functioning of particular organs or tissues; others affect physiological responses. In general, hormones are involved either in the body's metabolic processes or its sexual cycles. Another group of hormones known as *neurohormones* are produced by specialized nerve cells and are secreted from nerve endings into the bloodstream or particular organs or tissues. These also affect the growth and functioning of organs and tissues.

The therapeutic use of hormones has a very long history. As early as the 2d-century B.C.E., the Chinese were extracting hormones from human urine and using them to treat various sexual dysfunctions. The science of endocrinology (the study of hormones) was initiated by the research of two English physiologists, William Bayliss (1860–1924) and Ernest Starling (1866–1927). In 1902, while studying the release of digestive fluids by the pancreas, they discovered that a substance was secreted by the walls of the intestine when food entered into it. When injected into an animal's blood, this substance caused pancreatic secretions even when no food was present. They called the substance *secretin* and classified it under the term *hormone* (from the Greek, "to rouse to activity").

A few years earlier, Jokichi Takamine (1854–1922), a Japanese chemist working in the United States, had discovered that the adrenal glands secreted a substance that raised blood pressure when injected into the body. He gave it the name *adrenaline*, a term that is commonly used today, although it is now a trade name (the scientific name is *epinephrine*). It too was recognized to be a hormone.

The next hormone to be isolated was *thyroxin*, which was extracted in 1915 from the thyroid gland by the American biologist Edward Kendall (1886–1972). The importance of thyroxin in regulating metabolism was discovered in the ensuing years. This discovery gave rise to a therapeutic application, the administration of thyroxin to individuals with defective thyroids. An even more important contribution to medical practice was the discovery that the absence of a certain hormone was the cause of diabetes, the inability to properly metabolize sugar. In the 19th century, the German physiologists Joseph von Mereng (1849–1908) and Oscar Minkowski (1858–1931) had discovered that the removal of the pancreas resulted in diabetes. In 1916, Albert Sharpey-Schaefer (1850–1935), a Scottish physician, suggested that groups of cells on the pancreas (known as the *islets of Langerhans* after their discoverer, Paul Langerhans [1847–1888]) produced a substance that he named *insulin* (from the Latin word for *island*). Insulin seemed to offer a means of combating diabetes, but extracting it from the pancreas proved very difficult. It was finally accomplished in 1921 by two Canadians, Frederick Banting (1891–1941) and Charles Best (1899–1978). It was later learned that the islets of Langerhans secrete another hormone, glucagon, which acts in a manner opposite to that of insulin. The two hormones work through a kind of pull-push process to keep the level of glucose in the blood at a stable level.

Another set of hormones known as *steroid hormones* control physiological processes having to do with sexual activity and reproduction. In 1927, the German physiologists Bernhard Zondek and Selmar Aschheim found that an extract of a pregnant woman's urine produced estrous when it was injected into rats. Within 2 years the hormone causing the effect had been isolated by Adolf Butenandt in Germany and Edward Doisy in the United States. *Estrone*, as it was called, is one of a group of hormones known as *estrogens*. In 1931, Butenandt extracted the first male sex hormone, *androsterone*, and this was followed by his extraction in 1934 of a second female sex hormone, *progesterone*. Female hormones are now widely used for the treatment of physiological problems associated with menopause.

Another important steroid hormone is produced by the cortex of the adrenal gland (another part of this gland, the medulla, is responsible for the production of *epinephrine*). The history of its discovery and application is somewhat circuitous. Four cortical hormones had been identified by 1940 and had been given the names Compound A, Compound B, Compound E, and Compound F. Spurred by an erroneous report that the Germans were using cortical hormones to improve the high-altitude performance of their pilots, Allied scientists engaged in a massive effort to synthesize them. By the time Kendall synthesized Compound A in 1944, it had become apparent that these hormones did nothing to improve piloting performance, but it was hoped that Compound A could be used to treat Addison's disease. This too proved to be a false hope. However, Compound E, which was synthesized in 1949 by Lewis Sarret, was found to be of considerable value for relieving the symptoms of rheumatoid arthritis. Today, this hormone is known as *cortisone*.

Many of the hormone-producing glands are governed by the pituitary, a small gland located at the base of the brain. The pituitary secretes a number of hormones that affect the supply of other hormones. For example, by secreting a thyroid-stimulating hormone, the pituitary stimulates the thyroid gland to produce thyroxin. The pituitary and the thyroid are linked by a feedback process, so the secretion of thyroid-secreting hormone ceases when there is an adequate amount of thyroxin in the blood. The pituitary secretes a number of other hormones, including the somatrophic hormone that controls growth.

The actual mechanisms through which hormones work are still not completely understood. In general, it is known that hormones affect the operation of enzymes. They also act directly on cell membranes, allowing certain substances to enter the interior of the cell while blocking others. This "gatekeeping" function either allows or prevents the entry of different enzyme substrates, thereby affecting enzymatic action.

See also CELL; ENZYMES; HORMONE REPLACEMENT THERAPY; HUMAN GROWTH HORMONE; NEURON

Horsecollar

In many ways, the horse is the ideal draft animal. As might be expected, it is capable of a sustained output of about 1 horsepower, and the largest varieties are capable of generating more than 3 horsepower for brief periods. In reasonable conditions, a horse can work at a rate of 1 meter (3.28 ft) per second, 30 to 50 percent faster than an ox. It can work steadily for 8 to 10 hours, whereas 4 to 6 hours are the norm for cattle. Finally, a horse has a working life of as much as 20 years, double that of an ox. Horses require that some farm acreage be sown to oats or other feed, for they cannot forage as efficiently as cattle and other ruminants, but the energy they supply greatly offsets the food energy that they consume.

Despite these many advantages, horses were not widely employed as draft animals in the ancient world. The main reason was the lack of an effective harness. In contrast, oxen were relatively easy to hitch to a plow or other implement. Although still leaving something to be desired, the simple yokes fastened to the head or neck of oxen did an adequate job of harnessing these animals. Early horse harnesses were also simple, but ineffective, being little more than straps that passed across the animal's throat and girth. This was an unsatisfactory arrangement on two counts. First, this kind of harness attached an implement to the horse at a high point of traction, partially negating the superior strength of a horse's front legs. Second, and more seriously, the strap passing across the throat pressed against the horse's windpipe, choking the animal as it strained against a load.

A partial solution was the breastband harness, introduced in China during the Han Dynasty (206 B.C.E.–220 C.E.) A harness of this sort was mounted lower down on the horse's body, concentrating force on the sternum rather than the throat. It eventually made its way into the Middle East, then reached Italy as early as the 5th century, and finally arrived in Northern Europe a few centuries later. An even better harness, one that in its basic form is still in use today, was also invented in China no later than the 1st-century B.C.E., and perhaps as early as 4th-century B.C.E. It consisted of a padded yoke that went all the way around the horse's neck and to which was attached the traces that connected the horse with a cart or farm implement. With this arrangement, the horse's pulling force was distributed evenly rather than being absorbed at the throat, allowing a threefold increase in pulling power when compared to the throat-and-girth harness.

Although the sources of inspiration for both in-

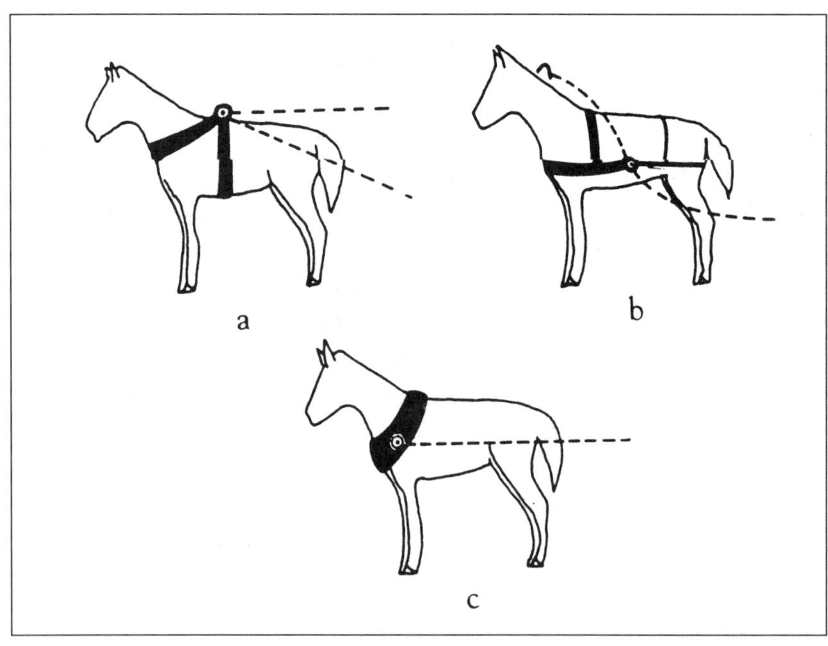

Three means of harnessing a horse: *c* is the superior method because the collar does not restrict the horse's ability to breathe (from J. Needham, *Science and Civilization in China*, 1965; used with permission).

ventions will never be certain, it may have been the case that the Chinese, who made considerable use of human motive power, recognized on the basis of personal experience that a harness around the chest and shoulders made much more sense than one around the neck. Although it is only speculation, it is possible that the padded horsecollar was derived from pack saddles used for camels. These fixtures were U-shaped wooden frames, padded with felt. Perhaps after some modifications they were used for harnessing horses.

As with many other inventions that originated in China, the breastband harness and the padded collar slowly diffused across central Asia and then made their way to Europe. The breastband harness might have been brought by the Avars when they invaded Hungary in the 6th century, and by the 8th century it was probably well known throughout the continent. The padded horsecollar made a similar journey from China to Europe, appearing in the 9th century and then diffusing over the course of 3 centuries until it became the standard means of harnessing horses. It is still in widespread use in many parts of the world, especially where horses continue to be an important source of power for agriculture and transportation.

Further Reading: Vaclav Smil, *Energy in World History*, 1994.

Horseshoes

Until rather recently, one of the main sources of horsepower was horses. The effective use of horses rested on a number of key inventions: the stirrup, the horsecollar, and the wheel. Also of considerable importance was the lowly horseshoe. A horseshoe is nothing more than a U-shaped piece of metal that is directly nailed to the horse's hoof. There are no nerves in the hoof, which is made of hardened collagen, so no pain is felt when the horseshoe is nailed to it.

As with many other seemingly obvious inventions, the horseshoe did not make its appearance until relatively late in human history. The nailed horseshoe was unknown in Greek and Roman times, and for many centuries afterwards. The best protection given a horse's foot was a kind of sandal attached with wires or thongs, and it was primarily used when a hoof was injured. An unshod horse got along tolerably well in dry Mediterranean climates, but in damp northern Europe some protection for the hoof was imperative, for wet roads and fields caused hoofs to become soft and easily damaged. Exactly when and where the horseshoe first made its appearance is still a matter of considerable disagreement. It can be said with confidence, however, that nailed horseshoes were common by the early 10th century, although some archeologists have argued for an earlier date. Whatever the exact date, horseshoes aided greatly in the effective use of horses and made possible the employment of heavier, more powerful breeds.

Horses continued to be important beasts of burden well into the industrial age. Horse-drawn carriages, carts, and even streetcars filled the roads of 19th-century cities. When roads were built with flat-surfaced stone blocks known as *setts*, the horseshoes worn by urban horses often had small protuberances called *calks* to catch the joints between the blocks to improve traction. Today, the shoeing of horses is not an obsolete occupation; it is still taught in agricultural schools under the name "farrier technology."

See also HORSECOLLAR; STIRRUP; WHEEL

Hovercraft

A hovercraft makes no direct contact with the surface, riding instead on a cushion of air. For this reason it is also known as an *air cushion vehicle*. Craft of this sort are also referred to as *ground effect machines*, but strictly speaking a ground effects machine requires forward motion to produce the cushion of air, whereas a hovercraft generates its own air cushion. Some hovercraft can travel over land, and some are amphibious, but the main commercial application of the hovercraft has been for overwater travel. Here they have an advantage over conventional ships, for they are not hindered by the resistance of the water against a hull or by wave formations at the bow and stern.

The idea of air cushion craft goes back to at least 1877, when John Thornycroft of England received a patent for such a vehicle, but he was never able to actually build one The hovercraft became a reality due to the efforts of Christopher Cockerell (1910–), a British engineer who had taken up boat building in the years following World War II. One of Cockerell's major insights was that the air forming the cushion should be pumped to the periphery of the craft and not its entire underside. Cockerell tested his idea by building a small model that consisted of a small can inside a larger one; When a hair dryer blew air through the cans, the resulting cushion of air moved the model above the surface of the table on which it had rested. Cockerell then went on to build a number of experimental air cushion vehicles, culminating in the launch of the first commercial hovercraft, the SR-N1, which began ferry service between the Isle of Wight and the English mainland in 1959.

Small hovercraft make use of air blown into the craft's concave underside from a central aperture. The air then moves outwards and is retained by a flexible skirt around the craft's periphery. Larger hovercraft may also use this basic design, but more sophisticated versions ride on air cushions produced by a ring of air jets arrayed around the craft's circumference and directed slightly inwards. The placement of these jets helps to prevent the leakage of air from the cushion, although most designs continue to make use of flexible skirts. Air cushion vehicles are propelled by gas turbine, diesel, or spark-ignited internal combustion engines that drive aircraft-type propellers or ducted fans. Directional control is achieved by reducing the pressure exerted by the air jets on one side, which causes the craft to bank. Vertical rudders may also be fitted; these come into play at relatively high speeds. Changes in direction and speed can also be effected by changing the rotational speed and pitch of the propellers.

Hovercraft are used as recreational vehicles, military craft, and commercial transportation. The largest hovercraft currently in service transports cars and people across the English Channel. This vessel, the British Hovercraft SR-N4, carries 60 cars and more than 400 passengers, and has a maximum speed of 120 km/h (75 mph). Power for forward movement and for maintaining the air cushion is provided by four gas turbine engines of 3,800 hp (2.8 MW) each. The success of cross-channel hovercraft produced considerable enthusiasm for these vehicles, but high initial and running costs, along with limited range compared to conventional ships, have prevented the widespread use of hovercraft at sea. Hovercraft have found a number of land applications, especially in the military, but maneuvering difficulties makes them poor substitutes for wheeled and tracked land vehicles for most purposes.

See also ENGINE, FOUR-STROKE; TURBINE, GAS

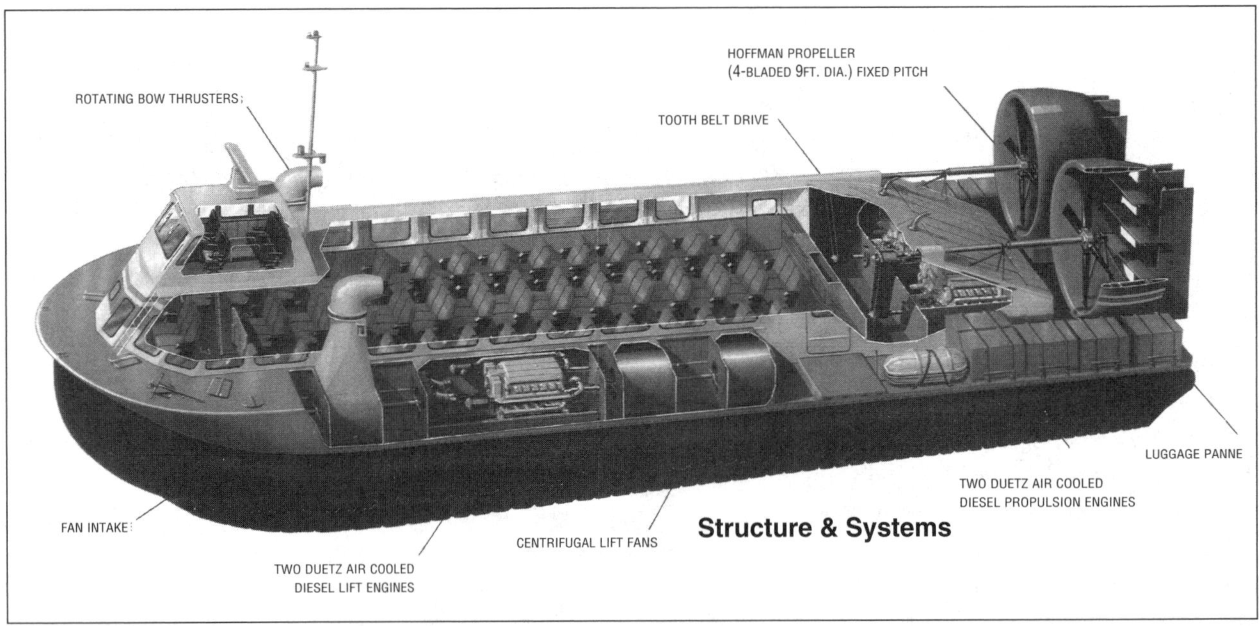

The basic features of a modern hovercraft (courtesy Hoverwork Ltd.).

Hubble Space Telescope

Since its invention in the early 17th century, the telescope has revealed much about the nature of the universe. But all the telescopes on Earth are handicapped because the atmosphere affects the light they receive. One manifestation of this is the apparent twinkling of the stars that is caused by the motion of the atmosphere. The Earth's atmosphere is also responsible for the blurring of objects seen through a telescope. To make matters worse, the atmosphere glows slightly due to the light of the moon and stars, as well as the light produced when charged particles from space collide with the air. Also, infrared and ultraviolet radiation, the segments of the electromagnetic spectrum that lie on both sides of visible light, are strongly affected. The Earth's atmosphere prevents much of the infrared radiation from reaching the ground, while the ozone layer does an even more thorough job of blocking ultraviolet light.

Astronomers have tried to reduce these constraints by situating their telescopes at high elevations where the air is thinner than it is at lower levels. In the 1980s, the success of the U.S. space program created the possibility of avoiding these problems altogether by putting a telescope into orbit above the Earth's atmosphere. After a delay of several years caused by the *Challenger* disaster on Apr. 25, 1990, the space shuttle *Discovery* deployed the Hubble space telescope 610 km (380 mi) above the Earth.

Named for the American astronomer Edwin (1899–1953), the space telescope is a massive object about the size of a railroad tank car. Weighing 11.2 metric tons (12.3 tons), it is 13.3 m (43.5 ft) long and has a diameter of 4.3 m (14 ft). Its electrical system is powered by two arrays of solar cells that extend its width to 12 m (40 ft). The main mirror has a diameter of 2.4 m (94.5 in.) and reflects light back to a .3-m (12-in.) convex secondary mirror. Light from this mirror is sent back through a hole at the center of the main mirror, where it is picked up by one of five scientific instruments. One of the most complex scientific instruments ever built, it is comprised of more than 400,000 parts, and has 42,000 km (26,000 mi) of electrical wiring. It is also one of the most expensive scientific instruments ever built. The telescope itself cost about $2.4 billion to build and has an annual operating cost of $270 million. Putting it into orbit cost an additional $450 million, as did a later repair mission.

Operational control over the space telescope is administered by NASA's Goddard Space Flight Center in Greenbelt, Md. Astronomical observations are the responsibility of the Space Telescope Science Institute, which in turn is administered by the Association of Universities for Research in Astronomy. Although the latter is a consortium of 21 American universities, the Hubble telescope has a strong international component. Some of its equipment has been provided by the European Space Agency, so foreign astronomers have been able to use the space telescope for their research. Data and images that have been gathered by the telescope is "owned" for a period of 1 year by the scientist who conducted the research, after which the data enter the public domain.

The space telescope makes a complete orbit around the Earth in about 1½ hours. Even if it were stationary, keeping the telescope trained on a region being observed would be a major technical challenge. Aiming the telescope is done by locking on to selected guide stars that provide a navigational fix for the telescope. Changes in the telescope's orientation are effected by four flywheels; powered thrusters cannot be used, for their exhaust would adversely affect viewing conditions. The telescope's position is monitored by gyroscopes that sense when it is properly located in three dimensions. Because the telescope is unmanned, all of the data it gathers have to be sent back to Earth electronically. This is done through the use of relay satellites that orbit at an altitude of 35,800 km (22,400 mi).

Contrary to common belief, the Hubble telescope does not "see" appreciably farther than the most powerful Earth-based telescopes. Rather, its advantage lies in its greater resolution. This refers to the ability to discern faint points of light that are in the presence of other sources of light. The best Earth-bound telescopes are capable of a resolution of about 1 arc-second; the Hubble can resolve down to 0.1 arc-second. This is equivalent to discerning one of the periods on this page at a distance of about 1.6 km (1 mi).

At first, the great potential of the space telescope was not fully realized due to an error in the grinding of the main mirror that left it too flat. Although the devi-

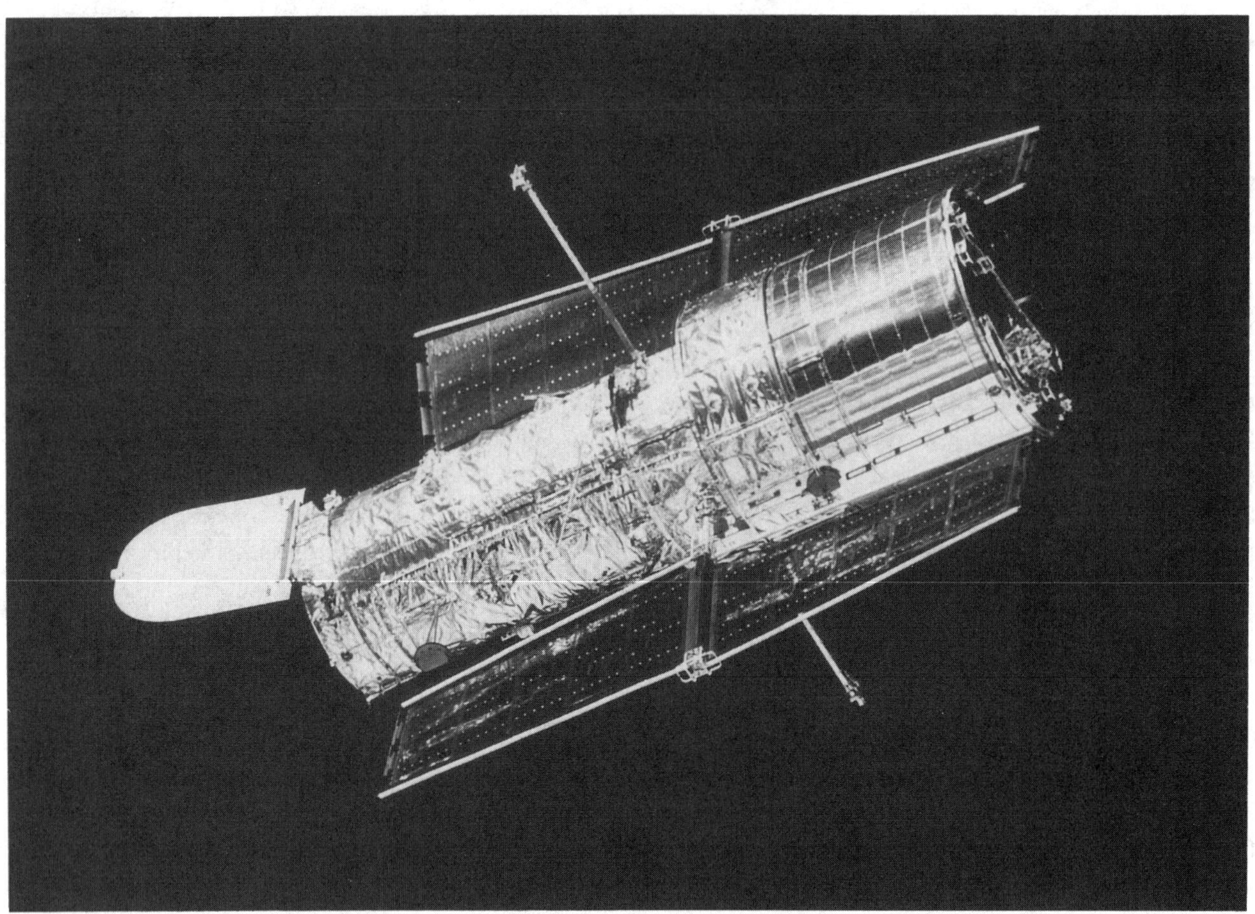

The Hubble space telescope shortly after its release from the cargo bay of the space shuttle *Discovery* (courtesy NASA).

ation is only .0001 inch, it resulted in a spherical aberration that significantly blurs the images collected by the telescope. When this imperfection became evident during the first months of the telescope's operation, many astronomers were understandably despondent, but computerized imaging technologies removed much of the bluriness.

On Dec. 4, 1993, the space shuttle *Endeavor* rendezvoused with the space telescope in order to effect some much-needed repairs. The most important of these entailed fitting the telescope with a system of mirrors that partially compensated for the spherical aberration caused by the main mirror. This significantly improved the telescope's sensitivity to light from faint objects, but at the cost of a reduced field of vision and the loss of one of the main instruments, a high-speed photometer. The repair mission also included the fitting of new solar panels and a rebuilt camera that also helped to compensate for spherical aberration.

The Hubble space telescope is expected to have a useful life of about 15 years. The problems that plagued it during its first years of operation delayed the momentous discoveries that many had predicted. Today, however, observations made by the orbiting telescope have provided insights into the age and size of the universe, how it was created, and even its ultimate fate.

See also *Challenger* DISASTER; COSMIC RAYS; EXPANDING UNIVERSE; FLYWHEEL; INFRARED RADIATION; NATIONAL AERONAUTICS AND SPACE ADMINISTRATION; SPACE SHUTTLE; TELESCOPE, REFLECTING; TELESCOPE, REFRACTING; ULTRAVIOLET RADIATION

Further Reading: Eric J. Chaisson, *The Hubble Wars*, 1994.

Human Biological Variation—see Biological Variation, Human

Human Genome Project

An international effort begun in 1988, the Human Genome Project seeks to determine the DNA sequence of all of the genetic information present in the 23 human chromosomes. The completion date has been set for the year 2003, the 50th anniversary of the date when James Watson and Francis Crick first proposed their model for the structure of DNA.

The DNA molecule is a double-stranded helix in which each strand is a linear array of the four nucleotide building blocks: adenine, guanine, cytosine, and thymine. Genetic information is determined by the specific sequence of these four nucleotides as they occur in a DNA chain. Genes are DNA sequences that, for the most part, encode proteins and include nucleotide sequences that define the first and last amino acids. Some sequences that function by regulating gene expression also are known, but at this point, much of the blueprint for human life remains to be deciphered. We know that only about 5 percent of the DNA in the human genome is expressed. In other words, the estimated 50,000 to 100,000 human genes comprise only 5 percent of the human genome. One purpose of the Human Genome Project is to determine the function of the other 95 percent of human DNA.

The human genome is composed of approximately 3 billion nucleotides organized into 23 chromosomes. The Human Genome Project will determine the identity and order of each of these 3 billion nucleotides. Once the DNA sequence is known, storing and analyzing the huge mass of information is an equally challenging undertaking that will require high-capacity computers and sophisticated software. The data will be stored at a few central depositories. An estimate of the magnitude of the project is illustrated in Fig. 1. Using this analogy, the stack of books required to record the sequence of a single human genome would reach to a height of 15 stories.

Determining the nucleotide sequence of 3 billion nucleotides is a daunting technical task. Even sequencing a single chromosome with its millions of

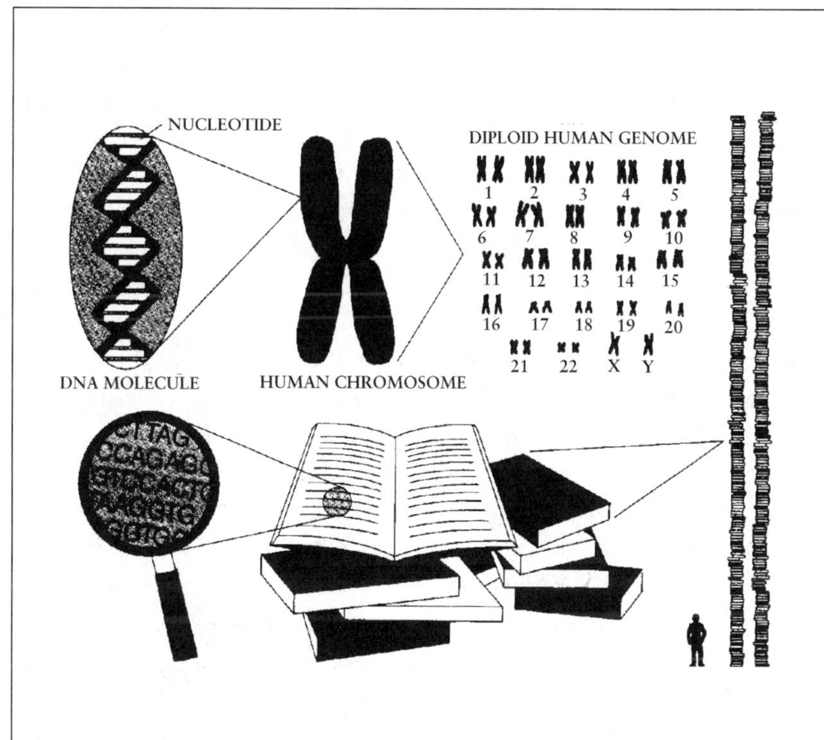

Fig. 1. The Genomic Book of Life If each nucleotide of the genome is equated to a single letter in this textbook and the nucleotide sequence is written *without spaces, punctuation, illustrations, or tables,* then a gene of 4,000 nucleotides could be represented on one page of type. At this rate, the DNA sequence of the haploid human genome (23 individual chromosomes) would fill 750,000 pages, 50 meters high. Twice this amount would be needed to represent the genetic information of diploid genome (23 pairs of chromosomes) (from D. A. Miklos and G. A. Freyer, *DNA Science,* 1990).

nucleotides is a stiff challenge, for only a few hundred nucleotides can be determined at a time. Moreover, sequencing requires an amount of DNA that vastly exceeds the amount that can be easily recovered from single human cells. Amplification of small segments of DNA by cloning makes it possible to recover enough DNA to sequence, and sequencing just small segments at a time will permit the entire sequence eventually to be determined. When the total DNA from an organism is cloned in bacteria, a large number of these bacteria should contain all of the DNA sequences from the organism of interest. Such a collection is called a *genomic library*.

A human genomic library is the starting point for the Human Genome Project. The library contains all human DNA sequences, isolated on restriction fragments of manageable size. Sequencing can begin on these fragments. The problem with a genomic library is that the library randomizes the fragments: No longer are individual fragments of DNA aligned with neighboring sequences. One big task is to fit the pieces into the jigsaw puzzle created when large DNAs are cut into fragments by restriction enzymes.

The first step is to construct a map of each of the human chromosomes, on which are noted the location of known genes and physical markers such as recognition sites for restriction enzymes. Genetic linkage maps are constructed using techniques developed around 1900 by Thomas Hunt Morgan (1866–1945) for his studies of *Drosophila*, the fruit fly. Genes are placed on a genetic map by studying the inheritance patterns of specific genes. Those genes that are always inherited together are located on the same chromosome. Genetic distance, however, is immense in molecular terms, as genes that are closely linked can be a million nucleotides apart. More refined maps are made using restriction enzymes, whereby DNA is cut with restriction enzymes and the relative order of recognition sites for different enzymes determined. Genes can be characterized by their restriction sites, and thus they can be placed on a physical map. Figure 2 is the map of human chromosome 12.

Mapping genes with restriction enzymes is facilitated by the study of inherited diseases. Family studies reveal patterns of inheritance for these diseases, and analysis of family members' DNA can reveal restriction patterns that are diagnostic of the disease. When an inherited disease is associated with a particular restriction fragment, the related gene is said to show "restriction fragment site polymorphism" (RFLP). RFLP mapping has produced crude genetic maps such as the map in Fig. 2. To determine the sequence of a mapped region of a chromosome, the DNA is cut with the appropriate restriction enzymes, the fragments are subcloned in bacteria, and then that DNA is sequenced.

The DNA double helix consists of two single strands in which each nucleotide in one strand is paired with a nucleotide in the other strand. The DNA molecule thus is somewhat like a ladder twisted into a helical shape. Nucleotide pairing follows the rule that A is always paired with T, and G with C. Because the sequence of the opposite strand can be predicted, it is necessary only to sequence a single strand. Determining the sequence of a single strand of DNA gives a pattern that looks like four lanes of bar codes on a sheet of X-ray film (see Fig. 3). Starting with the bottommost band, the DNA sequence is read upward across all four lanes, in the order in which bands appear. To get to this point formerly required a laborious process that cost an estimated $1 to $2 per nucleotide. Recently, however, the procedure has been automated, reducing the cost per nucleotide to pennies.

The Human Genome Project has been controversial, raising questions about its necessity, whose DNA will be used as a standard, the cost of financing this project at the expense of others, and questions about how the information will be used. Still, the potential benefits are vast. Identification of the molecular basis for human disease is a key goal of the Human Genome Project. By comparing DNA sequences from normal and diseased individuals, the basis for such diseases as sickle-cell anemia and cystic fibrosis have already been identified. In sickle-cell anemia, there is a single nucleotide difference between normal individuals and those with the disease, and in cystic fibrosis, a block of three nucleotides is missing. In Duchenne muscular dystrophy, where the affected gene (encompassing a DNA sequence of 2 million base pairs) is one of the largest identified, patients are missing portions of the gene. Deletions may occur in different regions, the severity of the disease related to the specific location of the defect. In addition to advancing medical practice,

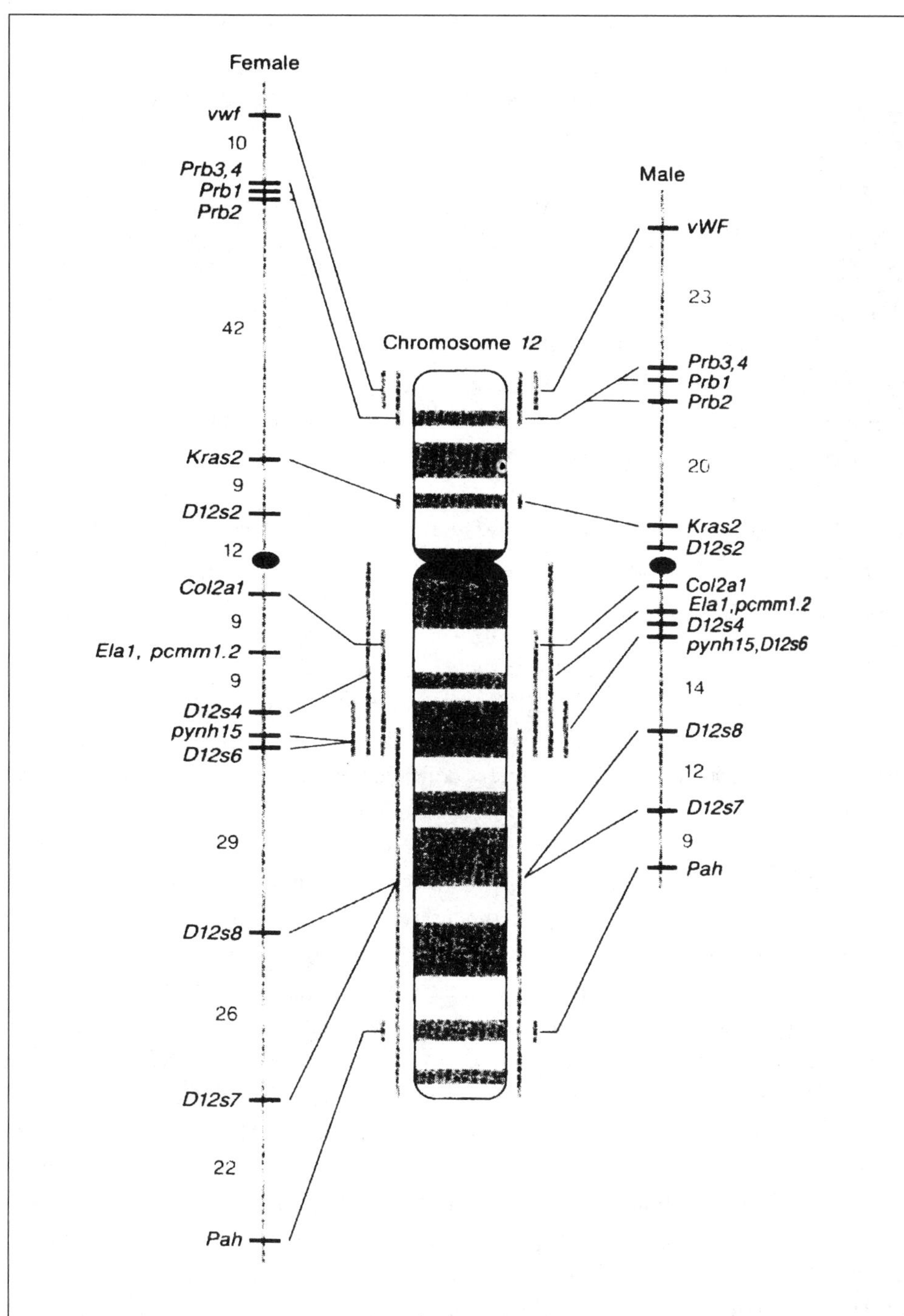

Fig. 2. The human chromosome 12, showing the location of RFLP marker loci that have been detected in various pedigrees (from A. J. F. Griffiths et al., *An Introduction to Genetic Analysis*, 1993).

analysis of the sequence of the human genome should give insights into the organization of genetic information itself. By comparing the sequence of the human genome to other organisms' DNA, scientists will be able to follow evolution on a molecular level.

The question of whose DNA sequence will serve as the standard is answered by the nature of the project: No single person's DNA will be the subject of the study. The project is an international, collaborative venture, in which DNA sequences determined by researchers all over the world will be collated. Many genes of research interest will undoubtedly be se-

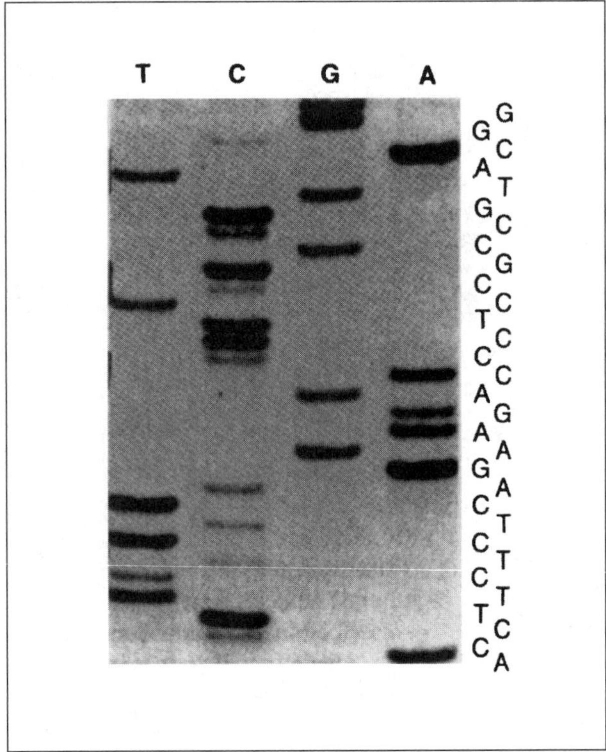

Fig. 3. A portion of the sequence data of a DNA strand (from D. A. Miklos and G. A. Freyer, *DNA Science*, 1990).

quenced with several different sources of DNA, and a consensus sequence derived. For many genes, it has already been established that there is strict conservation of DNA sequence. Other regions of the genome are polymorphic; those sequences will clearly not be the same for all individuals.

Possible abuses of DNA sequence information are also of concern. Scenarios have been painted wherein employers, lending agencies, or insurance firms might use knowledge of the "normal" human DNA sequence when assessing risk, and conceivably require genetic testing as requirement for eligibility. For example, if an individual's DNA revealed a sequence diagnostic of Alzheimer's disease, or of schizophrenia, could an insurer deny coverage to that person? Could lending institutions deny loans, even college scholarships, to those whose DNA indicates that they carry known genetic defects? Such ethical issues will undoubtedly come to the fore as the genome project proceeds.—C.E.

See also CHROMOSOME; CLONING; DNA; GENE

Human Growth Hormone

In human beings and other mammals, the growth of the long bones of the arms and legs is controlled by a hormone secreted by the pituitary, an endocrine gland located at the base of the brain. Secretion of *somatotrophin*, as this hormone is known, is governed by the interaction of two hormones produced in the hypothalamus, a part of the brain that connects with the pituitary by a thin stalk. One of these hormones (somatocrinin) promotes the release of somatotrophin, while the other (somatostatin) inhibits it.

On occasion, the human growth hormone (HGH) will be overproduced, resulting in gigantism in children, and acromegaly (excessive bone growth that leads to coarsening of the facial features) in adults. More common is its underproduction, a condition that causes dwarfism. The possibility of using HGH to combat dwarfism emerged in 1956, when a team led by Li Choh-hao (1913–), a Chinese-American biochemist, isolated the hormone. HGH was found to be a complex molecule that consisted of 256 amino acids and differed substantially from the growth hormones of other animals. In 1970, Li and his associates succeeded in synthesizing the hormone.

Therapeutic use of HGH to counter abnormally low levels in the body began in 1958. At that time, the hormone had to be extracted from the pituitary glands of cadavers, a process that was tedious, costly, and on occasion led to fatal brain infections. The production of HGH was given a major boost when genetic engineering technology developed to the point that bacteria could be used to make the hormone. In 1985, HGH was approved for sale by the U.S. Food and Drug Administration, making it the second genetically engineered drug sold in the United States.

HGH has been widely used for the successful treatment of dwarfism. It is also used to treat short children who do not suffer from a deficiency of naturally produced HGH. Although an estimated 15,000 short children were being treated in the United States during the mid-1990s, the hormone's effects on stature are uncertain. One Italian study of short children with normal HGH levels found that treatment with the hormone did not cause them to become taller adults. In contrast, a study conducted in Israel found a gain of as much as 5 cm (2 in.) after 4 years of HGH therapy. Like many therapies, the use of HGH may produce serious side effects. One Japanese study found that children treated with HGH have twice the leukemia rate of untreated children. A controlled study currently being conducted under the auspices of the National Institute of Child Health and Human Development is expected to provide a definitive assessment of HGH by the late 1990s.

Clinical studies of elderly people treated with HGH found that it can cause increases in muscle mass and bone mineral density, while at the same time enlarging organs that have begun to atrophy. Also at the same time however, HGH treatment sometimes has been accompanied by fluid retention, joint pains, and carpal tunnel syndrome. There is also some possibility that the hormone may stimulate the growth of cancerous tumors. Few medical researchers believe that HGH will prove to be a magic elixir that reverses the aging process. Even so, the attraction of HGH has stimulated the emergence of a black market in the hormone, as had already happened with anabolic steroids.

See also BIOTECHNOLOGY; FOOD AND DRUG ADMINISTRATION, U.S.; GENE THERAPY; GENETIC SCREENING; HORMONES; STEROIDS, ANABOLIC

Hybrid Corn—see Corn, Hybrid

Hydroelectric Power

Hydroelectric power plants transform the energy of falling or flowing water into electric current. Most facilities consist of an artificial reservoir fed by a river; to produce electricity, operators release water through turbine generators located in a dam at the reservoir's lower end. Smaller "run-of-the-river" plants exploit a waterway's natural flow. Pumped storage facilities use electricity to pump water into elevated reservoirs when demand is low and power cheap, and then release it back through the generators when demand peaks.

At the beginning of the 1990s, installed hydroelectric capacity totaled 615 gigawatts, or 24 percent of world electric-generating capacity. The 2.1 million gigawatt-hours generated annually by these facilities represented 7 percent of the world's primary commercial energy and 20 percent of the world's electricity. North America and Europe contain over half of this capacity. Hydroelectric development gives an indication of the extent of industrial development; North America harnesses 60 percent of its hydroelectric potential, Europe 36 percent, Asia and Latin America 10 percent each, and Africa only 5 percent.

As those figures suggest, hydroelectric power often plays an important role in industrialization. Most obvious, it offers cheap, plentiful power. The world's first large hydroelectric plant, inaugurated at Niagara Falls, N.Y., in 1895, immediately attracted aluminum smelters; manufacturers of artificial abrasives, carbide, and chemicals; and firms involved in electric ore reduction. Significantly, all these concerns exploited newly invented electricity-intensive processes.

In some contexts, hydroelectricity represents a staple product. By 1957, for example, half the output of Quebec's dams went to electricity-intensive export industries such as pulp, paper, and aluminum. (Later, the province would export bulk power directly to the United States.) Thus, hydroelectricity continued the province's historical trend of exploiting virgin resources through mature techniques. However, hydroelectric power also altered Quebec's economy by facilitating secondary manufacturing and processing industries such as furniture making and food preparation; in 1957, this sector used 30 percent of Quebec's hydroelectricity. Nor did the cost and quantity of the power represent its only revolutionary qualities. In the late 19th century, hydroelectricity helped to free manufacturers from the webs of belts and rods that had transferred the mechanical power of waterwheels—until then the dominant industrial energy resource—to their machines. Electrification allowed them to restructure their plants according to the logic of the production process, rather than according to the constraints of the power source.

Hydroelectric power developed synergistically with electricity transmission technologies. By 1900, most unexploited high-energy water sources flowed in locations that were remote from sources of labor, transport, and consumption. Because the low-voltage direct current promoted by Thomas Edison (1847–1931) lost substantial power when transmitted more than a few miles, the lure of cheap power at remote waterways encouraged the adoption of the high-voltage, alternating-current system developed by George Westinghouse (1846–1914). Energy-poor California began its impressive record of energy pioneering in the 1890s by sending hydroelectrically generated power from the cascades of the Sierra Nevada to the state's populous western coast. On the way, managers connected their systems with the agricultural communities of the Central Valley, thereby extending service to farmers decades ahead of the rest of the nation. By the time of World War I, 49 of the world's 55 high-voltage systems carried hydropower; California boasted 8 of those.

Even more than other types of generating plants, hydroelectric facilities cost far more to build than to operate. Few private firms proved able to raise the necessary funds, particularly in developing countries. Thus, hydroelectric development often fell to foreign-aid agencies or to national governments seeking to associate dam construction in the popular imagination with progress and patriotism. For instance, in the 1950s Brazil's government seized responsibility for hydroelectric construction, while limiting the country's foreign-owned power franchise to the management of distribution services. Like railroads, national governments saw hydroelectric development as the key to regional growth, and therefore too important to leave to the market. Often, governments subsidized such projects for the sake of their social benefits.

Complicating the pursuit of hydropower as a "development project," many regions of the nonindustrialized world lack the infrastructure and management skills to deliver the power, and the economy to absorb it. Hydroelectricity supplies only 3 percent of sub-Saharan Africa's energy, even though, according to one estimate, the region's 1.1 million gigawatts of exploitable hydroelectric capacity dwarfs the world's total electric-generating capacity. Nations such as Nepal, India, and China have installed many run-of-the-river projects with capacities under 100 kilowatts. Still, economies of scale, energy independence, and the mystique of massive civil construction attract many engineers and planners of the developing world (and, often, the sponsors of international aid projects). Central and South America, for example, derive 70 percent of their electricity from hydropower.

In the United States, entrepreneurs initiated the first hydroelectric projects (frequently on public lands), but the federal government has undertaken the largest efforts. Initially, the government pursued river engineering to facilitate navigation, agriculture, and flood control, rather than to generate electricity. Federal planners presented power sales as a way to finance their projects' larger goals. Private electric utilities, which struggled early in the century to repel demands by social "progressives" that the federal government take over the utility system, vilified "multiuse" hydroelectric projects as the first encroachment of the federal government into their sector. Nevertheless, with the Depression of the 1930s, power generation played the "paying partner" of many such efforts. This trend reached its apex with the Tennessee Valley Authority (TVA), commissioned by Congress in 1933.

Supporters of hydroelectricity praise it as a "sustainable" energy resource, noting its atmospheric cleanliness and the ability of pumped hydro storage to "firm up" such intermittent energy resources as solar and wind power. But because of their remoteness, large hydroelectric projects in the developing world disproportionately affect poor, politically powerless populations, entailing forced relocations and disrupting indigenous cultures and livelihoods. Less notoriously, this problem affects industrialized countries such as Norway, which relies on hydropower to generate 95 percent of its electricity. Norway has encountered resistance to hydroelectric construction in the country's northern reaches due to its effect on Saami reindeer pastoralism. In the United States, the TVA project displaced some 12,000 people, who, some researchers suggest, had little choice in the matter.

Among energy resources, only nuclear plants pose a greater risk of catastrophic malfunction than hydroelectric dams. In 1995, a nonprofit agency charged that 230,000 people had perished after two 1950-era dams burst on China's Huai River in 1975. The organization accused the Chinese government and its Soviet advisors of sacrificing safety considerations for impressive size. (The Chinese government subsequently admitted the disaster but not the death toll.) Some observers fear that China's recently begun Three Gorges Dam on the Yangtze River will suffer from the same politically tainted engineering decisions, and that it poses immense environmental risks. In general, environmental effects of large hydroelectric projects can include the appearance of new disease vectors, altered salinity, changes in sediment and nutrient levels, barriers to animal movement (especially migrating fish) and loss of terrestrial habitat, and aesthetic degradation. A largely unaddressed problem concerns water loss from surface evaporation.

In the United States, the environmental furor over hydropower continues a long-standing debate. Early in the century, conservationists fought to check rapacious exploitation of waterways by instituting a "wise use" policy for federal resources. They collided over hydroelectric policy, not only with commercial interests but also with preservationists such as John Muir (1838– 1914), who opposed any human intervention in natural ecosystems. The battle continues, as different interests, each legitimately claiming the title of "environmentalist," clash over hydropower. Although the Energy Policy Act of 1992 applauds hydroelectric development's capacity to check the nation's contribution to the (purported) trend of global warming, laws such as the Endangered Species Act make problematic the licensing and relicensing of hydroelectric plants. In a famous 1978 episode, the Supreme Court upheld a lower court ruling to halt construction of the Tellico Dam on the Little Tennessee River due to its potential to destroy the habitat of the snail darter, a small rare fish.

Assessments of hydroelectric power probably systematically underestimate its social and environmental impacts. Yet, it is easier to localize such effects for hydroelectric facilities—and to assign the blame to a single owner or operator—than for other energy resources of equal capacity. For example, a similar assessment of coal would require investigation at least of mining, rail transportation, and long-term meteorological trends. Environmental considerations will probably not slow hydroelectric exploitation in the energy-hungry economies of the developing world. In industrialized countries such as the United States, however, the future of existing and planned hydro facilities depends on the improvement of comprehensive environmental assessment techniques and political factors such as attitudes toward global warming.—A.S.

See also ALTERNATING CURRENT; DIRECT CURRENT; ENDANGERED SPECIES ACT; TENNESSEE VALLEY AUTHORITY; TURBINE, HYDRAULIC

Further Reading: Martin V. Melosi, *Coping with Abundance: Energy and Environment in Industrial America*, 1985.

Hydrofoil

A hydrofoil is similar to a hovercraft in that it makes little contact with the surface of the water. Unlike a hovercraft, however, a hydrofoil does not generate a cushion of air for this purpose. Rather, it uses winglike devices (known also as hydrofoils) that lift the craft out of the water once it reaches a certain speed. The hydrofoils are attached to spars under the hull, where they act in a manner similar to the wings of an airplane, generating a negative pressure at the top that causes the craft to lift out of the water. Because water is 840 times as dense as air, the hydrofoils can be much smaller than aircraft wings. In most cases there are two sets of hydrofoils, one fore and one aft. In some cases the main hydrofoil is at the bow; in others it is aft. In the latter case, the bow hydrofoil is steerable.

Hydrofoils come in two basic varieties: surface-piercing hydrofoils and fully submerged hydrofoils. Surface-piercing hydrofoils operate on a simple principle: If lift is lost because the hydrofoil breaks above the surface of the water, the craft will sink deeper in the water, so more lift will be generated as the hydrofoil is immersed. In the fully submerged hydrofoil, lift is controlled by the hydrofoil's angle of attack, which in turn is governed by electrical signals actuated by height sensors and gyroscopes.

A patent covering the hydrofoil principle was issued in 1869 to Farcot in France, but no craft was actually built. Also in France, Count de Lambert did build one in 1897, but it failed to work as expected. The first vessel to successfully use a hydrofoil was built in 1900 by Enrico Forlanini of Italy. In 1905, it reached the remarkable speed of 80 kph (50 mph). The telephone pioneer Alexander Graham Bell helped to design a hydrofoil based on Forlanini's patent that reached 110 kph (70 mph) in 1918. Beginning in the 1950s, hydrofoils have been built in many sizes and configuration for commercial and military use. The first commercial hydrofoil

Artist's rendition of a turbine-powered, high-speed hydrofoil (courtesy Hydrofoils, Inc.).

went into service between Italy and Sicily in 1956, and a year later the Soviet Union put the first passenger-carrying hydrofoils into service. European countries (including Russia, the world's largest operator of hydrofoils) have concentrated on commercial hydrofoils, whereas the United States has been concerned primarily with military applications.

The low friction between the hydrofoil and the surface of the water gives hydrofoil craft a considerable speed advantage over conventional vessels. At speeds over about 50 kph (31 mph), however, lift is adversely affected by cavitation, the formation of air bubbles due to the turbulence of the water. Hydrofoils also face the same inherent limitation of aircraft wings: Their size cannot be increased indefinitely, because lift is a function of surface area. Surface area increases in square increments, but weight increases in cubic increments, so a very large wing would be impossibly heavy. Finally, advanced hydrofoils employ gas turbine engines, aircraft-type construction, and complex electronic controls, all of which increase their initial and operating costs. These craft are therefore best suited for applications where speed is a paramount concern and cost is secondary.

See also BERNOULLI EFFECT; HOVERCRAFT; PROPELLER, MARINE; TURBINE, GAS

Hydrogen Bomb

The first nuclear weapons were fission bombs, devices that gained their explosive power from "splitting" atomic nuclei. An even more powerful bomb works on the opposite principle: Its destructive force comes through the fusing of atomic nuclei. The nuclei that are fused are isotopes of hydrogen, hence the term *hydrogen bomb*. Weapons of this sort also are called *thermonuclear bombs* because the fusion reaction is caused by very high temperatures.

Fusion produces large amounts of energy because the protons and neutrons of an atomic nucleus are more tightly bound together as atomic mass increases. For example, the nuclear forces in a helium atom (atomic weight = 4) are greater than the nuclear forces in a deuterium atom, an isotope of hydrogen that contains a neutron in its nucleus (atomic weight = 2). Due to the difference in nuclear forces, energy is released when the nuclei of two light atoms are fused to form a heavier atom.

In a hydrogen bomb, energy is produced by fusing deuterium with another isotope of hydrogen, tritium (atomic weight = 3). This reaction produces only about 1/11 the energy produced by fissioning an atom, but because the mass of the elements being fused is much smaller, a fusion reaction converts a much larger fraction of nuclear mass into energy.

In the mid-1930s, physicists determined that energy could be released through the fusion of light elements, and in 1937 the German-American physicist Hans Bethe (1906–) solved a long-standing scientific puzzle when he demonstrated that the sun's energy results from fusion reactions that convert hydrogen into helium. Building a fusion bomb had been proposed by the Hungarian-American physicist Edward Teller (1908–) in 1942, but a decade passed before one was actually tested.

The impetus for America's development of a hydrogen bomb was provided by the Soviet Union's detonation of an atomic bomb in 1949. Up to that time, American military and civilian leaders assumed that their country would enjoy for many years the security that came from having a monopoly on nuclear weapons. Military capabilities became an issue of growing importance as relationship between the United States and the Soviet Union became increasingly tense due to the Soviet blockade of Berlin and the rise of communist governments in Eastern Europe and China. Under these circumstances, possession of a bomb of unparalleled destructive power seemed to offer an extra measure of security. However, more was involved than the issue of military capability; the American decision to proceed with the development of a hydrogen bomb also was affected by domestic politics. At a time when Senator Joseph McCarthy (1908–1957) was leading a hunt for alleged communists in the government, reluctance to support the development of hydrogen bombs was taken as an indication of disloyalty. One of the reasons given for removing the security clearance of Robert Oppenheimer (1904–1967), the scientific director of the program that built the first atomic bombs, was that he had opposed the development of a hydrogen bomb in favor of constructing more tactical nuclear weapons. Also pushing for the development of the hydrogen bomb was the U.S. Air Force, which saw it as a means of enhancing the importance of strategic bombing.

In early 1950, the U.S. Atomic Energy Commission was ordered by President Truman to begin a crash program to develop a fusion bomb. As with the atomic bomb, the basic principle was straightforward, but many scientific and technical problems had to be solved in order to produce a working bomb. Fusing deuterium and tritium nuclei requires the application of heat at temperatures of hundreds of millions of degrees Kelvin. This is accomplished by using a fission bomb as a source of heat. Since gaseous deuterium and tritium are insufficiently dense for efficient fusion, hydrogen bombs use lithium deuteride (LiH) as a source of deuterium. Neutrons generated by the fission explosion convert lithium nuclei into tritium, which is also produced by subsequent fusion reactions.

The first fusion bombs were based on a design created by Stanislaw Ulam (1909–1984) and Edward Teller that still remains classified. It is likely that the efficacy of their design resulted from the particular configuration of the bomb's fissile and fusible materials. On Nov. 1, 1952, the United States detonated a hydrogen bomb that literally vaporized an atoll in the Eniwetok chain of Pacific islands. On Mar. 1, 1954, the United States conducted another hydrogen bomb test; this one sent substantial amounts of radioactive fallout over a wide area. Inhabitants of the island of Rongelap and the crew members of a Japanese fishing boat suffered severe radiation sickness due to exposure to the fallout.

The production of radioactive fallout is not an inherent feature of a fusion reaction, which by itself produces only tritium. The fission "trigger," however, is the source of a substantial amount of radioactive materials. Moreover, hydrogen bombs are purposely designed to increase their destructive power by surrounding the bomb's core with a uranium casing that adds another fission reaction. The detonation of a hydrogen bomb therefore procedes in a fisson-fusion-fission sequence.

The second fission reaction is the source of radioactive fallout that can travel vast distances. Fission products from a l-megaton (the equivalent of 1 million tons of TNT) hydrogen bomb reach an altitude of about 20 km (12 mi), while the detonation of a larger bomb will put these products into the stratosphere, from which they are carried to all parts of the world. The fission products include radioactive isotopes like iodine-131, strontium-90, and cesium-137. The latter two have half-lives of 28 and 30 years, respectively, so they can be a problem for a number of decades. Strontium-90 is particularly harmful because it accumulates in bones where it damages bone marrow.

The Soviet Union tested its first thermonuclear device on Aug. 12, 1953. This was a very low-yield device, but on Nov. 23, 1955, a hydrogen bomb that probably had a yield of several megatons was tested. Some historians have speculated that had the United States delayed its development of the hydrogen bomb it might have been possible to reach an agreement with the Soviet Union to limit the spread of these weapons. This never came to pass, and the two countries went on to build bombs of enormous destructive power: The United States tested 20-megaton bombs, while the Soviet Union tested one with a yield of 58 megatons—more than 3,000 times the power of the atomic bomb dropped on Hiroshima. However, in the years that followed, the terrifying capabilities of fusion weapons helped to convince both sides that efforts had to be made to reduce the danger of a thermonuclear war, and a number of arms-limitation treaties followed.

See also ATOMIC BOMB; BOMBING, STRATEGIC; ISOTOPE; NUCLEAR FISSION; SALT I AND SALT II; STRATEGIC ARMS REDUCTION TALKS

Further Reading: Dietrich Schroeer, *Science, Technology, and the Nuclear Arms Race*, 1984.

Hypersonic Flight

Hypersonic flight is defined as flight at speeds above Mach 5, five times the speed of sound. Before World War II, the concept was explored through designs by German researchers. Some experimentation continued during the war, but the Germans opted for missile designs such as the V-2 in place of true aircraft. After World War II, American and British researchers, utilizing German data, continued work on high-speed aircraft with high-altitude launches of test vehicles developed by the laboratories of the National Advisory Committee on Aeronautics (NACA). A series of near-hypersonic vehicles, such as the X-7 series, furnished a great deal of information.

The X-15, which achieved Mach 6.72 in 1967, was the premier research airplane of a new generation of experimental aircraft. But by the late 1960s, the goal of putting a man on the moon had shifted much of the hypersonic emphasis from aircraft to spacecraft. While missiles played the central role in that new policy, designers at the National Aeronautics and Space Administration (NASA), which had supplanted the NACA, also considered a piloted aircraft capable of flying into orbit and landing like a traditional airplane. This would permit much more rapid use of the aircraft and would address crucial issues of supplying manned satellites. To that end, although the X-15 was phenomenally valuable, a different aircraft was needed. Consequently, NASA's Langley Research Center initiated studies for a hypersonic boost-glide aircraft called the Dyna-Soar (X-20).

Launch of the Dyna-Soar was to take place from a high-altitude bomber, such as a B-58 or B-52, and was expected to attain more than 27,400 kilometers per hour (17,000 mph). But by 1964, funding for the program ended. The Dyna-Soar generally was regarded as a well-conceived research program, but it fell victim to budget constraints. More significantly, the program showed that the United States had difficulty committing to long-term research projects that promised little immediate practical return.

Other experiments continued while the rocket-boost NASA moon missions received most of the funding and public attention. When the Apollo missions ended in the early 1970s, however, NASA had no clear long-term plan. Instead, it simultaneously pursued work on a space station design and on a lift vehicle known as the Space Shuttle. The shuttle incorporated much of the Dyna-Soar's lifting-body design and permitted piloted reentry. It still required large rocket boosters to attain orbit, however, leaving open the question of whether or not the shuttle was a true hypersonic aircraft.

In the United States, research almost disappeared on alternatives to rocket boost, such as supersonic combustion ramjets (scramjets), which would use air in the atmosphere to achieve orbit, thus reducing overall weight. At the same time, however, the concept of "airplane-like" hypersonic operations still captivated researchers around the world. By the 1980s, Germany, Japan, France, Great Britain, and the Soviet Union all had hypersonic research programs.

Once again, the United States took the lead in moving to produce an actual aircraft. In the early 1980s, designer Tony du Pont presented a concept for a scramjet-powered, 22,700-kg (50,000-lb) aircraft to the Defense Advanced Research Projects Agency (DARPA). Robert Williams of DARPA embraced the idea and secured funding for a series of studies. On the basis of those studies, further research funding was provided for a joint interservice/interagency program to validate the concept of a hypersonic, scramjet-powered piloted aircraft called the X-30. The U.S. Air Force, NASA, the U.S. Navy, and the Strategic Defense Initiative Office all joined DARPA to commit funds to the program.

In 1986, DARPA released contracts to five airframe contractors and two propulsion companies, and allowed a third engine company to compete using its own funds. After a year, the research had produced several breakthroughs, including new materials and modeling techniques absolutely necessary for analyzing flow paths in the hypersonic regime. However, the du Pont design had not proved feasible, and in fact the contractors' proposed airplanes ended up looking little like du Pont's. By 1988, two airframe contractors and one engine company had been eliminated from competition. By this time the program had been officially named the National Aerospace Plane (NASP), and the U.S. Air Force had taken over management of the entire program. Despite almost a $1 billion investment by the federal government, and over $800 million by the contractors, by 1990 it was clear that the technology was too challenging to maintain several competing designs. At the government's prodding, the NASP program contractors all consolidated their efforts into a single team design. Funding continued to wither, however, and the scramjet technology had not progressed to the point that the government was willing to maintain the program. In 1994, NASP was canceled, with low-level hypersonic research continuing at Air Force and NASA labs.

Other nations, such as Germany, Japan, and France, experienced similar difficulties with their programs. Funded initially at even lower levels than NASP, foreign designs had even less chance of producing an operational hypersonic aircraft. Attempts to form an international consortium failed, largely due to the fact that most private and public organizations had gained funding for their programs by promising an improved competitive position in the world's aerospace industries.—L.S.

See also Apollo project; National Aeronautics and Space Aministration; Space Shuttle; space station; supersonic flight

Further Reading: Richard P. Hallion, *The Hypersonic Revolution: Eight Case Studies in the History of Hypersonic Technology*, 1987.

Hypothesis

A hypothesis is a supposition or claim about something for which one has incomplete knowledge. A mother supposes that her children are hard at work at school. In fact, perhaps, they have played hooky and gone to the record store instead. If she insisted on testing her hypothesis, she could call the principal of the school or the attendance officer, just to check up. The response would either confirm her supposition or not.

In a general sense, then, a hypothesis is a description of something that is imperfectly known, perhaps even unknown. It is a description of an object, a state of affairs, or an event, a portion of which is not known directly (factually) by experience. The description might be based on nothing more than pure opinion or fantasy. For instance, one might make the hypothetical statement that there are currently 100 alien space vehicles of various kinds and sizes circling the earth, but in fact the person might have no grounds whatsoever for suggesting this particular number. However, hypotheses can also be based on background information. The mother in the first example might have a long background of consistent behavior from which to suppose that her children have actually gone to school. In any event, when a hypothesis is made, it is not important what its inspiration or basis was; it is a hypothesis because it seeks to describe something beyond the limits of our present factual knowledge.

In the context of science, hypotheses are fundamental to all research. Scientists very rarely engage in unguided investigations or pure, undirected observational experiences. Virtually all scientific experiments are guided by anticipations of what will be observed; in a very important sense, we always have to prepare ourselves to observe anything. A scientist who intends to measure a melting point, for instance, must begin with a hypothesis that the melting point will fall into a certain range of temperature so that the experiment can be designed with an appropriate thermometer. A mercury-in-glass thermometer is not appropriate for measuring the melting points of most metals, for instance.

Finally, one could say that a scientific theory is a very large and complex hypothesis. It is, at the very least, a hypothetical description of classes of things for which we have incomplete knowledge. When we study a theory by doing scientific research, it is broken down into specific limited hypotheses that direct the construction of suitable experiments.—T.B.

See also EMPIRICISM AND SCIENTIFIC METHOD; EXPERIMENT; THEORY

I

Icebox

People long realized that cooling retarded the spoilage of food. They did not know the reason—we now know that bacterial and enzymatic action decreases at cooler temperatures—but they nonetheless devised a number of ways to keep things cool. For most of the 19th century, natural ice provided the most effective means of cooling. A large industry centered on cutting ice from frozen ponds during the winter, storing it in icehouses, and shipping it to the consumer throughout the year. The United States and Norway were the leading producers of natural ice, and their products were widely used in Europe, North America, and the Caribbean. Under the stimulus of a growing market, inventors devised many improvements in the cutting of ice. The most important was a horse-drawn ice saw patented in 1829 by Nathaniel Wyeth that reportedly lopped off two-thirds of the cost of harvesting ice. Significant improvements were also made in the design of ice houses and the insulation of ice.

The rapid growth of commercial brewing in the late 19th century increased the commercial demand for ice, while rising incomes generated a growing household demand. Most of the ice consumed by households went into kitchen refrigerators (the term *refrigerator* is properly used for iceboxes as well as mechanical refrigerators). Icebox patents date back to the early 19th century, but the real growth in their use began in the 1870s. Early iceboxes had the serious flaw of having no moisture barrier between the ice and the food, resulting in a pervasive dampness that harmed food stored for any length of time. Some improvement was effected by separating the ice and food compartments, and adding vents and baffles that caused the cooled air to flow from the ice compartment to the food compartment.

By the turn of the century, mechanical refrigeration was rapidly supplanting cooling with ice cut from ponds. In the home, however, the ice box maintained its predominance well into the 20th century, for domestic mechanical refrigerators were expensive and not always reliable. Most kitchens received regular visits from the iceman. Although delivering ice provided employment for many, the work was arduous. Delivering ice to a fourth-floor walkup tenement required strength and endurance, while commercial deliveries could be even harder, requiring the lugging of blocks of ice that weighed as much as 135 kg (300 lb) or even 180 kg (400 lb).

In 1930s, the icebox began to go into a slow decline in the United States as sales of mechanical refrigerators began to take off. The production of mechanical refrigerators, which had totaled only 5,000 in 1921, jumped to 3 million in 1937, considerably surpassing the production of iceboxes. To an increasing degree, iceboxes were to be found only in the kitchens of people too poor to afford a mechanical refrigerator or lacking the electrical service to run one.

See also BREWING; REFRIGERATOR

Further Reading: Oscar Edward Anderson, Jr., *Refrigeration in America: A History of a New Technology and Its Impact*, 1953.

Illumination, Electrical

In the summer of 1878, a traveling circus brought electric lighting to cities of the American Midwest. Promoters of the circus boasted of "18 electric light chandeliers, yielding a volume of light equal to 35,000 gas jets." Because electric lighting had not been seen before, continued the promoters, "philosophers are bewildered . . . [by it] . . . and all scientists gaze in wonder at it." During the next 2 decades, Americans continued to identify electric lighting with magic and mystery.

Beginning around 1900, business and political leaders in a number of cities secured installation of numerous and often brilliant street lights, creating what were known as White Ways. Bright lights attracted customers, retailers contended, and political leaders believed that enhanced street lighting would improve traffic safety and reduce crime. Building owners associated brilliant lighting with increasing property values. In short, social, economic, and political concerns coalesced around the idea that bright lights created an unmitigated improvement. In turn, bright lights, especially those located among the great buildings of downtown areas, created the illusion of a nighttime city, a city different and somehow more magnificent than the industrial city lit by the sun and often shrouded in fog, dirt, and pollution.

Diffusion of electric lighting into homes, schools, factories, and offices proved a slower process. During the period before 1910, prices—as high as 20 cents per kilowatt-hour—was one factor limiting widespread installation of electric lights. Even declining prices did not bring rapid increases in the use of electricity for indoor lighting. In order to increase consumption, electrical manufacturers and salespersons launched marketing campaigns aimed at increasing demand for electric lighting and other electrical products such ad flat irons, coffee percolators, and wafflemakers.

During the late 1920s, most Americans residing in cities installed electric lighting in their homes and apartments. Yet many of those installations amounted only to one or two bulbs hanging from a wire, usually in the kitchen, the main family meeting room, and the room occupied principally by women. In the distant suburbs with names like North Shore and Country Club District, well-off householders purchased "all-electric" homes featuring abundant lighting and other electric appliances such as a stove, wafflemaker, and toaster oven. For most urban Americans, such homes were unaffordable and often unknown. Instead, ordinary residents of the nation's large cities could visit the downtown. Here, all could observe the great White Ways, department stores, and theaters. Bright lights guided motorists, shoppers, and fun seekers, creating the illusion of democratic participation in the "advance" of technology and of equal access to the city's wealth.

More rare during the 1920s and 1930s were bright lights in the nation's rural districts and homes. In general, farmers used electric lights less frequently than their urban neighbors. Lower incomes were sometimes to blame, but most farmers, no matter what their income, did not yet enjoy access to electric service. In many parts of the country, rural electrification required the intervention of the federal government from the 1930s onwards.

What took place during the 1920s and 1930s was a process of learning about electricity. That process might have started in the public schools, where architects, business executives, and political leaders insisted on fitting each classroom with adequate ventilation and electric lights. The process of learning about electric lighting and appliances extended from public schools to the fabulous department stores located downtown in virtually every American city. If many Americans could afford to purchase only a few of their products, they could nonetheless examine merchandise under bright lights and roam through electrical and appliance departments stocked with vacuum cleaners, washing machines, toasters, and coffee percolators.

As Americans of ordinary means learned about the day-to-day uses of lights and domestic appliances, still others during the interwar years helped create the framework for widespread use later. Public schools continued to teach basic skills in fields such as electric lighting. By the 1920s, moreover, educators had adopted the idea that improved lighting not only enhanced eyesight but also enhanced deportment and encouraged students to polish study skills.

Even during harsh Depression days, electric usage crept upward, increasing consumption for lighting and for new appliances. But rising consumption represented only a portion of a more fundamental change that was taking place in the realm of electric lighting. The period between the World Wars was one in which Americans created and taught one another about the uses and advantages of electric illumination. What took place during the interwar decades was creation of an educational, social, and technical infrastructure that underlay and guided purchases of electric lights and other electrical equipment and appliances. Post–World War II prosperity presented the final dimension in the assembly of that infrastructure and in the widespread diffusion of electric lighting.

Following World War II, the number of electric lighting installations skyrocketed. In 1939, the nation's lighting manufacturers had produced 517 million large incandescent lamps. In 1945, manufacturers produced 787 million lamps. By 1955, production jumped to more than 1 billion, and in 1970, production rose to 1,582 billion lamps, representing an increase in production since prewar days of 305 percent. Starting in 1945, moreover, manufacturers launched production of fluorescent lamps, turning out 37 million. By 1970, they had boosted output in that area to 267 million, an increase since 1945 of 721 percent.

Public policy facilitated this remarkable increase in the number of electric lights. During the Depression era, officials of the Federal Housing Administration stipulated construction standards for the homes on which they insured mortgages, including a stipulation of two electric circuits in every home—one for light and one for power. After World War II, leaders of the Veteran's Administration, which also insured home mortgages, adopted similar rules. However, most consumers gave little thought to the role of government agencies, assuming that their choices were based on personal tastes acted out through the market.

Whether in 1880 or 1980, rare were the Americans who understood the technical, social, or political aspects of electric lighting systems. In the earlier period, electric lighting seemed like a magic and mysterious science. In a later period, that magic and mystery, now yoked to large corporations and public policy, appeared as if by magic in virtually every public and private setting, whether inside or outdoors. Educators, business leaders, politicians, and ordinary consumers had created a system that at first encouraged and then later required vast amounts of electric lighting. One of the paradoxes of electric illumination in the United States, however, was the impression of near-sovereign Americans making choices about lighting in a regime that actually permitted only a few details to them.—M.R.

See also ELECTRIFICATION, RURAL; IRONS, DOMESTIC; LIGHTS, FLUORESCENT; STOVES, COOKING; TENNESSEE VALLEY AUTHORITY; VACUUM CLEANER; WASHING MACHINE

Further Reading: David E. Nye, *Electrifying America: Social Meaning of a New Technology, 1880-1940*, 1990. Mark H. Rose, *Cities of Light and Heat: Domesticating Gas and Electricity in Urban Americas,* 1995.

Immunization

Immunization, the process by which people develop resistance to disease-causing organisms, occurs both naturally and artificially, and starts during gestation. A mother passes immunoglobulin G to a fetus through the placenta, offering protection that lasts for about 6 months after birth. During the 18th week after conception, the fetus begins to develop its own immune system and afterwards can synthesize immunoglobulins M and A if challenged by an intrauterine infection. At birth, bone marrow "stem cells" flow through the umbilical cord, giving the baby its lifelong ability to produce an enormous array of immunogenic cells. Antibody-producing "B cells" alone can generate somewhere in the magnitude of 100 million different molecules. Natural immunization continues throughout life as foreign antigens, including those of pathogenic organisms, provoke challenges that result in a long-term response called *immunologic memory*.

For all but the last 200 years, natural immunization alone protected human beings in their struggle for survival. Artificial immunization began during the 18th century. In 1718, Lady Mary Montagu, wife of the British ambassador to Constantinople, described what the Turks called ingrafting, or inoculating people with dried smallpox pustules ("varioles") to prevent the full-blown disease. Variolation must have derived from observing that not everyone who contracted smallpox died. The practice increased, especially during epidemics and particularly after France's Louis XV died of smallpox in 1774, and Louis XVI immediately took variolation. Still, it is certain that many ingrafted people developed smallpox from this live-virus inoculation, unknowingly contributing to epidemics.

English physician Edward Jenner (1749–1823), a practicing variolator, noticed that implanting pustules did not produce the expected symptoms among people who showed evidence of having previously contracted the milder scab-raising disease, cowpox. In May 1796, he used a needle to transfer a cowpox pustule from the hand of a milkmaid, Sarah Nelms, to the arm of 9-year-old James Phipps. Two months later, he inoculated the boy with smallpox, who afterwards failed to show even mild signs of the dreaded disease. Jenner published the results of several *vaccinations*—named for the Latin word for cow—in 1798, and, within 20 years, millions followed his practice. Among other things, vaccination allowed immunized Europeans to push into areas where smallpox decimated the indigenous people. Ultimately, though, vaccinia immunization accounted for humankind's only complete victory over a pathogen; in 1979, the World Health Organization announced that smallpox disease no longer existed.

Jenner's work appeared when humoral theory, which had prevailed for 2 millennia, still provided the main conceptual framework for understanding disease and health. Germ theory precipitated modern medicine in the late 19th century, with its first human trial coming in the summer of 1885, when Louis Pasteur used an attenuated virus to immunize against rabies. To accomplish this, he first inoculated rabies-infected tissue from a dog to a rabbit. Usually, rabbits developed hydrophobic signs in 15 days. However, by repeatedly transferring virus from one rabbit to another, through 20 animals, Pasteur reduced the reaction time to 7 days. In so doing, he had increased the virus's adaptation to rabbits and thereby weakened it for other species. On July 6, 1885, Pasteur inoculated 9-year-old Joseph Meister, an Alsatian boy who had been bitten by a rabid dog 2 days earlier, with spinal cord tissue of a rabbit that had developed rabies in 7 days. Over the next 10 days, he made 12 more inoculations into the boy's abdomen with virus that was progressively more potent. The last inoculation had the same virulence as a fresh "mad-dog" bite. Without knowing the actual response mechanism, Pasteur built up the boy's serum antibodies before the rabies virus did irreparable damage to the central nervous system. Three months later he announced the results, and by the end of 18861 nearly 2,500 people had been treated similarly for rabies, laying the footing for germ theory in medicine.

As bacteriologists studied microbe-caused disease, they discovered that in some cases germs localized in certain tissue, but damage extended to distant organs. This raised suspicion that pathogens might be harmful because of toxins they produced. Emile Roux (1853–1933) and Alexander Yersin, Pasteur's colleagues, proved this hypothesis with the diphtheria bacillus. Then, in Germany, Emil Behring (1854–1917) and Shibasaburo Kitasato (1856–1931) demonstrated that if diphtheria toxin was injected into horses or sheep, the animals produced "antitoxins" (i.e., immunoglobulins) that could be extracted from their blood and used as serum therapy in human beings. On Christmas night 1891, Behring used serum immunization for the first time, dramatically rescuing a child stricken with diphtheria. Kitasato showed similar findings with tetanus. Before the 19th century ended, researchers learned they could use formalin to render toxins of diphtheria and tetanus harmless while retaining its immunizing effect, though these "toxoids" were not used for human beings until the 1920s.

During the 20th century, the list of vaccines gradually lengthened: for pertussis (1925); yellow fever (1937); influenza (1943); polio virus—Salk, inactivated (1954) and Sabin, attenuated (1956); measles, (1960); rubella (1966); mumps (1967); and hepatitis B (1975). In 1986, the first recombinant vaccine (for hepatitis B in human beings) and the first gene-deleted vaccine (for pseudorabies in swine) were developed, followed in 1993 by the first "naked" DNA vaccine (for human influenza A). This extended the full range of *active* immunizing agents from live virus, to attenuated virus and bacilli, toxoids, killed pathogens, and several genetically altered organisms. So-called "herd immunity" built up as more individuals become vaccinated, further protecting populations from epidemics.

The serum therapy that Behring and Kitasato originally used in 1891 actually was a type of *passive* immunization: the transfer of antibodies from one individual to another, in that case, from a horse to a human being. The transit of immunoglobulin G across the placenta from mother to fetus also conveys immunization passively. In 1940, Edwin Cohn, professor of biochemistry at Harvard Medical School, developed the cold ethanol process for fractionating blood plasma into its components. Five classes of immunoglobulins (designated G, A, M, D, and E) comprise about 20 percent of the plasma proteins, with "IgG" making up the largest component. With Cohn fractionation, it became possible to separate these classes of antibodies from blood-donor pools for intramuscular injections against such things as snake and black widow spider bites, and *Clostridium botulinum* poisoning. After U.S. Army pediatrician Ogden Bruton identified the first immunodeficiency disorder in 1952, X-linked hypogammaglobulinemia, passive immunization, with intravenous immunoglobulins, remained the only treatment for about 80 such diseases, until the first bone marrow transplantation (1968) and the first gene therapy (1990) opened other alternatives.—G.T.S.

See also EPIDEMICS IN HISTORY; GENE THERAPY; GERM THEORY OF DISEASE; SMALLPOX ERADICATION; VACCINES, POLIO

Inductive Reasoning

To reason inductively is to reason from individual cases to a general principle. We call it "enumerative induction" when we are able to enumerate a number of situations in which the same initial situation was invariably followed by the same results. Event B has followed event A in 100 separate observations; the inductive reasoner concludes that B will always follow an observation of A, making this a general principle.

Mathematicians reason by induction in a similar way. Suppose that a mathematical proposition can be proved for a case where the variable $n = 1$, as well as for the case where the variable $n = 2$. If the mathematician can also prove, in general, that the proposition is true for the case $n = n + 1$ whenever it is assumed to be true for the case of $n = n$, then the proposition is considered to be proved in general for all cases of n.

What mathematicians possess and most inductive reasoners lack is a general formulation of a proposition allowing them to prove that one case is related to another in a certain manner. Toward the end of the 18th century, David Hume (1711–1776) demonstrated powerfully that thorough analysis of experience does not offer us observed relations between events except for the mere relation of succession. In other words, in the example above, we merely observe the succession of B after A, but we do not observe anything more powerful linking the relationship of a B event to an A event. Physical scientists, then, are not so privileged as mathematicians and have no logical basis for proving the general principle of B following A.

The riskiness of inductive reasoning can be seen if we imagine a child who observes a white powder being dissolved in water and connects it with a sweet taste. This is a good connection of taste with ingredient, until someone replaces the sugar with table salt. In spite of the risks, physical science was able to make progress in inductive reasoning by categorizing materials and events with greater and greater accuracy. In the example above, the child would not have assumed that a sweet taste would follow if he had understood that sugar and table salt are different white, granular substances. By classifying materials carefully, scientists have determined many more general principles with success. Nevertheless, error is always possible because of something that we have not yet observed. Understanding this, the philosopher Karl Popper (1902– 1994) suggested that the true principle of inductive reasoning is contained in his thesis of conjecture-and-refutation. We conjecture a generalization from the experience we have and, then, we subject that generalization to crucial tests that may refute it. Only what survives is safe.

In a sad story of American farm life, a turkey follows inductive reasoning and decides that the farmer comes through the gate, each morning, to feed him and make him comfortable—until Thanksgiving morning when, alas, he comes through the gate with an axe.—T.B.

Industrial Revolution

The word *revolution* conjures up images of rapidly unfolding events that result in massive changes. There can be no denying that the Industrial Revolution was the source of vast economic, social, and cultural changes. Only the agricultural revolution of Neolithic times equaled the Industrial Revolution in the way it transformed human life. At the same time, however, the Industrial Revolution was not a single event but a complex process that unfolded over a span of many decades. In the course of those decades, industrialized modes of production took hold, but at first they did not produce a greatly accelerated pace of economic growth. From 1760 to 1820, British industrial output grew at a rate of 1.5 percent per year, about the same growth rate for the economy as a whole. Still, in its long-term consequences, the Industrial Revolution was of greater significance than the French or Russian political revolutions, although it had nothing equivalent to the dramatic events that occurred in 1789 and 1917. Historians sometimes use the year 1760 as the beginning of the Industrial Revolution, but this date is rather arbitrary. It is equally difficult to discern a precise end point for the Industrial Revolution; indeed, it can be argued that this is a revolution that continues to the present day.

The term Industrial Revolution goes far back in time. In 1788, the British writer and economic commentator Arthur Young summed up current economic and technological changes with the observation that "a revolution is in the making." By the 1820s, the expression *Revolution Industrielle* was commonly used in France. The posthumous publication of Arnold Toynbee's *Lectures on the Industrial Revolution* in 1884 helped to popularize the term, which subsequently appeared in a vast literature that surveyed and analyzed the economic, social, and technological changes that transformed the world.

Despite an immense amount of writing about it, the Industrial Revolution remains an elusive subject. It was an economic revolution, a technological revolution, a social revolution, and a cultural revolution. The changes produced by these revolutions interacted with one another in an exceedingly complex fashion. The long span of time occupied by the Industrial Revolution meant that there was no abrupt break with the past, and industrializing nations contained paradoxical mixtures of the traditional and the modern. In mid-19th century England, decades after the onset of the Industrial Revolution, half of the population still lived in the countryside, and half of the labor force continued to use preindustrial production methods.

Although the timing of the Industrial Revolution does not lend itself to precise dates, its place of origin is evident. The technological and organizational changes that transformed the traditional economy first emerged in Great Britain, reaching their most concentrated form in the Midlands of England where the growth of the textile industry provided the model for industrial modes of production. Why the Industrial Revolution first appeared in Great Britain is a question that continues to challenge economic historians. Great Britain was favored by abundant deposits of coal and iron ore, adequate stocks of capital, a fair number of people with mechanical ability and/or an entrepreneurial spirit, and a government less intrusive than most, yet it cannot be said for certain that these advantages put Great Britain in a class by itself. In any event, the Industrial Revolution soon moved beyond the confines of Great Britain into Western Europe and North America. By the latter part of the 19th century, an industrial revolution was beginning to take place in Japan, the first non-Western nation in which this occurred. In the Soviet Union and later the People's Republic of China, an industrial revolution took place under government aegis and amidst a great amount of needless suffering. Today, there are many countries that have only recently begun to experience their own industrial revolutions. Although Great Britain provided the first model of industrialization, history never repeats itself exactly. Every industrialized country (indeed, every region) has had an industrial revolution with its own distinctive features, and any attempt to find a single, unvarying pattern of industrialization is doomed to failure.

Many different enterprises were profoundly changed in the course of the Industrial Revolution, everything from steel mills to breweries. But most emblematic of the Industrial Revolution was the large factory with machines run by an external source of power. Of these, the most significant were those producing textiles. The mechanization of many phases of textile production, most notably spinning and weaving, completely transformed an industry that for centuries had resided in small-scale enterprises based on handiwork, simple technologies, and human muscle power. A plethora of inventions, many of them based on new sources of power, contributed to this transformation. Although the Industrial Revolution is sometimes thought to have been the creation of the steam engine, in actual fact the majority of mechanized operations were powered by waterwheels and water turbines. It was not until the middle of the 19th century that steam engines collectively supplied more industrial power than falling water. Steam, however, made long-distance railroad transportation possible. In conjunction with canals and improved inland waterways, the railroad greatly enlarged the size of the market served by individual firms. This in turn stimulated the growth of individual enterprises and attendant economies of scale.

According to one famous assessment, "science owes more to the steam engine than the steam engine owes to science." More generally, it often has been claimed that science had little to do with the key innovations that made the Industrial Revolution possible. This statement is true in regard to mechanical devices like spinning jennies and power looms; these contrivances could be, and usually were, designed and built by artisans with no scientific training. In similar fashion, the construction of early steam engines owed little to scientific principles. But in a fundamental sense, the steam engine owed its existence to the scientific discovery that air has weight, and that this weight could be used for practical purposes. Also, James Watt's invention of the separate condenser, while not directly motivated by a scientific principle, nonetheless took place in an environment strongly influenced by scientific approaches to understanding the world. Finally, although the machines that industrialized textile manufacture did not directly result from scientific ideas, other aspects of the cloth-making process were deeply indebted to scientific discoveries. Most notably, chlorine bleaching, a process directly descended from chemical research, opened up a major bottleneck in textile production.

While the Industrial Revolution was propelled by a series of technological innovations, no less important were changes in the way production was organized. In the place of a few workers weaving cloth or making cutlery in a small cottage, there were now hundreds of people employed in a single factory. The high cost of centralized sources of power and specialized machinery contributed to the concentration of the labor force, but no less important was the desire to keep workers under close supervision. Hired through an increasingly impersonal labor market and lacking capital of their own, workers had no personal connection to the enterprise in which they worked. The face-to-face contact and personal attachments found in the traditional craft shop gave way to impersonal managerial methods. Hierarchical authority, division of labor, formal rules, and strict work schedules became the typical means of organizing work.

Many of the workers who labored in these early factories were women. The presence of women workers was especially evident in the textile industry; in the cotton industries of Europe and the United States, half of the labor force was comprised of women, while in Japan it was even larger. Often confined to routine operations and barred from many skilled or managerial positions, women workers bore a disproportionate share of the costs of industrialization. Children too were an important segment of the Industrial Revolution workforce. Children as young as 6 were put to work in mines and factories where they put in long hours for very low wages. Although many children were hired as adjuncts to their parents, their working lives were often marked by exploitation. In 1833, this deplorable situation began to be addressed in Great Britain by the first of a series of child labor laws.

Reference to child labor forces a consideration of some of the unfortunate aspects of the Industrial Revolution. Long working days were the norm: 12 hours a day, frequently including Saturdays, were by no means unusual. In many factories, the pace of work was unrelenting, leaving little time for relaxation and socializing. Work was not only long, monotonous, and fatiguing, it could also be dangerous. Among the prominent features of early industrialization were miners killed in underground explosions, workers maimed by machinery, and operatives debilitated by lung diseases caused by the inhalation of cotton dust. Time spent away from work was often unpleasant, marked by cramped living quarters, inadequate ventilation, poor diets, and periodic epidemics. At the same time, however, while the life of workers during the Industrial Revolution was harsh, not all of the problems were the result of industrialization per se. The growth of industrial production was accompanied by rapid urbanization; for example, the English textile town of Manchester grew from 25,000 inhabitants in the 1770s to more than 365,000 by the middle of the following century. The rapid growth of the urban population jammed people together, overwhelmed sanitation facilities, and created perfect conditions for the spread of disease. In some cases, government policies made bad situations worse, as when the British government levied a tax of window glass, needlessly making dwellings dark and badly ventilated. At the same time, the in-migration and high birth rates that generated the growth of industrial cities show that, for all its shortcomings, industrial life had its benefits. Life as a factory operative was hard, but life in the countryside was surely not the idyll it is often made out to be. For many former rural dwellers, a factory job and an urban tenement represented an improved standard of living, however bleak it may seem to us today.

The "classic" phase of the Industrial Revolution in Western Europe and North America was essentially completed by the middle of the 19th century. In the following years, what is sometimes termed the Second Industrial Revolution began to gather momentum, spurred by the development of the chemical and electrical industries. In our own time, there has been much speculation about the ebbing of industrial society and the emergence of a "postindustrial" society. Should this happen, it will have been made possible by the prior success of the Industrial Revolution in transforming the way people lived and worked.

See also AGRICULTURAL REVOLUTIONS IN INDUSTRIAL AMERICA; BLEACHING; FACTORY SYSTEM; MASS PRODUCTION; POST-INDUSTRIAL SOCIETY; POWER LOOM; RAILROAD; SEPARATE CONDENSER; SPINNING JENNY; STEAM ENGINE; TURBINE, HYDRAULIC; WATERWHEEL; WEAVING

Further Reading: Peter N. Stearns, *The Industrial Revolution in World History*, 1993.

Infant Formula

Throughout most of human history, infants' survival depended on a lactating woman, the child's mother or a wet nurse (a woman who breastfeeds another woman's child for payment, charity, or friendship). Available alternatives to human milk were inadequate, indeed often dangerous, and contributed to high infant mortality rates. Studies conducted in 19th- and early 20th-century Europe indicated the differences in the risks of serious infection and death between breast- and bottle-fed infants. We now better understand the natural superiority of human milk: for example, colostrum (the liquid in a new mother's breasts before her milk is produced) and breast milk provide the antibodies, acid-base balance, and intestinal flora that protect infants against staphylococcus infections, infant diarrhea and *E. coli* infections. Breast milk provides a natural immunity to most common childhood diseases for the early months; reduces problems with allergies, constipation, and indigestion; and supplies the perfect balance of nutrients for babies' growth. The unique biochemistry of human milk has yet to be artificially duplicated, although researchers have come closer in recent decades.

Still, societies have needed alternatives to maternal breast feeding due to serious illness or death of the mother, unusual inadequacy of her milk production, abandonment of infants, social customs that discouraged women of high status from breast feeding, abandonment of infants, or maternal employment that separated mothers from their newborns. Wet nursing was the most common alternative in such circumstances, and in some societies it was quite common. (As early as 1800 B.C.E., Hammurabi's code set forth regulations on the practice of wet nursing.) When wet nurses could not be found, or proved unreliable, caretakers turned to "artificial" feeding approaches that usually relied on the milk of animals, usually cows and goats, and/or pap, cereal, or bread thinned with water or milk. Spouted feeding cups have been found throughout Europe in the graves of infants dating from about 2000 B.C.E. Visits to museums today will reveal other specially designed implements for feeding infants, ranging from pottery infant feeding bottles of the ancient Greeks to the "cornets" or feeding horns and pewter feeding bottles of 17th-, 18th-, and 19th-century Europe.

In the United States, the overwhelming majority of infants were breastfed throughout the 19th and early 20th centuries, but by the 1950s babies were routinely bottle fed as their parents followed physicians' advice. This sweeping change in infant-feeding practices has been attributed to a number of factors: developments in bacteriology, nutrition, physiology, and chemistry that resulted in new "scientific" infant formulas: the "commoditization" of infant foods by commercial enterprises—such as Nestlé, Mellins, and Horlicks—that vigorously advertised their products to physicians and mothers; and the medicalization of infant care that increasingly encouraged parents' reliance on the "scientific" advice of physicians.

Manufacturers aggressively increased marketing of infant formula in Third World countries as the birth rates declined in the United States and Europe. Millions of mothers in poor nations began to bottle feed their infants; consequently, infants have been deprived of the immunity that is passed on by mothers' milk. The risk of contracting serious diseases is compounded further in these countries, because many people do not have access to a clean water supply, facility for proper storage, or income sufficient to provide their offspring with the undiluted formula needed for sufficient nutrients. According to some researchers, the result was 10 million cases of severe malnutrition and approximately 3 million infant deaths a year from diarrhea, vomiting, respiratory infections, malnutrition, and dehydration, sometimes called "baby bottle disease." In response, people from dozens of nations organized a major campaign to put a stop to some of the promotional practices of the infant formula industry, e.g., direct consumer advertising, "milk nurses" (women in white uniforms who were not nurses) in hospitals who provided formula and instruction in bottle feeding, and the provision of free samples to new mothers and medical personnel. In the 1970s and 1980s, the Infant Formula Action Coalition (INFACT) and the International Baby Food Action Network (IBFAN) organized a boycott against the Swiss company, Nestlé, the world's largest multinational food corporation, which accounted for about half of the infant formula sales in the poor nations. The international boycott had a significant economic impact, won an agreement from Nestlé to modify its marketing claims and practices, and helped to inspire the World Health Organization (WHO) to develop an International Infant formula Code of Marketing Behavior for the industry. The WHO Assembly adopted the code in May 1980, with the support of all voting countries except one: the United States, which has not changed its position to date. Though the boycott ended in 1984, many issues remain regarding infant feeding in the Third World and educating and empowering mothers so they can make good choices for their newborns.

Ironically, while reliance on bottle feeding and formula has grown in the Third World, where poorer families can least afford it, breastfeeding has increased somewhat in the United States, especially among more educated women. Physicians' groups such as the American Pediatric Association urge mothers to breastfeed, and advice is readily available from experienced mothers in the La Leche League and other support groups. Third World governments and healthcare providers are also dedicated to promoting breastfeeding while limiting the activities of companies that sell formula. One nation, Papua New Guinea, restricts the sale of the feeding bottles as well.

Continuing research has led to the development of formula that comes closer to human milk than did earlier formulations. Today, normal full-term infants being bottle fed generally get a conventional cow's-milk–based formula, but some have adverse reactions to the protein in cow's milk formula (vomiting, diarrhea, abdominal pain, and rash) or develop symptoms of lactose intolerance (excessive gas, abdominal distention and pain, and diarrhea). The main alternative to cow's milk formula is soy based; about 20 percent of the formula sold in the United States is soy. Another alternative for infants who cannot tolerate cow's milk protein is hydrolyzed-protein formula which, although made from cow's milk, has the protein broken up into its component parts.—A.H.S.

See also *E. coli*

Further Reading: Rima Apple, *Mothers and Medicine: A Social History of Infant Feeding 1890–1950*, 1987. Penny Van Esterik, *Beyond the Breast-Bottle Controversy*, 1989.

Influenza

Influenza, or flu for short, is caused by the presence of certain viruses known collectively as *adenoviruses*. Its symptoms include nasal and lung congestion, sore throat, fever, chills, sore muscles and joints, headache, and fatigue. Some forms of flu also cause gastrointestinal upsets. Flu is readily transmitted by breathing in the aerosol sprays that are produced by sneezing and coughing, and by bodily contact such as shaking hands. The incubation period is 1 to 4 days after first exposure.

The rapid spread of the viruses that cause flu has on many occasions resulted in massive epidemics of the disease. The most disastrous example of the present century was the great influenza pandemic of 1918–19. It began as a fairly mild outbreak in the spring of 1918, but by the summer it had turned

World War I soldiers with face masks for protection against influenza (courtesy National Library of Medicine).

deadly. It subsided for a while, only to resume late in the year. By the spring of 1919 it had taken a terrible toll in human life. Relatively few people died of the flu itself, but many of its victims were so weakened by it that they were highly vulnerable to other infectious diseases, pneumonia in particular. In the year that the pandemic raged, influenza claimed between 15 and 25 million lives. By way of comparison, about 8.5 million soldiers, sailors, and airmen died in the course of World War I. In the United States, influenza killed about 550,000 people, 10 times the number that were killed in action during the war.

At this time the cause of flu was a mystery. The disease was known not to be bacterial in origin, but it was not until the 1930s that the development of more powerful microscopes allowed the observation of flu-causing viruses. The subsequent development of flu vaccines provided a modicum of protection against the disease, but there are many strains of flu virus, and no single vaccine confers protection against them all. In addition, the viruses continually mutate, further complicating efforts to develop effective vaccines. On the other hand, infection by one strain of flu usually confers some immunity to strains that are prevalent during the next few years.

Although antibiotics have effectively addressed pneumonia and other flu-related diseases and prevented a repeat of the influenza epidemic of 1918–19, influenza continues to be a significant danger, especially to older people. From 1972 to 1988, flu caused an average of 20,000 deaths a year, more than 80 percent occurring among men and women aged 65 or more. In 1976, the United States initiated a massive vaccination program in response to an outbreak of "swine flu" that was thought to be imminent. In fact, no such outbreak occurred, but a very small number of vaccinated persons contracted Guillain-Barré syndrome, with fatal results in a few instances. The administration of a flu vaccine had never been associated with Guillain-Barré syndrome, but the vaccination program had to be quickly terminated, and ended up being one of the greatest public health fiascoes in recent times.

See also EPIDEMICS IN HISTORY; VIRUS

Infrared Radiation

Infrared is the part of the electromagnetic spectrum occupying wavelengths from approximately 0.7 microns to 1,000 microns (7×10^7 to 1×10^{-3} m). The existence of infrared radiation was discovered by the British astronomer William Herschel (1738–1822) in 1800. Herschel was taking the temperature of the different spectral bands produced by passing the light of the sun through a prism. He found that the temperatures increased as he moved towards the red end of the spectrum. He then moved his thermometer to the dark area adjacent to the red area, expecting to see a drop in temperature. Instead, the temperature rose, indicating the presence of radiation invisible to the eye. Since it was below the red part of the spectrum, he called it *infrared*.

Like all other parts of the electromagnetic spectrum, infrared radiation was encompassed by James Clerk Maxwell's (1831–1879) grand theory of radiant energy that was formulated in 1873. In a more empirical vein, 19th-century astronomers conducted a number of observations of infrared sources in the heavens. In 1881, Samuel Pierpont Langley (1834–1906), an American astronomer, invented an infrared detector known as a *bolometer*. He then used it to determine that the radiant energy of the sun extended well past the zone that Herschel had discovered. In effect, Herschel had found near-infrared, whereas Langley detected the longer-wavelength radiation of the middle infrared band. In 1890, the British scientist Charles Vernon Boys (1855–1944) used very sensitive thermocouples and galvanometers to measure the infrared radiation given off by the planets.

Attempts to go much beyond this were thwarted by the atmosphere's blockage of most infrared radiation from space. This was partially remedied by sending measuring instruments aloft in balloons, rockets, and a converted C-141 cargo plane. A major breakthrough in infrared astronomy came in 1983 with the launching of the United States–Dutch Infrared Astronomical Satellite (IRAS). This was a major undertaking, for the infrared-measuring devices on board had to be kept at a temperature of 2 Kelvin through the use of liquid helium. IRAS used a 60-cm (24-in.) mirror to focus the radiation. Weighing 1,076 kg (2,372 lb), IRAS orbited the Earth for 11 months at an altitude of 900 km (559 mi). During its time in orbit, it discovered more than 250,000 sources of infrared radiation, including eight previously unknown galaxies. In 1989, the U.S. Cosmic Background Explorer (COBE) satellite detected infrared radiation at very low temperatures (around 3 Kelvin). This radiation is thought to be the red-shifted remnant of the Big Bang that had created the universe 15 million years ago. The Wide Field and Planetary Camera carried on the Hubble space telescope is capable of detecting radiation in the near-infrared portion of the spectrum, and at some point the telescope will be fitted with a more sensitive infrared detector. The European Space Agency also plans to orbit its Infrared Space Observatory (ISO) some time in the future.

A somewhat more mundane use of infrared-detecting satellites is remote sensing, which can be used for everything from the mapping of thermal currents in oceans to detecting the movement of ground troops. One common application is the mapping of cloud patterns by weather satellites, made possible by the fact that high-altitude clouds are appreciably colder than the surface of the Earth.

Observations of the infrared part of the spectrum is also the basis for closer-to-home observations. Infrared imaging plays an important role in medicine, where it is widely used to screen for the presence of tumors and other pathological conditions that are revealed by their warmth relative to the rest of the body. Infrared can also be used for therapeutic purposes: by directing infrared radiation on an affected area, it is possible to relieve many ailments, including sprains, bursitis, arthritis, and muscular aches. Another important application is a procedure used in chemistry known as *infrared spectroscopy*. This procedure allows the identification of unknown substances through an analysis of the infrared spectra that they absorb or emit.

Some modes of infrared imaging use special films that are sensitive to infrared radiation with a wavelengths up to 1.2×10^{-6} m. However, most forms of infrared imaging do not use film; they produce their images electronically. In these devices, infrared radiation causes a change in the voltage of a detector, which then can be used to generate an electronic signal that is converted to a visual image.

See also BIG BANG THEORY; DOPPLER EFFECT; HUBBLE SPACE TELESCOPE; MAXWELL'S EQUATIONS; REMOTE SENSING; SPECTROSCOPY; TELESCOPE, REFLECTING

Injector, Steam

Early steam engines used some sort of pump to bring water into their boilers. In 1858, Henri Giffard (1825–1882), a Frenchman who also pioneered the building and flying of dirigibles, invented a device to replace the pump the stream injector. Instead of being mechanically driven like a conventional pump, the injector used some of the steam produced by the engine to lift water into the boiler. Its operation was an example of the Bernoulli principle; the expanding steam accelerated as it passed through the injector, causing a drop in pressure—a partial vacuum that sucked water into the boiler.

By making it possible to do away with energy-consuming pumps, the injector increased the efficiency of steam engines. But more than this, the injector contributed to the emerging science of thermodynamics by helping to lay to rest an erroneous conception of the nature of heat. At the time of Giffard's invention, many scientists believed that heat was an actual substance known as *caloric fluid* or simply *caloric*. According to this theory, an object became hot when it absorbed caloric, and cooled when it lost caloric. Also, caloric particles were supposedly mutually repellent: consequently, steam was created when these particles came into contact with water and vigorously repelled one another.

The caloric theory explained a number of things about heat, but it was at a loss to explain the operation of Giffard's injector. If heat really was a substance, the injector seemed to operate as a perpetual-motion machine, yet by this time most scientists had come to the realization that such as device was an impossibility. In fact, the injector was no perpetual-motion machine; it used a quantity of heat equal to the quantity of work expended in lifting water into a boiler, plus the losses caused by radiation and contact with surrounding surfaces. Its operation therefore made sense only when the interconvertability of heat and work was understood. This concept in turn rested on the kinetic theory of heat, the idea that heat was the result of molecular motion. The kinetic theory of heat had been around for a number of years prior to Giffard's invention, but acceptance had come slowly at best. The apparent anomaly of the steam injector helped to discredit the caloric theory and replace it with the kinetic theory.

The intellectual consequences of the steam injector shows that the relationship between science and technology is not a one-way process, as is often assumed. The steam injector was devised for a very practical purpose, but efforts to understand its operation also contributed significantly to the advancement of scientific knowledge.

See also BERNOULLI EFFECT; CALORIC; PUMPS; ENERGY, CONSERVATION OF; ENERGY EFFICIENCY; HEAT, KINETIC THEORY OF; PERPETUAL-MOTION MACHINES; STEAM ENGINE

Further Reading: Eda Fowlkes Kranakis, "The French Connection: Giffard's Injector and the Nature of Heat," *Technology and Culture*, vol. 23, no. 1 (Jan. 1982): 3–38.

Ink

People have used various kinds of paint for tens of thousands of years. The first use of ink—that is, something used for writing and printing rather than painting—goes back to around 3200 B.C.E., when the Egyptians combined lampblack (nearly pure carbon), water, and vegetable gum. The Chinese were also early users of ink, which was usually produced as small blocks or sticks that were converted to ink by the user. Calligraphy was practiced as an art and recreation in China, and part of the pleasure of writing lay in preparing the ink prior to doing the actual writing. Ink in China was made from iron sulfate mixed with varnish obtained from the sap of the sumac tree. It was thicker than inks used in other places—rather like today's "india ink"—to accommodate the use of a brush rather than a pen.

Our word *ink* comes from the Latin *encaustum*, a purple-hued ink commonly used in Rome. The Romans also wrote with a sepia-toned ink that was obtained from the ink sacs of cuttlefish. In the 17th century, a new kind of ink came into use in Europe; it was made from an iron salt and tannic acid derived from the bark of oak trees. Colored inks were made in the years that followed through the use of various dyes. The ink used in today's ballpoint pens is different from the ink used for fountain pens, for it has to be viscous, resistant to premature drying, and capable of rapidly penetrating the fibers of the paper.

The invention of printing required the parallel development of new kinds of inks that stuck to the typeface before it was pressed onto the paper. In the early days of printing, this was done through the use of an ink made from lampblack and linseed oil. The requirements of today's high-speed printing presses have led to the development of many specialized inks. In general, these inks consist of four basic components: colorants, vehicles, additives, and solvents. Colorants are the dyes and pigments that give inks their color. Vehicles carry the colorants and cause them to bind to the surface being printed. Additives are used to affect such things as drying speed, and solvents are used to control the viscosity of the ink. Many inks do not set because they dry (that is, the vehicle evaporates) after they are applied to the object to be printed. Instead, they may be oxidized after being exposed to air, undergo polymerization, or simply be absorbed into the surface like water in a sponge. The latter process is commonly found in the inks used for the printing of newspapers; the print sets but never really dries, and is easily smudged. Some kinds of inks literally dry, but this happens as a result of exposure to ultraviolet light. This process, which saves energy and reduces air pollution, is used for glossy book and magazine covers and for packaging materials.

The production and testing of inks has become a complex and sophisticated process. Not only must inks meet special technical needs, they also must meet environmental guidelines, for the manufacture and use of inks is now heavily regulated. In the United States, the Environmental Protection Agency, the Occupational Safety and Health Administration, and the Consumer Product Safety Commission, as well as state and local agencies, all oversee various aspects of ink production and usage.

See also DYES; ENVIRONMENTAL PROTECTION AGENCY; PAINT; PEN, BALLPOINT; PRINTING WITH MOVEABLE TYPE

Innumeracy

Innumeracy is an ignorance or disability with numbers and mathematical concepts. In some ways, innumeracy is the arithmetic equivalent to illiteracy, an ignorance or disability to connect written letters with sounds and concepts. Innumeracy, however, is wider than illiteracy in that it extends to the failure to grasp mathematical concepts as well as ignorance of visual representations. It is every bit as disabling and in some ways more dangerous than illiteracy. Moreover, people who think of themselves as understanding arithmetical matters frequently prove innumerate in special areas. For example, claims by cancer patients that they have been cured by unorthodox forms of treatment may well be attributed to innumeracy. To illustrate: a male patient treated in established ways does not respond well to treatment. His oncologist tells him that there is nothing more that medical science can do. The patient has a 10 percent chance of surviving beyond 6 months. He goes to a practitioner of alternative medicine, who prescribes megadoses of vitamins, mountain stream water, and organically grown vegetables. The cancer patient recovers and is convinced that the practitioner of alternative medicine cured him. The medical establishment argues that an earlier treatment may have cured the patient. It is also pointed out that megavitamin treatments are nothing new and are usually harmless but ineffective.

The recovered cancer patient as well as the medical establishment are exhibiting features of innumeracy. The first thing that must be pointed out to the recovered patient is a numerical fact: 10 percent of patients who suffer from his form of cancer experience spontaneous remission. It is therefore premature to attribute recovery either to the patient's regimen or to prior medical treatments.

Perhaps an investigation of shared characteristics or activities of that group of 10 in every 100 who survive can move medicine closer to a cure for that form of the disease. In any event, a proper assessment of the efficacy of any variable requires not just the experience of one patient but also the experiences of a properly constructed sample of cancer patients and perhaps the use of a control group.

Innumeracy is a relative rather than an absolute condition. In 1994, the president of the Southern Regional Education Board, Mark Musick, averred that almost all students of high school age could do simple arithmetic and money exchanges, but in a 1991 test of mathematical achievement spanning 15 nations, the highest average score was Korea's and Taiwan's 73 percent, while 55 percent for the United States and Ireland placed them above only Jordan.

Another way of gauging innumeracy is by the percentage of high school students enrolled in second-semester algebra, geometry, trigonometry, and calculus. In 1982, fewer than 1 percent were enrolled. In 1987, the number had increased to 2.4 percent, then dipped back to 2.2 percent in 1990. Competency at this level of mathematics is of course essential for advanced work in mathematics, science, and technology.—A.Wa.

See also CANCER; SAMPLING; STATISTICAL INFERENCE; VITAMINS

Further Reading: John Allen Paulos, *Innumeracy: Mathematical Illiteracy and Its Consequences, 1988.*

Insecticides

A considerable amount of the world's agricultural products is destroyed by insects. The Food and Agriculture Organization (FAO) of the United Nations estimates that one-sixth of the world's crops are eaten by insects. Over the centuries, farmers have employed a number of methods and substances to reduce losses caused by insects and other pests. Sometimes nothing more was involved than picking harmful insects from the plants they were attacking. A number of substances were used to kill insects, but none of them was of much value. The first effective insecticide was paris green (copper acetoarsenite), which was used in the 1860s to stop the eastward spread of the Colorado potato beetle. In the years that followed, farmers used a number of other arsenic compounds, as well as everything from kerosene to nicotine, in their efforts to reduce insect infestation.

Towards the end of the 19th century, Othmar Zeidler in Strasbourg synthesized a chemical named dichlorodiphenyltrichloroethane. No particular significance was attached to this until 1939, when Paul Müller (1899– 1965), a chemist at the Swiss firm J. R. Geigy, A.G., discovered that the substance was highly toxic to moths. Before long it was apparent that dichlorodiphenyltrichloroethane, or DDT for short, effectively killed a variety of insects. The United States took a particular interest in DDT when the outbreak of World War II led to sharply diminished imports of commonly used insecticides. The War Food Administration tested a DDT-based substance known as Geserol, and found it to be 10 times more potent than other insecticides when used on cabbage aphids, thrips, mealybugs, whitefly larva, flies, carpet beetles, and German cockroaches. DDT was extensively used during the war to suppress malaria-bearing mosquitoes and other insect pests. Its effectiveness was dramatically illustrated in January 1944 when it abruptly ended a typhus epidemic that had broken out in Naples, Italy. Although not originally intended for use on crops, DDT was widely used for exactly this purpose in the postwar years.

At first, DDT was thought to be harmful only to insects. However, it eventually became evident that DDT posed a long-term hazard. DDT is a volatile chemical that is easily transported through the air; as a result, it soon began to appear in rivers, lakes, and ponds. Because it is more soluble in fatty tissues than in water and is not easily broken down in the body, DDT accumulates in organisms that have been directly exposed to it or who have ingested it with their food. Concentrations of DDT tend to be greatest in animals at or near the top of the food chain. In a lake, for example, plankton store up DDT, which then gets into the systems of the small fish that eat the plankton. These small fish are then eaten by larger fish, which accumulate even greater concentrations of DDT. At this point, concentrations of DDT may be 10,000 larger than the water in which the fish live. Finally, human beings at the top of the food chain risk taking in substantial amount of DDT when they eat these fish.

The long-term effect of DDT and other chlorinated phosphates is still a matter of some controversy. It is known that it disrupts the calcium metabolism in some birds, resulting in broken eggshells and consequent depleted populations. In 1962, the hazards of DDT and other insecticides were brought to public attention in Rachel Carson's book, *Silent Spring*. The hazards of DDT led to severe restrictions being placed on its use in the United States in 1972. However, DDT is still used in many countries, where it is assumed that crop reductions caused by insect infestations outweigh the dangers posed by the use of DDT.

The other major family of insecticides are the organophosphates, of which malathion and parathion are the best known. Organophosphates were first developed in Germany during the 1930s for possible use as nerve gases. As it turned out, gas was never used as a weapon during World War II, but in 1947 researchers began to experiment with the use of organophosphates as insecticides. Within a few years they were widely used for this purpose. Organophosphate insecticides are not as potent as chlorinated hydrocarbons like DDT, but they have the significant advantage of decomposing soon after application, and they do not accumulate in animal tissue.

Even relatively benign insecticides may be hazardous when they are not used properly. There is also a deeper problem with the use of any insecticide, one that was not anticipated when powerful insecticides were first employed. Initial applications of an insecticide are likely to be quite effective, but their effectiveness often is a temporary phenomenon. Insecticides are indiscriminate, and they will kill off the beneficial insects that prey on the harmful ones. The net gain of insecticide use may therefore end up being less than anticipated. At the same time, some of the harmful insects will survive the application of an insecticide because they are naturally resistant to it. With most of their natural enemies gone, and their nonresistant relatives also killed off, these insects will enjoy a favorable ecological niche that allows them to multiply rapidly. The result is a population of insects that are largely resistant to the pesticide that was used in the hope of eliminating them.

Although pesticides have their place in modern farming, it has become evident that they are used to excess in many places. One of the reasons for the overuse of insecticides is that in many countries their purchase is subsidized by the government. Moreover, insecticide production is a big and profitable business, and pesticide-producing firms have naturally sought to expand the use of their products. In the 1970s, some international organizations attempted to rein in the use of insecticides and other pesticides by drawing attention to the problems that surround their use. In the 1980s, the FAO sponsored the drafting of a code of conduct to govern commerce in pesticides. Its provisions included informing importing nations of the health hazards of specific pesticides, along with reforming labeling and advertising practices. An effort also was made to include a "prior informed consent" provision in the code of conduct. This would have forbidden the export of pesticides that had been banned or sharply restricted by the exporting country unless the government of the importing country was informed of the reasons for banning or restriction of the insecticide. Opposition from pesticide-exporting nations resulted in watered-down strictures that only called for notification of possible hazards after a sale had been made. Pressure from environmental organizations continued, however, and in 1989 a prior informed consent provision was incorporated into the FAO Code of Conduct for member nations. These guidelines are only voluntary, but they serve to indicate the importance of preventing the indiscriminate use of insecticides and other pesticides.

Growing concerns about excessive insecticide use have spurred a growing interest in integrated pest management, the use of biological controls and other natural agencies to reduce populations of harmful insects and other pests. These procedures will not eliminate the need for pesticides, but they will allow them to be applied only in situations where their use is appropriate.

See also FOOD CHAIN; GAS, POISON; INTEGRATED PEST MANAGEMENT

Instant Coffee—see Coffee, Instant

Instrument Flying

In the early days of flying, pilots had to rely on their eyesight, sense of balance, and perhaps a few primitive instruments that gave approximations of their altitude and flight attitude. These were usually adequate when the distances to be covered were short and there was a clear view of the ground below. Such was not always the case, however, and many pilots met disaster because they were unable to orient themselves while flying at night or after flying into clouds. Under these circumstances, turning maneuvers, often unintended, could cause a great deal of trouble, for the banking of the airplane produced centrifugal force that mimicked the force of gravity. This made the pilot think that the ground was directly beneath him, even though the plane might be almost on its side. The misperception was often fatal, for a turning airplane loses altitude and gains speed as it descends. Trying to recover by pulling back on the stick made a bad situation worse, for it tightened the turn and caused the plane to drop at a faster rate. Only a sighting of the ground would cause the pilot to take the proper corrective action by leveling off. Many pilots did not realize what was happening until it was too late, and their planes either crashed or broke up in the air.

This tragic situation could be remedied only by developing instruments that allowed pilots to orient their planes when the view of the ground was obscured. One important instrument that made this possible was Paul Kollsman's improved altimeter, which used precision gearing to convert barometric pressure into an accurate reading of an airplane's distance from the ground. Another crucial device was the turn indicator; it first appeared in 1917, although it was not used in large numbers of airplanes until the 1930s. Turn indicators are built around a gyroscope, a spinning disk or ring that resists forces that would cause it to tilt on its axis. Gyroscopes also exhibit the important property of "precession": If a force is applied to the side of a rotating gyroscope, it reacts with a slow rotative movement that is at a right angle to the point where the force was applied. Consequently, when an airplane turns, the reaction of the turn indicator's gyroscope can be used to move a needle on the instrument. By observing the needle, a pilot can hold to a straight course or make a precise turn without ever looking out the window. Moreover, turn indicators are calibrated so that a pilot can make a turn of a particular number of degrees by keeping the needle at one point on the gauge for an appropriate time period. In modern airplanes, the turn indicator is combined with a turn-and-bank indicator, a steel ball inside a liquid-filled curved glass tube. By keeping the ball centered in the tube, the pilot can more easily coordinate the movement of the ailerons and rudder in order to produce a balanced turn, that is, one with no skidding or slipping.

Another instrument necessary for "blind" flying is the artificial horizon. It too is based on a gyroscope. This instrument contains an indicator shaped like an airplane seen from the rear. The indicator moves relative to a line representing the Earth's horizon; when the real airplane is banked to one side, the airplane indicator appears to bank. The instrument also indicates pitch, the up-and-down movement of the plane's nose.

The third gyroscope-based device essential for instrument flying is the gyrocompass. Conceived by the French physicist Jean-Bernard-Léon Foucault (1819–1868) in the middle of the 19th century and developed in Austria by Hermann Anschutz-Kaempfe in the first decade of the 20th century, the gyrocompass is used to keep an airplane on a proper heading. By itself, a magnetic compass is of little use in an airplane, for the turning, pitching, and yawing movements prevent accurate readings. A gyrocompass is not affected by the motion of the airplane, although the ones used in small airplanes have a tendency to creep, and therefore must be recalibrated through reference to a magnetic compass during level flight. Gyrocompasses used in larger aircraft are automatically corrected by a magnetic compass.

Much of the early development of gyroscope-controlled instruments was the work of Elmer Sperry (1860–1930) and his son Elmer Sperry, Jr. In 1929, with the financial sponsorship of the Daniel Guggenheim Fund for the Promotion of Aeronautics, an airplane equipped with a radio homing device, an accurate altimeter, a Sperry gyrocompass, and a Sperry artificial horizon was flown blind by James H. Doolittle (1896–1993). Although the plane also had a pilot who could take over in the event of trouble, all of the flying was done by Doolittle, even though his outside vision was completely obscured by a hood over the cockpit. The plane took off, gained altitude, made two 180-degree turns, and then landed. Although the flight's duration was only 15 minutes, it demonstrated that it was possible to fly with reference only to one's instruments. Many pilots, however, had great difficulty in putting greater trust in their instruments than their bodily sensations. All too often, pilots were unwilling to believe that they were in a turn just because the turn indicator said so. This could be a dangerous delusion, for on some occasions, the willingness to trust one's instruments could literally be a matter of life or death.

In conjunction with other flying aids such as the automatic pilot, radio navigational signals, and eventually radar, aircraft instrumentation made aviation much more safe and reliable. Aided by aircraft instrumentation, commercial airlines could adhere to regular schedules, and military aircraft could carry out their missions despite adverse weather conditions. Except for recreational pilots with no instrument training, the days of seat-of-the-pants flying were over.

See also GYROSCOPE; MAGNETIC COMPASS; RADAR

Further Reading: T. A. Heppenheimer, "Flying Blind," *American Heritage of Invention and Technology*, vol. 10, no. 4 (Spring 1995): 54–63.

Integrated Circuit

The electronic circuits that are used for everything from missile guidance systems to garage door openers are comprised of transistors, capacitors, resistors, and diodes. In the past, these components had to be produced individually and then wired together with the other components. In the late 1950s, it became possible to produce a circuit in a single unit. This is known as an *integrated circuit*, sometimes known as the *microchip* or just "the chip." It has been one of the century's most revolutionary advances in electronics and, indeed, technology in general.

In some ways, the integrated circuit was a natural outgrowth of ongoing trends in the design and manufacture of electronic components. It was also the result of a rising demand for smaller componentry. In the late 1950s, the electronics industry was plagued by what was sometimes called "the tyranny of numbers," solid-state devices like transistors made all sorts of new electronic devices possible, but more sophisticated electronic devices required the cramming of more and more interconnected components into a circuit. Anything that allowed more components to be added and at the same time eliminated the need to wire them together would be a major step forward. This seemed feasible, for the electronics industry had already made some headway in this area. The development of printed circuit boards provided an easier and cheaper way to connect components, while transistors were being produced in large batches by making a single wafer that was then cut into the individual transistors. Still, the problem of manufacturing components as disparate as transistors and capacitors in a single unit remained.

The first working integrated circuit was invented by Jack Kilby (1923–) at Texas Instruments. He began with an idea that came to him in the summer of 1958, that while transistors were commonly based on silicon, the same thing could be done with resistors and capacitors. Accordingly, a complete circuit could be produced from silicon. The idea was realized in the form of a phase-shift oscillator that Kilby demonstrated in September 1958. Still, a problem remained: Kilby's silicon-based circuit provided for the simultaneous production of separate components on a single chip, but the components would then have to be interconnected with wires, a time-consuming and error-prone process.

The connection problem was solved by Robert Noyce (1927–1990) at Fairchild Semiconductor in what is known today as "Silicon Valley" in Northern California. In January 1959, Noyce sketched out what he called his "monolithic idea," putting all the components on a chip of silicon and connecting them with copper lines that were printed on an oxide layer. This was an idea that owed much to Jean Hoerni (1924–), another Fairchild employee, who had conceived the "planar process" (putting a layer of silicon oxide on the top of a silicon chip) for transistor production.

Both Kilby and Noyce received patents for their integrated circuit designs in 1961. Kilby had clearly been the first to come up with the idea, but Noyce and Fairchild Semiconductor believed that their integrated circuit had a better claim to patent protection, for their application had been much clearer about how component interconnection would be accomplished. The conflicting claims led to a patent-rights battle that eventually ended up in the U.S. Supreme Court. Noyce's claim was upheld over Kilby's, but by that time their two firms had agreed to share in the rights to make integrated circuits and to split the licensing fees that were paid by other manufacturers.

Sales of integrated circuits were slow at first, but in the early 1960s they began to receive a substantial boost from the military and the space program, both of which used many integrated circuits for their missile-guidance computers. By the time Apollo 11 landed on the moon, the Apollo program had used more than a million integrated circuits Government funding was also responsible for about half the research and development that had been directed at integrated circuits for the first $1\frac{1}{2}$ decades of their existence.

In 1964, the first commercial product using an integrated circuit, a hearing aid amplifier, went into production. As more integrated circuits were made for space, military, and commercial applications, more was learned about their properties, and especially about how to produce them. Manufacturers of integrated circuits traveled up a steep learning curve that resulted in plummeting prices for integrated circuits. In only 10 years, the price of a chip went from $32 to $1.27, and new commercial possibilities emerged as prices fell. In 1971, two of the staples of the consumer electronics industry appeared, the pocket calculator and the digital watch.

Also in 1971, an integrated-circuit–based device was introduced that was to have vast future consequences, the microprocessor. Sometimes described as a "computer on a chip," the microprocessor was invented by Marcien E. "Ted" Hoff at Intel, a firm that Noyce helped to found after leaving Fairchild Semiconductor. Originally designed for a Japanese business machine company, the microprocessor rapidly developed into a multipurpose integrated circuit that could be programmed for specific purposes. Today it is an essential part of personal computers and many other products.

The integrated circuit has been responsible for a staggering increase in the number of electrical components that can be packed in a given area. In 1964, Gordon Moore, a chemist who had been one of the cofounders of Fairchild Semiconductors, facetiously suggested that the number of electronic components on a single chip would double each year. What began as something of a joke subsequently was enshrined as "Moore's law," and to the amazement of its author it has held up quite well. In the early 1970s, a chip might contain a few thousand transistors; 20 years later, well over a million transistors could be put on a chip about 1.3 mm (.5 in.) square.

The ability to pack large numbers of transistors and other electronic components into a small space has been a major contributor to the development of increasingly powerful computers. Stripped to its essentials, a computer works by performing vast numbers of switching operations very rapidly. The switching is done by transistors, so in general, the more transistors, the faster and more powerful the computer. Transistors are also the basis of a computer's memory, so again, a computer's performance will depend to a significant degree on the number of transistors it contains. The integrated circuit is the perfect complement to a computer, for transistors can be made very cheaply in an integrated circuit, while other components such as resistors and capacitors are relatively more costly to make. But since a computer uses relatively few of the latter, this is not a significant problem. Older electronic devices, such as radios and televisions, use far fewer transistors and proportionately more capacitors and resistors, so integrated circuits are useful but not essential. The integrated circuit has improved existing electrical and electronic devices by enhancing their performance and reliability, and often lowering their cost. But the real significance of the integrated circuit is that it has given rise to devices like the personal computer that were scarcely envisaged at the time it was invented.

See also APOLLO PROJECT; CALCULATORS; COMPUTER, MAINFRAME; COMPUTER, PERSONAL; MICROPROCESSOR; MOORE'S "LAW"; PATENTS; PRINTED CIRCUIT; RESEARCH AND DEVELOPMENT

Further Reading: T. R. Reid, *The Chip; How Two Americans Invented the Microchip and Launched a Revolution*, 1985.

Integrated Pest Management

Insecticides and other pesticides are an essential part of modern agricultural practice. Without the application of pesticides, a sizable portion of agricultural products would be destroyed before it could be harvested. In similar fashion, farm yields would be significantly lower if crops had to compete with the weeds that are now controlled by herbicides. At the same time, however, herbicides and pesticides have been imperfect solutions to the problems they address. The health hazards of many pesticides and herbicides are evident and must necessarily limit their use. Moreover, in some cases pesticide application has actually resulted in a pest problem becoming more intractable. This occurs because over time many insect species build up an immunity to pesticides. When this happens, larger and larger quantities of pesticide must be applied, until the point is reached where the pesticide has lost most of its effectiveness.

The shortcomings of herbicides and pesticides have sparked a growing interest in an assemblage of techniques known collectively as integrated pest management. Through better methods of crop management, selective breeding, and the use of biological controls, it is often possible to substantially cut down on the amount of pesticides and herbicides while retaining high crop yields.

One element of integrated pest management is checking the growth of weeds through the use of improved cultivation practices. Crop rotation is a good weed control strategy, for each crop requires different cultivation practices, which decreases the ability of weeds to gain a foothold. Planting different crops in close proximity, a practice known as *intercropping*, also reduces weed growth. Each crop takes up space that otherwise would have been claimed by weeds, and the crops will not compete with each other if they are properly chosen. For example, experiments have shown that the intercropping of corn with mung beans may increase the yield of corn by 20 percent, because the beans suppress the growth of weeds and do not compete with the corn.

The breeding of pest-resistant crops is another component of integrated pest management. Contemporary agriculture uses large amounts of pesticides because monoculture leaves vast expanses of crops highly vulnerable to particular pests or disease. This vulnerability can be reduced by interbreeding these crops with disease- and pest-resistant varieties. These varieties are often found in a wild state, for their survival over long periods of time required them to build up a fair degree of resistance. Of course, new pests and diseases will eventually come along, making it necessary to once again find resistant strains for breeding. Unfortunately, the loss of wild areas to cultivation has diminished the number of wild species available for breeding; this may cause serious harm to agriculture in the long run.

Natural biological processes can be effectively used for the control of insects and other pests. The most basic of these is the introduction of natural enemies. This method has been successfully used to control the spread of many types of insect, and it also has been effective in checking the spread of undesirable plants. Infestations of insects can also be checked by radiation if these insects are of a species that mates only once. After being irradiated, the insects are released into crop-growing areas, where they mate with indigenous insects. Few offspring result from such unions, and the overall population is considerably diminished.

Natural biological processes also can be disrupted by spraying certain hormones on insects before they have completely developed. This causes the prolongation of the juvenile stage of the insect, and it dies before becoming an adult. Another set of natural substances used in integrated pest control are pheromones. These are naturally occurring substances that certain animals secrete in order to affect the behavior of other animals of the same species. Pheromones play a vital role in the mating practices of some insects. For example, in some insect species the female releases a pheromone in order to indicate that she is ready to mate. Accordingly, pheromones that have been synthesized in a laboratory can be sprayed over a wide area in order to prevent males from finding females. Although it is in its early stages, the use of pheromones for pest control is very promising, for it is specific to one species of insect, requires the application of only minute amounts of chemicals, and causes only a minimal amount of environmental disruption.

In sum, integrated pest management in general and biological controls in particular offer effective alternatives or complements to chemically based methods of pest control. There is considerable potential for further advances in the use of integrated pest controls, but far more money and effort have been put into the development of conventional insecticides and herbicides. Research in biological controls, for example, receives less than 1 percent of the funding that is provided for conventional pesticides.

See also CROP ROTATION; HERBICIDES; HORMONES; INSECTICIDES; MONOCULTURE; RADIOACTIVITY AND RADIATION

Further Reading: Maarten J. Crispeels and David Sadava, *Plants, Food, and People*, 1977.

Intelligence, Measurement of

For over a century, researchers have attempted to define the nature of intelligence. Although there is little agreement among experts as to the precise definition of intelligence, three common themes run through definitions proposed. Intelligence is reflected in the capacity to learn from experience, to adapt to the surrounding environment, and to analyze problem-solving or decision-making strategies.

Central to the study of intelligence is the study of

differences between human beings. Such differences are measured using a variety of tests; hence the measurement of intelligence and the study of intelligence are inseparably intertwined. The oldest tests of individual differences in mental abilities were probably the government examinations developed in China during the Qin and Han dynasties (221 B.C.E.–220 C.E.). The modern era of intelligence testing, however, began just over a century ago with the research of Charles Darwin's cousin, Francis Galton (1822–1911), in the mid-1800s. Inspired by Darwin's theory of natural selection, Galton proposed in *Hereditary Genius* (1869) that intelligence, like physical characteristics, is inherited. In *Human Faculty* (1883), he proposed that individual differences in mental ability could be accounted for by two general qualities: capacity for labor and sensitivity to physical stimuli. Galton's ideas were brought to the United States by the American psychologist, James McKean Cattell (1860–1944). In 1890, Cattell proposed a series of 50 psychophysical tests, including the rate of arm movement over 50 cm (20 in.), the squeeze strength of the hand, and the span of letters that can be recalled from memory. Cattell's approach, however, was seriously compromised a decade later by research demonstrating that performance across these numerous tests was largely unrelated, and hence they could not be tapping a similar central ability.

A different conception of intelligence was advanced at the turn of the century by the French researcher Alfred Binet (1857–1911) and his collaborator, Theodore Simon. Subsequent to a law mandating mandatory education for all children, Binet was contracted in 1904 by the French Ministry of Public Instruction to develop a test to identify "mentally defective" children who were not suitable for public school. Binet and Simon believed that the core of intelligence was mental judgment, as demonstrated by problem solving and common sense. They argued that people with sensory handicaps could nonetheless demonstrate high levels of intelligence, citing the deaf and blind author Helen Keller (1880–1968) as a case in point. Binet and Simon further proposed that intelligent thought or judgment was composed of three distinct elements: *direction* (correctly identifying a goal), *adaptation* (customizing and monitoring a strategy to achieve the goal), and *criticism* (critically analyzing one's thought processes).

Binet and Simon are responsible for the first reliable and valid intelligence test, which was first published in 1905. In 1908, they introduced the term *mental level* to describe performance on their revised intelligence scales. This revised test was very similar to modern-day IQ tests in that it consisted of a series of tests arranged according to increasing age levels. If 75 to 90 percent of children of a given age could solve a test, the test was assigned to that age level. This was the first normed intelligence test; that is, children's mental age was determined relative to the performance of their same-age peers. Binet and Simon preferred the term *mental level* (rather than *mental age*) to emphasize change and fluctuation in performance. They believed that a child's mental level was not fixed but could be changed with proper training. By 1914, Binet and Simon's test had been translated and was in use in at least a dozen countries. Henry H. Goddard (1866–1957) translated the test into English and introduced the Binet-Simon scales into the United States in 1908. In 1916, Louis Terman (1877–1956) revised and restandardized the test for use in the United States, constructing the first version of what has come to be known as the Stanford-Binet.

The notion of a flexible mental level was rapidly replaced by that of a mental age. However, this measure did not allow for easy comparison of intelligence across chronological ages. The German psychologist L. W. Stern (1871–1938) proposed that an *intelligence quotient* (IQ) be defined in terms of the ratio of mental age (MA) divided by chronological age (CA), multiplied by 100 (to make a whole number). This ratio is expressed mathematically as:

$$IQ = \frac{CA}{MA} \times 100$$

Although this ratio, today termed *ratio-IQ*, was useful for children, it proved problematic when applied to adult intelligence. Because changes in mental age slow down in mid-adolescence, the relationship between mental age and chronological age changes in adulthood. Today, IQ scores are defined in terms of a normal distribution curve, where the average score for a given age is assigned the score of 100, and each standard deviation is assigned 15 points. Thus, a person who scores above 75 percent of her or his peers, or 1 standard deviation above the mean, has an IQ of 115.

Although genetics no doubt plays a role in intelligence, Galton appears to have overemphasized the role of heredity. However, he set the stage for the "nature-nurture" controversy that is still central to the intelligence debate. Galton's hereditarian position was widely accepted at the beginning by many of the first generation of psychologists through the first decades of the 20th century, but was then widely rejected in the 1930s by the behaviorist movement. In the 1960s, however, it was revived amidst much controversy by two prominent researchers, Arthur Jensen and Hans Eysenk, and the controversy continues today.

The controversial nature of intelligence tests and intelligence research stems in part from the fact that intelligence tests are used both for research ("What is intelligence?" and "How do we measure it?") and for classifying abilities in real-world situations ("Should I hire this person?"). The social implications of intelligence testing originate largely in how tests are developed and how their results are used. In the United States, where widespread use of intelligence tests caught on rapidly, the early use of IQ tests and the research on which their use was based was frequently flawed and politically motivated.

Among the first uses of intelligence tests was the identification of "mentally defective" people subsequent to state sterilization laws that permitted eugenical sterilization. The first law was passed in Indiana in 1907, and by 1928, an additional 20 states had passed such laws These laws were based on a series of flawed studies that claimed to trace widespread mental retardation and related personality defects (e.g., criminality) through several generations of a family. The first and most important of these studies was Goddard's *The Kallikak Family*, published 1897. Based on his findings and the conclusion that intelligence was hereditary, Goddard publicized widely what he considered to be a serious threat to the American people: bad genes.

By the end of the 19th century, another threat to the integrity of the American genetic constitution was seen in the flood of immigrants from Eastern and Southern Europe. In 1882, Congress passed a law forbidding entry to the United States of lunatics and idiots; however, there was no easy screening process. In 1913, Goddard offered the solution of intelligence testing, which produced a sixfold increase in the number of deportations within a year. To assess the overall intelligence of the arriving steerage passengers, Goddard's staff tested 178 arriving immigrants over 3 months in 1914. Through untrained interpreters, immigrants were given IQ tests that included culturally biased questions, such as "What is Crisco?" and "Who is Christy Matthewson?" Goddard's flawed results showed that between 80 and 90 percent of the Eastern and Southern European immigrants tested were feebleminded. These findings were used as the basis for restrictive immigration quotas.

The widespread testing of army recruits was introduced (Army Alpha and Army Beta tests) during World War I. The validity and reliability of the testing process, however, is called into question by the Army Testing Project report that the average intelligence of the army soldier was 13 years of age, and that half the soldiers tested were retarded (described as "morons"). This report was also used to justify a variety of racist and classist claims about intelligence, which although later refuted, remained a part of common IQ lore.

Researchers have developed a number of approaches or models to study the nature of intelligence. The *psychometric* model, the oldest model, uses factor analysis to represent intelligence in terms of its components, or factors. Charles Spearman proposed in 1904 that intelligence could be understood in terms of a general factor (*g*) that pervaded performance on all tests and a set of specific factors (*s*) that were specific to a single type of ability (e.g., arithmetic). The debate over the existence and nature of a *g* continues today. The number of factors proposed ranges from Louis Thurstone's model (1938) of seven primary mental abilities to J. P. Guilford's model (1967) with 120 factors.

A more recent approach to intelligence, *information processing*, emphasizes processing abilities. This model focuses on differences in the complexity and efficiency of cognitive processes that are used, including understanding the problem, accessing memory, constructing strategies, and applying solutions. This model focuses on the speed with which problems are solved and the nature of the solutions, rather than simply overall test performance.

The *biological approach* to intelligence has benefited from increasingly sophisticated technologies. This approach studies physical changes in the brain (e.g., electrical and chemical activity) while it is involved in intellectual activity. Some research suggests that IQ scores may be linked to the speed of neural impulses. However, this research is still controversial and limited. At the opposite extreme, is the *contextual approach* to intelligence, which argues that intelligence must be wholly or partly defined in terms of the cultural values. That is, what might be considered intelligent behavior is relative, rather than fixed, and varies across cultures. This approach has important implications for the construction of culture-free (unbiased) intelligence tests.

A century of research has produced much information about the nature of intelligence and has led to more sophisticated measures and methods for studying intelligence. However, the question "What is intelligence?" still remains an active area of research. The complexity of the problem suggests that no given model will by itself be able to define and describe intelligence fully.—M.M.

See also BEHAVIORISM; EUGENICS; INVOLUNTARY STERILIZATION; NATURAL SELECTION; NEURON; NORMAL CURVE

Intelligent Transportation Systems

Throughout most of the 20th century, designers of ground transportation systems have relied on a combination of traffic signs, regulations, and the judgment of individual drivers to improve travel mobility, safety, and route selection. Rapid growth in the number of on-road vehicles, vehicle trips, and average vehicle miles traveled, however, have strained system capacity in many communities, creating severe traffic congestion, added driving hazards, and increased pollution. Now, thanks to continuing revolutions in information technology, robotics, and telecommunications, the transportation system is undergoing a transformation that may eliminate or reduce many of these problems. Vehicles are being fitted with onboard sensors, navigation devices, and other "smart" technologies that promise to enhance future travel efficiency and accident avoidance, and possibly reduce environmental damage from vehicle emissions. These advanced transportation technologies, collectively known as Intelligent Transportation Systems (ITS), include innovations for traffic control and congestion relief, en-route driver and transit information, personal security and safety monitoring, electronic payment of tolls and other travel fees, advanced emissions testing and mitigation, automatic vehicle control and collision avoidance, and commercial fleet management.

In contrast to the past emphasis on road building and other supply-side approaches to transportation development, ITS strategies seek to enhance the capacity of existing facilities, while reducing the rate of growth in travel demand. For example, automatic vehicle identification and electronic toll collection can serve as enabling technologies for congestion pricing, a market-based scheme through which travel on heavily used roads is priced according to demand. By making the cost of driving reflect the cost of congestion—estimated at $100 billion per year (U.S.) in lost productivity, wasted fuel, and accidents resulting from stop-and-go traffic—travelers will have a financial incentive to avoid or reduce trips by private vehicles during peak use periods. Along with telecommuting, these ITS measures are viewed by many as promising ways to reduce on-road travel and associated emissions, energy consumption, and accident risks.

ITS applications already benefiting today's traveler include advanced traffic signalization to smooth traffic flow and antilock braking to reduce traffic accidents. Before long, many highway users may have the benefit of precise route guidance, using global positioning satellite systems (GPS), tiny transponders attached to vehicles, and roadside beacons for automatic vehicle identification and location. Drivers and their passengers are likely to travel with greater safety, due to collision avoidance radar, special windshield vision enhancement systems, and driver alertness monitors that can detect drowsiness and erratic steering.

Eventually, the electronics and robotics available for on-road vehicles may permit fully automated travel in platoons of private cars, much like a train, but without the fixed couplings. Automated highway systems (AHS) that will steer, accelerate, and execute braking of vehicles without driver involvement are being designed and tested in the United States and Europe. Moving at high speeds with only a few feet of headway separating them, driverless vehicles are considered by many to be the ultimate goal of ITS development. Others caution that cost, legal liability, and consumer acceptance issues will preclude any widespread applications of AHS in the foreseeable future. For many drivers, giving up control of their vehicle may be unacceptable. Over time, however, concerns about control, safety, and reliability of ITS technologies are likely to dissipate as the public becomes more familiar with their general operation.

The principal barriers to ITS deployment are likely to involve issues of cost and equity more than technical feasibility. Some critics, noting the sheer growth in the number of single-occupancy vehicles and vehicle trips, argue that a fully deployed ITS infrastructure will offer only temporary relief from congestion, and that these benefits will accrue mostly to drivers who can afford expensive in-vehicle technologies. While the cost of building new roads and transit systems to accommodate increased travel demand appears to be much higher than the cost of making existing systems more efficient, clean, and safe through applications of ITS, the technology needed to improve the existing infrastructure will still require heavy public and private investment. ITS America, the public-private partnership created to coordinate ITS development in the United States, has estimated that more than $100 billion of investment will be needed over the next 20 years, roughly 80 percent coming from the private sector.

Economics aside, there are also concerns about whether ITS can be optimized to produce simultaneous benefits for mobility, air quality, energy conservation, and social equity. ITS is like an amplifier. It amplifies both what is good and what is bad about our transportation system, and extends the consequences to other domains, such as energy, environment, and land use. In some configurations, ITS technologies could boost mobility and safety dramatically, while at the same time increasing vehicular pollution, demand for foreign oil, and inequality in transportation access. This is so because by themselves capacity enhancements might only free up space for more dirty, fuel-inefficient cars to use the road, while raising the cost of private vehicle travel beyond the limits of many low-income drivers.

Optimizing the configuration of ITS technologies to satisfy simultaneously the goals of mobility, equitable access, air quality, and energy conservation will require both engineering foresight and flexible management. Estimates of ITS benefits depend on fundamental assumptions about latent travel demand, modal choice, and the capabilities of vehicles and route guidance systems operating in the future. which in turn involve assumptions about such unknowns as the rate of technological innovation, government policies regarding emissions and fuel-efficiency standards, and extent of transit use. Many transportation experts argue that ITS has the potential to double and perhaps even triple vehicle capacity on existing roadways. Any rapid expansion of capacity, however, is likely to induce some additional demand, raising the question of whether the added vehicles emitting pollution and consuming gasoline will be more than offset by the emissions savings and fuel savings gained from reduced congestion. Also pertinent is the question of whether and to what degree transit applications of ITS will mitigate potentially negative social equity, energy, and environmental consequences of enhanced mobility for private automobiles.

Mounting evidence suggests that ITS is a promising means for enhancing mobility, but that it is unlikely to contribute greatly to the goals of improved transportation equity, environmental quality, and energy security. It is primarily a bridge to other solutions in these areas, and its potential for indirect contributions is what matters most. Accordingly, ITS can be thought of as a set of enabling technologies for expanding travel choices with real-time information, assisting drivers in navigation and accident avoidance, and eventually, perhaps, internalizing the social and environmental costs of driving through road pricing and other measures.—L.H.

See also BRAKES, ANTILOCK; TRAFFIC SIGNALS

Interferon

Interferon is a protein produced in the body by T lymphocytes (commonly known as T cells) and other bodily cells. Interferon appears when cells are brought into the proximity of bacteria, viruses, and other foreign cells. It plays a vital role in the body's defense mechanism by stimulating surrounding cells to produce proteins that slow down the reproduction of viruses, retard the growth of cells (especially tumor cells), and regulate the immune response. Interferon was discovered in 1957 and became the focus of much medical research after 1980, when one form of interferon, alpha-2, began to be produced in large quantities through the use of recombinant DNA technology. In 1986, the U.S. Food and Drug Administration approved the use of interferon in clinical trials. In the ensuing years, it has been used as an experimental treatment for a number of disorders.

One of the most noteworthy applications of interferon has been its use as a treatment for cancer. Although not effective for the treatment of advanced cancers, interferon has been successfully used to treat certain kinds of leukemia, and in clinical trials it has been shown to prevent the recurrence of melanoma, a serious skin cancer. Interferon is effective in the treatment of some cancers because it inhibits the growth of cells, particularly malignant ones, and it inhibits cell motility, which is a critical part of metastasis (the spread of a cancer through the body). Interferon does not act alone in inhibiting the growth of cancer cells. Although the exact mechanisms are still not understood, it probably works in concert with tumor-suppressing proteins normally found in the body. Interferon also has been used to treat Karposi's sarcoma, a vascular tumor often associated with HIV infection. In addition, good results have been obtained through the use of interferon as a treatment for hepatitis-B and hepatitis-C.

As with many other modern medical technologies, the benefits of interferon therapy are not obtained cheaply. For example, an 18-month interferon treatment for hepatitis-C costs $10,000 to $15,000, not including the costs of medical staff and hospitalization. Interferon treatment also produces unpleasant side effects. Therapeutic doses of interferon cause flulike symptoms, which is to be expected, for when a person has the flu the accompanying fever, body aches, and other symptoms are caused not by the flu virus but by the interferon as it goes on the attack.

Interferon was the subject of intense interest when it entered public consciousness in the 1980s. At that time, extravagant hopes were expressed that interferon would be a "magic bullet" against cancer and many other diseases. It has not lived up to these unrealistic expectations but has shown its worth in the treatment of a number of illnesses. Interferon will likely take on even greater importance as more is learned about how it acts in conjunction with other organic substances.

See also ACQUIRED IMMUNE DEFICIENCY SYNDROME (AIDS); BIOTECHNOLOGY; CANCER; CLONING; DNA; FOOD AND DRUG ADMINISTRATION, U.S.; INFLUENZA; VIRUS

Internet

The Internet is a vast system of interlinked computers and local computer networks. It allows the sending and receipt of text, pictures, and sound that move along leased telephone lines in the form of binary digits. Individual computers are able to send and receive these bits by means of a modem or through direct connections with a telephone line. The Internet has no central administrative core, only a small agency that makes sure that each Internet site has a unique address (known as a Uniform Resource Locator, or URL).

Many educational institutions, government agencies, business firms, and nonprofit organizations provide Internet access for their personnel. People lacking these connections can gain access to the Internet by subscribing to a commercial online service. In addition to providing the usual Internet access, these services usually provide extra benefits, such as home shopping, electronic encyclopedias, and up-to-the minute stock prices.

The Internet originated in 1969 as a system of linked computers known as ARPAnet (for Advanced Research Projects Agency Network) and was intended to facilitate communication and data transfer by the military services and their suppliers. During the mid-1980s, ARPAnet became the Internet as the original network was linked to the computer networks of universities, laboratories, and business firms. At first confined to members of the defense establishment, the network was gradually opened to other users. By the late 1980s, it was growing at an explosive rate as the number of users doubled each year. By 1996, approximately 40 million users in 159 countries were connected to the Internet. The result has been the formation of an "information highway" of great power and utility that has gone well beyond the intentions of its original creators.

Individual and institutional users have been able to employ the Internet for a variety of purposes. One of its most common uses is for the transmission of electronic mail. Electronic mail can be sent one-to-one, or it can be used to send communications to large numbers of people who subscribe to a particular online mailing list. The Internet also contains thousands of "chat rooms" where individuals can engage in real-time discussions with like-minded individuals. They also can use the Internet to post and read messages on electronic bulletin boards. Many chat rooms and electronic bulletin boards are "moderated" by one or more individuals who try to keep the discussions on a particular topic, or perhaps screen out contributions that are deemed offensive. The Internet also enables users to tap into a vast number of online databases, allowing them to retrieve information about everything from airline schedules to up-to-the-minute sports scores. Many organizations have a "home page" that presents material that can be easily accessed by Internet users. For example, most colleges now have home pages that provide information about courses, student services, faculty, and library holdings.

The retrieval of information has been facilitated by the development of the World Wide Web (WWW), a graphical interface that allows easy access to a vast number of Internet sites. Through the use of programs known as *web browsers*, it is possible to move from one Web site to another with relative ease. Most Web sites contain highlighted words or phrases known as *hypertext links*: By clicking on these links with a mouse, the user can quickly move to another part of the site that contains more information about the subject. Further adding to the utility of the Web are programs known as *search engines*. These allow the user to simply type in a word or set of words designating a particular subject; the search engine will then look for Web sites that contain material relating to the word or words that had been typed in. Search engines help users tap into a huge reservoir of information, but a word or a phrase given to a search engine may yield thousands of entries, forcing the user to sift through a great amount of useless material. To counter this problem, many search engines allow the use of Boolean operators so a search can be limited to a more specific topic.

The domain of the Internet, sometimes known as *cyberspace*, is at times difficult to navigate. Connecting with a Web site can be slow, particularly when graphics are transmitted or when a user's modem does not allow rapid transmission. Also, some Web sites use obsolete computer systems that can be slow to respond. Accessing a Web site can entail typing in dozens of largely meaningless characters that comprise a URL, and the connection will not be made if any one of them is missed or incorrect. The task of accessing a particular address is made much easier when Web browsers have a "bookmarking" function that allows a URL to be saved for future use.

A major concern regarding Internet use centers on the presence of Internet sites that contain pornography or other materials that are inappropriate for children and adolescents. In 1996, the U.S. Congress passed the Communications Decency Act, which made it a criminal offense to provide "indecent" materials via the Internet to people under the age of 18. This legislation was vigorously attacked by civil libertarians and others opposed to governmental control over communications media. In 1997, the U.S. Supreme Court ruled that the act violated constitutional protections of free speech. Although the act was struck down, parents can still restrict their children's access to objectionable material by equipping their computers with special programs that block out material deemed unsuitable.

Authoritarian governments that seek to limit their citizens' access to information would like to have a similar ability to restrict Internet access. At present, however, material on the Internet cannot be selectively blacked out in order to prevent its being received by particular countries or geographical areas. Like other modern communication devices, the Internet can be a powerful force for social and political change by contributing to the free flow of information. Repressive regimes may attempt to deny Internet access to most of their citizens, but the widespread availability of personal computers and modems makes this difficult. And when they are successful at restricting Internet access, authoritarian governments deny their country a major source of economic advance in a world increasingly dependent on the rapid transmission of information.

The Internet also poses challenges for free-market economies. Up to now, the Internet has been a low-cost alternative to other modes of communication such as telephones and fax machines. In the immediate future, it is likely that telephone companies and other providers of communications services will seek to add to their revenues by appropriating parts of the Internet and integrating their services with it. The potential market for an augmented Internet is vast, for despite the explosive growth of the 1990s, use of the Internet is by no means universal. Of American households, 40 percent had personal computers in 1996, but only 15 percent of households were connected to the Internet. Of this latter group, only about 15 percent could be considered heavy users of the Internet. This means that, at most, only 3 percent of American households made significant use of the Internet. The commercially motivated integration of the Internet with other services—televised movies on demand, for example—would greatly expand Internet usage. This would have a social cost, however. For much of its history, the Internet and its constituent sites have been available at very low cost. Commercialization of the Internet would in all likelihood make it less accessible by putting more of it on a fee-for-service basis. Should this happen, economically disadvantaged groups would be unable to make use of many Internet-based services, and society would be further divided between the information-rich and the information-poor.

See also BINARY DIGIT; BOOLEAN LOGIC; COMPUTER NETWORK; ELECTRONIC MAIL; FAX MACHINE; MOUSE; TELEPHONE

Interstate Highway System

In April 1939, General Motors Corporation opened "Futurama" at the New York World's Fair. Visitors to the exhibit rode in two-person cars on a track and looked down at a gigantic diorama of the United States in the year 1960. They saw automobiles speeding through city and countryside at high speeds and without the hazards and delays of intersections. City and countryside appeared prosperous and at peace, a marked change for visitors who had endured 10 years of Depression and news of war around the world. If Americans built majestic express highways and improved the flow of traffic, it appeared that they could also restore prosperity and remain at peace. Futurama was the most popular exhibit at the New York Fair. So exciting was this promise of quick-flowing traffic and its many economic and social benefits, that President Franklin D. Roosevelt hosted a formal dinner at the White House for Norman Bel Geddes (1893–1958), Futurama's designer, and for leading members of the U.S. House and Senate. At this time, however, disagreements about highway finance and location proved impossible to overcome. By late 1941, preparation for war took precedence over planning for expressways.

Between 1942 and 1956, political leaders and ordinary Americans learned how improved traffic would lead to the renewal of cities and rural areas, and help to create a more prosperous economy. Countless articles in newspapers written by architects, engineers, and journalists explained and celebrated the promised benefits of urban and rural highways free from intersections. Although these writers largely disseminated ideas that were well known and time honored among road engineers, the net result was to produce images about traffic, economic growth, and urban and rural improvements that were widely shared and most likely widely approved.

After World War II, however, congestion on the nation's roads and streets only grew worse. Between 1946 and 1956, Americans doubled their ownership of cars and trucks, adding to traffic congestion in retail and manufacturing districts near downtown and in new and distant suburbs. Whatever motorists and truck operators thought about highways in the abstract, their own day-to-day encounters with traffic were often irritating, costly, and sometimes deadly.

In 1944, members of Congress and President Franklin D. Roosevelt had approved legislation creating the Interstate Highway System, but they failed to make a special appropriation of funds to begin construction. Not until 1952 did Congress and President Harry S. Truman approve limited funding, but the budgeted amount of $25 million was insignificant. In 1954, Congress voted $175 million to build the Interstate System, which still was a modest amount compared to projected expenses.

However much motorists, truckers, and political leaders endorsed the concept of building the Interstate Highway System, no such consensus surrounded the details of highway finance. Whether motorists or truckers would pay the largest share of taxes to finance construction comprised a major unresolved question. Within that question stood another point of dispute: If motorists and truckers in heavily populated states such as New York paid the bulk of taxes to construct costly roads in great cities, then who was responsible for construction expenses in sparsely populated areas like Montana? At the same time, members of Congress from farm districts would not increase the tax burdens of their constituents in order to construct roads in far-off cities, or at least they would not do so without providing a comparable benefit for their own constituents. Still others in and out of Congress wanted the federal government to pay up to 90 percent of the cost of building the interstate system, and do so through increased taxes on gasoline, diesel fuel, and trucks.

In 1955, members of Congress rejected several plans to finance construction. In particular, truck operators objected to a large increase in taxes on gasoline and diesel fuel. In all, federalism and a keen understanding of regional and industrial self-interest stood in the way of a project that enjoyed nearly universal approval. In the realm of postwar highway building, the politics of stasis took precedence over technological enthusiasm.

In 1956, President Dwight D. Eisenhower and Congress approved an expenditure of vast sums to launch construction of the Interstate Highway System. The keys to this new-found consensus lay in three of the arcane details of finance. In brief, Congress and the President agreed to limit the increase in gasoline and diesel taxes to 1 cent per gallon (3.8 liters). Next, all agreed that expensive sections of the Interstate—those running through downtown sections of the largest cities—would be paid for as a percentage of each year's total costs. Finally, Congress voted increased funds to build rural roads, even going so far as to promise that money to construct rural roads would increase each year that the costly Interstate System was under construction. In keeping with the emphasis on linking important and costly undertakings with the Cold War, Congress even changed the name of the proposed expressway system to the National System of Interstate and Defense Highways. In 1956, with knotty questions of taxes and funding formulas settled at last, state and federal road engineers and numerous contractors could get on with the work of constructing a limited-access highway system with a projected length of 66,000 km (41,000 mi).

The terms of the Highway Act of 1956 determined the framework of road politics for the next 20 years. Congress had set in motion a program that rewarded timely construction and accommodation of traffic, especially accommodation of urban traffic. By the early 1970s, state and federal engineers completed construction of most of the Interstate Highway System. Cost-benefit ratios and traffic counts guided engineers in locating roads. By the late 1980s, as engineers had long predicted, the Interstate System was carrying about 20 percent of the nation's traffic. Moreover, nearly 50 percent of tractor-trailer combinations moved on the Interstate System, the continuation of a trend that had been underway since the 1920s.

This emphasis on traffic flows and expedited construction failed to take account of emerging features on the urban landscape. Political and business leaders in cities such as New York and Chicago had been among the proponents of building the Interstate System. Free-flowing roads, it was asserted, would encourage shoppers to return downtown and help unclog central-city traffic. Instead the opposite happened, as the availability of limited-access roads and the boom in mall construction fostered the outward movement of jobs, houses, and retail sales.

Between 1939 and 1956, the politics of Interstate construction revolved around matters of finance. After 1956, engineers and their traffic-flow data took precedence. Once in place, users of the Interstate System such as truck operators and suburban commuters viewed the Interstate System in terms of their own needs. Although construction of the National System of Interstate and Defense Highways constituted one of the nation's largest and most expensive undertakings, no one idea, organization, or institution remained in charge for long.—M.H.R.

See also AUTOMOBILE; ROAD CONSTRUCTION; TECHNOLOGICAL ENTHUSIASM; TRUCKS

Further Reading: Mark H. Rose, *Interstate: Express Highway Politics, 1939–1989*, 2d ed., 1990.

Intrauterine Device

An intrauterine device (IUD) is a small device inserted in the uterus through the cervical canal in order to prevent pregnancy. Women have used various kinds of contraceptive devices for millennia, most of them of their own invention. In the 20th century, a few doctors developed rings and other shapes intended for this purpose, but it was not until the 1950s that the IUD began to be widely used. Two types of IUD gained widespread acceptance: a flat coil invented by Jack Lippes and a T-shaped spiral invented by Ralph Robinson. IUDs are made from plastic; some are coated with copper, which increases their effectiveness. Others release progesterone or a more potent steroid known as levonorgestrel, which also add to their efficacy. IUDs coated with copper are designed to remain in place for 4 years, while IUDs that release a hormone have to be replaced every year. The IUD has a number of advantages as means of preventing pregnancy. It is effective, inexpensive, and usually requires little attention after it has been inserted. Its actions last only as long as it is in place, so users experience no unusual difficulties in conceiving after the device is removed. On the other hand, the IUD does carry a number of risks, which are noted below.

The reasons for the IUD's contraceptive effect are still not completely understood. It is known that an IUD will prevent the implantation of a fertilized ovum on the walls of the uterus, probably as a result of irritating the lining of the uterus and causing secretions that prevent the development and implantation of an embryo. For those categorically opposed to terminating any fertilized ovum, the IUD may be viewed as an abortifacient, and therefore an unacceptable means of birth control.

Provided it is of the proper size and is correctly inserted, an IUD will be about 95 percent successful in preventing pregnancy. An IUD that is too small may be expelled (often with the user unaware that this has happened), while one that is too large may implant in the uterus, or even perforate it. Even an IUD of the proper size may cause problems, such as increased risk of pelvic inflammatory disease, spotting between periods, and pain and discomfort. In extreme cases, a pelvic infection may lead to permanent sterility. Women who have never given birth, as well as those who have multiple sexual partners or have a partner who is not monogamous, are more susceptible to infection. In the rare event of a pregnancy, the chance of a subsequent miscarriage is about 50 percent There is also a higher risk of an ectopic pregnancy, the implantation of the embryo in a fallopian tube. However, since pregnancies to women using IUDs are rare, the overall risk of this is quite low. There is also some indication that IUDs may cause an increased risk of contracting a sexually transmitted disease, including HIV infection.

Although in the late 1970s, 3 to 4 million women were using IUDs in the United States, in recent years their use has fallen off sharply due to their side effects and the threat of lawsuits against their manufacturers. IUDs continue to be used extensively in Third World countries, but as with all birth control methods, their successful use depends on the motivation of the couples involved, and whether the woman has sufficient power in the family to be able to limit her fertility.

See also ACQUIRED IMMUNE DEFICIENCY SYNDROME (AIDS); CONTRACEPTIVES, ORAL

Inverter

An inverter is an electrical device that converts direct current into alternating current. This makes it the opposite of a rectifier, which converts alternating current into direct current. Rectifiers are used much more frequently than inverters, for electrical power is almost always transmitted as alternating current (AC), while many electrical devices require the use of direct current (DC). However, inverters are sometimes necessary when a direct current source such as a fuel cell or battery is used to run an AC device. Also, some telecommunication equipment runs on AC, but batteries are used to maintain service in the event of a power failure. Under these circumstances, it is necessary to use an inverter to convert the direct current supplied by the battery to alternating current.

For many years, the inversion of DC to AC was done by rotary converters, a device that uses a direct-current motor to power an alternating-current generator (other kinds of rotary converters can be used to convert alternating current to direct current). Rotary converters were especially important at the end of the 19th century, when the "battle of the currents" between AC and DC was being waged in the United States and elsewhere, for they allowed systems using the different types of current to be coupled together. This in turn allowed for the operation of larger electrical power systems capable of serving wider areas and greater numbers of customers.

In the 1960s, solid-state semiconductor devices

began to be used for the conversion of direct current to alternating current, and vice versa. The most important of these is the silicon-controlled rectifier, also known as a *thyristor.* When used to convert DC to AC, several thyristors are gated together; when the gating is delayed by more than 90 degrees, the DC supply is inverted to AC. This form of alternating current differs from regular AC, for it has a square-wave pattern instead of the sine-wave pattern characteristic of the latter.

See also ALTERNATING CURRENT; BATTERY; DIRECT CURRENT; FUEL CELL; RECTIFIER; SEMICONDUCTOR

In-Vitro Fertilization

In-vitro fertilization (IVF) is the fertilization of an ovum and the initial development of an embryo outside a woman's body. The term literally means "fertilization in glass," and reference occasionally is made to "test-tube babies," although petri dishes and not test tubes are used for this purpose. IVF may be used when dysfunctional fallopian tubes prevent normal conception. It is also useful when a man's production of sperm is inadequate and the likelihood of normal fertilization low. IVF may be employed because it allows the selection of the healthiest spermatozoa. This process, which is known as *intracytoplasmic sperm injection*, is still considered to be an experimental procedure, but initial results have been promising.

In-vitro fertilization begins with the removal of an ovum from a woman's ovaries immediately prior to expected ovulation, as indicated by a surge of leutinizing hormone. In most cases, ovulation-inducing agents are used to guarantee proper timing. The ovum is removed through a laparoscope or through an incision, in which case ultrasound imaging is used to identify the location of the ovaries. The ovum is then fertilized by a spermatozoa, one of the 50,000 to 100,000 that had been rinsed in a solution that separates them from the seminal fluid. The fertilized ovum (zygote) is then allowed to develop in a culture medium, and after 48 to 72 hours, the two- to eight-cell embryo is planted in the uterus of the woman who will carry the developing fetus to term. The embryo is introduced into the uterus through the cervical canal or is directly implanted onto the uterine wall by means of a laparoscopy or laparotomy.

The practice of in-vitro fertilization has been used for many years as a means of producing superior farm animals. It was first applied to human conception by Patrick Steptoe and Robert Edwards in England; its successful outcome was the birth of Louise Joy Brown on July 25, 1978. Since then, in-vitro fertilizations have been performed thousands of times. IVF is a safe procedure; fetal and neonatal defects are no more likely than they are with normal conception. At the same time, however, the use of IVF raises a number of ethical issues. One of these has to do with fetal research. IVF can be the basis for innovative research into fertilization and the early stages of embryo development. But some researchers and laypeople consider the creation of embryos for research purposes to be unjustifiable, given their belief that human life begins at the moment of conception.

Another ethical problem with IVF is that the procedure makes it possible to fertilize one woman's ovum and then transfer the resulting zygote to another woman's uterus. This procedure, known as surrogate pregnancy, has been criticized as inherently exploitative and potentially psychologically damaging to the surrogate mother. Surrogate parenthood also raises troubling questions about what properly constitutes "motherhood," and it has generated some thorny legal issues, such as instances when a surrogate mother refuses to give up "her" baby.

Further complicating the use of IVF is the employment of cryogenics to freeze the resulting embryos at an early stage of development so they can be implanted some time in the future. This has given rise to more legal problems; one notable case centered on frozen embryos whose biological parents had been killed in a plane crash. At issue was not only their fate, but whether or not they were rightful heirs of the parents' estate. Another ethical dilemma emerged in 1996 when a British law mandated the destruction of 5-year-old frozen embryos that had not been claimed by their biological parents. Antiabortionists viewed the destruction of these embryos as tantamount to the willful ending of human life.

In the United States, the ethical dilemmas posed by IVF led to the creation of an Ethics Advisory Board under the Department of Health, Education, and Welfare. In 1979, the board recommended that IVF research not be denied federal research funds, although these have never been appropriated by the U.S. Congress. Opposition to some of the uses of IVF led to the dissolution of the board by the Reagan administration in 1980. More than 15 years later, many of the ethical dilemmas of IVF remain unresolved, and individuals, the legal system, and society as a whole continue to struggle with the issues posed by in-vitro fertilization.

See also SELECTIVE BREEDING OF ANIMALS; ULTRASOUND

Further Reading: Lawrence J. Kaplan and Rosemarie Tong, *Controlling Our Reproductive Destiny*, 1994.

Involuntary Sterilization

In the late 19th and early 20th centuries, there was considerable enthusiasm for eugenics, a doctrine that promised the "improvement" of humanity through controlled reproduction. One manifestation of the spirit underlying eugenics was a vogue for the sterilization the "feeble minded" and "mental defectives" so that their presumably inferior genes would not be passed on to succeeding generations. Not only was the sterilization of these people encouraged by eugenic theory, it also was actually mandated by law in more than 30 states in the United States. The constitutionality of these laws was challenged, but in 1927 the Supreme Court upheld them in the case of *Buck v. Bell*. The verdict was marred by inaccuracies in the factual information that had been presented to the court. In fact, Carrie Buck, the women whose sterilization was at issue, was not retarded or mentally ill. Nevertheless, the case gave legal sanction to involuntary sterilization, and approximately 60,000 Americans met this fate in the 3 decades following the court's decision.

Revelation of the weak scientific basis underlying eugenics along with the Nazi atrocities that were committed in the name of eugenics eventually soured politicians and the citizenry on these practices. Even so, forced sterilizations continue to be performed, although for a different reason. Girls and women in mental hospitals, group homes, and other facilities for the mentally handicapped may be subject to rape or sexual exploitation. Under these circumstances, sterilization may be preferable to an unwanted pregnancy. Although *Buck v. Bell* has never been formally overturned, authorities are naturally hesitant to have people sterilized without their consent. At the same time, however, many state courts have upheld the right of a parent to order the sterilization of a mentally incompetent daughter. Such a course of action is allowed only when it can be shown to be of benefit to the person to be sterilized and when proper procedures have been followed.

In recent years, improvements in contraceptive technology have obviated the need for many involuntary sterilizations. In particular, Norplant, a set of contraceptive capsules implanted under the skin of the upper arm, can be used to prevent pregnancy. Its great advantage is that, unlike sterilization, its effects are reversible.

See also EUGENICS

Iron

Iron may be one of nature's most common elements, but it is not easily transformed into a metal. Once the secrets of its smelting and use were learned, however, ironmasters produced a relatively inexpensive metal with superior hardness, durability, and unparalleled utility.

The earliest iron objects (from the 3d-millennium B.C.E.) were fashioned from iron found in meteors, for metallic iron is never found naturally. Rather, iron exists as an ore that must be smelted (heated in a furnace) to remove the other minerals with which it is found. The resulting metal was almost never pure iron but an alloy (mixture) of iron and carbon. This makes iron a special material, for the amount of carbon determines its basic properties. Most ancient iron contained almost no carbon (less than 0.15 percent) and was called *wrought* or *malleable* iron. It was relatively soft and could be worked into many shapes (e.g., plows, nails, bars). But when iron was later melted (at 1,535°C [2,800°F]), it had a significant carbon content (1.5 to 5 percent). This "cast iron" was very hard, but brittle, suited best for pots. With luck and ores of very high quality, steel (iron with 0.15 to 1.5 percent carbon) also was produced in antiquity. Supremely durable, hard but workable, able to hold an edge, this rare metal was treasured for weapons and edge tools like razors and axes.

As early as 1900 B.C.E., experiments with iron had begun near the Black Sea in what is today Turkey. By 1500 B.C.E., the Hittites of eastern Turkey routinely produced wrought iron. How this breakthrough occurred is unclear, for compared to other metals (gold, silver, lead, copper, bonze), iron is difficult to smelt. It must be worked at higher temperatures, yet because it did not melt, it was not obvious that heating the ores produced a useful metal. Clues may have come from potters using iron ores as a glaze, a suggestion in keeping with metallurgist and historian Cyril Stanley Smith's argument that the development of metallurgy owed much to aesthetic considerations and art.

After 1000 B.C.E., iron-making technology diffused across Europe, Asia, and Africa. For more than 2 millennia, the basic process for producing wrought iron was unchanged in most places. Smiths used a direct process, going from ore to metal in a small furnace or forge, called a *bloomary*, that had a natural or forced draft of air. Iron ore was mixed with charcoal in a small hearth and heated to more than 1,000°C (1,830°F). Charcoal provided both heat and an atmosphere that released the oxygen to which iron is chemically united. Iron was not melted in antiquity; it became a pasty mass that had to be hammered to remove other impurities (slag). The result was a "bloom" that was worked into bars and shapes. Enormous skill was involved, as smiths were completely lacking in metallurgical and chemical knowledge; they achieved marvelous results with their eyes and other senses, trained by long experience. They read subtle clues about the nature of the ore, the appropriate mixture of materials, the temperature of the iron, and the state of the process. Not surprisingly, ironmasters were accorded special status in many cultures, and some considered them magicians.

China, which was making iron by the 6th-century B.C.E., produced the first cast iron about 2 centuries later. Slightly larger furnaces were required to attain the higher temperatures needed. For centuries, China's metallurgical capabilities were the best in the world. Iron was used for buildings, including a pagoda at Dang Song in Hebei that stood 21 m (70 ft) high and used more than 48 metric tons (53 tons) of iron; temple bells, which required special quality control; numerous statues; iron-chain suspension bridges; plows that fed a growing population; and weapons. Only China enjoyed such a liberal supply of iron in 1000 C.E.; the 113,000 metric tons (125,000 tons) made in 1078 included 16 million iron arrow heads. But other developments occured in India, where a material called *wootz steel* was made after 400 B.C.E. and used to make world-famous Damascus swords, renowned for their beauty.

Europe did not equal China's abilities until the end of the Middle Ages, when European ironworks slowly mechanized by harnessing waterpower to the bellows and the hammers. By the early 15th century, Europeans adopted the blast furnace, so named because of the blast of air blown into it, which produced temperatures high enough to make cast iron. They called it *pig iron* because the molten liquid cascaded into trenches in a bed of sand and then into rows of molds that resembled suckling piglets. This hard, brittle iron had to be refined into wrought iron, a step accomplished by heating the pigs in a charcoal-fired forge to remove carbon and then hammering the bars. This two-stage *indirect process* was disseminated through technical treatises like Vannocio Biringuccio's *Pirotechnia* (1540) and Agricola's *De re metallica* (1556). The latter, with illustrations covering all facets of mining, smelting, and assaying metals, remained a standard reference for about 2 centuries.

Thanks to these changes, both production and consumption of iron in Europe rose sharply. In 1500, a typical blast furnace produced 1,500 kg (3,300 lb) of iron per day; by 1700 output stood at 2,500 kg (5,500 lb). Total European consumption in 1500 was about 54,500 metric tons (60,000 tons). In 1739, Sweden alone produced 46,300 metric tons (51,000 tons) of iron. Indeed, blessed with high-quality ores, the Swedes developed a reputation for superior iron. But the British for a time monopolized the production of cast-iron cannon, a key to the Royal Navy. British production was only 22,700 metric tons (25,000 tons) of pig iron and 10,900 metric tons (12,000 tons) of wrought iron in 1720. But for a century, experiments had been underway to substitute coal for charcoal as the fuel for various smelting and refining processes. The main obstacles were impurities in coal like sulfur, which made iron even more brittle. British ironmaster Abraham Darby (1677–1717) first smelted iron in a blast furnace by using coke (coal burned in the absence of air to produce almost pure carbon) in about 1712. Darby's process spread slowly because it seemed to give a lower-quality iron. But after 1750, much less-expensive coke-smelted iron dominated British production. A crucial reason was the development by another Englishman, Henry Cort (1740–1800), of a process called *puddling*, which refined cast iron into wrought iron while using coal. Puddling called for melting iron in a reverbatory furnace that reflected the heat from burning coal or coke onto the iron but kept impurities away. Once refined, the pasty mass of iron was passed through rollers rather than hammered to remove slag and other impurities.

These new production processes combined with other technical developments to drive Great Britain's Industrial Revolution. Darby's coke-fueled furnaces allowed the casting of large cylinders for Newcomen atmospheric engines (introduced in 1712 or perhaps a bit earlier), used mostly for mine pumping. These engines burned coal and in turn allowed coal to be mined from deeper and wetter seams. Coal became available in larger quantities at lower cost for all. But ironmakers found that coke did not crush under the weight of iron ore as easily as charcoal, permitting taller furnaces that raised output. The use of coal for this purpose, and many others, was a crucial element in the Industrial Revolution.

Iron had given ancient empires and modern nations alike crucial military advantages; now it was central to Great Britain and other nations that developed industrial economies. Iron was essential, for example, to railroad development, being used for rails, locomotives, and the train sheds at every large station. It became widely used in construction, including large textile mills. Darby's firm built the first European iron bridge in 1779; it still stands at Ironbridge in Shropshire. Eventually iron water mains and gas pipes were used to expand public utilities in growing cities.

The British recognized the economic importance of ironmaking technologies by jealously guarding their secrets, although process experience proved easier to protect than the hardware. But as important as the economic advantages it conferred, iron production came to be viewed as a measure of civilization. This attitude was firmly embedded in the 1819 decision of a Danish museum curator, C. J. Thomson, to categorize the museum's collection of prehistoric tools by dividing time into three ages: stone, bronze, and iron. Accordingly, cultures like the Incas, Aztecs, and American Indians (Native Americans), which never developed iron metallurgy, were found wanting. And the decline of native iron industries in places like India as a result of a flood of iron imported from Great Britain seemed to offer further evidence of cultural as well as industrial standing. By 1850, British ironmasters produced 2.27 million metric tons (2.5 million tons) of iron. Interconnected to coal, steam, railroading, and machine tools, iron production was the essence of heavy industry, the base of what Lewis Mumford (1895–1990) labeled the *paleotechnic era*. And the technologies of coal-fueled iron production had spread to France, Germany, and the United States, where it was assumed that only by mastering this technology could they consider themselves rivals to Great Britain.—B.E.S.

See also CAST IRON; FURNACE, REVERBERATORY; INDUSTRIAL REVOLUTION; IRON, WROUGHT; LIGHTING, GAS; LOCOMOTIVE, STEAM; RAILROAD; STEAM ENGINE; STEEL, ALLOYED; STEEL, BESSEMER; STEEL IN THE 20TH CENTURY; WATERWHEEL

Further Reading: Theodore A. Wertime and James D. Muhly, eds., *The Coming of the Age of Iron*, 1980. J. R. Harris, *Industry and Technology in the Eighteenth Century, 1780–1850*, 1992.

Iron, Wrought

Iron is among the most abundant elements on Earth. As a metal, it has been worked for thousands of years. But as late as the 18th century, it was used selectively to make tools, weapons, or other articles when wood or another more easily worked material would not suffice. To make iron, ore must be converted by smelting. In ancient times, the task was accomplished in a bloomery with a simple hearth furnace. Iron ore and charcoal were combined and heated in a small furnace, stoked by a manually operated bellows. Because iron ore is an oxide, the oxygen united with the carbon during heating and escaped as a gas, carbon dioxide. In a few hours, the process produced a bloom, a spongy ball of malleable wrought iron that was hammered into shape to remove rocky particles. Later, a blacksmith reheated the material and forged it into a variety of tools and hardware for local consumption.

The conversion of ore into metal was a craft requiring experience and judgment. Ores are not uniform; they contain varying amounts of carbon, manganese, silicon, sulfur, and phosphorous. The nature of iron depended especially on the amount of carbon it contained, with slight variations having significant consequences.

Wrought iron contains little or no carbon. As a consequence, it is malleable and easily forged. When heated, quantities of wrought iron can be welded together by simple hammering. Cast iron contains up to 4.5 percent carbon. It is brittle and cannot be welded. The production of cast iron requires higher temperatures to melt the iron and allow it to flow from the furnace into molds. Steel contains about 2.5 percent carbon. It can be easily forged when red hot, and when tempered and cooled it becomes extremely hard. But steel was much more difficult and costly to produce than wrought iron, which remained the dominant metal until the middle of the 19th century.

The chemical properties of iron were not understood until well into the 18th century. Accordingly, manufacture resulted from trial and error based on craft processes devised to either add or eliminate carbon from iron. Significant advances in the manufacture of iron followed the introduction of water-powered machinery, as older processes were adapted to allow for production on a larger scale. By the 15th century, water-driven bellows produced a stronger blast, allowing for larger furnaces with more intense heat. In addition, water-powered machinery crushed ores and operated hammer forges, pumping apparatus, hoists, and rolling and slitting mills. During the 16th century, the spread of printing led Vannoccio Biringuccio, Agricola (Georg Bauer), and Lazarus Ercker to publish comprehensive works describing the processes of mining, smelting, alloying, and assaying metals. These works remained the standard texts of metallurgy for 2 centuries.

During the 18th century, three advances increased the quantity of iron, and in so doing revolutionized industrial culture. First, coal replaced charcoal as a fuel for smelting. Because charcoal was easily crushed in a furnace, the size of the furnace and the quantity of iron produced were limited. In 1709, at Coalbrookdale in Shropshire, Abraham Darby (1677–1717) smelted iron with coke, a form of baked coal. Coke burned at a much higher temperature, melting the iron and producing cast iron. But the process was troublesome, for coke contained impurities that contaminated iron. Fortunately for Darby, he employed coke that was low in sulfur, an element that made iron extremely brittle. However, the use of coke spread slowly, since other manufacturers did not employ ores with the same degree of purity. Also, greater supplies of cast iron were of limited value since the greatest demand existed for the more malleable wrought iron.

In 1784, Henry Cort (1740–1800) patented a process for *puddling*, which converted large quantities of cast iron into a malleable wrought iron. Cort placed bars of coke-produced cast iron in a coke-fired, reverberatory furnace that contained separate chambers for the iron and the fuel. After heat passed over and melted the iron, it could be stirred. The carbon in the iron combined with oxygen in the air until the pool of iron became a pasty puddle ball that could be squeezed into shape. During the 19th century, this process was improved by heated blasts, by the use of fluxes, and by improvements in the refractory linings of furnaces. These processes allowed for the production of wrought iron in much greater quantity. Still, puddling always required highly skilled workmen who were among the aristocrats of labor.

Finally, steam engines altered fundamentally the source of power available to iron manufacturers. These engines powered apparatus that produced a stronger and more continuous blast of air, necessary because coke burned less readily than charcoal. Steam engines allowed for consolidated manufacturing sites developed near sources of ore. These were populated with steam-powered forge hammers, slitting mills, and rolling mills that turned out various shapes of iron. By the middle of the 19th century, machine tool building was an art that demonstrated mastery in the use of metals. In turn, machine tools and the production of great quantities of malleable iron allowed for the creation of the railroads that transformed modern culture.

Steel began to replace wrought iron in the latter half of the 19th century. The process for making steel involved heating the best-quality wrought-iron bars packed in pulverized charcoal for several days at a very high temperature. Carbon hardened the shell of the wrought iron, resulting in what was known as *blister steel*. The major advance in production came with the introduction of Bessemer steel in the mid-19th century. The process presented insurmountable competition for wrought and cast iron. By the end of the 19th century, steel had become the key material of the industrial age, although wrought iron is occasionally used today for applications where resistance to corrosion and shock are of paramount importance.—J.C.B.

See also CAST IRON; FORGING; FURNACE, REVERBERATORY; IRON; PRINTING WITH MOVABLE TYPE; RAILROAD; STEAM ENGINE; STEEL, BESSEMER; STEEL IN THE 20TH CENTURY; WATERWHEEL

Irons, Domestic

Ironing devices have existed for thousands of years. In many cultures throughout the world, heated metal rods or flat pieces of metal were employed to press clothes. Ancient Chinese irons consisted of an open pewter or brass bowl with a wood handle. The reservoir was filled with hot coals, and then the bowl was moved in circles over the clothing. Until the 19th century, very little change occurred in the design and technology of such ironing devices. Colonial Americans, for example, used wood-handled "box" irons, so-called because the irons had a sliding drawer, door, or hinged opening that was filled with hot coals or charcoal bricks.

Introduced in the 1800s, the earliest flatirons or sadirons (*sad* meaning *heavy*) were cast iron with a handle attached at only one point. Sometimes the handle was detachable so that it could be switched from a cold iron to a warm iron that had been heating while the other was in use. During this century, specialized irons were also developed for specific purposes, such as for fluting, sleeving, and polishing, and were shaped accordingly.

Gasoline was briefly experimented with as a substitute for standard coal, but to no great success since it was smelly and expensive, as well as very dangerous. Another option was the use of gas, once houses began to be equipped with this fuel for heating and lighting purposes in the mid-19th century. The iron, or other household appliance, could be attached to a gas outlet by a rubber tube. This rather inconvenient arrangement never caught on with consumers, especially because gas was soon superseded by electricity for many domestic tasks.

The electric iron, one of the first domestic electrical appliances, was patented by Henry W. Seely of New York in 1882 and went on sale 3 years later. Its base was made of heavy cast iron, and it was heated by an electric arc, a powerful spark that jumped between two carbon rods. The next major improvement in ironing devices occurred in 1903, when Earl Richardson of California developed an electric iron with as much heat in its tip as in its center, thus the origin of the Hotpoint trademark. This meant that the iron emitted more even heat over a larger surface area.

Because they were generally made out of cast iron, clothes irons weighed up to 6.8 kg (15 lb). Cast iron remained standard until the introduction of new, lighter materials in the early 20th century. For example, General Electric produced the first iron made with a stamped, nickel-plated steel shell in 1910. Twenty years later, Sunbeam introduced the first truly lightweight iron, which weighed under 2.7 kg (6 lb). Another reason for the increasingly light weight of irons was the inclusion of newly developed plastic, first in the handle and then in the body of the iron as well. Today, only the soleplate is still made out of metal, and it is now usually covered with Teflon or some other nonstick coating to help prevent scorching and keep the surface clean.

As with other domestic appliances, the clothing iron benefited from a general consumer boom after World War I. Increased demand meant increased competition among manufacturers to meet it, and this caused numerous improvements to be introduced during the 1920s and 1930s. One of the most important developments in irons was the addition of automatic thermostats to better regulate their heat, especially since they had to cope with the new demands of synthetic fabrics requiring lower ironing temperatures.

Another major development was the invention of the steam iron, first introduced in 1927. Early steam irons had only one or two holes from which steam could emerge. By the 1950s, competition among iron manufacturers had led to a sudden escalation in the number of perforations on iron soleplates. More of a marketing ploy than a practical improvement, steam irons were produced with more and more holes in their soleplates, from 8 to 80 within 2 decades.

One of the more recent innovations in ironing devices is the "self-cleaning" steam, spray, and dry iron introduced by General Electric in 1972. Other improvements have included more temperature options specifically tailored for the growing range of synthetic fabrics, as well as the addition of a device that automatically turns off an iron that has been left on for too long. For anyone who has ever gone on a long trip and wondered "Did I leave the iron on?", this addition may be the best innovation yet.—M.S.

See also ACRYLICS; CAST IRON; NATURAL GAS; NYLON; RAYON; TEFLON; THERMOSTAT

Further Reading: Earl Lifshey, *The Housewares Story: A History of the American Housewares Industry*, 1973. Susan Strasser, *Never Done: A History of American Housework*, 1982.

Irradiated Food

Many techniques have been used to preserve food. One of the most recent technologies for food preservation entails the use of radiation. In 1898, it was noted that radiation killed bacteria, and the first patents for preserving food by radiation were granted a few years later. But many years passed before irradiated food became available. An important step was taken in 1948, when the U.S. Navy awarded a contract to the Massachusetts Institute of Technology to explore the use of radiation for food preservation. In the years that followed, techniques were developed for preserving a variety of foods through the use of radiation.

Foods are irradiated by exposing them to beta and gamma rays given off by the radioactive isotope cobalt-60. Radiation causes the formation of free radicals from the water and other substrates in the food (free radicals are atoms or molecules with one unpaired electron; as a result, they are usually highly reactive chemically). The presence of free radicals serves to kill insects, to inhibit the sprouting of tubers, to destroy bacteria and other microorganisms, and to inactivate some enzymes. Different levels of radiation are required for each function. Inhibiting the sprouting of tubers requires 4×10^3 to 4×10^4 rads, while the destruction of microorganisms requires 2×10^6 to 5×10^6 rads. The inactivation of enzymes requires the most intense applications of radiation—up to 10^7 rads.

Anything having to do with radiation is likely to be a source of anxiety, especially when it involves the food we eat. The use of radiation for food preservation has always been a matter of governmental concern. In the United States, the Food, Drug, and Cosmetic Act of 1958 specifically defined irradiation as a food additive; this meant that the process had to be demonstrably safe before it could be certified for use. In 1966, the U.S. Food and Drug Administration certified irradiated canned bacon, white potatoes, wheat, and wheat flour, but approval was withdrawn in 1968 when one study found cancers and cataracts in laboratory animals that had been fed irradiated meat.

Subsequent research seemed to indicate that food irradiation presented no undue dangers. There was never any question of food being rendered radioactive; this does not occur with the amounts of radiation used for food preservation. Rather, the concern has been with the creation of possibly toxic radiolytic products. In 1976, a joint meeting of the Food and Agriculture Organization (FAO), the International Atomic Energy Agency (IAEA), and the World Health Organization (WHO) concluded that the radiolytic products found in the irradiated foods that had been investigated did not appear to present any toxicological dangers. In fact, research conducted before and after the 1976 report indicated that the radiolytic products were no different from the substances found in foods preserved by other methods. Bolstering the safety of irradiated foods were animal studies that found no adverse effects from the eating of these foods. Due in large measure to these studies, 32 countries have granted limited or unconditional approval to more than 40 irradiated foods. By 1990, the U.S. Food and Drug Administration had approved the irradiation of many food items, including meat, poultry, fruits, and vegetables. At the same time, FDA regulations stipulate the maximum levels of radiation that can be used and require that irradiated foods be labeled as such.

Government approval notwithstanding, the irradiation of food remains controversial. Even if irradiated foods are completely harmless, the process of irradiation does require an increase in the volume of radioactive materials, and all the problems of transportation, disposal, and possible theft that this entails. The irradiated-food industry also faces formidable economic challenges, for a substantial increase in food irradiation would require substantial capital investment. This is not likely to be forthcoming, given the widespread consumer resistance to buying and consuming irradiated food.

Despite these obstacles, food irradiation may eventually play an important part in feeding a continually expanding world population. Spoilage is a major reason that many countries cannot provide enough food for their people, and irradiation offers the possibility of preserving food at a considerably lower monetary and energy cost than canning and refrigeration.

See also ANIMAL RESEARCH AND TESTING; BETA PARTICLES; CANNED FOOD AND BEVERAGES; FOOD AND DRUG ADMINISTRATION, U.S.; FREEZE DRYING; FROZEN FOOD; GAMMA RAYS; ISOTOPE; PASTEURIZATION; RADIOACTIVITY AND RADIATION

Further Reading: Jacques Leslie, "Food Irradiation," *Atlantic Monthly* (Sep. 1990).

Irrigation, Center-Pivot

Among the requirements for productive farming is an adequate supply of water. When rainfall is not sufficient, it is necessary to bring water to the fields. In the past, this was done through the use of irrigation channels. This, however, can be a labor-intensive process, both for construction and maintenance, as well as for the distribution of water. Furthermore, there are places where the water lies deep underground, not easily accessible to conventional irrigation technologies. In 1952 a Nebraska farmer named Frank Zybach received a patent for a system of irrigation that overcame these drawbacks.

Zybach's system is known as center-pivot irrigation, for it uses a long pipe equipped with sprinklers that rotates around a central pivot. The length of the pipe ranges from one-half mile to a mile (0.8–1.6 km) and is made of numerous short pipes joined together. The pipe is supported at regular intervals by wheeled towers and equipped with an alignment device to keep the pipe reasonably straight. The towers are equipped with wheels that are hydraulically or electrically powered. So powered, the pipe can traverse a complete circle in as little as 12 hours, although 3 to 4 days is the usual time for one revolution, during which time about an inch (2.5 cm) of water is distributed.

In the United States, center-pivot irrigation systems are often set up to service a quarter of a section, 160 acres (64.75 ha). Because the circular area covered leaves the corners unwatered, only 133 acres (53.8 ha) are irrigated, although there are special attachments that allow even the corners to be reached. In addition to providing water, the system can be used to dispense chemical fertilizer, insecticides, and herbicides. The gradual application of water is particularly useful for the irrigation of the sandy soils typical of the high plains, for if water is more rapidly applied it quickly passes through the soil, leaving less for the crops.

In some ways, center-pivot irrigation typifies American agriculture. It allows high crop yields from soil that might otherwise be unsuited to cultivation, but at the same time it requires large amounts of energy for its operation. A typical yearly application of 22 inches (56 cm) of water requires 50 gallons (189 liters) of diesel fuel, 10 times the amount used for tilling, planting, cultivating, and harvesting a crop on the same acreage. Water is also used in large quantities: In operation, a center-pivot irrigation system uses as much water as a town of 10,000 inhabitants.

Heavy consumption of water is particularly problematical when it takes place in semiarid regions where the major source of water is an aquifer hundreds of feet below the surface of the ground. Center-pivot irrigation has been used extensively in the region served by the Ogallala aquifer, an area comprising parts of Colorado, Oklahoma, Nebraska, Kansas, and Texas. This is a region that resisted decades of efforts to produce crops on a reliable basis, ruining many farmers in the process. It began to burgeon as a crop-growing area when deep-drilling technologies began to tap the aquifer and center-pivot irrigation made effective use of the water. The aquifer is immense, more than 3 billion acre-feet (an acre-foot is a foot of water on 1 acre; it equals 325,851 gal or 1,233,480 liters). But more than a half-billion acre-feet have been used since 1960, and very little new water has been making its way to the aquifer. Under these circumstances, the sustainablity of agriculture in this region is in doubt. New farming techniques employing sensors and computers hold out the promise for a more efficient use of the water, but they may only postpone the inevitable. In this region at least, the high crop yields made possible by center-pivot irrigation may prove a temporary success.

See also AGRICULTURE, IRRIGATED; HERBICIDES; INSECTICIDES

Further Reading: William C. Splinter, "Center-Pivot Irrigation," *Scientific American* (June 1976): 234–36.

Isotope

The discovery of radioactivity at the end of the 19th century generated a host of questions about the nature of matter. One major puzzle was posed by the discovery of more radioactive elements than there were places for them on the periodic table of the elements. But were they really elements? As far as their chemical properties were concerned, each one seemed no different than an existing element or one (or more) of the newly discovered ones.

The solution to this puzzle lay in a reconceptualization of the atom and its nucleus. From the inception of the modern atomic theory in the early 19th century, it was assumed that elements of the same substance all had the same atomic weight. But research conducted in the early 20th century demonstrated that some elements exhibited variable atomic weights. For example, samples of lead taken from radioactive ores had atomic weights as low as 206.4 and as high as 208.4.

An understanding of the nature of radioactivity provided a reason for this variability. Frederick Soddy (1877–1956) in England and Kasimir Fajans (1887–1975) independently observed that radioactive substances turn into another kind of atom when they lose alpha particles and beta particles. In the first case, the loss of an alpha particle (which carries a charge of +2) results in an atom with a charge that is two less than its original value. The atom's position on the periodic table is thus shifted two places to the left of the place it had occupied prior to the loss of the alpha particle. The loss of a beta particle (which carries a charge of -1), puts the atom one place to the right of its original position on the periodic table. Depending on the element, this process of radioactive decay might continue several times until it terminates with the production of a nonradioactive element.

According to Soddy's account of this process, which he called *radioactive displacement*, a place on the periodic table could be occupied by an atom that had undergone alpha decay or by a different one that had undergone beta decay. But while their atomic numbers would be identical, their atomic weights would differ. Soddy gave them the name *isotope*, meaning "the same place." Today, the term is used to describe elements that have the same atomic number (determined by the number of protons in their nuclei) but different atomic weights (due to different numbers of neutrons in their nuclei). The identical chemical characteristics of isotopes is the result of their having the same number of electrons orbiting their nuclei.

Although isotopes were originally identified through research into radioactive substances, it was soon determined that isotopes of nonradioactive elements also existed. In 1919, Francis Aston (1877–1945) examined ionized neon using a scientific instrument he had invented, the mass spectrograph. What he found were two forms of neon, neon-20 and its isotope neon-22. Aston went on to discover 212 of the 287 stable isotopes now known to exist.

Isotopes began to be used for scientific research within a few years of their discovery, the first example being Gyorgy Hevesy's (1885–1966) employment of a radioactive lead isotope to study plant growth. Research of this sort continues to the present day, with isotopes being extensively used to gain a better understanding of biological processes. Isotopes can also be used to determine the age of anything containing carbon, a procedure that has many applications in paleontology and archeology.

In the 1940s and thereafter, an isotope of uranium, U-235, was essential to nuclear research and the production of nuclear weapons. U-235 is the only naturally occurring isotope of uranium that can be fissioned, and a great deal of effort was devoted to increasing the concentration of U-235 in uranium, a process known as *enrichment*. On a more peaceful note, radioactive isotopes have also been used extensively for the treatment of various forms of cancer.

See also ALPHA PARTICLES; ATOMIC BOMB; ATOMIC NUMBER; ATOMIC THEORY; ATOMIC WEIGHT; BETA PARTICLES; CANCER; CHEMICAL BONDING; PERIODIC TABLE OF THE ELEMENTS; RADIOACTIVITY AND RADIATION; RADIOCARBON DATING; SPECTROSCOPY

Jet Aircraft, Commercial

The development of the turbojet began to revolutionize aviation in the 1940s. The first applications were military, and by the 1950s most front-line combat fighters and bombers were powered by turbojets. By this time it was evident that jet engines conferred advantages that also made them well suited to commercial aircraft. Jet engines promised a new generation of airliners that traveled at higher speeds then before, climbed to altitudes that put them above the weather, and subjected their passengers to less noise and vibration.

The first commercial airliner to be powered by jet engines was Britain's De Havilland Comet. First flown in 1949, the Comet was initially powered by four 2,018-kg (4,450-lb) thrust engines buried in the wings (later versions had engines producing 2,495-kg [5,500-lb thrust]). Carrying 44 passengers and cruising at a speed of 790 kph (490 mph), in 1952 the Comet went into regular service that connected London with Johannesburg, Singapore, Tokyo, and Colombo. Unfortunately, four fatal crashes occurred within 2 years of service, causing the grounding of the airplane until the cause—metal fatigue—was discovered after extensive testing.

In 1952, Pan American World Airways placed an order for a larger version of the Comet, but the crashes and subsequent grounding of the airplane upset the scheduled delivery date of 1956. By this time, American aircraft manufacturers were at work on their own jet airliners. This first of these was Boeing's 707, which took to the air in 1954. The development of the 707 was greatly aided by Boeing's prior experience in the building of the B-47 and B-52 jet bombers. Boeing also received a crucial boost from the U.S. Air Force, which provided financial backing for an air tanker version of the 707 known as the KC-135. The 707 underwent an extensive period of testing and was finally put into service on Pan American's New York–to–London route in late 1958.

Douglas Aircraft, a firm that had produced a series of highly successful commercial aircraft beginning with the DC-2 in 1934, entered the jet airliner business with its DC-8. A four-engine airplane like the 707, the DC-8 first flew in 1958 and began carrying paying passengers a year later. Both the 707 and the DC-8 were commercial as well as technological successes. Boeing built 800 707s; Douglas produced 550 DC-8s. A much less-successful venture was the Convair (now General Dynamics) 880. Although it was a capable aircraft, the 880 appeared on the scene when the market for jet airliners was temporarily saturated. Only 102 examples of the 880 and its successor the 990 were sold, causing Convair to rack up some of the worst financial losses ever suffered by a private firm up to that time.

In Europe, a number of shorter-range jet passenger aircraft made their appearance in the 1950s and 60s: the French Aerospatiale Caravelle (first flown in 1955); the British BAC One-Eleven (first flown in 1963); BAC VC10 (first flown in 1962); Hawker Siddeley Trident (first flown in 1962); and the Soviet Tupelev-154 (first flown in 1968). Europe also led the way with the first supersonic airliners, the Soviet Tupelev-144 and the Franco-British Concorde. The latter two aircraft failed as commercial ventures, however. As it turned out, a much more important step in the development of jet travel was the widebody or "jumbo jet." In 1969, Boeing rolled out its 747, an airplane that represented a quantum leap in the passenger-carrying capabilities of jet aircraft. The 747 was Boeing's

A Boeing 707, the first jet aircraft used by U.S. airlines (courtesy Boeing).

response to a elongated version of the Douglas DC-8, which was capable of carrying as many as 259 passengers. At the same time, the 747 benefited from earlier work on a proposed high-capacity military cargo plane, even though the contract eventually went to Lockheed's C5A. Capable of carrying 500 passengers in a spacious cabin, the 747 brought significant operating economies to the operators of heavily traveled long-distance routes.

Two widebodies with less passenger capacity and shorter range followed, the Lockheed L-1011 (first flown in 1970) and the McDonnell Douglas DC-10 (also first flown in 1970). Another important entry was the Airbus, a 300-passenger jetliner that was designed and built through the collaborative efforts of firms in Britain, France, and Germany. The Airbus also benefited from financial subsidies from the governments of these countries, although in fairness it should be pointed out that American aircraft manufacturers had received crucial assistance in the form of military projects that had civilian applications. While American aircraft manufacturers grumbled about what they considered unfair competition, they were able to produce new widebodied aircraft, the Lockheed MD-11 (an updated version of the DC-10) and the Boeing 767 and 777. At this point the cost of developing a new airplane had become so astronomical that these projects required international cooperation. Boeing, for example, used an Italian firm, Aeritalia, to do a considerable amount of airframe development and manufacture for the 767.

By this time, jet aircraft had changed the nature of travel, and many other things as well. The jet age had forced fundamental alterations to the operation of the air travel system. The airplanes themselves required longer, better-surfaced runways, while the increased number of passengers carried put great strains on airports and air traffic control systems. Jet aircraft also changed the spatial economy by allowing businesses to be more decentralized; it became commonplace for a business executive to fly into a city, conduct a meeting, and then depart on at the end of the day. Airports became more than places of arrival and departure; they became hubs of commerce with their own hotels, restaurants, and meeting sites.

Jet travel also increased the volume of recreational travel, making possible brief vacations in Europe, Hawaii, and many other destinations that in the past had been the domain of the small minority who had sufficient time and money for leisurely travel. At the same time however, jet travel was not without its discomforts. In addition to having to cope with jammed airplanes, airports, and the highways leading to them, long-distance passengers had to contend with a new physiological problem, "jet lag," the inability of the body to adjust rapidly to having been transported to a different time zone. More ominously, jet travel facilitated the rapid spread of influenza and even more serious illnesses.

In the nearly 50 years that have passed since the flight of the first jet-powered passenger airplanes, there have been great improvements in comfort, quietness, and operating efficiencies. What has been lacking has been a concomitant increase in performance. The first generation of jet airliners cruised at 880 to 1,000 kph (500–620 mph), about the same speed that today's jets travel. Efforts to operate supersonic airliners at a profit have not met with success, and most technological efforts now center on improving the efficiency of subsonic airliners. Visionaries look to a future with passenger-carrying aircraft capable of flying at several times the speed of sound, but many decades will elapse before these become a reality.

See also HYPERSONIC FLIGHT; METAL FATIGUE; SUPERSONIC TRANSPORT; TURBOJET

Further Reading: Robert J. Serling, *The Jet Age*, 1982.

Juke Box

Thomas Edison's invention of the phonograph in 1878 made the reproduction and subsequent playback of sound possible. Although Edison did not foresee it, the primary application of his invention turned out to be musical entertainment. This use was not restricted to the home. In 1889, the first coin-operated phonograph made its appearance in a San Francisco saloon. Before long, machines known as *nickelodeons* were in widespread use; 2,000 of them were in operation by the end of the century.

During the 1920s, a new entertainment medium, radio, grew in popularity. One consequence was a severe cut in record sales, from 100 million in 1921 to 2 million in 1933. Given the ready availability of radio broadcasts, the purchase of records was hard to justify, especially in the hard times of the Depression.

A 1947 Wurlitzer jukebox (courtesy Rohm and Haas).

Still, not everyone could afford a radio. Coin-operated record machines continued to find a market in bars and dance halls patronized by poor people, especially those living in rural areas. As a result, they came to be identified with sleazy diversions. In fact, the word *jukebox* may stem from a African term for unsavory partying and dancing. With the repeal of Prohibition in 1933, the audience for jukeboxes expanded with the reopening of bars and other sites. By 1940, 250,000 jukeboxes were collectively devouring 5 million nickels every day.

By then, they had expanded into hotel lobbies, beauty parlors, bus stations, and even restrooms.

Jukeboxes could not have satisfied the demand for high-volume music had it not been for the invention of the paper cone speaker around 1920 and the development of sound amplification. During the 1940s and 50s, the design of jukeboxes went well beyond the technical requirements of a coin-operated record player. The most elaborately designed were ornamented with multicolored lights, and tubes through which bubbles continuously circulated.

At present, 225,000 jukeboxes are in operation, a far cry from the 700,000 in use in the 1950s. A number of things contributed to the decline of the jukebox. Highway projects bypassed many of the roadside bars and cafes that housed them, while most fast-food franchises deliberately refrained from installing jukeboxes because they encouraged people, especially teenagers, to hang out and take space from paying customers. At the same time, musical needs were being satisfied by the emergence of the transistor radio and home stereo. Even so, jukeboxes from a bygone era are eagerly collected and command high prices. It is also possible to buy modern reproductions of such classics as the Wurlitzer 1015.

See also AMPLIFIER; RADIO, TRANSISTOR; RECORD PLAYER

Kidney Dialysis

In a healthy individual, the kidneys regulate the body's acid-base balance, maintain proper amounts of water, and concentrate metabolic wastes that are subsequently excreted as urine. Complete or substantial kidney failure (known in medical jargon as "end-stage renal disease") results in incapacitation and premature death. Many people suffer from defective kidneys; urinary diseases are the fourth leading killer in the United States, right behind cardiovascular diseases, cancer, and pneumonia.

In the early 1940s, Willem Kolff, a Dutch physician, used a bathtub and parts salvaged from a foundry to construct a device that performed the functions of the kidneys, a process known as *dialysis*. He later emigrated to the United States, where he and others developed more refined versions of his invention. These machines succeeded in cleansing the blood, but they only could be used for short periods of time. Permanent dialysis became possible in the early 1960s as a result of improved equipment and the invention of a connecting tube that obviated the need to open a new artery and vein every time the machine was hooked up. Long-term dialysis was now possible, but it brought a number of economic and ethical problems along with it.

When dialysis first became an accepted medical practice, the number of patients who could benefit from it far exceeded the number of available machines, making it necessary to use some kind of selection procedure. The pioneering dialysis institution, the Seattle Artificial Kidney Center employed an Admissions and Policy Committee to screen applicants and determine who would get dialyzed and who would not. The committee, the membership of which was intended to reflect the community as a whole, made use of criteria that included the age of prospective patients as well as their personal characteristics, such as income, marital status, occupation, and educational background. As might be expected, to some critics this smacked of the "the bourgeoisie sparing the bourgeoisie," of people being evaluated in accordance with the committee members' own middle-class values.

In due time the problem of controlling access diminished as more machines were put into service. Dialysis techniques also became easier, allowing many patients to dialyze themselves at home. But formidable cost problems remained. In 1970, even home dialysis entailed start-up expenditures of $9,000 to $13,000 and annual operating costs of $3,000 to $5,000.

Most patients could meet these expenses only with great difficulty or not at all, but in 1972 the U.S. Congress lifted patients' financial burdens when it authorized payment for dialysis treatments through Medicare. Congress's willingness to have the government assume these costs was the result of intense lobbying, and not a great deal of deliberation. The final vote was preceded by only a few minutes' debate in Congress, and even less discussion, while the Senate and House conference committee was drafting the final legislation. Medicare's assumption of its costs made dialysis available to all who needed it, but in 1987 the government's bill was $2.4 billion. These treatments absorbed 4 percent of total Medicare disbursements, yet dialysis patients comprised only one-quarter of 1 percent of Medicare beneficiaries.

About one-third of dialysis patients are over the

One of Willem Kolff's early dialysis machines.

age of 65. An aging population will be accompanied by a need for greater expenditures for dialysis, putting further pressure on a healthcare system already under a great deal of financial strain. In some countries with national health insurance programs, access to dialysis is restricted, and people over the age of 55 are often turned away. The official reason is that people in this age group do not have the physical constitution that allows them to endure a regimen of lifetime dialysis, but in fact there is greater variation in levels of health among the elderly than there is in any other age group. Using age as a means of limiting access to an expensive medical technology is administratively convenient, but it dodges the need to make choices according to more relevant criteria. At the same time, in a world of fiscal limits, a policy of open access to dialysis absorbs money and personnel that might be used for other medical procedures and programs, such as improved prenatal care. The problems of distributive justice posed by the use of artificial kidneys mark only the beginning of a host of dilemmas that will emerge as new and even more expensive medical technologies become available, and medicine gains the ability to address hitherto untreatable disorders.

See also ORGAN TRANSPLANTATION

Kilns

For thousands of years, the task of baking, or firing, ceramic products in a kiln was one of the potter's greatest technical challenges. A carefully controlled fire is among the potter's key tools, enabling ceramists to create objects that are hard, durable, water-resistant, and often beautiful. Kilns come in all shapes and sizes, depending on the time, place, and product.

The simplest firing method, which is still used by some African and Native American craft potters, consists of open pits that are lined and covered with combustible materials, such as corn stalks, grass, twigs, wood, or dung. Once pots are stacked inside the pit, a fire is lighted and allowed to burn very gradually. Using this elementary method, potters can achieve temperatures of 700° to 900°C (1,300°–1,650°F), a range that is high enough to harden porous earthenware but too low to melt glazes or create waterproof products.

Higher temperatures are reached in specially designed ovens, or kilns. Early kilns were simple improvements on open pits; they featured air holes at their bases to improve heat circulation along with clay walls that were rebuilt around each new stack of ware to retain heat. In many cultures, temporary ovens gradually yielded to abode, brick, and sandstone structures. The earliest permanent kilns in the Near East, Far East, and Mediterranean region were beehive-shaped updraft kilns, which featured separate fireboxes beneath perforated floors for the burning of fuels such as wood and grass and flues at their apexes for the exiting of heat.

Chinese, Korean, and Japanese potters, the first to master the art of making stoneware and porcelain, understood that controlling the intensity of the heat and the character of the atmosphere inside a kiln was an essential prerequisite to the uniform production of hard, refined ceramics. In their quest for high temperatures, these potters ultimately developed large, multichambered, wood-fired, hillside kilns, which remained standard until modern times. Asian potteries were the earliest ceramics enterprises to employ kiln specialists, who assumed responsibility for all aspects of oven management, from kiln building and maintenance to loading, firing, cooling, and unloading products.

While Asian potters were making significant advancements in kiln technology, European potters continued to use relatively simple firing techniques that dated from ancient times. During the Renaissance, potter Cipriano Piccolpasso wrote a three-volume treatise on earthenware manufacturing, and his illustrated text is a window on pottery production in 16th-century Italy. Piccolpasso used a rectangular updraft kiln that was equipped with separate chambers for ware and fuel and that could reach a maximum temperature of 1,000°C (1,830°F). Although Europe's most advanced potters took pride in the brilliance and flawlessness of their ware, they nonetheless lagged behind their Asian counterparts in kiln design.

In the 18th century, Western potters began thinking seriously about kiln construction. The rediscovery of porcelain at Meissen, Germany, in the first decade of the century provided the impetus for significant advances in firing technology, for it was impossible to make china without the requisite kiln temperatures that fused kaolin and feldspar. At Meissen, alchemist Johann Frederich Böttger and his assistants modified the design of the standard European updraft kiln in their quest to create porcelain. They altered the kiln's shape to improve air circulation, used heat-resistant fireclay bricks in construction, and fueled the oven with coke and coal rather than wood. British potters, who were industrializing in the 18th century, in turn modified the designs of their bottle-shaped ovens to take advantage of these improvements, which increased production, lessened fuel costs, and improved product quality.

During the 19th century, kiln builders in the East and West relied on long-established technical traditions, making at best small changes to oven designs. In the early 20th century, European and American ceramists who were seeking increased production, fewer ware losses, and improved product quality embraced the new technology of tunnel kilns, long refractory chambers equipped with railway tracks. In this continuous-flow process, railway carts stacked with ware entered the cold kiln and were gradually propelled through the tunnel's hot middle section before emerging from a cool exit. Today, ceramics factories throughout the world are equipped with fast-firing tunnel kilns, designed to meet the needs of quick-response merchandising programs.—R.B.

See also ALCHEMY; BRICKS AND BRICKMAKING; PORCELAIN; POTTERY; STONEWARE

Knot Theory

A mathematical knot is a curve in space that begins and ends at the same point and otherwise does not intersect itself. Intrinsically, all knots are just (topological) circles. What is of interest is the extrinsic nature of the knot, i.e., how it is embedded in the ambient space; this is what defines its "knottedness." Tying a knot in a real piece of rope and then fusing the ends together gives a good model for a mathematical knot. The rope is flexible and yet cannot pass through itself—two properties that we allow mathematical knots to have as well. Two different knots in space are called *equivalent*, and considered to represent the same knot if the first can be moved continuously through space, never passing through itself, until it coincides with the second. This kind of continuous deformation is typical in topology.

Although the great 19th-century mathematician Johann Karl Friedrich Gauss (1777–1855) and his student J. B. Listing were known to have considered knots, it is not until the end of the 19th century that the mathematical study of knots began in earnest. In 1867, the great thermodynamicist William Thomson (Lord Kelvin) suggested that atoms were knotted vortices in the ether. Thus a careful understanding of knots and their properties should lead to an understanding of atoms and the physics of matter. One of the fundamental problems of knot theory surfaced immediately: Given two knots, how can one decide if they are equivalent or not? Urged on by Kelvin, the Scottish physicist P. G. Tait, in collaboration with mathematicians T. P. Kirkwood and C. N. Little, spent the next 20 years creating a table of knots organized by the number of under- and over-crossings that appear in pictures of knots. They tabulated knots through 11 crossings, working almost entirely by intuition and experimentation. They had no formal proof that the knots in their tables were indeed distinct, or that their list, in so far as it went, was complete. It was not until the advent of modern topology that their table could be checked for accuracy and, incredibly, only a few errors were present.

Despite this early interest in knots, it was not until the 1920s and 1930s with the development of topology, and especially algebraic topology, that real progress in knot theory was made. Various algebraic topological measures of knots, known as *knot invariants*, were developed which could sometimes distinguish pairs of knots. Most of these invariants were based on the topology of the complement of the knot, the space surrounding, but not including, the knot.

Knot theory plays an important role in low-dimensional topology, and in particular the still-unanswered question of what are all possible three-dimensional spaces, or manifolds. This is because every three-dimensional manifold can be constructed from ordinary three-dimensional space by a process involving knots. Briefly, a knot can be removed from one manifold and the resulting "hole" filled back in an interesting way so as to produce a new manifold. Thus understanding knots and their properties leads naturally to an understanding of how three-dimensional manifolds can be constructed.

Knot theory also has important applications outside of mathematics. Chemists are now able to synthesize knotted molecules and linked polymers. Knot theory can often be used to answer important questions regarding these molecules, such as whether or not the molecules can exist in both right- and left-handed forms. Recombinant DNA research has also benefited from knot theory. The DNA molecule is an extremely long strand usually wadded into a compact tangle inside the cell. Enzymes present in the cell are able to selectively cut the strand, perform interesting maneuvers, and then rejoin the pieces. By first inducing the ends of DNA strands to join up so as to form knots, and then inspecting what knots are present before and after a reaction caused by a certain enzyme, knot theorists can determine the most likely action of the enzyme. Understanding how the various enzymes work will ultimately allow us to "operate" on DNA, repairing, for example, damaged sections that contribute to hereditary birth defects.

Knots also arise in the study of dynamical systems, where the solutions to systems of nonlinear differential equations can sometimes be knotted curves in space. Applications such as these, together with discoveries made in the last 15 years of powerful new knot invariants, have led to a renaissance in knot theory. The new knot invariants have surprising connections to physics—to statistical mechanics and to the newly developed topological quantum field theories. Ironically, after more than 130 years, knot theory is turning once again to physics for inspiration and application.—J.H.

See also DNA; ENZYMES; POLYMERS; TOPOLOGY

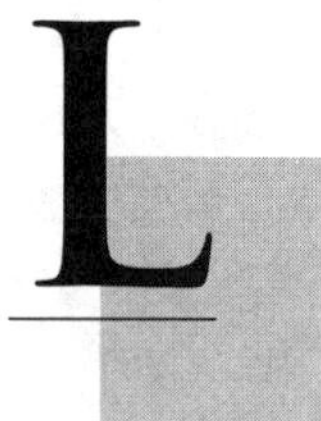

Lacquer

Lacquer is a generic designation for any finish that dries as a result of a volatile solvent evaporating from a dissolved resin, rather than a designation of chemical changes such as oxidization or polymerization, as with paints and varnishes. Originally, lacquer was made with the resinous incrustation formed by small parasitic insects on the twigs of many kinds of trees indigenous to China, India, Siam, and certain East Indian islands. The Hindu *lakh*, meaning 100,000, refers to the insects that produce lac, or shellac. Until recently, lacquer was primarily associated with ornamental finish on wood or porcelain art objects. China first excelled at lacquerware but was later surpassed by Japan, where for a time lacquered articles were accepted in payment of taxes.

Eventually the name was given to any transparent coating, whether or not made with lacquer or a distilled spirit lacquer (shellac). It was a name also given to glossy pigmented varnishes and ultimately to quick-drying finishes made with nitrocellulose or cellulose acetate, or acrylic, melamine, or other synthetic resins—collectively the basis of the modern industrial product. Beginning in the latter 19th century, cellulose lacquer was applied to all sorts of materials including wood, metal, glass, and even fabric. It was more expensive than enamels, and not as weather resistant, but harder and tougher. The first important industrial use for cellulose lacquers was as a photographic coating for glass plates. In the early 20th century, it was often used to coat brass objects such as bedsteads and to stiffen the fabric covering airplane wings (there are U.S. patents for coating fabrics with plasticized lacquer dating from the 1850s). During World War II, it was applied as a protective coating to all sorts of war material. Other uses have included insulation, vaporproofing, and fingernail polish, as well as automotive finishes.

Automotive lacquers were pioneered by the DuPont Chemical Company, which introduced its "Duco" finishes in 1924, when that year's Oakland was available in "True Blue." At that time DuPont was a major stockholder in General Motors, and the availability of GM cars in a variety of hues gave General Motors a marketing advantage over Ford, whose Model-T was available only in black. Beginning in the 1930s, aluminum particles were added to automotive finishes to give a "metalflake" appearance, and more recently a pigment material called ferric hydroxide has yielded automotive finishes that are exceedingly rich and translucent, particularly when applied in multiple coats, each one rubbed out before the next.—R.C.P.

See also PAINT

Land Mines

Land mines are explosive devices that are detonated when pressure is applied to them. In the case of large antitank mines, a weight of several hundred kilograms is required to set them off. Most of the mines that have been laid around the world are much smaller antipersonnel mines. These divide into two general categories: blast mines and fragmentation mines. In the former, the mine's explosion does all the damage; in the latter, the detonation of the mine causes the scattering of metal fragments at very high velocities. Blast mines are usually detonated when they are directly stepped on, whereas fragmentation mines are usually set off by a trip wire.

Blast mines are quite small, with diameters ranging from about 9 to 11 centimeters (3.5–4.3 in.) and contain from 40 to 240 grams (1.4–8.5 oz) of TNT. They generally do not kill their victims directly; depending on their size, they may blow off part of a leg or the entire leg, and will often inflict serious injuries to other parts of the body. Fragmentation mines cause widespread injury and usually result in death. Some of these mines are of the "bounding" variety: After being triggered they jump into the air before exploding, causing the dispersal of fragments over a larger radius and producing a more lethal effect.

Mines were originally developed as weapons against military forces. They served to close areas to enemy advance, to channel enemy troop movements, and to guard military installations. As such, they posed a threat to combatants but not to most civilians. In recent decades this pattern has changed, and to an increasing extent mines have been employed as terror weapons directed at noncombatants, part of a general trend: In the course of the 20th century, many civilians have been engulfed by military violence, and increasing numbers of civilians have been victimized by land mines. An estimated 15,000 people are killed or maimed by land mines every year; of these, 80 percent have been civilians, many of them children (one type of mine, the so-called "butterfly mine," is especially appealing to children, who think that it is a toy).

Mines are an inexpensive way to wreak havoc on a civilian population. Their cost ranges from $3 to $15, and they can be "sown" by helicopters and aircraft in large numbers. It cannot be said for certain how many mines exist in the world today, but informed estimates put the number at more than 100 million, and 64 countries have areas that have been mined. In many places the number of land mines runs in the millions. To take one extreme example, Cambodia, a country with a population of about 7 million, has 10 million land mines within its territory, a ratio of 142 mines for every square mile of land. A sizable portion of good agricultural land can no longer be farmed due to the mines that have been planted. Especially high concentrations of land mines are found also in Croatia, Bosnia-Herzegovina, Iraq, Egypt, and Afghanistan.

Many questions have been raised concerning the military value of land mines that have as their sole purpose the killing or grievous wounding of individuals, especially noncombatants. Several international conferences have taken up the issue of antipersonnel weapons, including land mines. No consensus has emerged about their use, although a United Nations–sponsored protocol limits the sowing of land mines by aircraft or artillery, and requires that the location of minefields be recorded. However, no nation is bound to accept the provisions of the protocol, which also includes an escape clause that notes that the provisions must be honored "unless circumstances do not permit."

In addition to international efforts, individual countries have attempted to limit the spread of land mines. In 1992, the U.S. Congress enacted a 1-year moratorium (subsequently extended for 3 additional years) on the sale of antipersonnel mines to other countries. Several other nations—including France, Germany, Israel, Italy, and Sweden—have adopted similar bans.

Even if efforts to stem the spread of land mines are successful, there will remain many parts of the world where land mines pose a constant danger. And should the world's leaders come to an agreement that antipersonnel mines have the same status as poison gas removing them will require many years of effort and substantial financial outlays. In the meantime, tens of thousands of people will be killed or crippled for life as a result of their encounter with a land mine.

See also GAS, POISON

Further Reading: Gino Strada, "The Horror of Land Mines," *Scientific American*, vol. 74, no. 5 (May 1996): 40–45.

Laser

In the 40 years since lasers were first developed, they have been used for an astounding variety of purposes. Originally little more than a laboratory curiosity, lasers have transformed many aspects of the everyday and scientific worlds. They now occupy an important niche in fields as diverse as surgery and long-distance communications.

A laser is a device that produces radiation that has a unique combination of properties. These properties are widely useful and all but impossible to produce by any other means. A common incandescent lightbulb operates with a power of approximately 100 watts, emits light that travels in all directions, and is composed of a broad range of visible and infrared frequencies in random phases. In contrast, lasers emit an intense beam of radiation at powers up to the terawatt (10^{12} watts) level. Laser radiation is monochromatic (contains one specific frequency or wavelength only), nondivergent (travels a single direction with very little spread), and coherent (the waves of the radiation are all in phase with each other). The first laser, originally called a *maser* because it operated in the microwave region, was demonstrated in 1954 by Charles Townes (1915–). The first laser producing visible light was constructed by Theodore Maiman (1927–) in 1960. Since then, scientists have developed lasers that produce radiation from all of the regions of the electromagnetic spectrum, up to and including X rays.

The word *laser* is an acronym that stands for "light amplification by stimulated emission of radiation," an expression that accurately describes the production of radiation in a laser. *Light* usually refers to visible radiation, but the term *laser* is now used for devices that produce radiation from microwave to X rays. The theory underlying the laser was first formulated by Albert Einstein in 1917, and was further developed as quantum mechanics advanced in the 1920s and 1930s. Physicists realized that atoms and molecules cannot possess all energies, but that they exist in specific, well-defined energy states. The process of stimulated emission therefore can be described as follows: An atom or molecule in an excited energy state (one that is higher than the lowest possible "ground" state) can be stimulated to a lower energy level by radiation whose energy matches the transition to the lower energy state. In the process, radiation is emitted at the same frequency and in the same phase as the original stimulating radiation; the stimulating radiation is thus amplified by the emitted radiation. The original, stimulating radiation is produced in a process called *spontaneous emission*, whereby excited atoms or molecules spontaneously de-excite and emit radiation simultaneously. In order for stimulated emission to occur to any extent, there must be a large number of atoms or molecules in the excited energy state, a condition called a *population inversion*. The population inversion is produced by "pumping" the atoms or molecules with energy in the form of electrical discharge, radiation, collisions, or chemical reactions. The atoms or molecules (the "lasing medium") may be in the form of large single inorganic crystals, solutions of organic dyes, mixtures of gases, or semiconductor chips. The frequency of the emitted radiation depends on the lasing medium. Lasers can be designed to operate continuously or in pulses.

Originally, lasers were used only in scientific research. At present, however, lasers are used for many everyday products and processes. Recently, lasers have been improved to the point that people untrained in laser physics or engineering can routinely use them as tools. They have also become cheap enough that they can be used for many diverse applications. Each of these applications takes advantage of one or more of the properties of laser radiation noted above. Lasers vary greatly in their properties, so no one laser is suitable for all applications.

One of the most commonly used lasers is the HeNe laser (so called because the lasing medium is a mixture of helium and neon gases), an exceptionally stable laser that emits a continuous, narrow beam of visible red light. It is used as an alignment guide in surveying (HeNe lasers were used to guide the excavation of the Bay Area Rapid Transit tunnel under the San Francisco Bay), in machine shop tools, and to align patients in medical X-ray units. It is also used in retail bar code readers, laser pointers, and compact-disk readers.

Pulsed lasers are used in a light radar (LIDAR) to perform precise distance measurements: Through the use of pulsed lasers, the distance to the surface of the moon can now be measured with an accuracy of several centimeters. LIDAR is also used to monitor pollution and trace gases such as ozone in the upper atmosphere.

The high coherence of laser light is useful in devices called *laser interferometers* that can measure distances and small displacements with high precision by allowing two laser beams to interfere with each other and produce the interference patterns typical of waves. These instruments have been used to measure displacements in the Earth's crust and to monitor automated machine tools. Laser light coherence is also used to produce holograms.

The high output power of the laser has been used

for the removal of material, for example, drilling holes in hard substances. Lasers have also been used as high-tech graffiti removers, and ancient artifacts have been restored by laser removal of superficial encrustations on stonework

A major application of laser technology is the general area of communications. The high frequency of laser light makes it capable of encoding very complex signals. These signals may be passed through space by means of satellite communications or through fiber optics cables. Lasers also are used to store and read video as well as audio information on disks.

In medicine, the use of lasers for diagnostic and therapeutic medical purposes is increasing at a dramatic rate. Many of the medical applications of laser technology involve serious or even life-threatening conditions. Laser beams are used to clear arteries and dissolve blood clots, to surgically weld detached retinas, and to control bleeding in peptic ulcers. Selective absorption of laser radiation by certain tissues has been used to remove pigmented tumors. Lasers also are increasingly being used to correct vision defects such as myopia and for elective cosmetic surgery.

There are a few dangers that are associated with lasers. Damage to the eye is the most common of these: Laser light can be focused into the back of the eye by the lens and the retina severely damaged. Precautions must therefore be taken to protect the eyes. High-powered lasers, especially those that emit ultraviolet radiation, are capable of burning the skin.

Lasers have been used to improve the quality of results for tasks that were performed in other, less technologically advanced ways. At the same time, however, the development of the higher-tech methods is likely to be accompanied by higher costs and job displacement. For instance, lasers are key elements of fully automated machine tools. With respect to medical applications, it has been speculated that as the American medical system makes increasing use of HMOs and other cost-cutting institutions, many physicians will shift their focus from the treatment of serious medical conditions to laser-based procedures such as elective cosmetic surgery for which individuals may be more willing and able to pay.—B.L.

See also COMMUNICATIONS SATELLITES; COMPACT DISK; DISK STORAGE; FIBER-OPTIC COMMUNICATIONS; HOLOGRAM; INFRARED RADIATION; OZONE LAYER; QUANTUM THEORY AND QUANTUM MECHANICS; SEMICONDUCTOR; SIGNAL-TO-NOISE RATIO; TECHNOLOGICAL UNEMPLOYMENT; ULTRAVIOLET RADIATION; UNIVERSAL PRODUCT CODE; WAVE THEORY OF LIGHT; X RAYS

Further Reading: Joseph H. Eberly and Peter W. Milonni, *Lasers*, 1988.

Latent Heat

Elements and compounds commonly exist in one phase—solid, liquid, or vapor—under ordinary conditions of temperature and pressure, for example, 25°C (77°F) and 1 atmosphere. The term *state* is also used for phase, e.g., water in the liquid state. When elements or compounds undergo a change in phase, energy is either absorbed or released, even though the temperature remains constant. In going from a solid to a liquid phase (melting or "fusion"), an input of energy is required. This is known as the "latent heat of fusion," an expression coined by the 18th-century British scientist Joseph Black (1728–1799). Likewise, when a liquid boils, the latent heat of vaporization is needed to change the substance from a liquid to a gas. The values of these energy changes are usually determined at the "normal" temperatures for fusion and vaporization, when the pressure is 1.00 atm. The value may be reported either "per gram" or "per mole" of the substance in question. The latter description allows for direct (per molecule) comparison among all types of substances. Normally, vaporization requires much more energy than melting, because the atoms or molecules must be distinctly separated in the vapor phase.

Conversely, when substances go from a less condensed phase to a more condensed phase, energy is released in the form of heat. The magnitude of the energy change is the same as for the opposite process. That is, when steam condenses at 100°C (212°F) and 1.00 atm, the heat emitted is the same as that absorbed in boiling an equal mass of water. This explains why it is much more harmful to be scalded with steam at 100°C compared to water at the same temperature. Not only is the steam very hot, but it also releases the latent heat of vaporization (2,259.4 joules per gram) when it condenses on the skin.—A.Z.

Further Reading: J. C. Kotz and K. F. Purcell, *Chemistry and Chemical Reactivity*, 2d ed., 1991.

Lathe

A lathe rotates an object between two pivots so that the object can be shaped by a cutting tool. For example, a lathe can be used to turn a plain wooden cylinder into an ornamental table leg or a baseball bat. Lathes are the original machine tool and continue to be of prime importance for many types of manufacturing.

The lathe is a very old device. Although the historical evidence is inadequate, it is likely that it was in use 3,000 years ago, perhaps even earlier. The first lathes were driven by a reciprocating motion; this was supplied by a cord wound around the piece being worked, the two ends of the cord being alternately pulled by an assistant. A bowstring wound around the workpiece also could be employed to allow one-person operation, but controlling the cutting tool while at the same time turning the piece was no easy matter. In a more advanced application, which first appeared in Medieval times, an elastic pole attached to a cord was used to twist the stock back after it had been pulled in the opposite direction by the operator. A refined version used a foot treadle to pull the cord, leaving both of the operator's hands free.

By the second half of the 15th century, lathes gained a true rotary motion, provided by a crank turned by power from human beings, animals, or water. At about the same time, lathes began to appear with slide rests to hold the cutting tool, which enhanced the force and accuracy of the cut. By the late 16th century, the lathe also had been improved through the fitting of a lead screw that moved the slide rest and cutting tool alongside the object being turned. In this way it was possible to accurately cut spiral grooves, making possible the production of large numbers of screws and bolts.

Lathes of this type were useful for making small, relatively lightweight objects, but they could not be used on heavy stock made from iron or steel, for they were not rigid enough for this purpose. This began to change as a result of the employment of all-metal lathes designed by Harry Maudslay (1771–1831). Although individual features of Maudslay's lathe had been anticipated several decades earlier by Polem in Sweden, Vaucanson and Senot in France, and Wilkinson in the United States, Maudslay's lathes put everything together. First made in the late 18th century, the lathes were very rigidly constructed and built to high levels of precision. The lead screws of Maudslay's lathes were turned through the use of a master lead screw that itself had been painstakingly fashioned for utmost accuracy. Maudslay's approach to lathe design and construction was continued and taken to new heights by his student Joseph Whitworth (1803–1887). Whitworth made some important improvements to the lathe, but his major contribution was to make stronger, more accurate tools. For a half century, his lathes were considered the world's finest, and they played a key role in bringing industrialized manufacture to many parts of the world.

The versatility of the lathe was enhanced by the turret lathe, which was first used in the mid-19th century at the Colt armory in Hartford, Conn. It employed a number of cutting tools mounted on a revolving drum, or turret, that held special tools for making different kinds of cuts; these could be quickly deployed simply by rotating the turret. Another highly useful device was the copying lathe invented by Thomas Blanchard (1788–1864) in 1819. This was an ingenious collection of levers and cams that formed wood into the duplicate of a metal prototype, making it possible to turn out large numbers of gun stocks, axe handles, shoe lasts, and other wooden objects. It greatly speeded up production; Blanchard's first copying lathe could produce a gunstock in 22 minutes, and with subsequent improvements the time was reduced to 5 minutes.

The lathe in its various forms was essential to the development of mass production, for it made possible the large-scale manufacture of standardized, interchangeable parts. At the same time, however, the development of the lathe and other machine tools often was oriented toward putting more skill into the machine and taking it from the machine's operator. Improvements in the lathe boosted productivity and lowered costs by replacing skilled workers with unskilled ones. Larger volumes of goods at lower prices was one result, but so was a reduction of skills and wages for one group of workers. According to some critics, this process continues today, exemplified by the development of numerically controlled machine tools.

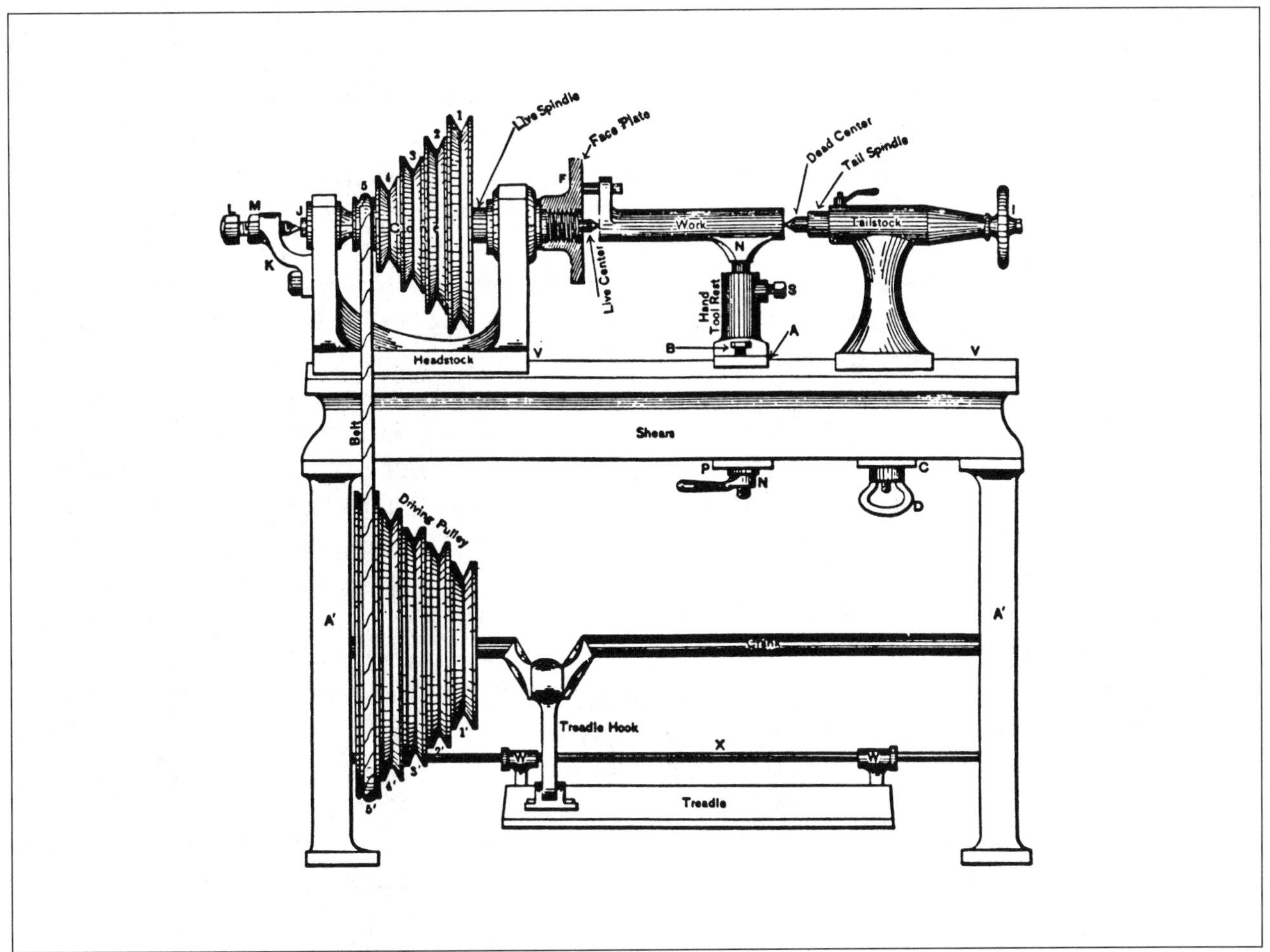

A 19th-century foot-powered lathe. The rotational speed was changed by shifting the belt between the two sets of pulleys (from C. C. Cooper, *Shaping Invention*, 1991).

See also DESKILLING; MASS PRODUCTION; MACHINE TOOLS, NUMERICALLY CONTROLLED

Further Reading: Robert S. Woodbury, *Studies in the History of Machine Tools*, 1972.

Lawn Mower

Although they are now ubiquitous in most parts of the United States, lawns have not always been essential features of residential life. Until the second half of the 19th century, a few well-to-do Americans lived on estates that emulated parklike English country homes, but most people lived in urban dwellings with at best a small front yard characterized by wild plants and grasses, and in many cases beaten earth was deemed sufficient. In the years following the Civil War, lawns began to take on increasing prominence as a result of suburbanization, higher incomes, and the desire to take on some of the trappings of wealth and status.

In most regions of the country, a green lawn is an unnatural entity; to flourish it requires regular applications of fertilizer, herbicides, insecticides, and water. Lawns also require frequent mowing during the seasons when the grass is actively growing. At first, grass was cut by wielding a scythe, a tedious, labor-intensive process. Larger lawns were often cropped by grazing sheep. For obvious reasons, neither method was suitable for suburban lawns tended by individual homeowners. Accordingly, lawn-mowing machines began to be developed just when lawns were growing in popularity. The first reel-type mower was invented in 1830 as an adaptation of a machine used for cutting the naps of carpets, but with few lawns to be mowed, the invention was ahead of its time. As more residences included lawns, mowers became the subject of increasing attention; 38 patents for lawn mowers were issued by the U.S. Patent Office during the years 1868 to 1873. In 1881 alone, 138 patents were granted.

Although steam-powered mowers were produced in the late 19th century, it was the gasoline engine that made the power mower a practical proposition. Although power mowers were manufactured during the early decades of the 20th century, the great boom in power mowers did not occur until after World War II, when accelerated suburbanization and rising living standards resulted in more and more land being devoted to lawns. In the 1930s, fewer than 35,000 power mowers were produced annually in the United States; in 1947, the number had increased more than 10-fold. In 1951, more than 1.6 million power mowers were produced, and 7 million were made in 1974, with the number declining to 5 million in 1989.

A 19th-century lawn mower (from Edward de Bono, *Eureka, An Illustrated History of Inventions*, 1974; used with permission).

The majority of the power mowers sold were rotary types that use a single horizontal blade to cut the grass. Finicky gardeners assert that rotary mowers do not produce as smooth a cut as reel-type mowers, but their real problem is safety. In 1971, the National Commission on Product Safety rated rotary mowers second only to the automobile as the most dangerous household devices. Despite voluntary safety standards instituted by the mower industry, an estimated 140,000 people were injured by mowers in 1969 alone. In 1977, the Consumer Product Safety Commission mandated certain safety features to be installed on all new mowers manufactured after July 1982, including a brake that stopped the blade when the operator's hands left the mower's push handle. Costs went up by as much as 40 percent, but injuries went down. Riding mowers, which began to be sold in significant numbers in the 1950s, are even more dangerous when not prudently operated, for they can overturn and pin the operator. In 1985, more than 20,000 people were treated in emergency rooms for injuries suffered in riding mower and garden tractor accidents. All in all, power lawn mowers have made grass cutting easier and faster, but they require more care and attention than some operators are willing to give.

See also ENGINE, FOUR-STROKE

Further Reading: Virginia Scott Jenkins, *The Lawn: A History of an American Obsession*, 1994.

Lead Poisoning

People have used lead since about 3500 B.C.E. A useful material that resists corrosion and is easily molded, it can be highly toxic when ingested or breathed in. Lead has no known biological function in the human body, and it interferes with many important processes such as the synthesis of hemoglobin.

Lead was extensively used by the Romans in their water supply systems. Lead pipes distributed great quantities of water within Roman cities, but they also exposed large segments of the population to lead poisoning. As the Roman architect Vitruvius warned:

> . . . water is much more wholesome from earthenware pipes than from lead pipes. For it seems to be made injurious by lead, because white lead is produced by it, and this is said to be harmful to the human body.

This may not have been a universal problem, however, for in many localities calcium carbonate from the hard water formed a protective deposit on the inner walls of the pipes. More dangerous was the Roman custom of using lead-coated copper and bronze vessels for food preparation and storage. The presence of lead prevented the corrosion of the vessels and slowed down enzymatic activity, keeping the food reasonably fresh. But the ingestion of lead also produced the classic symptoms of lead poisoning: colic, weakness, lethargy, neurological disorders, digestive upsets, and, in many cases, death. A few scholars have even implicated lead poisoning as the chief cause of Rome's collapse, but there is not enough information on the actual extent of lead poisoning to validate this claim.

Although other places experienced occasional outbreaks of lead poisoning from contaminated food and drink, few people were exposed to toxic quantities of lead until the Industrial Revolution, when large quantities of lead came to be used for battery plates, solder, and pipes. World production of lead increased from 100,000 tons per year in 1750 to over 5 million tons per year in the 1980s. In the 1920s, even more lead began to be put into the environment when tetraethyl lead was added to gasoline to raise its octane number, thereby allowing an increase in the compression ratios of automobile engines. The use of lead as a gasoline additive began to be sharply curtailed in the 1970s, but substantial amounts of lead from tailpipe emissions remain in the soil.

The other major source of lead is the paint found in older buildings, because a mixture of white lead powder and linseed oil was commonly used until the 1950s. The problems caused by children ingesting lead paint flakes was first documented in Australia at the turn of the century, and American studies began around 1910. Due in part to the fact that most of the victims were children from poor, inner-city families, the problem did not receive much attention until the 1960s. As the problem of lead poisoning became evident, the United States banned the use of lead-based paint for residential structures in 1978. Even so, an estimated 65 percent of the houses in the United States have surfaces covered with lead-based paint. If left undisturbed, painted walls and fixtures generally do not pose a hazard. The danger occurs when young children gnaw on surfaces covered by lead paint, or when the paint flakes and produces toxic dust.

High levels of lead have been blamed for a variety of physical and psychological disorders among children, along with lowered intelligence and other developmental problems. The actual extent of lead poisoning among children is, however, a matter of some dispute. The U.S. Center for Disease Control considers a level of 10 to 25 micrograms of lead per deciliter of blood to indicate some degree of lead poisoning, but not all medical experts agree with this standard. Many children with this concentration of lead in their systems perform poorly on tests of intelligence and psychological development, but it is not certain that lead is the cause. Being raised in a poor family is often correlated with both low test scores and relatively high concentrations of lead, making it difficult to determine if lead is really the source of the problem, or if it just happens to be associated with the real cause, poverty.

Determining the culpability of lead-based paint is more than an academic exercise, for the cost of removing this paint would be several hundred billion dollars. Moreover, the process of deleading houses, if not properly done, can throw a considerable amount of toxic lead dust into the immediate environment, making the situation far worse than it had been. Deleading also may have an adverse effect on the housing supply; in many cases, landlords may abandon a building rather than bear the expense of removing lead-based paint. Since a disproportionate share of these houses are rented by lower-income families, their loss would result in housing shortages and consequent higher rents. Finally, lead-free houses do not guarantee a lead-free environment. As was noted above, in many areas a great amount of lead stays in the soil, where it remains a hazard to all who come in contact with it.

The lingering problems caused by lead in the environment should not obscure the fact that exposure to lead is much lower today than it was in the recent past. The United States began to phase out tetraethyl lead in the 1970s because it destroyed pollution-reducing catalytic converters. Since then, world use of tetraethyl lead has declined by 75 percent, and in the United States concentrations of lead in the bloodstream have declined significantly.

See also AQUEDUCT; CATALYTIC CONVERTER; INDUSTRIAL REVOLUTION; OCTANE NUMBER; PAINT

Lean Production

Lean production is the name given to the assemblage of manufacturing techniques that began to be developed at Toyota Motor Company in the 1950s. It has as its goal the combination of the best aspects of craft production and mass production. Craft production allows customized production, but at a high cost. Conversely, mass production allows low manufacturing costs, but it results in a great deal of inflexibility. Putting together the best aspects of both through lean production requires a significant rethinking of manufacturing processes and the roles assumed by suppliers and shop-floor workers.

One of the basic assumptions of lean production is that it is impossible to completely plan production processes in advance and from afar. Consequently, the people who actually build the product have to be involved with manufacturing technologies from the outset. This in turn requires the hiring of educable workers who are given the opportunity to acquire and use new skills. This puts lean production at the opposite pole from the traditional treatment of workers that left little room for initiative, innovation, and the development of skills. The role of production workers is further expanded by having them serve as quality checkers, rather than delegating that task to people whose only job is inspection. Working in lean production regimen creates new stresses for workers in that they have to exercise problem-solving skills, but it is a different kind of stress than the kind engendered by continuously working on an assembly line

Another key aspect of lean production is an expanded role for component suppliers. In traditional manufacturing regimes, a firm might purchase a significant portion of the parts used in its products, but the relationship between the manufacturer and the supplier was one-dimensional. The manufacturer drew up exact specifications for a part, and then chose a particular supplier because it offered the lowest price. In lean manufacturing, suppliers are often involved with the actual design of the product. As a result, the manufacturer is able to take advantage of the suppliers' expertise. In addition, the product can be designed from the start so that all the components interact harmoniously with one another.

Supplier involvement is also crucial for another important facet of lean production: the "just-in-time method" of parts delivery. Sometimes known by its Japanese name, *kanban*, the parts provided by outside suppliers are delivered a few hours or even minutes before being incorporated into the final product. Just-in-time delivery obviates the need to build up large inventories of parts that take up space and absorb working capital. Equally important, a just-in-time delivery system requires that manufacturing operations be organized in the most efficient manner, with no wasted moments. In contrast, the buildup of inventories allows wasteful practices to go unnoticed and provides no incentives to eliminate bottlenecks and wasteful practices.

Lean production has been successfully adopted by a number of manufacturing and retail firms, which credit it with cutting their costs and improving their quality. At the same time, there are a number of issues that lean production does not address. It may be a superior way of making existing products, but it does not do much to help firms anticipate what kind of products they should be developing for the future. Also, the cooperative relationship between the firm and its workers and suppliers may come under considerable strain should the firm's profits be squeezed during cyclical economic downturns. Maintaining lean production under these circumstances requires a long-range vision and a willingness to suffer some short-term losses.

See also ASSEMBLY LINE; DESKILLING; MASS PRODUCTION

Further Reading: James Womack et al., *The Machine That Changed the World*, 1990.

Legumes

Legumes are plants belonging to the Leguminosae family, which encompasses approximately 18,000 separate species. The word also refers to their edible fruits: peas, beans, soybeans, lentils, peanuts, and chickpeas, to name the most common. Other legumes, notably alfalfa and various clovers, are widely used as animal foods. Leguminous fruits (strictly speaking, they are not vegetables, although common usage refers to them in this way) are important dietary items in many cultures. Legumes even played a significant role in European history; in the early Middle Ages the use of legumes in a three-field crop rotation system undergirded the advance of Western culture by substantially improving its nutritional base.

Legumes contain complex carbohydrates (starches) that serve as basic elements of an adequate diet. Grains provide more complex carbohydrates than legumes, but the latter have twice the protein of the former, while soybeans have three times as much. Additionally, legumes contain fiber, which has no direct nutritional value, but is nonetheless essential to digestion because it provides bulk. Fiber also reduces the absorption of fat in the body by helping it to move through the system before it can be entirely absorbed. Moreover, there is considerable evidence that fiber in the diet helps to prevent colon cancer. Legumes do have one negative quality. When eaten raw, they release certain chemicals that block trypsin, an enzyme that helps to covert proteins into amino acids so they can be absorbed by the body. Since legumes are rarely eaten in an uncooked or unprepared state, this problem rarely arises.

In addition to providing important nutrients, legumes can play an important role in the maintenance of soil fertility. The roots of leguminous plants have numerous nodules that are inhabited by *Rhizobium* bacteria. These bacteria help to fix atmospheric nitrogen in the soil, reducing or even eliminating the need to add nitrogen-based fertilizers.

See also CROP ROTATION; FERTILIZERS, CHEMICAL

Lifespan, Extension of

Human life expectancy (average lifespan) has essentially doubled during the past 500 years, from less than 40 years in 1500 to about 75 years today. This extension of lifespan has been most marked in the industrialized nations during the past century. This achievement gives rise to a number of questions: How long is the maximum potential human lifespan? Can lifespan be extended by pharmaceutical and genetic manipulations? Why, indeed, do organisms die?

Improved survival of the young during the past century has been the major factor in increased life expectancy. Improved sanitation and nutrition, as well as the use of vaccinations and antibiotics, have been the major factors contributing to decreased mortality among all but the oldest segment of the population. Except for antibiotics, these interventions prevent disease rather than cure it. Great epidemics and bacterial scourges—cholera, plague, typhus, tuberculosis—have been largely eradicated in industrialized nations. A century ago, such diseases were accepted with bitter resignation as a fact of life.

Vaccination was practiced before its mechanism of protection was understood. In 1798, Edward Jenner, a British physician, noted that milkmaids who had been infected with cowpox did not become ill with smallpox. He began preventive inoculations, a practice that soon spread around the world. In the 1880s, Louis Pasteur immunized farm animals against anthrax and cholera and, in his most famous case, protected a young boy against rabies. Following Pasteur's lead, vaccines against other major bacterial and viral killers were developed and brought into widespread usage. With promotion of vaccinations by public-health organizations, mortality among the young has been reduced to less than one percent of its previous toll.

Public-health organizations arose in the late 1800s in response to increased urban density and its accompanying squalor. The United States Public Health Service was established in 1878. Included among its responsibilities was quarantine of immigrants flooding into eastern seaports during this period, as well as housing condemnation and improved sewage disposal. As the bacterial causes of disease became better understood, government agencies also monitored water purification, food handling, and vaccination programs. Several women physicians (e.g., Florence Sabin and Alice Hamilton) were active in public-health efforts during the early 1900s.

Nutrition gained attention as a public-health issue in the early 1900s with the demonstration of specific nutritional deficiency diseases. Much earlier, James Lind (1716–1794), a Scottish naval surgeon, had shown that scurvy could be prevented with regular doses of citrus fruit. In the early 1800s the British Navy began providing lime juice for its sailors (hence the term *limey*). Some historians credit this practice with making possible the British mastery of the seas and expansion of the British empire around the world. Other nutritional diseases were not explored until after 1912, when Frederick Hopkins (1861–1947) postulated that diseases such as beriberi, scurvy and rickets were caused by a lack of trace nutrients called vitamins. Interest in nutrition is currently experiencing a revival, particularly as it relates to aging.

Antibiotics are products of the 20th century. In 1910, Paul Ehrlich (1854–1915) developed Salvarsan, an artificial arsenic compound used against syphilis. Nearly 2 decades later, Alexander Fleming (1898–1955) observed that the mold, Penicillium, was able to kill a disease-causing bacterium, Staphylococcus aureus. Subsequently, Howard Florey (1898–1968) and coworkers purified penicillin sufficiently for use against human bacterial infections and the U.S. government assisted with production of this antibiotic during World War II. Antibiotic development continues to this day in an effort to stay ahead of acquired bacterial resistance.

With the control of most nutritional and bacterial diseases (and many viral diseases) by a combination of vaccination, sanitation, nutrition, and antibiotics, life expectancy in the United States has increased steadily. It is now only a few years short of the estimated average maximum lifespan (about 85 years). This means that, on the average, human beings would probably live to about 85 years of age if everyone died of "old age" or diseases of aging, such as cardiovascular disease, cancer, and neurological diseases. Is it possible to retard or even cure such diseases? Gerontology (the study of aging) and geriatric medicine (a medical specialty) have emerged in the past quarter century to focus on what is sure to become a major social, medical and economic issue: the aging of the American populace. The U.S. Census Bureau predicts that the elderly—those over 65—will constitute about 20 percent of the U.S. population by the year 2030. Human beings live longer than most other multicellular animal species in terms of average maximum lifespan. Whales and elephants have an average maximum lifespan of about 80 years. At 120 years, the Galapagos tortoise (a cold-blooded reptile) is perhaps the longest-lived animal species. What factors correlate with extended lifespan in both individuals and species? Size correlates with lifespan (hence, the elephant's long life), whereas metabolic rate correlates inversely (metabolically active mice have short lifespans). Longevity also tends to correlate with later onset of reproductive age. Still, in terms of virtually every measure of longevity, human beings have already beat all the odds. Our maximum potential lifespan is twice that of chimpanzees, with whom we share about 98 percent of our genome and a similar metabolic rate.

How long can a human possibly live? That figure is usually given as 100 to 110 years. One gerontologist, however, suggests that the human lifespan might be extended beyond even Methuselahn figures: 2,000 years for males and 4,000 years for females! Most believe that the maximum potential human lifespan is limited, and probably genetically determined, but also that this limit can be extended by genetic and pharmaceutical intervention.

Successful intervention requires that we understand the basic causes of aging and senescence. Prevailing theories of aging fall into two broad categories: genetic and environmental. These are not entirely separable, since certain genetic conditions render individuals more susceptible to environmental insult. Most gerontologists attribute aging to one or more of the following factors: oxidative damage, bodily "wear and tear," decreased immunological competence, mitochondrial dysfunction, and genetic (programmed) senescence.

Oxidative damage is considered the prime suspect in the aging process. Ironically, the oxygen on which our lives depend is the major source of damage to tissue molecules. During metabolic processes ("burning" the body's fuel for energy and heat), several highly reactive oxygen derivatives are formed. These can react with cellular molecules, including DNA, sometimes creating damage that cannot be repaired. Lipofuscin, a type of pigment that clogs heart cells and neurons as we age, is a consequence of peroxidation of cellular lipids. The single most successful strategy for prolonging life, demonstrated repeatedly by experimental data, is the limitation of caloric intake (as long as nutritional balance is maintained). The resulting decreased metabolic rate apparently slows oxidative damage.

The wear-and-tear hypothesis suggests that we accumulate small mechanical, chemical, and radiation insults over the years, which cause the fabric of our bodies to deteriorate. Perhaps the most critical type of damage is the mutation of DNA in cells exposed to radiation (from the sun, for example), or to chemical carcinogens. An elevated cumulative exposure to damage increases the probability of mutations that can transform normal cells into cancer cells, which may then grow out of control and destroy the rest of the body.

Immune competence (the ability to mount a robust immune response to foreign molecules) increases from infancy to adulthood, remains high until about the age of 40, then gradually decreases with age. Since our body's immune competence is largely responsible for its ability to cope with the many microorganisms that inhabit our environment, aging increases susceptibility to disease. Immune competence is also involved in ridding the body of transformed cells that might otherwise develop into tumors. Mitochondria, the powerhouses of our cells, have their own special loops of DNA that code for some mitochondrial proteins. However, mitochondria lack the enzymes that repair DNA damage when it occurs in cell nuclei. Thus, as we age, our mitochondria accumulate an increasing load of crippling mutations. Some cells may no longer be able to produce enough energy to sustain life.

Genetic (programmed) senescence is an undercurrent of many theories of aging. In the early 1960s, Leonard Hayflick and coworkers made the mysterious and unexpected observation that normal human cells were able to undergo only a finite number of replications in tissue culture, after which the cells died. This became known as the Hayflick phenomenon, the disquieting reality that normal cells, even when grown in a protected environment and provided optimal nutrition, do not live forever. Such studies provided strong evidence for a genetic basis of aging and death.

Only recently has molecular biology provided a clue to the cause of this phenomenon. At the ends of chromosomes are segments of DNA, called *telomeres*, that apparently protect the internal genetic material. Bits of the telomeres are normally lost during each cell division until eventually, aged cells are left with little or no telomere capping their chromosomes, and cells die. "Immortal" cells, such as cancer cells and stem cells, produce an enzyme (telomerase) that helps to regenerate the telomeres so that cells can continue to divide indefinitely.

Other evidence that length of life has a genetic basis comes from studies on fruit flies and roundworms. Fruit flies have been induced to double their maximum lifespan by selecting for late offspring. Evidence exists for at least four genetic sites that affect lifespan in the worm, including a specific longevity gene (age-1). Two putative genetic diseases in human beings, progeria and Werner's syndrome, are characterized by an acceleration of the aging process.

As the genetic basis of longevity is explored, thorny philosophical questions arise. Individual human beings, conscious of being unique and separate from all others, often wish for immortality. What if it were possible to live indefinitely? Who would be allowed that privilege? What would be the biological and social costs and consequences? Most multicellular organisms, both plants and animals, reproduce sexually and have limited individual lifespans. Sexual reproduction allows much more genetic variability than occurs in asexually reproducing organisms. Genetic variability, in turn, allows for a broad species repertory of adaptive responses to environmental change. An important function of death is to free up resources for subsequent generations. It is clear that our planet is already approaching its maximum human carrying capacity, the Malthusian danger zone. As a species, can we afford to sacrifice our genetic flexibility? As individuals, would we forego the joys of parenthood for the indefinite extension of lifespan?—J.S.

See also ANTIBIOTICS, RESISTANCE TO; CANCER; CELL; DNA; ENZYME; EPIDEMICS IN HISTORY; EVOLUTION; GERM THEORY OF DISEASE; IMMUNIZATION; NATURAL SELECTION; PENICILLIN; PUBLIC HEALTH; RADIOACTIVITY AND RADIATION; SMALLPOX; VITAMINS

Light, Polarized

The wave properties of light are manifest in three phenomena: interference, diffraction, and polarization. *Polarization* refers to the orientation in space of the transverse electrical and magnetic fields that produce light. These fields, which vibrate in directions perpendicular to each other, are usually oriented at random in the plane perpendicular to the light's propagation. Light of this type is called *unpolarized*. However, it is possible to produce light that has a preferred orientation within this plane for the two fields; this type of light is said to be *polarized*. Scientists refer to the percent of polarization, which can range from 0 percent (completely random orientation) to 100 percent (completely polarized). In the latter case, the electromagnetic vibrations of every photon (i.e., a "packet" of light) are aligned in the same direction.

The phenomenon of polarization began to be explored in earnest when Erasmus Bartholin (1625–1698), a Danish physician, observed what he called *double refraction*. This is the double image produced when an object is viewed through a transparent crystal of calcium carbonate ($CaCO_3$), known as Iceland spar. Bartholin also noted that when the crystal was rotated, one of the images remained fixed, and the other appeared to revolve about it. For many decades, neither Bartholin nor anyone else could explain this phenomenon. In 1828, the Scottish physicist William Nichol (1768–1851) discovered that two pieces of Iceland spar could be cemented together with Canadian balsam to produce polarized light, although the term had not yet been invented. The dubious honor for coining the term *polarized light* goes to Etienne Malus (1775–1812), a French military engineer and amateur physicist. The word *dubious* is used because the word *polarized* is based on a faulty premise. Malus discovered that when sunlight reflecting from a window passed through doubly refracting Iceland spar, only one ray of light emerged. Malus adhered to the prevalent view that light was composed of particles, and that the particles had poles (an idea that originated with Newton), hence the term polarized "light."

In 1815, Malus's compatriot Jean Baptiste Biot (1774–1862) found that when polarized light passed through quartz crystals its plane of polarization twisted. This effect also was produced by liquid or in-solution organic substances. He correctly surmised that this phenomenon was caused by asymmetry in the molecules themselves. In effect, he had founded the technique of polarimetry, which was subsequently put to excellent use by Louis Pasteur (1822–1895) in the early phase of his scientific career. By this time, the development of the modern wave theory of light by Agustin Fresnel (1788–1827) provided an adequate explanation of polarization.

Iceland spar continued to be used for subsequent experiments with polarization. It was found that an unpolarized light beam entering the crystal emerges as two perpendicular beams, each at half the intensity of the incoming beam. Other materials, tourmaline for example, will greatly attenuate one of the two beams. This phenomenon is known as *dichroism*, and is caused by the alignment of the molecules in the crystal. In the 1930s, the American inventor Edwin Land (1909–1991) hit upon the idea of embedding small crystals of a polarizing material in sheet plastic. Today, polarizing sheets often are manufactured from polyvinyl alcohol, whose long chain molecules are carefully aligned by stretching in order to produce the effect of a large single dichroic crystal. These sheets have largely supplanted naturally occurring crystals for technical and commercial applications.

Some substances such as sugar consist of molecules with a corkscrew structure that can rotate the plane of vibration of the light passing through them. These substances are said to be "optically active." An optically active material placed between two filters that polarize light will often produce spectacular colored patterns that can be used for special effects, as well as for such practical applications as stress analysis and the measurement of concentrations.

When unpolarized light strikes a nonmetallic surface such as a body of water or the surface of a road, the reflected light is partially polarized, with the electric field vibrating in a plane parallel to the surface. At one special angle, called a Brewster's angle (named after the Scottish physicist David Brewster [1781–1868]), the reflected light is completely polarized. Polaroid sunglasses are built to take advantage of this effect. The Polaroid lens is oriented to pass light that is perpendicular to the horizontal surface; consequently, the sunglasses will eliminate much of the glare normally experienced when driving or boating. In a similar manner, light scattered from air molecules is partially polarized at right angles to the direction of the sun; some animals and insects, notably bees, appear to take advantage of this phenomenon to help determine directions.

Many materials, including some that are biologically interesting, are *bifringent*; i.e., the two polarized waves travel through the material at different speeds. The emerging light is then said to be "circularly" or "elliptically polarized," while light passing through a calcite crystal or Polaroid sheet is said to be "plane polarized."

Although light emitted by the sun and most lightbulbs is either unpolarized or at most a fraction of a percent polarized, light emitted by the synchrotron process is highly polarized. This distinction has been important for astrophysicists, for it allows them to distinguish between thermal emissions, as seen in stars, and synchrotron emission as seen in quasars and some galactic nuclei.—S.N.

See also CRYSTALS; QUASARS

Further Reading: Hugh D. Young, *Fundamentals of Waves, Optics, and Modern Physics*, 2d ed., 1976.

Light, Speed of

Everyday experience tells us that sound has a finite speed. When we observe an action taking place a distance way, we see what is happening before the sound reaches us. Light, however, appears to travel instantaneously, or at least at a very great speed. It was a major accomplishment of physics to determine that light traveled at a finite speed and to measure what it was.

The first attempt to measure the speed of light was conceived by Galileo (1564–1642). For this experiment, one person stood on top of a hill with a lantern, while another person stood on another hill with his lantern. After one of them uncovered his lantern, the other, having seen the flash, uncovered his lantern. The speed of light could then be calculated by dividing twice the distance separating the two lanterns (since the light made a round trip) by the time interval between the two flashes. The experiment was performed several times, with the expectation that the time interval would increase as the distances lengthened. Instead it stayed about the same, for the speed of light was much too fast to be measured by Galileo's procedures.

The first successful attempt at measuring the speed of light was done in 1676 by the Danish astronomer Olaus Roemer (1644–1710). Roemer based his calculation on the time it took the satellites of Jupiter to make a revolution around that planet. Knowledge of these times allowed the determination of the precise moment that a given satellite would be occluded by Jupiter (i.e., appear to go behind the planet). Roemer found to his surprise that the occlusions came earlier when the orbits of Earth and Jupiter brought the two planets closer together, and that the opposite happened as they diverged. Roemer correctly attributed these differences to the greater or lesser distances traveled by the light emanating from Jupiter and its satellites. Knowing the various distances between Earth and Jupiter, Roemer calculated that light traveled at a speed equivalent to 227,000 km/sec (141,051 mi/sec). The word *equivalent* is used because the kilometer did not exist as a unit of measurement in Roemer's time.

Oddly, Roemer's discovery did not raise much of a stir in the scientific community. In 1728, a calculation of the speed of light performed by the British astronomer James Bradley (1693–1762) met with more interest. While seeking to measure stellar parallax (the apparent shift in a star's position caused by changing viewing angles as the Earth moves through its orbit), Bradley found that the stars seemed to shift position due to the speed of the Earth's orbit. From this he was able to calculate the speed of light as 176,000 mi/sec (283,245 km/sec).

The first attempt to measure the speed of light through the use of Earth-bound experimental apparatus was made by the French physicist Armand Fizeau (1819–1896) in 1849. Fizeau used a toothed disk that was set in the path of a beam of light. As the disk was rotated, the light passed through the gaps between the teeth. The resultant pulses of light then bounced off a mirror located 8.67 km (5.39 mi) distant. If the disk was turned at the proper speed, the returning pulse of light passed through the gap next to the one through which the initial pulse had passed, and it could be seen by an observer located behind the disk. When this happened, the time it took for a pulse of light to pass through a gap in the wheel, reflect off the mirror, and then pass through the next gap in the wheel was the same as the time it took for the wheel to rotate from the first gap to the next, 1/18,000 sec. Since the light had traveled 17.34 km (10.78 mi), i.e., 2 × 8.67 km, the speed of light was approximately 312,000 km/sec (194,000 mi/sec).

A year later, another French physicist, Jean Foucault (1819–1868), improved on Fizeau's technique by using a rotating mirror in the place of the toothed disk. The elapsed time for a pulse of light to the mirror and back was determined by measuring the small displacement of the light caused by the mirror's rotation. Foucault continued to work with this apparatus, and in 1862 he came up with a figure of 297,728 km/sec (185,000 mi/sec). He also used this technique to measure the speed of light as it passed through various fluids, finding that these speeds were appreciably slower than the speed of light in air.

In 1878, the American physicist Albert Michelson (1852–1931) began to measure the speed of light using an improved version of Foucault's apparatus. After many trials, in 1882 he came up with a figure of 299,853 km/sec (186,320 mi/sec), a figure that held sway for the next 45 years. Even more important than precision, Michelson conducted experiments that profoundly affected scientists' perception of how light is propagated. At this time, most physicists considered light to be a wave. A wave required a medium through which it could travel (air is the medium for sound waves), so they hypothesized the existence of a mysterious substance dubbed "the luminiferous ether."

The ether was thought to remain stationary while the Earth rotated on its axis and revolved around the sun. Consequently, light that was beamed in the direction of the Earth's motion should travel more rapidly than light moving at right angles to it. To check this, Michelson designed an *interferometer*, a device that would detect differences in velocity by showing interference patterns made by the two beams. From 1881 to 1887, he tried to find these interference patterns. The most sophisticated of the experiments were conducted with Edward Morley (1838–1923). No interference patterns were ever found. Physicists had to conclude that the ether did not exist. Moreover, the Michelson-Morley experiment indicated that the speed of light in a vacuum never varied, no matter what the relative motion of an observer. This discovery had profound implications, for it reinforced Albert Einstein's revolutionary vision of the nature of space and time.

Michelson returned to the measurement of the speed of light in 1923, and worked at it until his death in 1931. His coworkers continued to work with the apparatus he designed, and their figures are very close to the currently accepted speed of light, which was determined in 1972 by a research team headed by Kenneth M. Evenson: 299,792.4561 km/sec (186,282.3959 mi/sec).

See also MAXWELL'S EQUATIONS; RELATIVITY, SPECIAL THEORY OF

Further Reading: Bernard Jaffe, *Michelson and the Speed of Light*, 1960.

Light-Emitting Diode

A light-emitting diode (LED) converts an electrical current to visible light. All diodes emit some electromagnetic radiation when they are forward-biased (that is, when a voltage is applied across the semiconductor junction, resulting in the flow of current through the junction). When certain semiconducting materials are used, the electromagnetic radiation from the diode may fall in the part of the electromagnetic spectrum that manifests itself as visible light. Diodes also differ in the amount of radiation they emit; a diode made from certain semiconductors like gallium arsenide phosphide will emit much more radiation than a silicon-based diode.

The light from a light-emitting diode is produced when an electrical current exceeds a threshold voltage, exciting electrons and causing them to cross the semiconductor junction and emit a photon. Unlike light from an incandescent bulb, the light produced by an LED consists of a narrow range of wavelengths because the electrons are all excited to about the same level.

The first LEDs became commercially available in the 1960s. Capable of emitting only red light, they had 1/100 the luminous performance of an incandescent lightbulb, itself no model of efficiency. LEDs were crucial components of digital watches and handheld calculators when these devices were introduced in the early 1970s. Since then, their LEDs have been replaced by liquid-crystal displays, which consume less power and are easier to read in sunlight.

While LEDs have lost a segment of their original market, they have gained many others. In recent years, it has become possible to use LEDs as sources of illumination in the place of conventional lamps. This has happened because their luminous performance (the output of visible light divided by the input of electrical power) is 20 times better than it was 2 decades ago. One notable example of LED use is taillights for automobiles. An ordinary incandescent taillight bulb emits about 15 lumens (a measure of light output) per watt of electrical power, but three-quarters of the light is absorbed by the red filter, so only 3 to 4 lumens of light is actually emitted, less than the light provided by an LED. Under these circumstances, an LED may be preferable, since it operates with greater reliability while allowing more stylistic freedom. For example, it is easy to fit a long, thin row of LEDs in the rear edge of a spoiler, where they can serve as auxiliary brake lights. Another potentially vast market for LEDs is traffic signals. Conventional bulbs are inexpensive, but failures can create serious traffic problems and even endanger lives, while the labor costs of replacing them are high. Since an LED can shine for a million hours before it loses half its brightness, it promises greater reliability and lower maintenance costs.

More than 20 billion LEDs are produced each year at an average cost of less than 10 cents each. Their greatest defect is that efficient LEDs are capable of producing light only in the red through green part of the spectrum. Blue LEDs are being manufactured, but most have relatively poor performance (less than 0.1 lumens per watt). Their primary use is to provide white light for moving-message panels by combining their light with light from other LEDs. A Japanese firm is currently producing a blue LED that puts out considerably more illumination, but it is much more expensive than other LEDs. Improvements in the price and performance of blue LEDs will likely lead to the development of a number of new applications.

See also CALCULATOR; DIODE; ILLUMINATION, ELECTRICAL; LIQUID CRYSTAL DISPLAY; SEMICONDUCTOR; TRAFFIC SIGNALS

Further Reading: Klaus Gillessen and Werner Schairer, *Light Emitting Diodes: An Introduction*, 1987.

Lighting, Gas

Until the 19th century, the only sources of artificial illumination were oil lamps and candles. The light they gave was dim, and usually accompanied by smoke and foul odors. By the end of the 19th century, electric lighting had become a practical and commercially successful technology, but before that occurred, greatly improved levels of illumination were being provided by gas lighting. Gas lighting has been characterized as the great unsung invention of the Industrial Revolution, and rightfully so, for it made streets safer, encouraged literacy, and allowed round-the-clock operation of factories.

The use of gas as a source of light has a long history; subterranean gas deposits were tapped and used for heating and illumination in some parts of China as early as the 4th-century B.C.E. In 1688, a communication to Great Britain's Royal Society noted that gas could be extracted from coal and oil, and then burned as a source of light, but nothing came of it. The use of gas for lighting was again noted in the late 1780s by German pharmacist J. G. Pickel and Belgian physicist Jean Pierre Minkelers. This time, more notice was taken. In 1792, William Murdock, a machinist-engineer in the employ of James Watt's firm, used gas derived from coal to illuminate his home in Cornwall. Within 10 years, it was being used to light up the Boulton and Watt factory in Birmingham. In the following decades, the use of illuminating gas spread throughout the industrializing world. In England, Samuel Clegg set up a central generating plant and a network of cast-iron pipes that carried the gas to individual consumers. In 1819, his son-in-law John Malam devised a crucial component of a commercially viable gas distribution system: a meter that indicated the amount of gas used by individual customers. Gas lighting was further improved by filtering the gas through quicklime (calcium oxide, CaO), which removed the noxious odor caused by sulfurous impurities in the coal.

By the second half of the 19th century, gas lighting was in widespread use. Gasworks had become sizable enterprises, and their vertical distillation towers were prominent features of many urban skylines. Yet by the end of the century, illuminating gas was threatened with extinction by the spread of electric lighting. Even so, electricity's triumph did not come as quickly as might be imagined. Old technologies do not always roll over and die when confronted by new, apparently superior ones. Sometimes the presence of a rival technology serves as a stimulus for significant improvements that prolong the life of the old technology. Illuminating gas is a case in point. Just when electricity was beginning to make significant inroads, gas lighting was given a substantial boost by Carl von Welsbach's invention of the incandescent gas mantle, which was patented in 1885. The Welsbach mantle, as it is still known, gave brighter illumination with a lower consumption of gas; it is used today in outdoor lanterns fueled by propane or butane gas. Of course, except for camping excursions, electricity became the sole source of illumination, but many decades passed before this happened. In the interim, gas continued to light up many offices, factories, streets, and homes.

Ironically, the staying power of an apparently obsolete technology facilitated the development of one of the 20th century's most important technologies, for illuminating gas was used to fuel the first internal combustion engines. Had illuminating gas not been available, the development of the internal combustion engine would have surely slowed while inventors struggled with liquid fuels and the devices necessary for their effective use.

See also CARBURETOR; ENGINE, FOUR-STROKE; ILLUMINATION, ELECTRICAL; INDUSTRIAL REVOLUTION; STEAM ENGINE

Lightning Rod

That lightning is a form of electricity was dramatically demonstrated in 1752 by Benjamin Franklin's (1706–1790) famous kite-flying experiment. This experiment used a kite that had a pointed wire attached to its frame; when flown in a thunderstorm, the wire attracted electricity, which then traveled down the wet kite string to a key. Franklin was then able to charge a Leiden jar with electricity drawn from the key. Contrary to common belief, lightning did not strike the kite; Franklin might have been electrocuted had this occurred. On the basis of this experiment, Franklin wrote an essay entitled "How to Secure Houses, etc., from Lightning." Franklin proposed the use of roof-mounted electrical conductors to collect the electrical charges in the air and conduct them to the ground before a dangerous bolt could be formed. The purpose of a lightning rod was, as he put it, to "draw the electrical fire silently out of the cloud before it came nigh enough to strike."

Not everyone welcomed Franklin's invention. Some members of the clergy were of the opinion that lightning was a manifestation of God's wrath, which was not to be thwarted. The argument was also made that the electrification of the Earth would ultimately result in the generation of earthquakes. Despite this opposition, lightning rods soon became standard fittings in lightning-prone areas. A few years after Franklin's invention of the lightning rod, one English visitor wrote that:

> I believe no other country has more certainly proved the efficacy of electrical rods than this: before the discovery of them these gusts [i.e., lightning bolts] were frequently productive of melancholy consequences; but now it is rare to hear of such instances.

Although the exact nature of lightning's propagation is not fully understood even today, it is known that a properly located lightning rod will provide an area of protection in the shape of a cone, with the top of the lightning rod at it apex and a base with a radius equal to the height of the rod. Moreover, lightning will only occasionally strike within a cone that has a base radius equal to twice the height of the rod. Although folk wisdom asserts that "lightning never strikes twice in the same place," lightning will in fact strike a tall structure repeatedly during a season. The installation of lightning rods keeps these strikes from doing serious damage.

See also CAPACITOR

Lights, Fluorescent

The incandescent lamp made the electrical lighting of homes, shops, and offices a practical proposition. Although incandescent lamps are still in widespread use, they remain poor ways of converting electrical energy into light, for most of the power fed into them is converted to heat and not light. A much better proposition is the fluorescent lamp. Usually tube shaped and anywhere from 15 cm (6 in.) to 2.4 m (8 ft) feet in length, they are also available in shapes that allow them to replace conventional incandescent bulbs. The inside of a fluorescent lamp is covered with a phosphorescent (light-emitting) coating such as calcium tungstate, zinc sulfide, or zinc silicate. The tube contains mercury vapor at very low pressure that produces ultraviolet radiation when electrically stimulated. The ultraviolet radiation hits the phosphorescent materials, causing the emission of light waves in the visible spectrum. The color of the light varies according to the particular phosphorescent coating employed; most fluorescent lamps give off a cool, greenish-blue light, but many tones are possible. Fluorescent lamps are started in a number of ways. The most common method uses a "ballast," which is actually an electronically controlled transformer that provides proper voltage for starting and the subsequent operation of the lamp.

Many inventions have long gestation periods that extend from their first demonstration to successful application. The fluorescent lamp is such an invention. The principle underlying fluorescent lighting was discovered in 1859 by Henri Becquerel (1852–1908) when he used a Geissler tube (a device that demonstrated the glow caused by an electrical discharge in a rarefied gas) that had been coated with a phosphorescent material. Although Thomas Edison (1847–1931) patented a lamp based on this principle in 1896, it was not commercially viable. A fluorescent lamp that worked through the vaporization of mercury was demonstrated in 1901 by Peter Cooper-Hewitt, but again a successful laboratory demonstration was all that was accomplished. Lamps energized by the electrical stimulation of a gas appeared in the 1920s in the form of neon lights, but these operated according to different principles and were used for signs and decorations, not for illumination.

The fluorescent lamp began to emerge in its present form in the 1930s as a result of research conducted at General Electric under the direction of Arthur Compton. At that time, GE had a patent-sharing arrangement with Westinghouse Electric, so in 1938, when the commercial fluorescent lamp was introduced, both firms were engaged in its manufacture and distribution. At first fluorescents were marketed as tint lamps for specialized applications, such as lighting display cases, and not as lamps for general illumination. Nor was the efficiency of fluorescent lamps stressed, for GE and Westinghouse did not want to antagonize the electric utility companies by encouraging consumers to use less electricity by replacing their incandescent lamps with fluorescent ones. When fluorescents began to be marketed as general-illumination devices, emphasis was placed on high-intensity lighting that required as much, if not more, electricity as lighting with incandescent lamps.

Still, there was no ignoring the energy-saving capabilities of incandescents, a fact that came into prominence with the energy crises of the 1970s. The first commercial fluorescent lamps more than doubled the efficiency of incandescent lamps, and today's fluorescence are even better, consuming 75 to 85 percent less electricity than incandescent lamps for the same output of light. In addition, they last 9 to 13 times longer than incandescents. Since lighting consumes about 25 percent of the electricity used in the United States (20 percent directly, and the remainder for the additional cooling required to offset the heat given off by lights), the conversion from incandescent to fluorescent lamps can result in substantial energy savings. For example, the replacement of a single 75-watt incandescent bulb with an 18-watt fluorescent lamp will save 350 kg (770 lb) of coal or 235 liters (62 gal) of oil over the course of the fluorescent lamp's 10,000-hour operating life.

See also ILLUMINATION, ELECTRICAL; NEON; TRANSFORMER; TUNGSTEN FILAMENT

Further Reading: Wiebe E. Bijker, "The Social Construction of Fluorescent Lighting: Or How an Artifact Was Invented in Its Diffusion Stage," *in* Wiebe E. Bijker and John Law, *Shaping Technology/Building Society: Studies in Sociotechnical Change*, 1992.

Linotype Machine

The invention of first the printing press and then the rotary press greatly increased the speed at which books and periodicals could be produced. But one important bottleneck remained: the setting of type. Compositors had to select and set pieces of type one at a time; at a rate of about two characters per second, manual typesetting slowed down the publication process and increased costs. In the 19th century, many inventors tried to create machines that allowed the rapid setting of type. Quite a few inventors created machines that selected, set, and justified lines of type, but that was the easy part. The real problem centered on returning the individual pieces of type to their proper place after the printing had been done.

On many occasions the solution to a problem lies in reconceptualizing the nature of the task to be done. So it was with mechanized typesetting. Instead of attempting to design a machine that returned the type to its place of origin, the successful inventor devised a process that melted down the type after it had been used. New characters then could be formed from the molten metal, set, used for printing, and then remelted so that the process could begin anew. The inventor was Otto Merganthaler (1854–1899), a German who had immigrated to Baltimore. After a decade's work, his "Linotype" machine began to set newspaper type in 1886, and its basic principles continued to be used for many years afterwards.

A linotype machine from c. 1890 (from M. Kranzberg and C. W. Pursell, Jr., *Technology in Western Civilization,* vol. I, 1967, p. 640).

Characters were selected by punching the appropriate key, which caused a type matrix (a small rod with a die on one end) to be released from a magazine. After a line of matrices was set, the type was cast into a solid line of type, hence the machine's name. The matrices were then returned to their respective magazines. This was done by equipping each matrix with combinations of teeth unique to each character. These teeth engaged with corresponding grooves in the distributing mechanism, which routed each matrix back to its proper place in the magazine.

One important variant of the Linotype machine was the "Monotype." This machine used keystrokes to produce perforations on a roll of paper. The perforations in turn controlled the release of matrices from which pieces of type were cast one at a time (the source of the name Monotype). Pages of type cast in this way could be more easily corrected, since it was not necessary to redo an entire line, as was the case with the Linotype machine. Moreover, the actual typesetting did not have to be done in the same place where the perforated rolls were created. It was even possible to make the perforated rolls by sending electrical impulses over telephone lines.

These technologies were rendered obsolete for the most part by the introduction of computerized typesetting in the 1970s. Typesetting machines of this kind used digitized characters that were generated on a cathode-ray tube and then exposed on a photographic plate. In this way the conventional typesetting phase could be eliminated altogether. A reporter could compose a story on a computer screen, edit it, and then electronically send it to a photocomposition machine. This made the process much faster. Whereas the operator of a modern typesetting machine using punched paper could manage no more than 15 lines per minute, computerized photocomposition equipment is capable of setting more than 4,000 lines per minute. As with many technological advances, computerized typesetting equipment did not distribute its benefits evenly. It has been a boon to the producers and consumers of newspapers, books, and magazines, but the new technology also eliminated the jobs of many skilled typesetters.

See also PRINTING WITH MOVABLE TYPE; TECHNOLOGICAL UNEMPLOYMENT

Liquid Crystal Display

Liquid crystal displays (LCDs) are among the most popular forms of electronic information displays, second only to cathode-ray tubes (CRTs). The technology of LCDs matured with and complemented the development of integrated circuits in consumer microelectronics. Their use has steadily grown since the 1970s, when they first appeared on wristwatches and pocket calculators. In the 1980s, they became standard features on measuring instruments, and today they are common components of fax machines, photocopiers, telephones, and videogames. They are beginning to replace CRTs for flat-panel screens in computers, televisions, and car instrument panels—a projected multibillion-dollar market.

Scientists have known about liquid crystals since the end of the 19th century, but applications only began to appear in the 1960s. Austrian botanist Friedrich Reinitzer first noted the phenomenon in 1888. When he heated a solid organic compound, cholesterol benzoate, it appeared to have two distinct melting points. It became a cloudy liquid at 145°C (293°F) and turned clear at 179°C (354°F). Otto Lehmann, professor of physics at the Technische Hochschule in Karlsruhe, Germany, learned of Reinitzer's experiment and continued research into the phenomenon. Using a microscope fitted with a heating stage, he determined that certain molecules do not melt directly but first pass through a stage in which they flow like a liquid while still retaining the molecular structure and optical properties of a solid crystal. This property led Lehmann to coin the term *liquid crystal.* While engaged in their basic research, Lehmann and Reinitzer never dreamed of an application of their findings.

European laboratory scientists came to understand the physics and chemistry of liquid crystals during the 1930s, but it wasn't until the 1960s that both basic research and efforts to find practical uses for liquid crystals began in the United States. Especially important in stimulating that research was G. H. Brown's founding of the International Liquid Crystal Conferences at Ohio's Kent State University in 1965.

The first published suggestion for using liquid crystal materials for display came in 1963 from Richard Williams and George Heilmeier at the David Sarnoff Research Center, RCA's laboratory in Princeton, N.J. Heilmeier's research group hoped to use liquid crystal displays for a future TV-on-a-wall, a long-standing dream of RCA head David Sarnoff. The challenge was to find a liquid crystal that would provide a display at room temperature, and by 1968 the RCA group had a display based on the dynamic scattering mode (DSM) of liquid crystals. In this mode, an electrical charge applied to the liquid crystal rearranges the molecules so they scatter light. But it was evident that large-screen, LCD-based TVs were many years off, and the group accordingly set its sights on displays that could be incorporated more immediately into commercial products. A number of Sarnoff lab pioneers left to form Optel Corporation in Princeton, where they perfected techniques for the manufacture of displays for digital watches.

Because the DSM LCDs suffered from relatively high power consumption, limited life, and poor contrast, the search continued for a workable LCD. James Fergason invented an improved display based on a phenomenon known as the "twisted nematic effect." Displays based on this principle are the kind found most frequently in today's LCD products.

Fergason founded a display-manufacturing company, International Liquid Crystal Company (ILIXCO) in Kent, Ohio. He did not make his patent application public at the time, and Wolfgang Helfrich and Martin Schadt of Hoffman LaRoche of Basel, Switzerland, published a paper on the same effect in 1971. Hoffman LaRoche eventually purchased Fergason's patent rights. In 1973, G. W. Gray of Hull University in England made synthetic liquid crystals (cyanophenyls), which were widely marketed in England. Three years later, scientists at Merck synthesized still another kind of liquid crystal, phenylcyclohexane.

Although liquid crystal research continues in the United States and Europe, Japan dominates the commercial market for LCDs. This has been a matter of considerable concern in the United States, for LCDs have many defense applications, and the absence of indigenous manufacturing facilities could be a problem should supplies from Japan be cut off for any reason. However, the United States is not completely lacking the ability to manufacture LCDs, as U.S. manufacturers specialize in the production of custom displays.—C.S.

See also CALCULATOR; CRYSTALS; FAX MACHINE; INTEGRATED CIRCUIT; PHOTOCOPIER; TELEPHONE; VIDEOGAME; XEROGRAPHY

Lithography

Lithography (literally "stone writing") is a printing process that uses a plate that has been treated to hold ink in some places and repel it in others. Its inventor was Aloys Senefelder (1771–1834), an aspiring playwright who had been born in Prague but was living in Munich while writing his plays. Having little success in getting them published, he printed them himself, using the standard medium of the time, engraved copper plates. Copper being quite expensive, in 1796 he turned his attention to limestone slabs. After writing on a slab with a grease pencil, he washed it with nitric acid (HNO_3). The grease resisted the acid, as the blank parts of the stone were etched, leaving the letters in prominent relief. He then was able to produce a mirror image of what he had originally written by applying gum arabic and ink to the stone, and then pressing it against a sheet of paper.

Subsequent experimentation revealed that successful printing did not require the letters to be raised above the surface of the slate. As Senefelder discovered, the image left by a grease pencil accepted greasy ink, while the blank portions of the slate would repel the ink if they first had been moistened. By 1798, Senefelder had created a press for printing with the slabs, and by 1803 he was using specially prepared metal plates in the place of the stone slabs. It should be noted, however, that to speak of a lithographic "press" is a bit misleading, because the slab is slid or scraped across the paper, rather than pressed down upon it.

Publishers were quick to take up lithography. One of the first was Mozart's publisher, but lithography's greatest impact was in the visual arts. In the mid-19th century, renowned artists like Francisco de Goya, Théodore Géricault, Eugène Delacroix, and Honoré Daumier created drawings that were lithographically reproduced in books, newspapers, and magazines. Many of these lithographs had a distinct political edge to them. While reflecting the social ferment of the times, they also provided a strong impetus for social reform.

Color lithography (sometimes known as *chromolithography* or *oleography*) goes back to 1808, but the process did not become a commercial proposition until about 1860. Prior to the widespread use of color lithography, publishers commonly employed teams of colorists, most of them women and children, where each member added a single watercolor tint to a lithograph that initially had been printed with black ink. Commercial color lithography employed a separate printing stone for each color. As many as 30 of these stones was used, one after the other, for each colored lithograph.

The mass production of lithographs also was advanced by the development of steam-driven rotary presses similar to the ones used for letterpress printing. In Austria, Georg Sigl constructed a cylindrical lithographic press in 1851, an invention that was perfected in England in the mid-1860s. Rotary lithography used rollers to moisten and ink the stones, and another revolving cylinder brought them into contact with the paper.

Today, the most important application of lithography is lithographic offset or litho-offset printing (usually rendered simply as *offset*). First patented in England in 1853 by John Strather, it did not come into widespread use until the 1870s, when it was first used for printing onto metal surfaces. In the early 20th century, offset began to be used for printing on paper. As with conventional lithography, offset printing is based on the principle that greasy ink will adhere to another greasy surface but will be rejected by a clean, wet one. The process begins with a thin metal plate (made from zinc, aluminum, or a combination of metals) that is treated to make it porous, and then coated with a photosensitive material. A photographic negative (bearing print, illustrations, or both) is placed on the plate, which is then exposed to a strong light. This gives the exposed areas of the plate a slightly greasy surface. The unexposed coating is washed away, so these areas of the plate will shed any ink applied to them. The plate is then attached to a rotating cylinder (the plate cylinder). As it rotates, the unexposed surface of this cylinder is moistened by other rollers in order to prevent ink from adhering. A series of inking rollers then pass over the plate cylinder, and ink is deposited on the exposed areas but rejected by the other areas. The inked image is then transferred to another roller, this one covered with rubber (the blanket cylinder). This roller then prints the image on paper being carried by yet another cylinder (the impression cylinder). The process is fairly simple in conception but complicated in its actual operation, so skilled operators are needed to make the process work as it should.

Offset lithography is by the far the most commonly used printing process. It is employed for a broad range of products, including books, packages, letterhead stationery, stamps, newspapers, magazines, labels, and maps. Indeed, just about anything that can be photographed can be lithographically reproduced. Adding to the versatility of offset lithography is its ability to print on a wide variety of materials: leather, wood, cloth, metals, and rough or smooth paper.

See also MASS PRODUCTION; PRINTING PRESS, ROTARY

Locomotive, Diesel-Electric

A diesel locomotive is properly called a diesel-electric locomotive because a diesel engine is used to turn an electrical generator that runs the electric motors supplying power to the wheels. This arrangement may seem needlessly complicated, but the direct application of power through some sort of transmission presents a number of mechanical problems. Some diesel locomotives have used hydraulic transmissions in place of the electrical generators and motors, but this arrangement has been largely confined to small industrial locomotives.

The use of railroad diesels began with a diesel-electric railcar (a single-unit, self-propelled passenger vehicle) that was built in 1913 by the Swiss firm Sulzer, and the first diesel-electric locomotives were produced in small numbers in the 1920s in the United States and Europe. These locomotives drew on technologies that had been developed in other sectors. The submarines used during World War I stimulated the design of compact, relatively powerful diesel engines, while advances in electrical generators and motors were also transferred to locomotive design. Early diesel-electric locomotives were primarily used for switching (shunting), but by the mid-1930s newly constructed diesels were pulling glamorous passenger trains, and by the end of the decade a few were in freight service as well.

Almost from the beginning, the diesel demonstrated its superiority over the steam locomotive. There were many reasons for this. Despite a century of development, the steam locomotive did a poor job of converting energy, operating at a thermal efficiency of 10 percent at best. In contrast, a diesel came close to quadrupling this number. In addition to saving on fuel, the greater efficiency of the diesel allowed it to go longer distances between refueling, whereas steam locomotives had to stop every 100 miles (160 km) or so just to replenish their water supplies. Other disadvantages attended the operation of steam locomotives. The pounding action of their driving wheels and side rods took its toll on the tracks, and the locomotives themselves required a great deal of maintenance and repair work. Making matters worse was a lack of standardization that necessitated a great deal of custom-making of repair parts. Steam locomotives could produce a great amount of power, but ensuring an even delivery of that power required a great deal of skill on the part of the engineer. When more power was needed, the addition of a helper locomotive entailed the services of another crew; in contrast, diesel-electric units could simply be strung together and operated by a single engineer. Finally, steam locomotives were dirty beasts, covering themselves and their surroundings with soot and ashes; some municipalities even deemed it necessary to prohibit their entry into the city.

Although railroads tended to be highly conservative and traditionally minded organizations, the manifest advantages of diesel locomotives could not be ignored indefinitely. The hard economic times of the 1930s slowed their adoption, while the demands of war production during the first half of the 1940s limited the resources that could be devoted to the manufacture of new locomotives. At the same time, however, the immense transportation demands of the war literally wore out locomotive fleets and guaranteed a large market for replacements.

In the United States, the dominant producers of diesel-electric locomotives were not the entrenched manufacturers of steam locomotives, but General Motors and General Electric (the latter initially combining its efforts with the American Locomotive Company [Alco]). Unlike the manufacturers of steam locomotives, these firms were committed to the principles of standardization and mass production that had served

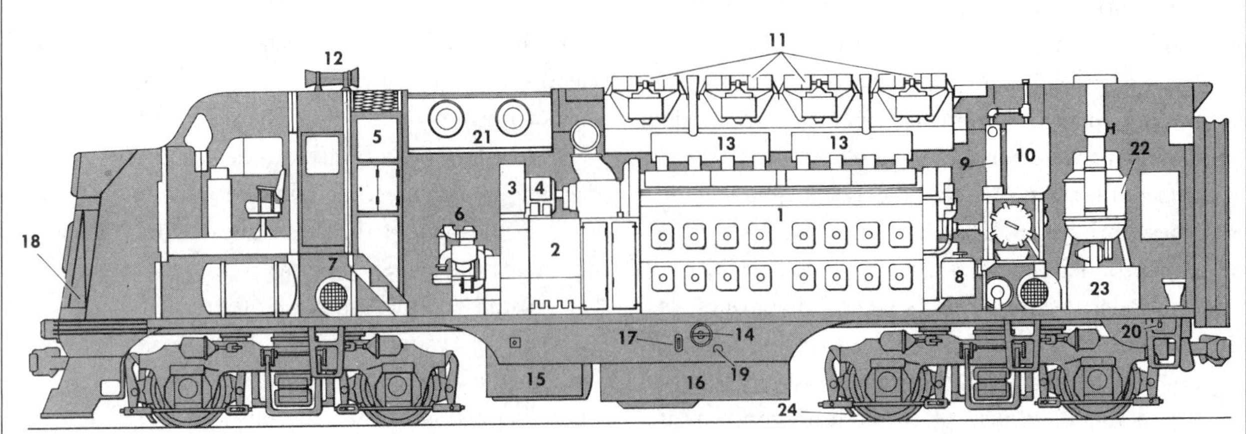

INTERIOR OF EMD F3 CAB UNIT

1 Engine: EMD model 16-567B
2 Main generator and alternator
3 Generator blower
4 Auxiliary generator
5 Control cabinet
6 Air compressor
7 Traction motor blower
8 Lubricating oil filler
9 Lubricating oil cooler
10 Engine water tank
11 34″ fans and motors
12 Horn
13 Exhaust manifold
14 Fuel filler
15 Batteries
16 Fuel tank: 1200 gallons
17 Fuel tank gauge
18 Door (plain)
19 Emergency fuel cutoff
20 Engine water filler (both sides)
21 Dynamic brake grids and blowers
22 Boiler
23 Boiler water tank: 200 gallons
24 Sanding nozzle

The major componments of a first-generation diesel-electric locomotive (from *The Second Diesel Spotter's Guide,* 1973, p. 7; copyright © Jerry A. Pinkepad, by permission of Kalmbach Publishing Co.).

so well in the production of automobiles and electrical appliances. The combination of high demand and rationalized production methods resulted in the rapid adoption of diesel-electric locomotives and a parallel decline in the utilization of steam locomotives. The United States in 1941 had nearly 42,000 steam locomotives and only about 1,500 diesels; after the passage of 2 decades, the diesel population had increased to more than 30,000, while the ranks of operating steam locomotives had been thinned to about 200.

Although diesel-electric locomotives brought many advantages to the railroads, not everyone benefited from them. The demands of steam locomotives for fuel, water, and repairs served as the basis of many small communities. Diesel locomotives undercut their reason for their existence and left some of them virtual ghost towns. Many of the skilled trades that had evolved alongside the steam locomotive were rendered obsolete. At the same time, those whose jobs had centered on the operation of steam locomotives tried to preserve work rules that had little relevance to the operation of trains pulled by diesels. In the United States, the labor agreements that had been forged in the days of steam were vigorously defended. In the steam locomotive era, 100 miles (161 km) was considered a whole day's work, and labor agreements stipulated that any haul in excess of this distance required a change of crew or the payment of overtime, even though a trained pulled by a diesel could easily travel four times that distance in 8 hours. Diesel locomotives did not require the services of a fireman, but the union insisted on his presence. For decades, railroad management fought these stipulations, which they dubbed "featherbedding." They did not succeed in eliminating them until the 1960s, and then only after legal battles that went all the way to the U.S. Supreme Court.

By this time the diesel-electric locomotive had become the dominant form of railroad motive power. Today's diesel-electrics use the same basic technology pioneered by their forebears in the 1930s, but their power and reliability have made impressive gains. The first-generation road diesels made by General Motors in the 1930s produced 1,350 horsepower (1,007 kW); today's locomotives easily triple this figure. Diesel locomotives also have benefited from computerized traction control systems that ensure that all of this power is effectively harnessed. Many diesel electrics are also equipped with dynamic brakes; these slow a train down by causing the traction motors to operate as electrical generators, converting a train's kinetic energy into an electrical current that is dissipated through resistors.

Because they represented such a vast improvement over steam locomotives, diesels were a technological fix that made possible the survival of the ailing American railroad industry. As with many technological fixes, however, they obscured the need for more thoroughgoing changes. Many railroads remained in precarious economic circumstances until deregulation in the 1970s allowed a fundamental restructuring of the industry and an eventual return to profitability.

See also ENERGY EFFICIENCY; ENGINE, DIESEL; FEATHERBEDDING; LOCOMOTIVE, STEAM; MASS PRODUCTION; SUBMARINES; TECHNOLOGICAL FIX

Locomotive, Steam

During the early years of railroading, there was no consensus on how trains were to be pulled. The earliest railroads used horses or mules, and to many that seemed a quite satisfactory arrangement. The first common-carrier railroad, Great Britain's Stockton and Darlington, supplemented its horses with a cable arrangement that pulled trains up a steep gradient. The great British engineer Isambard K. Brunel (1806–1859) took the most radical approach when he built a pneumatic railroad. This used a tube between the rails that was evacuated by pumping stations strung out along the line. Trains were propelled by suction pulling on a piston located underneath one of the cars. None of these methods had any lasting value, and from the 1820s to the early 1950s, the steam locomotive was the dominant source of motive power on most of the world's railroads.

Credit for building the first steam locomotive goes to Great Britain's Richard Trevithick (1771–1833). In 1804, his one-cylinder locomotive hauled five cars loaded with 9.1 metric tons (10 tons) of iron and 70 men a distance of 15.7 km (9.75 mi). In a bold departure from conventional steam engine practice, Trevithick used high-pressure steam for his locomotive, considerably increasing its efficiency. Trevithick built a second locomotive, but then moved on to other things, leaving George Stephenson (1781–1848), another British engineer, to do much of the pioneering work in the development of the steam locomotive. After constructing a few locomotives for use in collieries, Stephenson and his son Robert (1803–1859) designed the highly successful *Locomotion* for the Stockton and Darlington. Built in 1825, it could haul a load of 81.6 metric tons (90 tons) at speeds of up to 13 km/hr (8 mi/hr). This was followed in 1829 by Stephenson's *Rocket*, which incorporated all the essential features of a steam locomotive: a horizontal boiler, forced draft, and outside cylinders that were directly connected to one set of driving wheels that in turn were connected to another set of driving wheels by side rods.

American railroads at first imported British locomotives or bought ones built in the United States to British design. This was not always a satisfactory situation, for the trackwork of American railroads did not come up to British standards. The sharp curves of many American railroads caused regular derailments, making it necessary to equip locomotives with pivoted four-wheel leading trucks that guided them into curves. First used by John Jervis (1795–1885) for a locomotive called *Experiment*, the four-wheel truck soon became standard. For the better part of the 19th century, the most common steam locomotive in the United States was the 4-4-0 (indicating four leading wheels, four driving wheels, and no trailing wheels) "American type." Many of these locomotives were ornately decorated, and fitted with prominent oil-burning headlights and large smokestacks containing a spark arrestor, a necessity when wood was the most common fuel.

The most general trend in the evolution of the steam locomotive was the steady growth of size and power. Beginning in the 1840s, locomotives were built with six and even eight driving wheels. In the 1880s, the 2-8-0 Consolidation locomotive began its reign as the most popular type of locomotive for freight service. In similar fashion, the 4-6-0 Ten-Wheeler began to supplant the 4-4-0 as the most popular locomotive for hauling passenger cars. The next step entailed the design of locomotives with larger fireboxes, allowing the generation of more steam by larger boilers. Larger fireboxes necessitated the addition of trailing wheels to support them. Consequently, the 4-4-0 evolved into the 4-4-2 Atlantic (which in turn was followed by the 4-6-2 Pacific and the 4-8-2 Mountain), and with the addition of a trailing truck the 2-8-0 configuration became the 2-8-2 Mikado.

As locomotives got larger, many other improvements also contributed to their ability to haul longer, heavier trains. Beginning in 1880 and for about 20 years thereafter, locomotive designers often applied the principle of compounding to their locomotives. Compounding is a process whereby steam pushes a pair of pistons and then is exhausted into a second set of cylinders, where it pushes another pair of pistons. This arrangement did increase efficiency, but at the cost of higher maintenance expenses. Consequently, compounding was not used for most locomotives after the early 1900s. It was retained, however, for a special kind of locomotive known as a Mallet, which was named after its French inventor, Anatole Mallet (1837–1919). Mallets were articulated locomotives; that is, they had two sets of driving wheels, with the first set free to pivot. This design allowed the operation of very large locomotives without the need to widen existing curves. In a Mallet, high-pressure steam went into the two cylinders that served the rear drivers, and then was exhausted into the two that served the front ones.

Another key improvement was the mechanical stoker, which fed more coal than was possible with hand firing. Locomotives also benefited from improved valve gear and the use of superheated steam. From the mid-1920s on, steam locomotives were fitted with even larger fireboxes, which necessitated the use of four-wheel trailing trucks. The result was a generation of locomotives that represented the apogee of steam, engines like the 2-10-4 Texas and the 4-6-4 Hudson.

After serving America's railroads for more than a century, steam locomotives began to be rapidly replaced by diesel-electric locomotives in the 1940s. Relatively few steam locomotives had been built after 1925, and the demands of World War II left the fleet much the worse for wear. In addition to coming along when America's stock of locomotives was ripe for replace-

A Norfolk and Western 0-8-0 switcher, one of the last steam locomotives produced in the United States (courtesy Robert C. Post).

ment, the diesel-electric had a number of inherent advantages. Its engines operated a thermal efficiency that was much higher than that of the best steamers, which rarely exceeded 7.5 percent. Steam locomotives also needed large amounts for water, which was difficult to supply in arid regions. Steam locomotives also required intensive regular maintenance, and their wheels and side rods did considerable damage to track and roadbed. Finally, each steam locomotive needed an engineer and fireman, whereas diesels could be coupled together in multiple units that were controlled by a single crew.

Efforts to prolong the use of steam power through the development of steam turbine locomotives were unsuccessful, and by the early 1960s, steam locomotives had vanished from U.S. railroads, although a few are still used by tourist railroads. Steam locomotives are not completely irrelevant for ordinary railroading however, as many of them are still pulling trains in a number of Third World countries.

See also ENERGY EFFICIENCY; ENGINE, DIESEL-ELECTRIC; LOCOMOTIVE, DIESEL; RAILROAD; STEAM ENGINE; TURBINE, STEAM

Further Reading: Alfred W. Bruce, *The Steam Locomotive in America: Its Development in the Twentieth Century*, 1952.

Logarithm

A logarithm is the exponent of a number that indicates the power to which that number must be raised in order to produce another number. For example, when using a base-2 system, the logarithm of 32 is 5; that is, 2 raised to the fifth power (2^5) equals 32. Logarithms were devised as a way of taking the drudgery out of the multiplication and division of large numbers. This is done by using a table to find the numbers' logarithms, adding them (for multiplication) or subtracting them (for division), and then referring to the table to determine the number corresponding to the sum or difference. Again using a base-2 system, 32 can be multiplied by 8 by adding their logarithms: 5 + 3 = 8; 8 in turn is the logarithm of 256 (2^8), the same as the product of 32 × 8. Manipulation of logarithms is of little practical importance in an era of cheap electronic calculators, but in times past the use of logarithms saved a great deal of computational time and effort.

The invention of logarithms is generally attributed to John Napier (1550–1617), a Scottish lord who was more interested in Protestant theology than mathematical research. A Swiss watchmaker named Jobst Burgi (1552–1632) came up with the idea at about the same time, but did nothing to promote it. Intended as a means of simplifying calculations for navigation and astronomy, Napier's logarithms were published in 1614 in a book containing 56 pages of text and 90 pages of tables, complete with a dedication to the Prince of Wales, who later had his head cut off while reigning as the unfortunate Charles I. Napier constructed his tables to encompass the logarithms of sines, in right triangles the ratio of an angle's opposite side to the hypotenuse. *Logarithm* in fact comes from the Greek, meaning "number of the ratio." Napier's associate Henry Briggs (1561–1631) took the basic idea and simplified it somewhat by creating a logarithmic table based on exponents of 10.

Base-10 logarithms, also known as *common logarithms*, were used to construct a slide rule. Logarithmic scales are printed on the different parts of the rule, which are aligned to give a product, quotient, square, or root.

The other widely used logarithmic system uses a base of 2.718 . . ., an irrational number that is the sum of the infinite series:

$$1+\frac{1}{1}+\frac{1}{1\times 2}+\frac{1}{1\times 2\times 3}+\frac{1}{1\times 2\times 3\ldots}+\cdots$$

The number 2.718 . . . is signified by the character *e*, and base-*e* logarithms are known as *natural logarithms*. Although *e* is an irrational number, the system is quite rational, for it has many applications in the sciences, such as the flow of electrical current, population growth, and the decay of a radioactive element.

Logarithms remain an important part of the mathematical description of many processes in science. For example, acidity and alkalinity is measured by pH, the negative of the base-10 logarithm of the concentration of hydrogen ions in a solution. The magnitude of an earthquake on the Richter scale is defined in terms of the base-10 logarithm of the energy of the quake, so that $\log_{10} E = 11.4 + 1.5M$, where M is the Richter magnitude. The relative loudness of sounds is measured in decibels, where a bel is the logarithm to the base-10 of the ratio of the intensity of two sounds. Thus a difference of two bels means a 100-fold increase in loudness. Meteorologists study the way the average wind speed varies with height by means of a logarithmic velocity profile. Base-2 logarithms are used in computer science, especially to calculate the efficiency of computer algorithms.—J.G. and R.V.

See also ALGORITHM; RICHTER SCALE; SLIDE RULE

Longitude, Determining

To know one's position on the globe, two things are necessary: the latitude and the longitude. Latitude is the distance in degrees from an imaginary line, usually the Earth's equator. It can be determined without much difficulty by observing the position of the sun or the stars relative to the horizon. For centuries, navigators were able to do this with a fair degree of precision through the use of a simple instruments like a cross-staff or the more complex astrolabe. Determining longitude, the imaginary lines that run in a north-south direction and pass through the Earth's poles, is a far more difficult task that occupied generations of astronomers and others.

One way of determining longitude takes advantage of the fact that the Earth acts as a kind of clock. If the circumference of the Earth is defined as 360 degrees, after the passage of 1 hour the Earth will have rotated 15 degrees (360 divided by 24, the number of hours in a day), which isn't much distance near the poles, but it is a considerable amount of distance near the equator. Accordingly, to convert degrees into miles or kilometers, it is necessary to know one's latitude, which, as noted above, isn't much of a problem. If one is able to tell the time where one is and then compare it to the time at some fixed point, it becomes possible to determine the distance between the two. For example, a difference of 3 hours translates into an angular distance of 45 degrees. If this method is used as a means of navigation, it is essential that time be kept with great accuracy; a deviation of a minute in the course of a day can result in an error of dozens of miles, with potentially fatal consequences on the high seas.

The difficulty of determining longitude caused many a ship and its crew to come to grief. One such incident occurred in 1707, when four British warships got lost near the southwest coast of England, struck some rocks, and went down with the loss of virtually everyone on board. Goaded by tragedies like this one, the British government passed the Longitude Act of 1714, which offered a prize of £20,000 (the equivalent of several million dollars today) to anyone who could come up with a way of accurately determining longitude within a half degree.

Long before the announcement of the prize, astronomers had attempted to use observations of celestial bodies as a means of determining longitude. One of the most promising was put forward by Galileo, who after discovering four of the moons of Jupiter noticed that the regularity of their orbits allowed them to be used as clocks of a sort. The difficulties of observing these motions while aboard a ship sharply limited the usefulness of this technique, but by the mid-17th century it was being used for the compilation of more accurate maps of the Earth.

At this time, the possibility of using clocks for the determination of longitude was defeated by their inaccuracy. Moreover, the method of regulation that was commonly used, the pendulum, was manifestly unsuited for use on board a ship as it was tossed on the high seas. This situation began to change in 1730, when John Harrison (1693–1776) began work on a clock capable of accurate time measurement while operating at sea. After 5 years, he had produced a timepiece that met this requirement, but not satisfied, he went on to build two more clocks and a watch over a period of nearly 30 years. One of the clocks performed excellently in 1736, and a sea trial conducted from 1761 to 1762 showed that the watch worked even better, losing only 5 seconds in 81 days. Even so, the Board of Longitude did not award the full prize, but in 1765 it gave Harrison £10,000 with the stipulation that he hand over his timepieces and make two new ones.

Unfortunately for Harrison, the Board of Longitude that administered the prize included a number of astronomers who were little disposed to the use of clocks for determining longitude. Among them was Nevil Maskelyne (1732–1811), the Astronomer Royal, who had been actively engaged in the development of a rival means of determining longitude through obser-

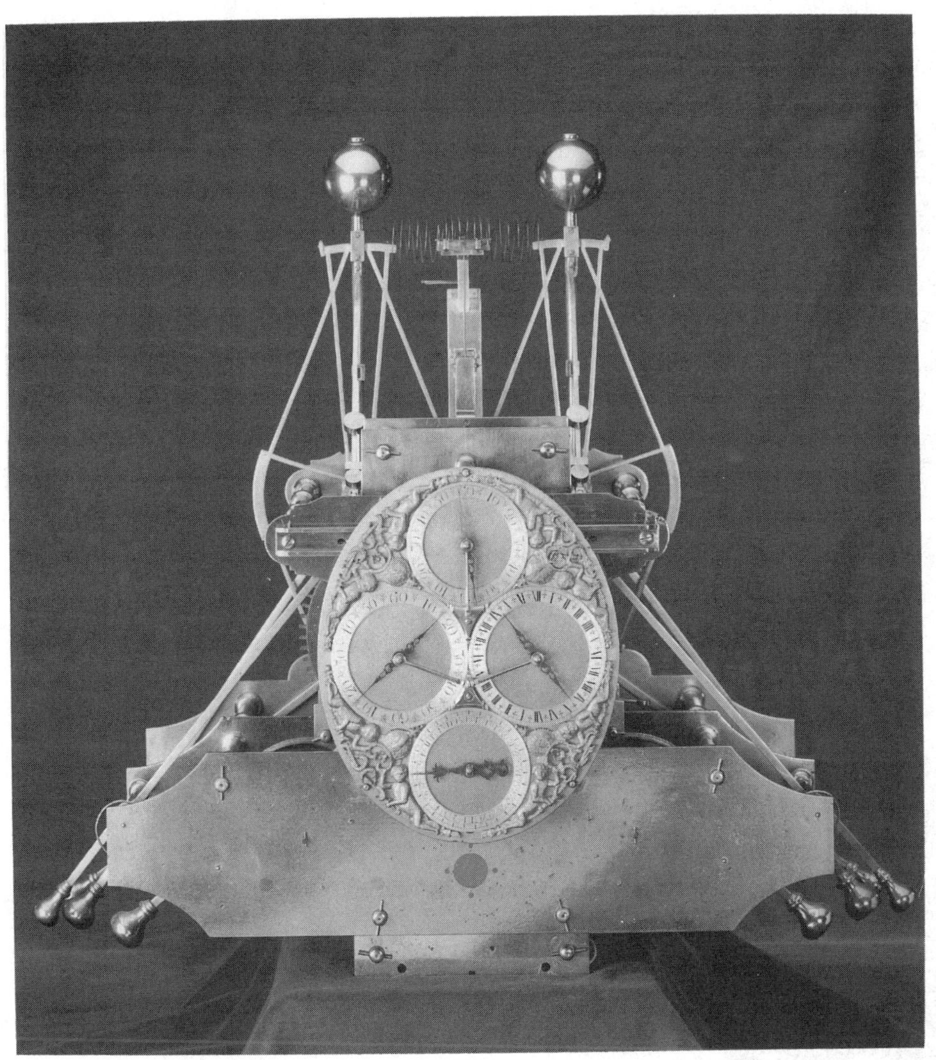

John Harrison's first marine timekeeper, 1751 (courtesy of National Maritime Museum, London).

a rival means of determining longitude through observations of the moon's angular distance from certain stars. Harrison's accomplishments were finally rewarded in 1773, when the intervention of king George III caused Parliament to award an additional £8,750. Harrison's chronometer had been expensive to produce, but by the end of the 18th century, watchmakers were able to get the price down to a point where all ship's captains could afford one. By this time, the astronomical method also had become well established, and the two methods of determining longitude complemented each other for many years. Today, longitude can be determined with the utmost precision through the use of a satellite-based global positioning system.

See also ASTROLABE; CLOCKS AND WATCHES; GLOBAL POSITIONING SYSTEM; TELESCOPE, REFRACTING

Further Reading: Dava Sobel, *Longitude*, 1995.

Loom, "Jacquard"

Weaving, a very old technology, benefited from a number of major improvements in the 18th and early 19th centuries. One of the most significant of these was the development of looms that automatically wove complex patterns. The mechanisms that made this possible were important not only for weaving but also for the future of automatic control. The loom that demonstrated the practicality of automatic weaving is known as a Jacquard loom, but it was only the latest stage in a series of preceding developments.

The crucial feature of the Jacquard loom is the use of binary coding to control the loom's operation. This was first accomplished in the 1720s by France's Basile Bouchon. In Bouchon's loom, the hooks that lifted the warp threads were controlled by prepunched holes in a long sheet of paper that rolled between two cylinders. Bouchon's use of perforated paper to control a mechanical process was much more significant than the loom itself, which suffered from a number of technical problems.

Bouchon subsequently was joined by Jean-Baptise Falcon, a silk weaver. Falcon was a resident of Lyons, the center of the French silk-weaving industry, and he was able to secure some financial assistance from that city. In 1728, the two produced an improved loom. To control the movement of the warp threads, the loom used a series of connected cards instead of a roll of paper; this made it the first device to be run by what we would now identify as punch cards. Operation was not automatic, however. An operator had to take each card as it moved into position, fit it over a perforated platen, then press the assembly against a set of needles that moved the warp threads. Although still requiring a considerable amount of operator intervention, the Bouchon-Falcon loom made it much easier to change the patterns being woven. About 40 looms with at least some of the features devised by Bouchon and Falcon were built.

The weaving of elaborate patterns was made more automatic by Jacques de Vaucanson (1709–1782), a French inventor who also distinguished himself by inventing a number of lifelike automata. Vaucanson's loom used

A Jacquard loom from the early 19th century. The pattern being woven is controlled by the roll of punch cards at the top of the frame (from C.A. Ronan, *Science: Its History and Development,* 1982)..

a rotating perforated cylinder instead of punch cards. This was a regressive step in that it limited the complexity of the pattern that could be woven into the cloth, but other details represented a substantial improvement over the looms designed by Bouchon and Falcon.

Vaucanson's loom never seems to have been put to productive use. The first commercially successful punch-card–controlled loom was the work of another French inventor, Joseph-Marie Jacquard (1752–1834). A former child worker in a silk mill, Jacquard had experienced firsthand the hard, monotonous labor of traditional weaving. With some financial assistance from the French government, he combined some of the key features of the Bouchon-Falcon and Vaucanson looms to produce an effective punch-card–controlled loom around 1804. Whether or not this constituted an original contribution has been a matter of considerable historical debate. Jacquard's loom was not an invention in the sense that it represented something truly novel, but by making previously invented elements work together effectively, it achieved the commercial success that had eluded its predecessors.

By 1812, no fewer than 10,000 Jacquard looms were in operation in France, and the looms began to be used in the British textile industry in the 1820s. The basic features of the Jacquard loom were carried over into powered looms, which were operated by perforated cards for many decades thereafter. Even more important, punch cards found their way into other sectors. The Jacquard loom provided inspiration for the computer pioneer Charles Babbage (1791–1871) when he designed his *analytical engine*. Another English engineer, Richard Roberts (1789–1864), employed a Jacquard-type control mechanism for drilling rivet holes in the wrought-iron plates used in the construction of the Britannia Bridge. By the end of the 19th century, the punch card was being used for rapid tabulation of census data, and it went on to be a key component of early digital computer systems.

See also AUTOMATA; BINARY DIGIT; PUNCH CARDS; WEAVING

Lubricants

Friction is the great enemy of all machinery. Although a well-designed bearing can reduce friction, in most cases it still requires some sort of lubricant. Lubrication is especially important when the wearing surfaces are metal, for unlubricated metals have more friction than nonmetals, but this situation is reversed when lubrication is introduced. Moreover, lubrication prevents metal-on-metal contact when one part rotates inside another, as with a crankshaft journal turning inside a plain bearing; when correctly adjusted and supplied with the proper lubricant, the bearing surfaces ride on a thin film of oil that is formed by the rotation of one of the components.

Lubricants do more than reduce friction. They often play an essential role in transferring heat away from the machinery, protecting against corrosion, and helping to seal gaps between moving parts. A lubricant can be a gas, liquid, or solid, but is most commonly found as a liquid in the form of oil or grease. Most of the greases used as lubricants are oils to which a thickening agent has been added, most commonly some kind of soap. Solid lubricants such as graphite and molybdenum disulfide are commonly used when temperatures are very high or it is difficult to renew the lubricant.

People understood the need for lubricants long ago. The invention of the wheel marked the greatest advance in reducing friction, but to perform well, wheel bearings required some sort of lubrication. Animal fats and vegetable oils were the primary source of lubricants from about the middle of the 2d-millennium B.C.E. and for many centuries afterwards. Whale oil was commonly employed for lubricating machines and shafts in the 19th century, but it was largely supplanted by another 19th-century innovation, lubricants derived from petroleum. Today's petroleum-based lubricants are a blend of liquid hydrocarbons that have been refined from paraffinic crudes, hydrocarbons with as many as 40 carbon atoms in each molecule.

Lubricating oils have a high boiling point, more than 400°C (750°F) and high viscosity. The latter refers to the "thickness" of the oil, or more properly the amount of force needed to move a layer of oil film over another layer. The extent to which a lubricant's viscosity changes with temperature is indicated by the viscosity index, an arbitrary scale that compares oils with a particular Pennsylvania crude oil that has an assigned value of 100. Commercial lubricating oils have a viscosity indexes ranging from 60 (equivalent to 10W motor oil) up to 3,300.

In general, a low-viscosity oil is preferred because it has lower friction and transfers heat at a faster rate. But when high operating temperatures are encountered, a higher-viscosity oil is employed because it is less volatile. Modern "multigrade" oils combine the beneficial characteristics of both by resisting viscosity loss as operating temperatures increase. Their range is indicated by two numerals separated by a W and a hyphen; 10W-30 motor oil, for example, has the viscosity of a light, 10-grade oil at room temperature, but at high temperatures it has the viscosity of a heavier, grade-30 oil. This characteristic is imparted by the addition of special polymers to the oil; at higher temperatures their molecules uncoil, increasing the oil's viscosity in the process.

In addition to viscosity extenders, modern motor oils have a number of additives that improve their overall performance. Antioxidants resist the formation of harmful acids and sludge through oxidation. Pour-point depressants are used to counteract the tendency of oil to thicken over time due to the presence of high melting-point paraffins. Dispersants (sometimes called *detergents*) are used to clean out combustion products like partially burned fuel and water, which form lacquers and sludge in critical places like piston ring grooves.

Because grit and other contaminants inevitably end up in lubricating oil, many engines and other machines are equipped with oil filters. These generally consist of a canister containing a filtering element made from paper, wire mesh, or some other substance. The folded-paper filters used in automobiles trap all particles with a diameter of more than 15 microns (a micron is one-millionth of a meter), 95 percent of the particles over 10 microns, and 90 percent of the particles over 5 microns. Oil filters can be of the "bypass" or "full-flow" variety. A bypass filter passes only a portion of the oil through the filter as it exits from the pump. In a full-flow filter, all of the oil pump's output passes through the filter unless it is blocked or the oil is too cold to flow properly, in which case a pressure relief valve returns the oil to the sump.

The development of improved lubricants is one of the major reasons that machines have made enormous strides in performance and reliability. Early internal combustion engines, for example, turned a few hundred revolutions per minute and developed 2 or 3 horsepower; a modern engine of the same displacement may turn 6,000 rpm while producing 100 hp. At the same time, however, lubricants can be a major source of environmental damage. The motor oil used in today's automobiles is one of the most serious sources of groundwater contamination; in the United States alone, 1.9 billion kg (2.1 million tons) of used motor oil ends up in rivers and streams every year. Just one liter (1.06 quart) can pollute 950,000 liters (250,000 gallons) of drinking water. In many communities this threat has been addressed by the development of recycling programs. To a certain extent these programs pay for themselves, for used motor oil can be reprocessed into fuel oil and lubricants.

See also BEARINGS; DETERGENTS; OIL REFINING; POLYMERS; RECYCLING; WHEEL

Luddism

During the early years of the 19th century, the economy of England was being transformed by the use of machinery. Although labor-saving machines increased productivity, not everyone benefited from their introduction. On occasion, dissatisfied groups of workers destroyed the new machinery in various industries, primarily in the north and Midlands. The name given to them was Luddite, taken from a possibly mythical Ned Ludlum, who, on being reprimanded by his employer, smashed his stocking frames with a hammer.

The early Luddites were not simply protesting the introduction of new machinery, fearful that the new devices were replacing their jobs. The motivation for Luddite outbursts differed from industry to industry. The movement began among the skilled workers in the hosiery trade, where there had been a long history of opposition to the use of wider stocking frames that allowed the employment of poorly paid unskilled laborers. The situation was exacerbated by the hard economic times caused by the Napoleonic Wars and the closure of many of England's export markets. A series of bad harvests compounded the problem as the price of food outstripped workers' wages.

In these difficult times, Luddite disturbances spread within the ranks of handloom weavers and shearers in the textile industry. The weavers viewed the advance of steam-powered weaving machinery with understandable apprehension, and some of them attacked the factories containing them, along with the houses of their owners. Only in a few instances was the machinery itself directly attacked. In contrast, new devices were the target of Luddite attacks in the cropping trade. Wool cloth was traditionally finished by raising the nap and then leveling the surface through the use of heavy shears. The introduction of the gig mill, along with a device for mechanized cropping, threatened the livelihood of the handworkers. They responded with some of the fiercest attacks of the Luddite epoch.

Within a few years, the Luddite uprisings came to an end due to the deployment of government troops (12,000 were called out in 1812 to restore order to those parts of England affected by the movement); the execution, imprisonment, and transportation to Australia of a number of participants; and the generally improved economic climate following Napoleon's defeat. Over the longer term, the replacement of small manufacturing establishments by large factories helped to stimulate the growth of labor unions and collective bargaining. Machine smashing by riotous crowds had been a favored form of labor protest when workers were scattered and lacking in permanent organizational linkages, but less relevant when unionization became a possibility.

This did not mean, however, that disturbances of this sort came to an end. As the pace of technological change has quickened and we have become more aware of its consequences, numerous efforts have been made to stop the advance of technology when it is perceived as harmful to a particular group. To take one example, in recent years the spread of computers and computer-controlled machinery has threatened many established occupational roles and procedures, leading to significant resistance to their installation and use. In one case that received a good deal of national publicity during the 1970s, newspaper workers in Washington, D.C., demonstrated their opposition to computerized typesetting equipment by engaging in widespread industrial sabotage. Today, *Luddite* is occasionally applied to individuals opposed to specific technologies such as nuclear power.

See also TECHNOLOGICAL UNEMPLOYMENT

Further Reading: Malcolm I. Thomis, *The Luddites: Machine-Breaking in Regency England*, 1972.

Lysenkoism

One of the outstanding examples of the corrosive effects of a totalitarian regime on scientific advance is provided by the career of Trofim Denisovich Lysenko (1898–1976). Not only did Lysenko's influence retard science in the Soviet Union, it also bore some responsibility for years of stagnant agricultural output and food shortages in that country.

Lysenko had received some training in agronomy, but he could hardly be considered a bona fide scientist. However, his weak scientific credentials were compensated by his apparent ability to meet the needs of the Soviet leadership. In 1929, Lysenko gained national prominence through his advocacy of "vernalization," a process that promoted the early germination of seeds by first soaking them in water and then freezing them. This was in fact a traditional practice, but Lysenko gave it a radical twist by claiming that these seeds would give rise to successive generations of early-germinating seeds. This of course was in direct contradiction of Darwinian evolution and Mendelian genetics, and strongly resembled the long-discredited views of the French biologist Jean Baptiste Lamarck (1744–1829). In espousing the theory that acquired characteristics could be inherited, Lysenko associated himself with another Soviet pseudoscientist, I. V. Michurin (1855–1935), who claimed that his application of this theory allowed him to breed better fruit trees.

In the 1930s, Lysenko put forth his theory of "phasic development," which claimed that the fundamental units of heredity were not genes in germ cells but complete organisms. Lysenko may have been little more than a misinformed charlatan, but he enjoyed the active support of Josef Stalin (1879–1953), who had his own reasons for embracing Lysenko's ideas. Under Stalin's leadership, the Soviet Union had plunged into a program of collectivization that resulted in sinking agricultural yields and widespread famine. Cultivation practices based on Lysenko's theories offered a quick fix for a ruined farm economy. Although it flew in the face of scientific orthodoxy, Lysenko's theory promised rapid increases in agricultural yields; high-yielding seeds would give rise to more high-yielding seeds, with no need for slow and painstaking breeding programs. Even better, Stalin believed that Lysenko's principles would bring increased farm production with only minimal investment in fertilizer, pesticides, and irrigation works.

In reality, Soviet agriculture continued to languish, but Lysenko was still able to maintain his influence. In 1940, he was appointed director of the Institute of Genetics of the Soviet Academy of Sciences. Scientists who opposed his views faced the prospect of denunciation and even imprisonment. In 1948, Lysenko further strengthened his position when he presented a report that endowed him with the full support of Stalin and the Central Committee of the Communist Party. His control over biology was complete; orthodox genetics was removed from books and journals, and an order was issued to destroy all stocks of *Drosophila*, the fruit fly commonly used for genetic experiments.

A number of disasters accompanied the application of his theories, but Lysenko managed to maintain his influence, even after Stalin's death in 1953. The unrealistic plans for accelerated farm production made by Nikita Khrushchev (1894–1971) were based in part on the "discoveries" that Lysenko continued to make at his experimental plot in the Lenin Hills near Moscow. Only after Khrushchev fell from power in 1964 did Lysenko lose his position as director of the Institute of Genetics. He continued to propound his views to a shrinking audience until his death, while Soviet geneticists engaged themselves in a valiant effort to rebuild a science that had been all but destroyed when Lysenkoism dominated Soviet biology.

See also EVOLUTION; GENE; NATURAL SELECTION

Further Reading: David Joravsky, *The Lysenko Affair*, 1970.

Machine Gun

A machine gun is a self-loading firearm capable of firing serial multiple shots from a single activation of the trigger mechanism. The idea of a multishot firearm is almost as old as that of firearms themselves. Heavy, multibarrel weapons—variously called *battery guns*, *volley guns*, or *ribaudequins*—go back to the 16th century. Such weapons, however, were consistently outperformed by multiple projectiles such as grapeshot and canister, fired from the more versatile 3- to 12-pounder field artillery cannon whose bulk they rivaled.

The advent of percussion ignition began to change this situation, allowing more efficient rapid-firing weapons to be designed. By the mid-19th century, a variety of mechanically loaded and fired weapons had reached a state of sufficient maturity to be offered to Western armies. These ranged from conceptual descendants of the battery gun, such as the French *mitrailleuse* with its bundle of 37 serially fired barrels, to more sophisticated systems such as the Gatling gun, with its cluster of 6 to 10 revolving barrels. All were externally powered, their actions operated by hand cranking. Metallic-cartridge ammunition made these weapons much more practical; however, they still suffered from bulk, fragility, and complication relative to other weapons. Of these drawbacks, the greatest was bulk: Since early mechanical rapid-fire weapons were so large as to be mounted on field gun carriages, they were generally seen as a type of artillery by military officers, even though their capabilities were by no means similar. In the case of the *mitrailleuse*, this problem, aggravated by extreme military secrecy, was serious enough to be a factor in several French losses during the Franco-Prussian War of 1870. Nonetheless, such weapons gained a military niche, especially in colonial and expeditionary forces where small forces found themselves in the way of enormously large, if primitively equipped, opposition.

Yet even as the Gatling gun and such competitors as the Gardner and Nordenfelt guns gained limited acceptance, they were obsolescent. In the early 1880s, Hiram Maxim (1869–1936), a U.S.-expatriate inventor living in England, set out to develop a practical self-powering action. By 1885, he had one, the first true machine gun. It operated by using the power of the gun's recoil to extract and eject the spent round, and then load and fire a new one to begin the cycle again. This revolutionary innovation, together with Maxim's elegant solutions to such ancillary problems as ammunition delivery (by enclosure of the cartridges in a fabric belt) and barrel cooling (by use of a water jacket), led to a weapon much lighter, more robust, and faster firing than the earlier, hand-cranked generation. The advent of smokeless propellants, with their reduced residues, allowed another approach to automatic fire outside Maxim's patents, tapping combustion gas to operate the action. The result was a proliferation of designs, as Maxim guns were joined in the marketplace and the arsenal by those of Browning, Hotchkiss, and Lewis, among numerous others.

This technical flowering was not accompanied by a tactical one, however. Early experience in colonial campaigning showed the lethality of the machine gun, as at Omdurman in the Sudan in 1898, where a British expeditionary force with six Maxim guns killed 11,000 Dervishes, with a loss of only 48 of their own number. Similar bloodletting accompanied the more traditional infantry actions of the Russo-Japanese War, in which, for the first time, both sides had machine guns.

Nonetheless, conventional military wisdom continued to regard the machine gun as an auxiliary weapon, mainly suited to defense and thus incompatible with European doctrine, with its Napoleonic focus on the primacy of the offensive at all levels of warfare. As such, it was issued on a limited scale, generally two guns per infantry battalion in European armies.

The major exception to this trend was Germany. The Reich's strategic situation, which necessitated a two-front war, put a premium on manpower and firepower. As a result, compared with the Continental norm the German army carried over three times the number of machine guns per capita, and integrated them closely into their tactics and institutions. The wisdom of this policy was shown soon after the outbreak of World War I in 1914. As the Western Front stabilized, it quickly became evident that increased firepower in the form of the machine gun and improved artillery had radically changed the tactical picture. The Germans, with their more extensive adoption and integration of the machine gun, were in a better position as a result, which helped them hold most of the ground gained by their initial offensive, despite being outnumbered and blockaded, for most of the war. While the machine gun was not the sole cause of the bloody stalemate that characterized the bulk of the war in the West, it was a major factor.

The Allies, however, learned quickly under wartime pressure, ramping up production and increasing the numbers of machine guns within units. This proliferation in absolute numbers was accompanied by a proliferation of types on both sides. The new threats posed by aircraft and tanks—themselves introduced in partial response to the tactical problem posed by the machine gun—led to heavier weapons and to the development of specialized ones for air and armored warfare. Machine guns were attached to infantry units farther forward, down to platoon and section level. This led to widespread adoption of bipod-mounted, magazine-fed light machine guns—especially the Lewis gun, originally rejected by numerous armies for just these characteristics—where portability became more important than steadiness or sustained fire. This trend culminated in a special case, the introduction of the submachine gun, a handheld weapon firing pistol ammunition. It was intended for the raids and clearing assaults characteristic

The famous Gatling gun, one of the first successful machine guns (from M. Kranzberg and C. W. Pursell, Jr., *Technology in Western Civilization*, vol. I, 1967, p. 494).

of trench warfare where the resultant drastic loss of effective range was irrelevant. On both sides, but especially the German, these improvements in infantry firepower, together with tactics evolved to exploit them, helped restore a measure of mobility to the battlefield.

World War II continued these proliferative trends, to the point where individual infantry squads fielded multiple light machine guns, sometimes leading to the subdivision of this nominal smallest group into even smaller "fire teams," each with its own light machine gun or automatic rifle. The logical final step, made conceivable by a combination of improved logistics, economies of scale, and the problem of producing a mass army of trained riflemen under the stresses of wartime, was the fielding of fully automatic weapons as infantry rifles. Efforts to this effect were set in motion during the war. They foundered, however, on the mechanical problems of controlling a lightweight, handheld weapon firing the infantry rifle rounds of the day, powered to engage targets at ranges up to 1,000 meters (3,280 ft), and keeping such a rifle functioning under the associated stresses for prolonged periods in the field. As a result, most of the weapons of this generation, typified by the U.S. M14 and the FN FAL, were largely restricted to service as semiautomatic rifles. It took the adoption of the "assault rifle" firing a less powerful— and thus smaller, cheaper, and more controllable—cartridge to make fully automatic personal weapons the standard infantry armament.

The machine gun per se, the fully automatic weapon firing high-powered cartridges, continues as one of the fundamental weapons of modern infantry and as an important component of armored and air weapons systems. Mechanically, current machine guns have advanced only slightly relative to designs produced in World War II; most advances have taken place in their fire control. Indeed, the greatest advance in the weapons themselves may be seen almost as a regression, as designers have returned to externally powered actions, electrically or hydraulically driven rather than by human muscle, in an effort to achieve greater reliability and rates of fire.—B.H.

See also RIFLE; SEMIAUTOMATIC WEAPONS; TANKS

Further Reading: John Ellis, *The Social History of the Machine Gun* , 1975. Jim Thompson, *Machine Guns: A Pictorial, Tactical, and Practical History*, 1990.

Machine Gun, Synchronized

The outbreak of World War I occurred less than 11 years after the first flight of the Wright Brothers. Airplanes were flimsy and often unreliable, but they were soon pressed into service for artillery spotting and the observation of enemy troop deployments. While conducting reconnaissance missions, crews occasionally encountered enemy aircraft but lacked the means (and often the desire) to effectively attack them. Pistols, rifles, and even shotguns loaded with chain shot were taken aloft and fired at enemy aircraft, usually with minimal results. An early attempt to take a machine gun into aerial battle was frustrated when the weight of the gun prevented the airplane from climbing above 1,067 m (3,500 ft).

More powerful aircraft engines allowed the carrying of machine guns, but they could not be mounted in an optimal manner. The most suitable position for the machine gun of an attacking airplane was its nose, but this was the location of the propeller. Some aircraft designs addressed this problem by putting the engine behind the fuselage. These "pusher" designs gave an unobstructed field of forward fire at the cost of performance that was inferior to conventional "tractor" configurations with the engine at the front. In an effort to provide a clear field of fire, guns were sometimes mounted to the fuselage at an angle so that the trajectory of the bullets put them clear of the propeller. This put a heavy burden on the flying abilities of the pilot, who had to approach his target in a crablike fashion. Other designs put the machine guns on the top of the wing, where they fired over the propeller arc. This worked, but it caused difficulties when the pilot had to reload the machine gun. Since the drum of the lightweight Lewis gun then in use contained only 47 rounds, this was a frequent occurrence.

A straightforward, if crude, solution to the problem was provided by Roland Garros, a French pilot. In March 1915, Garros instructed mechanics to fit the blades of his fighter's propeller with wedge-shaped steel plates. These deflected the bullets that hit them, while enough bullets got between the blades to make it to the target. In the next 2 weeks, Garros shot down six German airplanes, but then an engine failure forced him to land behind German lines. At this time the Dutch aircraft designer Anthony Fokker (1890–1939) was preparing to put his new monoplane fighter into service, and he was ordered to fit its propeller with deflector plates. Instead, Fokker devised a synchronizer gear that prevented the machine gun from being fired when the propeller was directly in front of its muzzle. This was done by interposing a cam-and-pushrod arrangement between the engine's oil pump and the machine gun's trigger; in this way the cam caused the trigger to be pulled only at the proper time.

The design was quite straightforward, and in fact it did not originate with Fokker. Interrupter gears had previously been designed by Poplavko in Russia, Schneider in Germany, the Edwards brothers in Great Britain, and Saulnier in France. In fact, Saulnier's design had undergone field tests, although it was hindered by the use of an unsuitable machine gun. This failure led to Saulnier's design of propeller deflector plates that served as Garros's inspiration.

In any event, during the winter of 1915–16, Fokker airplanes fitted with one or two synchronized machine guns fed by belts holding 500 rounds established what would now be called "air superiority." The Allies called it the "Fokker scourge." Allied efforts to produce synchronizing gear were only moderately successful, as the guns so equipped had a tendency to jam when the pilots throttled down. This unfortunate situation was resolved by the invention of a hydraulic synchronizer by George Constantinesco, a Romanian mining engineer.

Fighter planes continued to be fitted with synchronized machine guns in the years after World War I. Many World War II aircraft were so equipped, even though the interrupter gear necessarily slowed down the rate of fire. Consequently, fighter planes put all or most of their armament in the wings, as many as eight machine guns or four cannon. Accuracy deteriorated somewhat, but this was offset by the greater firepower afforded by wing-mounted weapons.

Further Reading: Bryan Cooper and John Batchelor, *Fighter: A History of Fighter Aircraft*, 1973.

Machine Tools, Numerically Controlled

Machine tools are devices that cut away material in order to make something of a particular size and shape. Until the 1940s, all machine tools were run by skilled operators who made the critical decisions in regard to such things as running speeds and feed rates. In a numerically controlled machine tool, these actions are controlled electronically. At first, digitized instructions were coded on magnetic tape or punch cards. From about the 1960s onwards, solid-state electronics came into use, and in the 1980s machine tools began to be equipped with their own computers (hence the current term *computer numerical control* or CNC). Instead of performing only one operation, CNC machine tools use sophisticated programs that allow them to do a variety of tasks. For example, a set of internal programs may guide the machine tool as it performs the same set of actions on parts of different sizes.

Development of numerically controlled tools in the early 1950s was financially sponsored by the U.S. Air Force, with major R&D work done at the Servomechanism Laboratory of the Massachusetts Institute of Technology. The Air Force was particularly interested in numerically controlled machine tools because they made it possible to form metal into the complex shapes required for wing surfaces and jet engine compressor blades.

All numerically controlled machine tools require programmed instructions. As noted above, many early numerically controlled machine tools used punch cards, digitized instructions that date back to 18th-century "Jacquard" looms. An alternative programming method was known as *record playback*. Developed in 1946–47, record playback generated instructions by recording on magnetic tape the motions of a machine tool as it was guided by a skilled machinist. The tape could then be used to control the actions of other machine tools. In contrast, the preparation of punch cards was divorced from shop floor practice. The source of the instructions were not experienced machinists but engineers and mathematicians who usually had little or no hands-on knowledge of machine tools and machining.

Numerical control was superior to record playback in that it made possible the machining of more complex parts. It also did away with expensive jigs, fixtures, and templates, and it promised lower labor costs by reducing the need for toolmakers, patternmakers, and skilled machinists. The latter, in particular, was of particular interest to managers who wanted to extend their control over machinists. In this way, the use of numerically controlled machine tools had a kinship with Scientific Management, the package of administrative methods that sharply divided the tasks of managers and workers. Both Scientific Management and numerically controlled machine tools were used to diminish the intellectual role of the workers and leave them with few opportunities to exercise discretion as they went about their work.

However, as happened with Scientific Management, the intentions of some promoters of numerically controlled machine tools were not fulfilled. Numerical control does not guarantee the automatic operation of machine tools. Even when using CNC machine tools, skilled machinists are needed to overcome the problems caused by dull tools, worn machine parts, rough castings, programming bugs, and periodic breakdowns. Successful managers have come to recognize that numerically controlled machine tools require the presence of skilled operators who can take an active role in machining operations.

See also AUTOMATION; DESKILLING; LATHE; LOOM, "JACQUARD"; MILLING MACHINE; PUNCH CARDS; RESEARCH AND DEVELOPMENT; SCIENTIFIC MANAGEMENT

Further Reading: David F. Noble, *Forces of Production: A Social History of Industrial Automation*, 1986.

Magnetic Compass—see Compass, Magnetic

Magnetic-Levitation Vehicles

Nearly all present ground transportation systems rely on wheels to transfer power and produce motion. Magnetic levitation, or maglev, uses magnetic forces to support and propel a single vehicle or a train. Two basic systems exist, one based on magnetic attraction, the other based on magnetic repulsion. In magnetic-attraction systems, direct-current electromagnets are mounted to the vehicle, where they supply a magnetic flux that is either transverse to, or in a straight line with, the direction of the vehicle's travel. The electromagnets are located on struts that extend below the elevated guideway; the vehicle can thus be made to hover over a track on the guideway by controlling the strength of the magnetic force so it offsets the force of gravity. In magnetic-repulsion systems, magnetic forces in both the vehicle and guideway are used to keep the two separated. In both systems the vehicles are propelled by linear synchronous motors. These use alternating current to produce a magnetic wave that interacts with magnets in the vehicle; the changing magnetic field moves the vehicle by pulling and pushing it in rapid succession.

The idea of magnetic levitation goes back to the early 20th century, when a proposed maglev system was patented by Emile Bachelet, a French engineer working in the United States; it never got beyond the working-model stage. In the 1960s, interest in magnetic levitation was revived when James R. Powell and Gordon T. Danby published a paper outlining the use of superconducting magnets for magnetic levitation. This theoretical insight was followed by the construction of a test model using superconducting magnets. The U.S. Congress gave a boost to maglev trains in 1965 when it passed the High Speed Ground Transportation Act. This act provided funding for research into maglev systems until the money ran out in 1975.

Although a promising start had been made, the United States did not build upon it. Meanwhile, Great Britain, Japan, and Germany were moving ahead with maglev systems. In Great Britain, a maglev train is in daily operation, providing slow-speed shuttle service at Birmingham Airport. A more ambitious prototype of a high-speed maglev train has been tested in Germany, and a system using German technology is being built in Florida, where it will connect the Orlando Airport with a tourist area near Disney World. In Japan, magnetic-repulsion technologies have been used in a half-scale unmanned experimental train that has achieved speeds of 500 kph (300 mph). A system based on this technology may eventually be used for a high-speed line connecting Tokyo with Osaka.

Interest in maglev transportation systems was revived in the United States with the federal government's creation in 1990 of the National Maglev Institute, which was supported by a 2-year budget of nearly $30 million. This was followed by the passage of the Intermodal Surface Transportation Efficiency Act of 1991 that had among its provisions an appropriation of $500 million for the development of a maglev prototype. However, the Bush administration balked at spending the money that had been appropriated, and while the Clinton administration has been more enthusiastic, budgetary constraints have retarded maglev development.

While only a few maglev systems have been built, enthusiasts continue to point out the inherent advantages of a maglev-based transportation system. Since there is no contact between the vehicle and guideway, there is nothing to wear out. The use of magnetism for propulsion also means that maintaining traction is not an issue, as is the case with conventional wheeled vehicles traveling at high speeds. However, a maglev train is noisy; at speeds of 480 kph (300 mph) it generates 100 decibels of sound at a distance of 24 meters (80 ft). Also, as with any new technology, a practical maglev transportation system will require the solution of many technical problems, not the least which is finding ways to insulate passengers from the effects of rapid acceleration and changes of direction.

At the same time, the success of maglev transportation will not be determined solely by the ability to solve technical problems. Maglev technology is most promising as the basis of high-speed rail systems, but high-speed rail systems are very expensive propositions. One proposed set of maglev lines serving three transportation corridors in the United States came with an estimated price tag of $27 billion. High construction and operating costs can be recouped only when the trains serve areas with a high population density and a large flow of passengers between cities. Even this may not be enough; one study found that only a very heavily traveled route, such as Los Angeles–San Francisco, would be able to produce revenues in excess of capital and operating costs. Under these circumstances, the construction of an extensive maglev system might be undertaken only if social goals such as relieving congestion (both air and ground) were deemed more important than making a profit. In any event, the high start-up costs of maglev systems may put them beyond the financial capabilities of private enterprise, making government involvement essential.

See also SUPERCONDUCTIVITY

Further Reading: R. G. Rhodes and B. E. Mulhall, *Magnetic Levitation for Rail Transport*, 1992.

Magnetic Resonance Imaging

Magnetic resonance imaging (MRI) is a technology that has become the preferred method for observing anatomical detail in soft tissues. Unlike X rays, which provide pictures only of hard substances like bones, MRI can be used to obtain information on body fluids, soft tissues in organs, and brain and nerve tissue. It has been successful in diagnosing heart diseases, tumors, and brain disorders such as multiple sclerosis. Recently developed techniques have been used to analyze brain chemistry (magnetic resonance spectroscopy) and to monitor brain activity (functional imaging). MRI can eliminate the need for exploratory surgery and exposure to high-energy radiation. Advantages over other imaging techniques include superior contrast resolution, as well as the availability of information in three dimensions.

MRI was developed through the use of nuclear magnetic resonance spectroscopy (NMR), a technique used since the 1950s by chemists to identify molecules and determine molecular structure. NMR spectroscopy was combined with imaging techniques developed for computed tomography (CT or CAT) scanning to develop what is at present one of the most widely used medical imaging methods. The possibility of imaging human tissue using nuclear magnetic resonance was first proposed in 1970 by Paul Lauterbur and Raymond Damadian. At first, the proposals were met with considerable skepticism. The first human image was reported by Damadian in 1977, and the first commercial instrument became available in 1980. Over 4,000 MRI instruments are now in operation worldwide, at least three- quarters of them in the United States. Approximately 6 million scans are performed in the United States each year.

When first used in medicine, MRI was called "nuclear magnetic resonance imaging." The word *nuclear* was dropped because of its association with the ionizing radiation from nuclear decay, which has a negative connotation because of the biological hazards of high-energy radiation. The signals in NMR and MRI do come from the nuclei of atoms, but high-energy radiation is not involved.

The basis of NMR and MRI is a property called *nuclear spin*, which generates an infinitesimal magnetic field—a "magnetic moment"—in nuclei with an odd number of protons, or protons plus neutrons. Several biologically important nuclei—such as hydrogen, phosphorus-31, and carbon-13—have a nuclear spin and are observed in NMR. Because the hydrogen nucleus (simply a proton) gives the strongest signal and is present in most molecules, it is the nucleus most often observed in NMR spectroscopy and is the nucleus almost exclusively observed in MRI. Because water is by far the most abundant molecule in the human body, most of the signal observed in MRI comes from the protons in water, while a lesser amount comes from protons in fat.

In the absence of an external magnetic field, the magnetic moments of nuclei with spin point in random directions. However, when a sample is placed in a strong magnetic field, the magnetic moments align themselves along the magnetic field. When the aligned nuclei are excited with radiofrequency (rf) radiation of the right wavelength, they absorb energy from the radiation, and their magnetic moments change alignment. After the rf radiation is turned off, the moments return to their former alignment, emitting radiation in the process. This emitted radiation (also in the rf range) is the source of the signal for NMR and MRI. Nuclei in different magnetic fields emit radiation of different frequencies. In an MRI instrument, a magnetic field is produced that changes as a function of the position in the magnet. The frequency of the signal observed will therefore depend on the position of the nucleus, allowing an image to be constructed.

The imaging subsystem consists of several main components: the magnet, the gradient coils, the probe, and the computer system. In most cases, the magnetic field is created by a superconducting coil. An open bore along the central axis of the cylindrical magnet is horizontal and large enough to accommodate most reclining patients. The gradient coils, which produce the additional position-dependent magnetic field, are wound just inside the bore of the magnet. The probe might also be on the inner side of the magnet bore, or it might be in a smaller unit that fits over the region of interest on the patient, such as the head. The probe contains radiofrequency coils that both deliver the rf radiation to the patient and detect the signal coming from the patient. The computer controls the pulse program—a sequence like a musical score. During the pulse program, the gradient coils are turned off and on, the rf radiation is delivered in a series of pulses, and the signal is collected, all in a carefully timed sequence. The computer then processes the data collected and constructs an image.

MRI is thought to be extremely safe, but there are some questions of safety and other problems associated with the technique. No harmful effects have been observed in mammals exposed to static magnetic fields. The switching off and on of the gradient coils during data collection generates a banging noise that can be uncomfortable for the patient. Newly developed techniques used to monitor brain activity employ faster switching of gradient coils and are associated with nerve stimulation that may cause cardiac arrhythmias and optical flashes. The rf radiation is not energetic enough to cause tissue damage as in X rays, but some heating of the tissue might occur. The most common problem associated with MRI is claustrophobia, experienced by 5 percent of patients. MRI examinations are not recommended for patients with cardiac pacemakers or large metallic implants, or for those who are in the first trimester of pregnancy.

MRI is an extraordinarily useful diagnostic tool. At the same time, however, the technique has helped to drive up the cost of medical care. The instruments themselves cost approximately $1.5 million each and have substantial maintenance and personnel expenses. Many systems have been purchased with the idea that they will be profitable for institutions. In order to pay for the equipment and make a profit, each scan is done at a high price (often $1,000), and many scans must be done. Economic pressure keeps charges high, may encourage unnecessary use, and, as with other high-tech instrumentation, tends to shift control of medical care delivery from physicians to hospital or clinic management.—B.L.

See also CARDIAC PACEMAKER; COMPUTERIZED AXIAL TOMOGRAPHY; NEUTRON; PROTON; RADIOACTIVITY AND RADIATION; SUPERCONDUCTIVITY; X RAYS

Further Reading: Stewart C. Bushong, *Magnetic Resonance Imaging: Physical and Biological Principles*, 1988.

Mammography

Mammography is a radiological technique used to visualize breast tissue. It is employed both to evaluate the breasts of women with symptoms, such as a lump, and to screen for breast cancer in women with no symptoms. It is the single most effective method of screening for breast cancer because it can detect cancer well before a tumor can be felt by women doing self-examinations or by physicians doing clinical examinations. Early detection of cancer is important, as it is correlated with more favorable prognosis than is the case with cancers discovered in later stages of development. Mammography has been shown to reduce deaths from breast cancer by between 30 percent and 50 percent in women older than 50.

Results of the first U.S. study of the feasibility of mammography were published in 1930. Since then the technology has improved greatly, beginning with the use of general-purpose X-ray equipment in the 1960s to today's modern dedicated screen-film units that provide a much higher quality image with significantly lower X-ray doses. In the procedure itself, a woman's breast is compressed between two flat plates and pictures are taken from different angles. Because of their differential absorption by the tissues, X rays of low energy are effective in imaging the breast skin, fat, and calcifications. Detected abnormalities indicate *possible* cancer, so further tests, including surgical biopsy, are required to confirm a diagnosis of cancer.

One problem associated with mammography is the possibility of misdiagnosis, and users should be aware that, overall, mammography has a false-positive rate (indicating cancer where there is none) of about 7 percent. Estimates of false negatives (missing an existing cancer) range from 4 to 30 percent. These errors are more common among premenopausal women, as noted below, because their breast tissue is denser and does not show up as well in a mammogram. Another concern is the potential of the radiation to cause breast cancer.

Quality control is another concern that has received considerable attention in recent decades. The American College of Radiology began its voluntary Mammography Accreditation Program in the late 1980s, and in 1992 the U.S. Congress passed the Mammography Quality Standards Act. This legislation, administered by the Food and Drug Administration, requires facilities to meet specific standards of quality in order to offer mammography.

In the United States, there is widespread agreement that women 50 and over should be screened annually, while mammography is generally inappropriate for women under 40. There is less consensus regarding recommendations for women between the ages of 40 and 50. The American Cancer Society advises mammography every 1 to 2 years for women 40 to 49, with a baseline mammogram by age 40. The panel of experts participating in the 1997 NIH-sponsored Consensus Development Conference, however, concluded that it could not recommend the procedure for all women 40 to 50. After reviewing hundreds of papers and hearing evidence from 35 experts, the panel concluded that there was as yet no overwhelmingly persuasive evidence that regular mammograms of women under 50 saves lives, and that this potential benefit must be counterbalanced by risks of false-positive readings (which may be as high as 30 percent in women 40 to 50), as well as the risk that the radiation itself might cause a small number of cancers (estimated at 3 women for every 10,000 women screened annually from ages 40 to 49). Some experts praised the panel's decision; others, who believed that the data strongly supported screening in this age group, expressed shock and dismay.

Women in the United States are, in fact, screened considerably less than is recommended. The 1992 Cancer Control Supplement of the National Health Interview Survey found, for example, that only 30 percent of women 50 and over had had mammography within the past 12 months. Women with no regular source of medical care, women with low income and less education, and African-American women are less likely to have had a recent mammogram.

Unfortunately, breast cancer mortality rates have been relatively stable since 1950, although between 1989 and 1992, breast cancer mortality declined 4.7 percent, the largest short-term rate decrease since 1950. However, the improvement occurred in white women only; in contrast, the rate rose slightly among African- American women. The recent overall decline in breast cancer mortality has been attributed to increases in breast cancer awareness and screening, along with advances in treatment and changes in risk factors. Breast cancer remains the most frequently diagnosed cancer in women, apart from skin cancers, and is the second largest cancer killer after lung cancer. Efforts to increase accessibility to and regularity of screening for *all* women of appropriate ages are urgently needed.—A.H.S.

See also CANCER; FOOD AND DRUG ADMINISTRATION, U.S.; X RAYS

Further Reading: American Cancer Society, *Breast Cancer Facts and Figures, 1996*, 1995.

Many-Body Problem

In 1668, in response to a Royal Society of London competition, John Wallis (1616–1703), Christopher Wren (1632–1723), and Christian Huyghens (1629–1695) formulated laws that we know as the conservation of momentum and conservation of energy in order to solve the problem of determining the final velocities of two colliding perfectly elastic or inelastic bodies that had known velocities before impact. Mathematicians have shown that, in general, in order to determine *n* different quantities, one must have *n* independent equations relating those quantities to one another. Thus, the problem of two-body collisions was, in principle, solved when the laws of conservation of momentum and energy were formulated, because there are two unknown quantities (the final velocities of the two bodies) and two equations of motion to use in determining them.

No additional general laws governing collisions between bodies have been discovered since 1668, so we are still without any way of solving the general problem of determining the final velocities of three or more bodies that strike one another simultaneously with known initial velocities. In special cases in which additional information is available about the initial velocities—for example that they are all equal in magnitude or that the velocities of two are symmetrical about the path of the third—the problem can be solved for more than two bodies, but in general, the *n*-body or many-body problem has no definite solution.

In an 1873 paper with the unwieldy title "Does the Progress of Physical Science Tend to Give any Advantage to the Opinion of Necessity (or Determinism) over that of the Contingency of Events, and the Freedom of the Will," James Clerk Maxwell (1831–1879) pointed out that the many-body problem makes the classical physical world strictly nondeterministic if one assumes that all events are ultimately caused by the impacts of moving bodies and that all conservation laws are known. The situation still resists solution, for we cannot be absolutely certain that there are no simultaneous many-body impacts. If there are, and there are no additional formal relationships to be found, then there is no determinate set of velocities produced in a many-body collision. More generally, the many-body problem demonstrates the limits of an overly deterministic approach to the world and its contents. The workings of natural systems (including human societies) are vastly more complex than the dynamic interactions of a few bodies. Efforts to predict future events with even a modicum of accuracy are not likely to be successful.—R.O.

See also CONSERVATION OF ENERGY

Margarine

A significant problem in many industrialized nations is the overproduction of dairy products. One outstanding example is the European Union, which tries to maintain farmers' incomes by buying large stocks of butter, most of which is simply put into storage. This is an ironic shift from the conditions that prevailed during the last century, when shortages rather than surpluses of food products were the paramount concern. One manifestation of this problem was a competition to find a substitute for butter that was initiated by Emperor Napoleon III (1808–1873) of France. The winner was a French chemist named Hippolyte Megé-Mourièz. His successful method began with beef tallow, which was converted into oleo, a soft, fatty substance. This was then emulsified with water and milk. In 1869 Megé-Mourièz received a patent for the resulting product. Due to its pearlescent appearance, he called it *margarine* after the Greek word for pearl, *margarites*. Megé-Mourièz never benefited financially from his invention. But it was successfully employed in the Netherlands, where two butter merchants built up thriving businesses in the manufacture and sale of margarine. Much of it was exported to England under the trade name Butterine until an 1887 law required the use of the name margarine.

Animal fat was still expensive, so experimenters

turned to plant-based fats: Peanut, soybean, sunflower seed, coconut, and palm oils were all employed. Initial efforts were not notably successful until the discovery in 1899 that these oils could be hardened by hydrogenation, the addition of hydrogen in the presence of a catalyst. Around 1910, this process began to be used on an industrial scale. Margarine today is manufactured by blending two or more oils in order to obtain the desired texture, consistency, and resistance to melting. These are emulsified after being mixed with aqueous ingredients such as water or milk products.

Dairy farmers have at times resisted the marketing of margarine due to its price advantage over butter. In the United States and other countries, pressure from dairy farmers led to the enactment of laws and regulations restricting the sale of margarine. For decades, margarine could not be packaged to resemble butter; consumers had to add a small amount of food coloring to the margarine to give it butter's yellowish appearance. The last state to mandate this was Wisconsin, which presented itself as "America's Dairyland" on its license plates. Travelers from other states were advised of their "last chance to buy margarine" when approaching the Wisconsin border.

Margarine's greatest advantage over butter has been cost, but growing health consciousness has also motivated the substitution of margarine for butter. Servings of butter and margarine contain about the same number of calories, but butter is considerably higher in cholesterol and saturated fats, both of which may contribute to cardiovascular disease in some individuals. A tablespoon of butter contains 31 mg of cholesterol and 7.1 grams of saturated fat, while a tablespoon of stick margarine made from corn oil has no cholesterol and only 2.5 grams of saturated fat. The saturated fat content of a spread can be further reduced by using smaller amounts of hydrogenated oils; margarine packaged in tubs may contain only 1.6 grams of saturated fat. People whose diets require very low levels of saturated fat can use spreads based on margarine that has air and water whipped in. This produces a spread with 40 to 60 percent fat; 1 tablespoon will contain 1.1 grams of saturated fat. It cannot be called margarine, however, for U.S. law requires that margarine has to contain no less than 80 percent fat.

See also CHOLESTEROL

Masonry

Building in stone or brick has always connoted permanence and durability. Most cultures have turned to masonry for their most important buildings: temples, palaces, and government structures. Not subject to rot and decay like wood, masonry has been a highly desirable form of construction despite the considerable labor involved in quarrying stone or firing brick.

Although suitable building stone can sometimes be found scattered on the ground ready for use, such instances are rare. Normally, building stones must be quarried out of the ground by drilling a series of holes in the rock and splitting off manageable sizes of blocks. The most popular stone types for building have been limestone (carbonates of calcium and magnesium) and marble (recrystalized limestone). Both are freestones, meaning that they do not have cleavage planes and can be freely carved in any direction. Marble, however, can take a fine polish and was preferred by the Greeks and Romans whenever available. Both limestone and marble have good resistance to weathering and can be worked with iron tools. Sandstones have also been used, but they vary in quality and often do not weather well. Granite, on the other hand, is extremely durable but much harder to work, requiring more labor and the more rapid consumption of tools.

When building stone is not available, bricks are often employed. The earliest examples were unfired mud bricks, such as the ones used in the Mesopotamian region in the 4th-century B.C.E. Builders soon discovered that fired clay bricks withstood the elements and began using them on the outer surface of walls. Most bricks used before the 20th century were fired in big stacks called *clamps*. Depending on the location of the brick in the clamp, it could be vitrified to a greater or lesser degree depending on the heat. Most of these so-called "common" bricks were underfired; they do not have the strength and durability of modern bricks.

Equally important as the stone or brick is the mortar used between the masonry units. Mortar provides a weathertight caulking and smooth bedding plane for the masonry, avoiding sharp projections that might concentrate stresses and crack the masonry. Stone can be laid up without mortar, but the faces of the stone have to be carved smoothly enough to provide an even surface. Examples of this are Greek columns, which are made out of many cylindrical sections called *drums*. Each drum was ground against the drum below, providing a column with joints that were nearly invisible. Such laborious construction took time and was the exception rather than the rule.

Most mortars in history were made of lime, produced by burning limestone or marble in a kiln, and then grinding the resulting clinkers into powder. Lime mortars do no possess great strength, but their compressive strengths are adequate for most building applications. Lime mortars do not, however, glue bricks together. The little tensile strength that lime mortars possess is unreliable, and builders cannot depend on mortar holding masonry together; it can only provide a bedding surface.

The Romans found that by mixing lime mortar with volcanic deposits called *pozzolan* that hydraulic properties and greater strength were given to the mortar. In fact, pozzolan mortars have qualities similar to modern Portland cement. Builders took advantage of this, building walls and vaults that were only faced with brick and stone, while the core was composed of broken bricks and rubble with pozzolan mortar poured around it. The durability of this pozzolan mortar has kept many Roman ruins intact over the centuries. Unfortunately later medieval builders did not have access to these volcanic deposits, and walls built with rubble cores and lime mortar have not proved as durable.

Masonry's great mass makes it vulnerable to earthquakes. Modern practice inserts vertical steel reinforcing grouted into the open cores of concrete block, and thin rods mortared into the horizontal joints to give resilience to seismic forces. But many modern masonry buildings, especially taller ones, use masonry only as a veneer attached to a metal frame. The Romans started this practice, covering their buildings with a veneer of marble over brick-faced walls with pozzolan cores. However, in most cases today, there is very little behind the solid-looking stone surface of contemporary buildings. Indiana limestone was a favored veneering material for steel-framed buildings in the early 20th century and is currently enjoying a resurgence in popularity, along with durable granites. These stone veneers give the appearance of solidity and permanence while reducing the mass of the wall.—E.C.R.

See also ARCHES AND VAULTS; BRICKS AND BRICKMAKING; CONCRETE; EARTHQUAKE-RESISTANT BUILDING STANDARDS; EARTHQUAKES; BUILDING CONSTRUCTION, STEEL-FRAME

Mass Production

For most of human history, life's necessities and luxuries were made on a small scale by individuals or by small groups of workers using a few hand tools. Products were often made to order, and long production runs were infrequent. Output was meager and costs were high. During the second half of the 18th century, manufacturing entered a period of revolutionary change. New sources of power, new machinery, and new ways of organizing labor drove production volumes to new heights. The process accelerated in the 19th and 20th centuries, resulting in an outpouring of products in numbers that would have been inconceivable to earlier generations.

In many industries, the key to increased output was mass production: the mechanized manufacture of large quantities of goods from interchangeable parts. An early example of mass production was the manufacture of ships' pulley blocks at the Portsmouth, England, Navy Yard. The creation of Marc Brunel (1769–1849) and Henry Maudsley (1771–1831), this mechanized operation employed 10 workers who turned out 160,000 blocks in a year, a larger output than had been achieved by 110 men using traditional techniques.

The machinery used for mass production required external sources of energy. At first, this came from waterpower. Steam power became significant during the second half of the 19th century, and by the end of the 19th century the electric motor was taking on a growing importance. These sources of power ran the lathes, milling machines, drill presses, and other machine tools that were essential to mass production. For mass production to work, it was essential that the parts made by these tools were fully interchangeable. For example, all of the axles made in a factory could be readily fitted into all the wheel hubs produced in that same factory. Interchangeability was accomplished through the development of precision machine tools and techniques, and the constant use of industrial gauges. Among the most important interchangeable parts were standardized nuts, bolts, and other fasteners.

Although mass production can be said to have started with the Industrial Revolution in England, in the 19th century leadership passed to the United States. So dominant was the United States in industrial technology that what we now call mass production was for many decades known as "the American system of manufacture." According to popular belief, Eli Whitney's (1765–1825) armory in Connecticut was the first manufacturing enterprise to mass-produce goods—in this case, muskets—through the use of interchangeable parts. In fact, Whitney's contribution was minimal: The parts of his muskets were not fully interchangeable, the quality of the completed muskets was poor, and they were never delivered on time. Nor was Whitney the first to try; interchangeable parts had been used in a few European enterprises long before Whitney commenced operations. In Sweden, Christopher Polhem's (1661–1751) factory built sets of standardized clock gears in the 1720s. In the early 19th century, New England manufacturers were manufacturing complete clocks from interchangeable wooden components. Especially important for the early development of mass production in the United States were federal arsenals, especially the U.S. armory at Springfield, Mass., where significant improvements were made in the design and use of machine tools.

The United States in the 19th century was a particularly fertile place for the development of mass production, for chronic labor shortages stimulated the development of machinery that took the place of workers. Also, as a new country, the United States was not burdened with encrusted traditions that dictated the way that things should be done. The accomplishments of American manufacturers were clearly in evidence at London's Great Exhibition of 1851. There, displays of Colt revolvers and other manufactured items convinced British industrialists that there was something to be learned from the young country. In the years that followed, delegations of British engineers and factory managers crossed the Atlantic to see with their own eyes how mass production was done in the United States.

The years following the Civil War saw sharp advances in mass production that threw onto the market massive numbers of sewing machines, pocket watches, typewriters, bicycles, and other accouterments of an industrial civilization. Toward the end of that century, the greatest of all consumer items emerged, the automobile. Although the first automobiles were built in small quantities using simple production technologies, within a few years sizable numbers of cars were being built in factories equipped with the latest machine tools. As with mass production in general, the key to the volume production of cars was the use of interchangeable parts. A striking demonstration of American manufacturing prowess came in 1908, when three Cadillacs were shipped to Great Britain and then dismantled. The parts were mixed together and reassembled into complete cars, which were then each driven a distance of 800 km (500 mi) with no evident problems, a feat that merited the awarding of the Royal Automobile Club's Dewar Trophy.

Coincidentally, 1908 also was the year that saw the introduction of the Ford Model-T. Conceived as a low-priced car from the outset, the Model-T became the very embodiment of a mass-produced product. Indeed, before the term *mass production* came to be firmly established in the 1920s, the word in common usage was *Fordism.* In producing Model-Ts in ever-growing numbers, Ford made heavy use of many innovative production technologies. The assembly line is the best known of these, but also of great importance were specialized machine tools, elaborate assembly jigs and fixtures, and gauges that allowed the rapid determination of whether or not a part was of the proper size. Ford's version of mass production also entailed an extreme division of labor. Ford described the work in his factory this way:

> In the chassis assembling [room] are 45 separate operations or stations. The first men fasten four mudguard brackets to the chassis frame; the motor arrives on the tenth operation and so on in detail. Some men do only one or two small operations, others do more. The man who places a part does not fasten it—the part may not be fully in place until after several operations later. The man who puts in a bolt does not put on the nut; the man who puts on the nut does not tighten it.

The Model-T Ford was produced in larger volumes than any other car until the Volkswagen finally topped its record in the 1970s. At the same time, however, mass production, Ford style, contained an inherent flaw. The Model-T could be produced in vast numbers because it was a standardized product that changed very little from year to year. But while the Model-T remained technically and stylistically stagnant, more advanced cars from other manufacturers were bidding for the consumers' dollars. By 1927, Ford realized that a static design was no longer commercially viable. A new car, the Model-A, replaced the redoubtable Model-T, but by then Ford had permanently lost the sales race to General Motors.

In recent years manufacturers have tried to combine mass production with the ability to rapidly change products, and to manufacture products that have some degree of individuality. Production technologies that allow this to be done are known as "flexible manufacturing systems." Making this possible are recent developments in the design of products, as well as the computerized control of production processes. Computerization is the key to flexible manufacturing, as different parts can be made by reprogramming the computers that control the machine tools and other production apparatus.

Mass production has brought a cornucopia of goods to the consumer; after all, mass production implies mass consumption. But in assessing the significance of mass production, it is also necessary to consider how it has affected the working lives of the people who work in mass-production enterprises. There is no escaping the fact that work in many of these facilities can be repetitious, boring, and physically and emotionally fatiguing. As one autoworker plaintively noted:

> There is nothing more discouraging than having a barrel beside you with 10,000 bolts in it and using them all up. Then you get a barrel with another 10,000 bolts and you know every one of those 10,000 bolts has to be picked up and put in exactly the same place as the last 10,000 bolts.

Automated mass production systems have eliminated the need for much of this kind of labor, but while this has been happening, mass-production methods have spread to many nonmanufacturing enterprises. As anyone who has worked in a fast-food restaurant will attest, many service jobs have strong similarities to the kind of work that was formerly confined to factories.

See also ASSEMBLY LINE; AUTOMATION; AUTOMOBILE; BICYCLE; FACTORY SYSTEM; GAUGES, INDUSTRIAL; INDUSTRIAL REVOLUTION; LATHE; LEAN PRODUCTION; MILLING MACHINE; MOTOR, ELECTRIC; SEWING MACHINE; STEAM ENGINE; TURBINE, HYDRAULIC; TYPEWRITER; WATERWHEEL

Further Reading: David Hounshell, *From the American System to Mass Production, 1800–1932: The Development of Manufacturing Technology in the United States*, 1984.

Mathematical Modeling

Our common inclination when attempting to understand something complicated is to isolate what is relevant, strip away inessentials, and clarify relationships that seem to be important. This is the essence of modeling, whether by mathematics or by other means. For example, a plastic human figure is a model that captures enough reality for children to encourage and enable their imaginations. Maps are models designed for a variety of purposes related to geographical proximity. In using a map as a model we accept that we are not physically present in the locations depicted by the map. We nevertheless expect to gain a better understanding of certain geographical features by studying the map than we could by being on location. The same is true of mathematical models, which frequently express relationships that are inconvenient or impossible to study directly.

We all engage in very rudimentary mathematical modeling when we count the number of items in a collection and then use that number in a calculation, rather than manipulating the collection directly. Another commonly used and historically early mathematical model is the Egyptian calendar, based on observing 365 days between occasions when Sirius, the Dog Star, rose just before the sun, precursive to the annual flood of the Nile.

Modern mathematical modeling is more conceptual than the previous two examples and frequently is not even numerically based. Leonhard Euler (1707–1783), a Swiss-born mathematician, created the branch of mathematics called *graph theory* by mathematically modeling, and subsequently analyzing, a popular problem of his time. In the town of Königsberg, Prussia, over the Pregel River, there were seven bridges connecting two islands with each other and with the two banks. No one seemed to be able to find a way to traverse each bridge only once while going for a walk. At the same time, no one could understand why it couldn't be done. The configuration of bridges included two from one of the islands to each side of the river, one from the second island to each side of the river, and one between the two islands. Euler's model of the situation represents each of the two islands and each of the two river banks by a large dot on paper, and arcs between dots to correspond to each of the seven bridges, thusly:

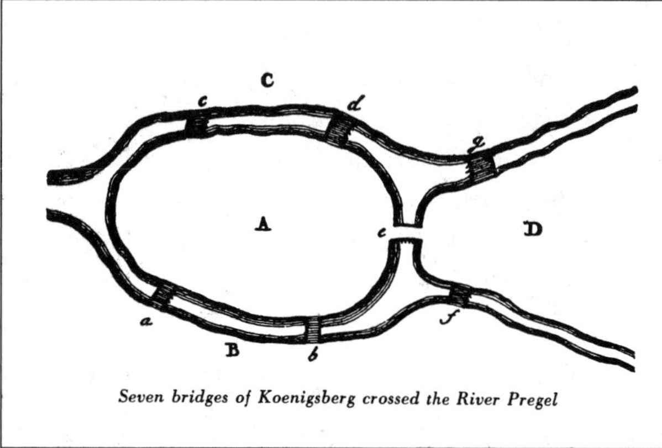

Seven bridges of Koenigsberg crossed the River Pregel

A contemporary diagram of the seven bridges of Koenigsberg.

This model is an example of what mathematicians have come to call a "combinatorial graph," with four "vertices" or "nodes" and seven "edges." Euler used it as an aid to his analysis: First, for a walker to be able to begin and end a successful tour of bridges (traversing each exactly once) at the same landmass—that is, at the same node—Euler reasoned that it would be necessary for each and every node to have a even number of edges attached to it. This is because the tour would have to include arrival at and departure from each node (landmass) along distinct edges (bridges). The node that is the common beginning and ending point for the tour is as subject to this reasoning as the rest. But Euler's model clearly shows that every one of the four nodes has an odd number of incident edges. (The nodes for the river banks and second island each have three, while the one for the first island has five.) So a successful tour must begin and end at different nodes, and *only* those two nodes could have an odd number of incident edges, since the walker would necessarily leave each of the other nodes after entering them, along a pair of distinct edges, as before. Since Euler's model has four nodes with odd numbers of incident edges, it must be the case that no successful tour is possible!

One sees in Euler's model of the Königsberg Bridge configuration the elements of a successful model: abstraction of just those elements of the situation that permit successful use of the model, with no frills. For example, his model contains no representation of distances between locations, which might make it useless for some other purposes, but has no bearing on the problem at hand. Other, more complicated, combinatorial graph models have been used to advantage in order to model communications networks, pipelines, and transportation (highway or rail) networks. Somewhat more surprising, perhaps, is the use of the models to describe progress in urbanization and civilization. In this situation, nodes of the graph correspond to various combinations of economic, technological, and political advance that describe the progress of civilization, such as the development of agriculture and metallurgy.

We can deal systematically with processes whose outcomes are uncertain by using probabilistic models. For example, when we bet on the spin of a roulette wheel with the typical arrangement of outcomes numbered from 1 to 36, and 0 and 00, we should realize that even with a fair table the house can win, on average (that is, probabalistically), about 5 percent of the time, which is the expected percentage of the 38 possible outcomes represented by the two outcomes, 0 and 00 ($^2/_{38}$ = 0.0526 = 5.26 percent). Queuing models use probabilities of waiting and servicing times to help us understand the differences, for example, between a supermarket situation where each checkout station has its own queue and a bank situation where many tellers serve a single queue. Complex systems like nuclear-generating stations are understood through probabalistic models of reliability. The probabilities of failure of individual components (valves, switches, containers, etc.) are balanced against redundancies of those components and their systems to determine whether the overall system has an acceptably small probability of catastrophic failure.

Statistical models help us understand large collections of data by analyzing relatively small samples. For example, the exit polls done on election days give a small sample of the outcomes and circumstances of the entire population of voters. Statisticians in the polling groups predict results of the election with reasonably well-understood margins for error, based on their analysis of the sampled data using statistical models.

In 1798, Thomas Malthus (1766–1834) wrote in his *Essay on the Principle of Population*, "Population, when unchecked, increases in a geometrical ratio." This is a mathematical understanding, or model, of a natural phenomenon. One may apply the point of view of the differential calculus to rewrite the increase by geometrical ratio as a differential equation. That equation expresses the consequence of assumptions that the population growth will be "unchecked," either by, for examples, contraception or by limitations that large populations tend to bring upon themselves, such as epidemics or wars. The modeling of populations in contemporary times includes not only consideration of animal and plant populations but has also been applied to consider such things as receptor sites on cells in order to study the immune systems of animals.

Mathematical models are especially helpful when a model, such as the Malthusian differential equation, applies to a completely different situation. For then the mathematical analysis that was done for the first application of the model will apply automatically for second and subsequent applications with little or no additional effort. We understand that in a radioactive decay, the mass of a substance decreases in a geometrical ratio. Despite "increase" (of population) being replaced by "decrease" (of mass), the results are very similar differential equations, whose analyses are very similar, except that the former predicts unmanageably huge populations while the latter predicts negligibly small quantities, after long times.

The differential equations of population growth and radioactive decay deal with the *time* rates of change of population and mass, respectively. Differential equations for heat and gas diffusion bring into play not only time rates of change but others also, such as *spatial* rates of change (e.g., the change of heat concentration per unit of change of distance, over very small distances). It turns out that these models of physical diffusion processes are very similar to the Black-Scholes model that is widely used to model financial markets, and thus our understanding of one is helpful to understanding the other.

We have seen mathematical models used to clarify not only numerical problems—such as calendars, populations, gas concentrations, and stock prices—but also geometric configurations such as the Königsberg bridges. The subject of *game theory* was developed in World War II to analyze human competitive relationships. It has become a powerful tool of economic analysis, as was recognized by the awarding of the 1994 Nobel Prize to John C. Harsanyi, John F. Nash, and Reinhard Selten, three economists who made important contributions to its development.

Today, spreadsheet software allows any moderately skilled user to use "what if?" analysis. This powerful, yet easily accessible, capability is a form of mathematical modeling. For example, one can undertake home budget analysis by investigating the effects of various percentages of income being allocated to current spending and saving for retirement, children's education, or major purchases. Then one asks "what if" the allocation percentages are changed, or the interest earned on savings changes.—R.E.

See also CALCULUS, DIFFERENTIAL AND INTEGRAL; "NORMAL ACCIDENTS"; RISK EVALUATION; STATISTICAL INFERENCE

Further Reading: Frank R. Giordano and Maurice D. Weir, *A First Course on Mathematical Modeling*, 1985.

Mathematics' Role in Science

Most students of late 20th-century science would probably agree with a comment made by Immanual Kant 200 years ago: A domain of knowledge is scientific principally to the extent that it has become quantitative and capable of formulation in terms of testable mathematical laws. It has not always been so. Aristotle, arguably one of the greatest scientists of antiquity, argued that though mathematics was a viable science in its own right, it was not applicable to understanding the natural motions of bodies. This was true, he claimed, because in arriving at the geometrical figures that constituted the subject matter of Greek mathematics, mathematicians had begun with the objects of ordinary experience and eliminated all of their features (such as their colors, hardness or softness, warmth or coldness) except for their figures. If mathematical entities were derived from abstracting figures from bodies and ignoring all else, how could one expect mathematics to answer questions about what had been ignored? As late as the 19th century, many students of medicine and the life sciences explicitly denied that the analytic techniques associated with mathematics were appropriate to describe the complex phenomena associated with living beings. But with the rise of molecular biology, population genetics, and medical fields such as epidemiology, mathematics has become pervasive as the principle language of scientific discourse even in biology and medicine.

The question of *why* so many successful explanations of natural phenomena are mathematical, is, however, likely to generate substantial disagreement. Over time, scientists and those who study science have answered this question in at least three fundamentally different ways, each of which has its advocates at present.

Perhaps the oldest self-conscious position with regard to why scientific knowledge should have a mathematical form is associated with the name of Pythagoras (c. 582 B.C.E.–497 B.C.E.), whose reputed answer was that mathematics works to describe the world because the world is, in its very being, mathematical. His followers, according to Aristotle, "became convinced that the elements of numbers are the elements of everything and that the whole Cosmos is harmony and number." A form of this position was expressed by many early modern scientists, including Galileo (1564–1642), Johannes Kepler (1571–1630), and William Petty (1623–1687) when they insisted that God created the world according to number, weight, and measure. For such scientists, the task of the scientist was and is to discover those mathematical relationships that are expressed in the natural world itself, usually by isolating key aspects of experience and ignoring complicating features. Galileo's discovery of the laws of free fall—that $v = kt$, and $s = \frac{1}{2}kt^2$ (where v = velocity at time t; if $v = 0$ when $t = 0$, k is a constant of proportionality, and s is the distance fallen)—by initially ignoring the resistance of the medium through which objects fall is the paridigmatic scientific discovery from this perspective. Today this point of view is most often found among theoretical physicists.

A second way of understanding why mathematics works in science is found among ancient astronomers such as Ptolemy (fl. 127–145 C.E.), medieval philosophers such as William of Occam (c. 1285–c. 1347), Enlightenment followers of Etienne Condillac (1715–1780), and even 20th-century positivists. None of these groups would claim that the true nature of things and their causal relationships are, strictly speaking, knowable. The best that human beings can do is to describe their experiences in clear and rigorous ways, using categories that they construct for the purpose. We choose descriptive categories that can be quantified because mathematical systems allow rigorous deductions of consequences, even though we may be unable to offer an essential causal explanation of them. Thus, in antiquity, any mathematical scheme for calculating the observed positions of celestial bodies that "saved the appearances" was understood to be satisfactory by astronomers, even though it did not pretend to offer an account of why the bodies moved as they did.

In the 19th century, the physicists Joseph Fourier (1768–1830) and James Clerk Maxwell (1831–1879) both held that scientists do not study what the world *is*, but what it *is like*. From this perspective the relationships among natural phenomena are held to be "like" the relationships among the elements in a mathematical theory. Both Fourier and Maxwell spoke of the situation as one in which there are analogies between the relationships found among phenomena in the natural world and the relationships among the elements in a mathematical system, which is a human creation. Today, most of those who hold such views would probably say that the mathematical system is chosen to "model" or "represent" the natural world.

Quite recently, students of science who focus on the social dimensions of scientific activity have become interested in asking whether there are social circumstances that might help to account for the preferences shown by some groups for mathematical laws and quantification. Good evidence now seems to exist to show that, especially in the social sciences, quantification functions as a strategy of impersonality or objectification that allows persons or groups in relatively weak social circumstances to compel assent to their views or at least to avoid criticism for them. For example, by treating the value of a human life mathematically as the present value of the future earnings of a person, bureaucrats have been able to offer assessments of the "costs" of various policies that create at least the illusion of being unbiased. Policies based on such calculations can then be implemented without being subject to claims of bias and unfairness. Some scholars have argued that quantification works in the natural sciences in much the same way. This view may be held along with either of the first two, or it may be held independently, but it is most frequently linked to the positivist or nominalist standpoints, which argue that our intellectual categories are human choices rather than natural categories imposed by a world independent of human needs.—R.O.

See also GRAVITY; MATHEMATICAL MODELING; RISK EVALUATION

Further Reading: Theodore M. Porter, *Trust In Numbers: The Pursuit of Objectivity in Science and Public Life*, Princeton: Princeton University Press, 1995.

Maxwell's Equations

In his paper "A Dynamical Theory of the Electromagnetic Field," which appeared in *Transactions of the Royal Society of London* in 1864, James Clerk Maxwell (1831-1879) first synthesized all of what was known of electrostatics and electrodynamics by means of a set of definitions of key concepts and four sets of partial differential equations. In his two-volume *A Treatise on Electricity and Magnetism* of 1873, he went on to explore the application of these equations to a wide range of phenomena. Maxwell's equations encompass some of the most fundamental concepts of physics. In their generality and significance, they are the equal of Newton's laws of motion. But unlike Newton's laws, Maxwell's equations have not required modification in connection with relativity and quantum physics as has been the case for classical mechanics.

For reference purposes, Maxwell's equations can be written in traditional vector notation as follows:

$$1 - V^2/c^2$$

Suppose that the electrical force acting on a unit charge at any point in any medium is D. D, which is called the electrical field, may be given by $D = \varepsilon E$, where ε is a function of the kind of medium present where D is acting, and E is the force that would be experienced by a unit charge in a vacuum at that point. Suppose also that B is the magnetic force acting on a unit magnetic pole at any point in any medium; then B, which we call the magnetic field, equals μH, where μ is a function of the medium at the point in question and of time, and H is the magnetic force that would act on a unit pole in a vacuum at that point. Finally, let ρ designate the charge density at any point and j designate the electrical current at any point, and let c be the speed of propagation of electrical and magnetic forces. Given these definitions of D, E, B, H, c, ρ, and j, the following four equations hold:

1. $\nabla \cdot B = 0$
2. $\nabla \cdot D = 4\pi\rho$
3. $\nabla \times E = -1/c\ dB/dt$
4. $\nabla \times H = 4\pi j + 1/c\ dD/dt$

The first two of these equations express the fact, articulated by Dennis Poisson (1781–1840) in 1813, that electrical and magnetic forces, like gravitational forces, are proportional to the charge or pole density, and decrease in intensity proportional to the square of the distance between two charges or poles. Also, opposite poles or charges attract and, conversely, like magnetic poles and electrical charges repel. In such cases, if all of the forces produced by some distribution of charges or poles is summed over a surface surrounding that distribution, the result will be proportional to the total charge or magnetic pole strength inside the surface. Equation 1 tells us that this value will always be zero for the sum of magnetic forces surrounding a distribution of magnetic poles. This is a consequence of the fact that magnetic poles always occur in pairs of opposite kinds, so that in any enclosed space, the number of unit south poles will be identical to the number of unit north poles.

The third equation expresses the fact, discovered by Michael Faraday (1791–1867) in 1831, that when a magnetic field changes over time, it produces a force on an electric charge that is proportional to the rate of change of the magnetic field, but is in a direction per-

James Clerk Maxwell (courtesy Institute of Electrical Engineers).

pendicular to that of the magnetic field. This phenomenon, called *electromagnetic induction*, makes it possible to produce electric currents in devices like dynamos or generators. And this fact, in turn, underlies the huge and economically vital electrical-power industry.

The fourth equation represents Maxwell's most original contribution to electromagnetic theory. In the spring of 1820, Hans Christian Oersted (1777–1851) demonstrated that when an electrical current is passed through a wire, magnetized needles nearby will align themselves at right angles to the direction of the current. By September 1820, two French experimenters, Jean Baptise Biot (1774–1862) and Felix Savart, had produced a mathematical expression for the magnetic force exerted on a unit pole by a small current. By the end of December, André Marie Ampère (1775–1836) had shown that Biot's and Savart's results could be expressed in a manner equivalent to $\nabla \times H = 4jB$, which is nearly, but not quite, the same as Maxwell's fourth equation (in 1821, Michael Faraday produced the first electric motor using this principle). Maxwell arrived at the final form of the fourth equation when he sought to provide a mechanical model to account for how magnetic and electrical forces produce and are produced by electrical currents in a solid medium. In working out the details of this model, he discovered that when a normal current of electricity was produced by a magnetic field, according to his model there was an unexpected change in the electrical field, which he called the "displacement current." This was mathematically expressed as $1/c\ dD/dt$, where c was the speed of propagation of the magnetic or electric field.

Maxwell's equations had implications that went well beyond the description of electrical and magnetic fields. Not only were electricity and magnetism interconnected; also the magnetic field produced by an oscillating electrical charge radiated outward at a constant speed. Calculations based on Maxwell's equations showed that the speed at which the magnetic field radiated was almost exactly the same as the speed of light, as it was known at the time. The inference was clear: Visible light was a form of electromagnetic radiation. Moreover, Maxwell formed the idea that there were hitherto unknown electromagnetic radiations on either side of the light spectrum. In 1888, Heinrich Hertz (1857–1894) endeavored to test this hypothesis, and in order to do so, he created the first radio transmitter and receiver.—R.O.

See also ALTERNATORS AND GENERATORS; QUANTUM THEORY AND QUANTUM MECHANICS; RADIO; RELATIVITY, GENERAL THEORY OF; RELATIVITY, SPECIAL THEORY OF; SPEED OF LIGHT

Mechanics, Newtonian

Sir Isaac Newton's (1642–1727) *Philosophiae Naturalis Principia Mathematica* ("The Mathematical Principles of Natural Philosophy," usually known simply as the *Principia*), published in 1687, contained in its introductory materials the foundations of a system for accounting for the motion in time and space of physical bodies. Such a system we call a *mechanics*, and that system of mechanics contained in the *Principia*, is, strictly speaking, "Newtonian mechanics." It has, however, become conventional among Anglophone scientists and historians of science to apply the term Newtonian to other and often more general formulations of mechanics that are logically consistent with, but not necessarily identical to, Newton's formulation. Though more properly designated "classical" to oppose them to the "relativistic" systems of mechanics associated with Albert Einstein, the systems of quantum mechanics associated with Werner Heisenberg (1901–1976) and Erwin Schroedinger (1887–1961), or to the Aristotelian systems that preceded them, we will accept the convention that identifies classical and Newtonian mechanics, offering occasional comments to suggest conceptual variations.

Newton's formulation of mechanics can be characterized in modern terminology as incorporating a set of basic assumptions, definitions, and laws derived from experience. First, it presumes the existence of an absolute and completely homogeneous and isotropic (with the same properties in all directions) space, which is essentially the space defined by Euclidean geometry. Bodies can be thought of as inhabiting absolute positions and having absolute motions within this space. However, Newton acknowledged that the positions and motions that we measure are always positions and motions relative to some particular body or collection of bodies. Similarly, Newton posits the existence of an absolute and uniform time that is completely independent of both place in absolute space and of bodies, yet he acknowledges that the times we measure are measured by the perceived and therefore necessarily relative motions of the heavenly bodies. Both the presumption of an isotropic absolute space whose characteristics are independent of the bodies in it and the presumption of time that is independent of place and relative motions must be discarded in "relativistic" mechanics.

Next, Newton defines the quantity of matter—which we now call "mass"—of a body as a product of its density and volume, and assumes that this is an invariable characteristic of matter, an assumption that must again be abandoned in relativistic mechanics. The quantity of motion—we say *momentum*—of a body is then defined as the product of the mass and velocity of the body, where the velocity v is a directed quantity with components along the three orthogonal dimensions of Euclidean space. Forces are then defined as actions exerted on a body in order to change the motion of the body and as proportional to the change in motion that they generate in a given time.

Suppose now that we designate the force acting on a body at any instant by the symbol F, where F is understood to be a directed magnitude with components along each of the three orthogonal dimensions of Euclidean space. We designate the mass of the body by the symbol m, where m is a scalar quantity (i.e., a quantity with a magnitude but no direction). And we designate the change in velocity of the body per unit time, or acceleration, by a, where a is again a directed (physicists usually say *vector*) quantity. If it is the case that no unforced changes of motion take place (Newton calls this his "first law of motion" and expresses it by saying that "Every body perseveres in its state of rest or of uniform motion in a straight line, unless it is compelled to change that state by forces impressed thereon."), then we can express the set of definitions above as $F = ma$, which is Newton's "second law of motion."

The final foundational law of Newtonian mechanics, "To every action there is always opposed an equal reaction: or the mutual actions of two bodies upon each other are always equal, and directed to contrary parts," is simply a reformulation of the discovery articulated simultaneously by John Wallis (1616–1703), Christopher Wren (1632–1732), and Christian Huyghens (1629–1695) in 1669, that in every interaction between two bodies, the total quantity of motion, or momentum, is conserved.

In principle, all mechanical problems capable of solution can be solved by the application of Newton's three laws (some problems, including that of what happens when three or more bodies simultaneously collide with one another have no unique Newtonian, or classical solution), assuming that one can discover the initial motions of the bodies under consideration and the forces acting on those bodies.

Newton proceeded in the *Principia* to solve a huge range of problems associated with celestial mechanics, especially the motion of the bodies contained within the solar system, by establishing that each particle of each body may be presumed to exert an attractive force on every other body that is proportional to the product of the two masses divided by the square of the distance between them. This force we call the *gravitational force*, and Newton established that all bodies, everywhere, may be considered to exert gravitational forces upon one another.

Some continental natural philosophers were disturbed by the fact that Newton offered no way of measuring forces that was independent of the accelerations that they produced. Given this situation, Newton's second law seemed to have no empirical content. Rather it seemed to be a tautology—i.e., true by virtue of the definitions of its terms. These natural philosophers sought a way to formulate mechanics that produced Newton's obviously correct results without introducing the concept of force. This was accomplished in its most general way in 1788 by Joseph Louis Lagrange (1736–1813), who was able to demonstrate that since Newtonian forces could always be derived from some function of interbody distances called a *potential function*, one could always bypass the use of forces by using potential functions alone.

Isaac Newton (courtesy Institute of Electrical Engineers).

Though Lagrange's formulation of classical mechanics is logically equivalent to Newton's formulation, some physicists prefer it because it does not posit the existence of quantities that are not directly measurable. Against this point of view, some natural philosophers point out that though force may not be easily measured, the use of this concept is valuable because it requires us to recognize that the very existence of bodies becomes known to us only because something resists our efforts to move ourselves through space. Thus, in a sense, our most primitive experiences are of forces, from which the very notion of body is only an inference.

Book 3 of Newton's *Principia* has had a particularly broad cultural importance. On the one hand, Newton's ability to find one single simple mathematical law of gravitational attraction that could solve a huge range of outstanding problems in celestial and terrestrial physics led many Enlightenment figures to the optimistic assumption that virtually any subject could be illuminated by applying Newtonian methods. Some came to the even more extreme notion that in almost every domain, phenomena could be accounted for through some single explanatory principle. David Hartley in England and Etienne Condillac in France, for example, openly sought to apply Newtonian techniques to ground all human psychology in a single law of the association of ideas, while Adam Smith sought to account for all economic phenomena through the single mechanism of the market. Such a vision of unifying simplicity remains in modern physics in the visions of theorists like Steven Weinberg and experimentalists like Leon Lederman, who continue to seek a grand unifying "Theory of Everything" that would bring together all of quantum mechanical theory and the general theory of relativity in order to explain all phenomena from the Big Bang at the origin of the universe to the present under a single 10-dimensional equation.

On the other hand, Newtonian mechanics as applied to celestial objects led to the development of some equations that could not (and cannot now) be solved in closed analytic form. And when Newton offered approximate solutions, they led to results that seemed to have important religious and political implications. Proposition 10 of Book 3 of the *Principia*, for example, strongly suggested that the solar system was not stable, and that given its then presumed age of approximately

4,000 years, Jupiter and Saturn would have crashed into the sun if God had not directly intervened in the natural course of events to adjust the planetary motions. Some religious figures argued that Newton had thereby provided incontestable evidence for God's miraculous and providential activity in the world. This argument also allowed Anglican clergy to justify the English civil wars and Glorious Revolution as providential interferences to keep England from becoming Catholic, while they remained political conservatives opposing all merely human rebellion against political authority. Gottfried Leibniz (1646–1716) found the whole providentialist argument from Newtonian mechanics appalling on the grounds that a God wise and powerful enough to create the world would not create it so imperfect that it demanded occasional repair.

Late in the 18th century, Lagrange was able to show that a more accurate approximate solution to the problems of Saturn and Jupiter's motions suggested that they might be perfectly periodic, thereby undermining both continental rationalist opposition to Newton and the support that Newtonian mechanics had seemed to offer providential Christianity. Finally in 1919, observations of the motion of Mercury, which had never been completely accounted for by Newtonian mechanics, supported Einstein's General Theory of Relativity, which has superseded Newtonian mechanics in accounting for cosmic scale phenomena.—R.O.

See also BIG BANG THEORY; MANY-BODY PROBLEM; QUANTUM THEORY AND QUANTUM MECHANICS; RELATIVITY, GENERAL THEORY OF; RELATIVITY, SPECIAL THEORY OF; SOLAR SYSTEM, HELIOCENTRIC

Memory, Computer

Memory, as in "computer memory," comes in many different forms and uses many different kinds of technology depending on the use to which it is put by the computer designer. Computers usually have at least three different memory devices that make up the total memory capacity of the machine.

The first is the smallest in capacity, the most expensive, and the fastest in terms of the machine's ability to access the information and to modify its contents. This is often referred to as the main memory, or often "core" memory—a term that comes from an older type of technology used to implement it. It is where the program, or software, and the data are usually stored while the machine is actually executing the instructions that make up the program. It must be extremely fast in order for the computer to find the next instruction, execute it, and store the results away at the electronic speeds available in the rest of the machine. It is not uncommon for this type of memory to be able to access any particular item of information or instruction in less than 100 nanoseconds (100 billionths of a second—the time it takes a beam of light to travel only 30 meters [100 feet]). While main memory capacities vary greatly depending on the price of the machine and the type of uses for which the computer was designed, the usual personal computer will have between 1 and 16 megabytes (1 million to 16 million bytes) of this very high-speed device.

The second type of memory on a computer is usually known as a *hard disk*. It is of much greater capacity, slower speed, and consequently lower cost than the main computer memory. It is usually used to store programs and data, not while actually being used by the machine but to have them rapidly at hand in case the user (or the computer) needs to use them (at which time it would copy them into the computer's main memory for actual use). This memory is usually constructed from one or more metal disks, coated with magnetic material, that rotate at high speed. The information is recorded on the magnetic coating of the disk in much the same way as music is recorded on the magnetic coatings of the tape in a music cassette. The fact that it is a disk and rotating at high speed means that the individual programs or items of information can be accessed in reasonably short periods of time, generally on the order of a few hundredths of a second or less (still many tens of thousands of times slower than the main memory). Hard disks, because they are used to store the programs and data items that may sometime be needed in the computer's main memory, are often of very high capacity. It is not uncommon to find hard disks capable of storing many hundreds of millions of bytes (hundreds of megabytes) or even billions of bytes (gigabytes) of information.

The third type of memory on a modern personal computer is usually the floppy disk. In principle it operates exactly the same way as the hard disk but is designed for long-term storage of programs and ease of carrying programs or data from one place to another. It is usually capable of storing between 1 and 2 million characters (depending on the disk and the disk drive it is used in) and is very slow when compared with other forms of memory—M.W.

See also BINARY DIGIT; BYTE; CD-ROM; COMPUTER, MAINFRAME; COMPUTER, PERSONAL; DISK STORAGE

Mercury Project

Early planning for an American manned space program began in October 1957 when a group from the National Advisory Committee for Aeronautics (NACA) began to discuss what might come after the X-15 experimental aircraft, which was capable of flying only to the edge of space. What began as a rather academic exercise moved into high gear after the successful launching of the Sputnik satellite on Oct. 4, 1957, made it seem that the United States was falling seriously behind the Soviet Union in "the space race." In 1958, the federal government transformed NACA into the National Aeronautics and Space Administration (NASA) and gave it the responsibility for organizing an intensified U.S. space effort. Even more governmental emphasis on space came with the presidency of John F. Kennedy, as demonstrated by Kennedy's successful request for an additional half-billion dollar funding for NASA for the 1962 fiscal year.

The initial round of space flights was given the name Project Mercury. Astronauts were trained, a space capsule was designed, and many problems associated with launching and then retrieving the capsule were addressed. The rockets that propelled the capsules into space were originally intended for military purposes. The only rocket capable of putting a man in orbit was the Atlas intercontinental ballistic missile, but concerns about its reliability dictated that the smaller and less-powerful Redstone rocket was used for the first suborbital flights, four in number. These began with an unmanned launch that got only a few inches off its launching pad, followed by the successful flight into space of a chimpanzee named Ham on Jan. 31, 1961. While NASA was making preparations for its first space launch with a human being aboard, the Soviets sent Yuri Gagarin into one orbit around the Earth on Apr. 12, 1961. On May 5, 1961, Alan Shepard in Mercury 3 made the first American flight into space, a 15-minute flight that reached an altitude of 100 miles above the Earth.

Another flight using a Redstone booster put Virgil (Gus) Grissom into space on July 21, 1961. This represented another step forward for the Mercury program, but it was quickly eclipsed on Aug. 6, 1961, when the Soviets put Gherman Titov into 17 orbits around the Earth, for a total time in space of more than 25 hours. The first American manned orbit on Feb. 20, 1962, was of considerably shorter duration, only three orbits, but it gave John Glenn the distinction of being the first American to orbit the Earth. Glenn's flight was the first to use the Atlas rocket as a launch vehicle. Three more Mercury launches were made, the last (May 15–16, 1963) putting Gordon Cooper into 22 orbits and a duration of 34 hours in space.

Project Mercury demonstrated that the United States was capable of putting people into space and bringing them back safely. Its total cost, $400 million, was substantial, but it was to be dwarfed by the more ambitious projects that followed, culminating with a series of lunar landings that began in 1969.

See also APOLLO PROJECT; GEMINI PROJECT; MISSILE, INTERCONTINENTAL BALLISTIC (ICBM)

Metal Fatigue

The widespread use of iron, steel, and other metals in the 19th and 20th centuries allowed the construction of larger and more complex structures. On occasion, however, one of these structures would fail for no apparent reason. The problem became much more evident in the 1930s, when significant numbers of aircraft began to be made from aluminum. With no warning, wings would snap off and propellers would shear, usually with fatal results. The cause, as it turned out, was metal fatigue.

Metal fatigue is the weakening of a metal part due to repeated cyclical movement such as bending or torsion (twisting). This usually happens as a result of the movement of a component in the course of its operation, such as the slight flapping of an airplane's wings when the plane is airborne. Another form of metal fatigue known as *thermal fatigue* can be brought on by temperature changes that cause cyclical expansion and contraction, as happens with the starting up and shutting down of gas turbine engines.

The cause of metal fatigue is still not fully understood; changes taking place in the microstructure of a metal are suspected, but exactly what happens is still a matter of conjecture. It is known that fatigue begins with tiny cracks at "nucleation sites," places where stresses are concentrated or the material is weak. From here, repeated loading causes the cracks to grow and consolidate into a fatigue crack.

A quantitative assessment of a metal component's vulnerability to fatigue is provided by subjecting a test specimen to different stress levels in order to find out how many cycles are required to bring on failure. The resulting data are then depicted on a graph where the amount of stress is plotted against the logarithm of the number of cycles. The result is an *S-N* curve, where *S* indicates the amount of stress and *N* indicates the number of cycles. For some metals, notably steels and titanium alloys, the graph will usually include a line parallel to the axis that shows the number of cycles; this is known as the *endurance limit*, for no fatigue occurs below it. In actual service conditions, variables in addition to amount of stress and number of cycles may have to be taken into account. These include the regularity or randomness of the occurrences of stress, the magnitude of peak stresses in the latter case, and operating temperatures. Engineers also have to take account of the fact that the performance of the metal is only one variable. Stresses set up in the course of fabrication may be subsequently manifested in the form of crack formation. Faulty welding may also have the same consequences.

The dangers of metal fatigue were dramatically highlighted in the early 1950s, when the first jet airliner, the British de Havilland Comet, went into service. In early 1954, less than 2 years after it had first begun to carry passengers, a Comet broke up in midair over the Mediterranean Sea. A short time later, a second Comet went down over the coast of Italy. Prior to these crashes, two other accidents with fatalities had occurred, but these had been attributed to structural failure brought on by severe weather conditions. In fact, the structural failures were the result of metal fatigue. This was demonstrated when the fuselage of a Comet was immersed in a large tank. The plane's cabin was filled with water and then emptied to simulate the cycle of pressurization and depressurization that it would experience in the course of actual operation. After 3,000 pressurization and depressurization cycles, the investigators found severe metal fatigue in the form of a 2.4-m (8-ft) crack that extended from an escape hatch window through the frame of another window. It was then surmised that similar cracks in the ill-fated Comets led to rapid structural failures and tragic loss of life.

In response to the hazards of metal fatigue, the aircraft industry substantially modified some of its design and construction practices, among them the use of thicker aluminum skins that were provided with crack stoppers welded in the fuselage interior. Static testing procedures were stiffened, including subjecting the airframe to 50,000 pressurization cycles in order to simulate the number of takeoffs and landings that might be expected in the course of an airplane's operational life.

A better understanding of the causes of metal fatigue can lessen its occurrence, but there is no way to prevent it altogether. Where a failure would be catastrophic or even just inconvenient, it is necessary to periodically ensure the integrity of a structure through the use of nondestructive tests. This can be as simple as visual inspection, or it can encompass sophisticated testing procedures that employ X rays, magnetic fields, or ultrasonic imaging. None of these is perfect, however, and engineers always have to include substantial safety margins when designing components or subassemblies subject to metal fatigue.

See also ALUMINUM; LOGARITHM; TURBINE, GAS; ULTRASOUND IMAGING; X RAYS

Further Reading: Henry Petroski, *To Engineer Is Human: The Role of Failure in Successful Design*, 1985.

Metal Stamping

Metal stamping is a common industrial practice for making parts out of sheet metal. Through stamping, a flat sheet can be formed into a car fender, the structural member of a building, or the housing of a domestic appliance. Stamping is distinct from forging in that the metal to be formed is not first heated. The sheet (steel, brass, aluminum—just about any malleable metal can be stamped) is placed between dies and pressure is applied. Depending on the kind of die used, the stamping operation can pierce or cut the metal, or it can give it a three-dimensional shape. In the latter case, enough pressure is applied to give the material a permanent set, but not so much as to rupture it. When something is formed through stamping, it is necessary to take into account some amount of "springback" in the formed part, as well as deformations, especially changes in the thickness of the material. A part may stamped in a single operation, or it may go through a series of stampings with different dies, getting closer to the final shape with each successive operation.

A stamping press consists of two major components. The lower part, which contains the "female" die, is fixed in place. The upper part holds the "male" part, which is pressed down by hydraulic pressure or through an electrically driven mechanical linkage. Depending on the size of the press, the pressure applied ranges from less than 900 kg (1 ton) to 45,000 metric tons (50,000 tons). A press operation may include a number of distinct functions, such as shaping, punching, and cutting. These functions are not usually done in a single pressing operation but in several distinct ones by means of a double- or triple-acting press.

The production of stamping dies is a complex process that calls for a high level of precision. Some stamping is done with commercially produced standard die inserts, but many applications require the manufacture of special dies. These dies may be cast to an approximation of their final pattern, and then cut to their final shape with a milling machine. Sometimes the milling process is guided by a tracer attachment that follows the shape of a wooden model die. The final finishing of the die often requires a fair amount of handwork with scrapers, grinders, and polishing cloths.

Metal stamping developed rapidly during the latter part of the 19th century to meet the needs of expanding manufacturing industries. One of the most important of these was the bicycle industry, where many manufacturers cut their production costs by substituting stamped parts for more expensive forgings. Although some manufacturers of high-priced bicycles derided stamped-metal parts as "cheap and nasty," many firms successfully used stamping for a wide range of components: frame lugs, sprockets, brackets, steering heads, wheel hubs, and handlebars. Manufacturers learned how to create complex components through the use of multiple pressing operations with intermediate annealing processes to soften the metal. Pressed-steel parts may not have been as strong as their forged counterparts, but they were strong enough to suit the needs of the buying public, most of whom were willing to sacrifice very high quality for low price. Stamping helped to create the mass-production bicycle, but in the long run what was more important was the transfer of this technology to the automobile industry. Automobile production would have been significantly lower and prices considerably higher in the absence of metal stamping.

See also BICYCLE; FORGING

Methadone

Since its creation in 1898, many people have been addicted to heroin. According to a survey conducted in the early 1990s of drug use among the American population, nearly 3 million people have tried heroin. The number of actual addicts is a matter of some debate, but given the highly addictive nature of heroin, it is certainly substantial. The personal and social problems caused by heroin use have motivated a number of strategies to end or at least diminish addiction. One of these entails the use of a methadone as a less-harmful substitute for heroin.

Methadone was synthesized in Germany in 1943 because World War II had disrupted the supply of natural opium from which opiate analgesics were derived. Its time and place of origin are revealed in the original name for the drug, Dolophine, a name bestowed in honor of Adolph Hitler. Oddly, one pharmaceutical company continues to use Dolophine as a trade name, and methadone tablets sometimes carry the street name of "dollies."

Methadone operates on the same neural receptors as heroin, but it does not provide the pleasurable sensations of that drug. It also blocks the feelings of euphoria brought on by the use of heroin. However, like heroin, methadone is an addictive drug, especially when injected. For this reason, methadone doses are usually administered orally, and the tablets are combined with insoluble substances so that they cannot be made into an injectable substance. The chief therapeutic benefit of methadone is that it allows an addict to get off heroin without experiencing painful withdrawal symptoms. For this to happen, methadone must be taken orally every day.

The use of methadone for the treatment of heroin addiction was initiated at Rockefeller University in 1964 by Vincent Dole and Marie Nyswander Dole. The success of their program led to the formation of methadone clinics throughout the United States. By 1987, methadone clinics in the United States were serving between 70,000 and 75,000 heroin addicts.

Methadone has achieved mixed success as a treatment for heroin. It has been estimated that 30 percent of clinical patients being maintained on methadone continue their use of opiates. Many former heroin addicts have stopped using heroin, but they are now addicted to methadone. In fact, there are methadone addicts whose addiction followed the use of illicit methadone that had been illicitly obtained. The unsupervised use of methadone can be fatal because therapeutic doses of 40 to 100 mg are close to toxic doses of 100 to 200 mg Even when taken under controlled conditions, methadone can produce some side effects, most notably nausea, constipation, and loss of sex drive.

Many heroin addicts have been successfully treated with methadone, but most drug experts agree that by itself methadone is of limited value. In successful methadone clinics, the administration of methadone is coupled with the provision of counseling and employment services. On a societal level, heroin addiction is symptomatic of a number of social and economic problems. To put excessive reliance on methadone is to fall into the trap of hoping that technology will solve problems that are nontechnological in nature.

See also HEROIN; TECHNOLOGICAL FIX

Microphone

A microphone converts sound waves into electrical signals that can be amplified and then converted back into sounds of higher volume. Microphones are used in telephones, tape recorders, public address systems, sound-recording studios, surveillance systems, and radio and television broadcasts. The part of the microphone that turns the sound waves into electrical signals is known as a *transducer*. Over the years, microphones have been built around hundreds of different types of transducers, although only a few have been commercially used.

Microphones fall into three basic categories: pressure, gradient, and wave. Microphones can also be categorized by the type of transducer they use: carbon (granules); magnetic (single-pole and double-pole armature); dynamic (moving coil, straight wire, or ribbon); electrostatic (condenser); piezoelectric (crystal); electrostrictive (ceramic); and electret (foil).

In a pressure microphone, changes in sound wave pressures drive the transducer. The transducer is activated by sounds coming from any direction; for this reason it is described as "omnidirectional." Pressure microphones may use any of the common transducers, including a subtype called a *contact* microphone. This type of microphone is fastened directly to an instrument, such as a guitar.

In a gradient microphone. the transducer reacts to a given function of the difference in pressure between two points in space; this allows the microphone to be aimed to pick up sounds from a specific location. One class is called *unidirectional* (often used for sound pickup on theater stages, motion picture sound stages, and in television studios). Another is known as the *bidirectional* or *velocity* type; it is preferred for difficult sound pickup at a distance or in noisy environments such as aircraft cockpits.

In a wave microphone, the transducer is usually a ribbon element, which is often activated by sound focused on it through a system of small tubes or pipes with open ends. This allows it to pick up speech sounds across long distances, while minimizing loud noises and reverberations in the area.

The term *microphone* was coined in 1827 by English physicist and inventor Charles Wheatstone (1802–1875) to identify an acoustical device he invented for hearing sounds when placed on "sonorous bodies." The discovery of the microphonic principle is generally credited to David E. Hughes (1831–1900). A paper written by Hughes was read before the Royal Society in London on May 8, 1878, describing what became known as the *carbon-pencil microphone*.

As sometimes happens in the history of technology, the invention of the microphone preceded a complete understanding of how it actually worked. In 1876, 2 years before Hughes's paper, Alexander Graham Bell (1847–1922) invented the first microphone. It consisted of a membrane diaphragm attached to a wire that was inserted into a metal cup filled with diluted sulfuric acid. This assembly was connected to an electric storage battery and, via a wire, to a tuned-reed receiver. Sound vibrations in the air caused the diaphragm to vibrate, moving the wire up and down in the liquid conductor. This varied the electrical resistance, so that a current passing through the cup and acid solution undulated in a pattern similar to the sound waves.

Although the liquid transmitter proved the principle of the telephone, it was not commercially feasible and was soon replaced by another Bell invention, the membrane transmitter. This microphone used a nonmagnetic diaphragm to which a piece of iron was attached. An electric coil produced a magnetic field around the iron, and movements of the iron within that field induced currents in the coil. The currents varied according to the motion of the membrane, producing a matching movement in the diaphragm of the receiver.

Thomas Alva Edison (1847–1931), employed as a consultant to a Bell competitor, the American Speaking Telephone Company (a subsidiary of the Western Union Company), created the first major application to telephony of the variable-resistance contact. In 1877, he filed patent applications for various transmitters using variable-pressure contacts between metal and electrodes made of, or coated with, graphite. Meanwhile, in 1878, Edison found a better transmitter that used highly compressed lampblack (carbon) in contact with metal.

Other inventors working for Bell developed even better microphones. Emile Berliner (1851–1929), who later invented the phonograph disk, worked together with Francis Blake in an effort that produced the Blake transmitter in 1878, the peak achievement of single-contact transmitters.

Edison and Berliner were working with microphones before the Hughes paper on the "microphonic principle" was published. However, Edison's patent application stressed the material rather than the principle, while Berliner seemed to lack as broad an understanding as Hughes of the underlying principles. In fact, the Berliner device worked poorly until Blake used the Hughes principles to make it a commercially useful microphone.

In 1878, Henry Hunnings, an English clergyman, came up with the idea of using loosely packed carbon granules instead of a tightly packed button of carbon. The granules were encased between two electrodes, one an electrically conducting diaphragm, the other a brass disk. Sound waves caused the diaphragm to flex, causing the granules to squeeze tightly together and then relax as pressures from the speech sounds changed. The loose granules formed multiple parallel contacts, allowing the microphone to carry higher currents than the single contacts used by Berliner, Edison, and Blake. It was subsequently marketed as the "long distance" transmitter because its higher output produced audible transmission over longer lines. Over 10,000 of these units were used by American Bell over the following years.

In 1886, Edison introduced the use of anthracite coal granules that had been carbonized by roasting. His multicontact transmitter was so superior to ordinary carbon granules that it became the standard material, and remained so for 80 years. However, callers had to bring their lips close the mouthpiece and speak loudly to generate enough sound volume for transmission. In 1890, another Bell engineer, Anthony C. White, developed an improvement on the Hunnings transmitter, using Edison's granules and a new "button" design that fastened the rear electrode to a metal bridge in the back of the apparatus, while the front electrode was free to move back and forth in response to the diaphragm's vibrations. A thin ring of mica insulated the two electrodes from each other, so that the current path was only through the carbon granules. This "solid-back transmitter" increased the volume level significantly and remained in production for 35 years.

Until the invention of the vacuum tube, no electronic means of signal amplification was available. The triode vacuum tube, invented in 1907 and improved by telephone engineers in 1913–14, provided microphone developers with a reliable means of connecting weak currents from transducers to loudspeakers.

The next significant step in microphone design occurred around 1917, when Edward C. Wente of Bell Telephone Laboratories invented the condenser microphone. It used two parallel plates that were insulated from each other; one plate was fixed; the other (a diaphragm of thin steel) moved in reaction to the changing pressures of sound waves. A battery and a resistor were connected to the plates in series, so that the apparatus produced a fluctuating current flow that varied in proportion to the changes of pressure in the sound waves. Originally intended for research projects since its electrical impedance was too high for use across even short wire spans, the condenser microphone became practical when it was combined with vacuum-tube amplifiers. Although not well suited to telephones, the condenser microphone found a ready market in the new radio and public-address applications emerging in the 1920s. By 1929, these microphones were being used on Hollywood sound stages.

One of the most important transducers now in use in microphones is the electret foil design. Invented at Bell Labs in the 1960s, it uses a precharged metalized dielectric film such as Teflon as a diaphragm. The nonmetallized side of the foil is placed next to a backplate that is metal or metal plated, and is ridged and perforated (and which is insulated from the case). A shallow airspace is left between the film and backplate. Sound vibrations cause the foil to oscillate, and these movements compress the air in the space between the backplate and the film, raising and lowering the electrostatic field generated by the electret charges. The electrical output of the electret microphone is taken between the backplate and the metal side of the foil and is then fed into a preamplifier.—R.Q.H.

See also AMPLIFIER; GUITAR, ELECTRIC; MOTION PICTURES, SOUND IN; RECORDS, LONG-PLAYING; TEFLON; THERMOIONIC (VACUUM) TUBE

Microprocessor

More than any other invention, the microprocessor is at the heart of the developing computer industry. Without it there would be no personal computers, and many of today's technological accomplishments, from sophisticated automobile engine management systems to automatic teller machines would be impossible. It is not an overstatement to say that the microprocessor has brought about a revolution in the way we think and do business, a revolution that is at least as profound as the one that followed the development of movable-type printing presses in the late 1400s.

Computers do not need microprocessors; indeed, when computers were first developed in the late 1940s, the only electronic device capable of serving as the on-off switch at the core of computer circuitry was the vacuum tube. These tubes were large, often 8 or 10 mm (3 or 4 in.) high and 2.5 mm (1 in.) in diameter, required large amounts of electrical power, were very prone to failure, and gave off considerable amounts of heat, which made them a potential fire hazard. Despite all these disadvantages, hundreds of them were the basic circuit elements of the first generation of computers (one famous machine, the ENIAC, required 18,000 vacuum tubes and used enough electrical power to light up a small town).

The next major advance was the development in 1948 of the transistor at Bell Telephone Laboratories. The transistor can accomplish the same basic electronic jobs as the vacuum tube, but is much smaller (it is usually put in a small metal container about the size of an eraser on the end of a pencil), requires almost no power to operate, is very reliable, and produces almost no heat. It was not until the mid-to-late 1950s that transistors were manufactured in sufficient volume, and with sufficient reliability, to allow them to become the electronic device of choice for the designers of computing equipment. Their small size and low power requirements meant that a computer that once filled a large room could be reduced in size to an office desk. At the same time, however, each transistor had to be individually designed into a basic electronic circuit, these circuits incorporated into larger modules, and these modules combined to make the finished computer. This not only entailed a lot of work in the manufacturing plants but also resulted in miles of wire being used to interconnect the thousands of transistors that went into the machine.

A transistor is made by taking a small sliver of silicon (the same material found in sand and glass) and carefully incorporating various impurities into it in order to create regions of conductive and nonconductive material. This is usually done by photographic means, in much the same way that a picture is printed on light-sensitive paper. While individual transistors could be made very inexpensively, the labor and materials that went into the interconnections between them kept computer prices beyond anything affordable by individuals.

By 1961, the techniques of depositing the impurities into the silicon were well enough known that it became possible to manufacture small collections of transistors on a silicon wafer that actually formed a small circuit, the interconnections between them being produced by thin deposits of metal on the silicon. These "integrated circuits" were then cut apart and incorporated into packages that significantly reduced the complexity of wiring together the individual transistors.

In 1969, a Japanese manufacturer of handheld calculators engaged the design staff at Intel, a California-based electronics firm founded only 2 years earlier, to design a series of integrated circuits. By this time it was possible to create a single silicon chip that contained about 1,500 interconnected transistors. Rather than produce several different sets of circuit chips for the different models of calculator, the engineers decided that by pushing the limits of the technology, they might be able to get the complete control circuitry of a computer on a single chip, and then simply use standard programming techniques to implement the functions required for the different models. While their concept was valid, the resulting 1971 design used 2,100 transistors, and the 2-year development delay resulted in the Japanese withdrawing their order. Intel was forced to look for alternative uses for this new product. This very elementary computer, known as the Intel 4004, because it was capable of dealing with four bits at one time, slowly became known in the industry and was used as the central control mechanism for devices that ranged from electronic meters to washing machines.

By 1978, Intel and other manufacturers had managed to create much more powerful microprocessors, setting the scene for the entry of the personal computer as we know it today. The creation of a complete computer on one chip eliminated the expense and problems associated with the wiring together of all the individual components. Moreover, the photographic techniques used in production allowed hundreds of these chips to be produced at once. As a result of these manufacturing efficiencies, prices fell to the point where computers became affordable by individuals.

The descendants of that 1978 chip (the Intel 8086) are the central processors in the vast majority of personal computers used today. The number of transistors that can be put on a chip the size of a fingernail has doubled each year. The production techniques went from using ordinary light to expose the sensitive material on the silicon wafer to the use of X rays and even beams of electrons, both of which allow much finer detail, and thus more transistors, to be created in the same space. As the process for creating the microprocessor changed, so did the terminology associated with the chip. From the simplest integrated circuits (or ICs), the process went to large scale integration (LSI) to very large scale integration (VLSI), and by the mid-1990s, simple adjectives could no longer be used to describe the next steps.

We are now close to the point where a single chip can contain over 50 million transistors and their interconnections. The resulting computer is many times more powerful that even the largest individual-component machine of a few years ago, and costs only a tiny fraction of the money that had to be invested in those machines. It also means that it has become relatively easy to combine these powerful microprocessors together to form supercomputers capable of processing many jobs in parallel.

While the most obvious places where microprocessors can be found are personal computers, other uses account for the majority of the chips produced. It is not uncommon to find new automobiles equipped with many microprocessors that monitor and control everything from gasoline usage to the operation of the brakes and suspension. It will soon be the case that virtually every device we use on a daily basis will have at least one, if not 10 or more, microprocessors. These will not only control the mechanical operation of everything from food blenders to airplanes but will also provide the basis for entirely new technologies, just as not too long ago they made possible the emergence of compact disk players and automatic teller machines.—M.R.W.

See also AUTOMATED TELLER MACHINES; BINARY DIGIT; CALCULATOR; COMPACT DISK; COMPUTER, PERSONAL; INTEGRATED CIRCUIT; MOORE'S "LAW"; PARALLEL PROCESSING; PRINTING WITH MOVABLE TYPE; THERMOIONIC (VACUUM) TUBE; TRANSISTOR; WASHING MACHINES

Microscope, Electron

One of the most powerful and important instruments of modern science, the electron microscope was first developed by Ernst Ruska (1906–1988) in Berlin in the early 1930s. A half century later, in the mid-1980s, Ruska was awarded the Nobel Prize in physics for his pioneering efforts. More than a straightforward tale of great achievements in science, however, the early history of the microscope reveals an interesting episode in both the development of technology and its relationship to scientific advance. The early chroniclers of the development of the electron microscope, apparently influenced by the definition of technology as "applied science," have viewed the microscope as a logical consequence of then-recent scientific developments. In this classic, and inaccurate, model of technology as applied science, the development of the electron microscope in the 1930s has been seen as a logical consequence of the application of Louis de Broglie's (1892–1987) scientific discovery of "matter waves" in the 1920s.

In the late 19th century, a consensus had been reached among scientists and microscopists that the wavelength of visible light limited the resolution of the conventional light microscope. According to the theory formulated in 1876 by Ernest Abbe, the resolution of a light microscope is limited to one-half the wavelength of the light used as the medium of the instrument. Thus, no matter how accurately lenses were ground, light microscopes were ultimately limited.

In the 1920s, de Broglie came upon his theory of matter waves, which argued that all matter emits a characteristic electromagnetic wave, the wavelength of which is determined by the mass of the object. An electron would have a matter wave significantly smaller than the wavelength of visible light, thereby making possible a resolution much better than that attainable with light microscopes. It would thus seem that the early builders of the electron microscope were inspired by the possibility of surpassing the resolution of the light microscope and peering further into the unseen microscopic world, a brilliant example of state-of-the art scientific theory inspiring and informing technological advance. In fact, Ruska had not even heard of de Broglie's work when he began work on the microscope. Rather, he wanted to evade the wavelength limitation of microscopy altogether, and put his hopes in the electron microscope because he was viewing the electron as a *particle*, a perspective derived from J. J. Thomson's 19th-century corpuscle theory of electrons. When Ruska had at last heard of de Broglie's research, his reaction was hardly one of elation or inspiration. As he recalled: "I was very much disappointed that now even at the electron microscope level the resoltion should be limited again by a wavelength." For now, he felt that he was once again up against the limiting wall of Abbe's theory. His spirits lifted, however, when he "became satisfied that these waves must be . . . shorter in length than light waves." Ruska and his colleagues then pressed on in spite of de Broglie's theory and not because of it.

In contrast to the model of technology as applied science, here was technology forging ahead of, or even away from, state-of-the-art science, indeed even basing its approach on outmoded science. One of the remarkable elements of this story is that, far from being an isolated and inconsequential artifact, the electron microscope played a key role in many of the scientific advances of the 20th century. Without the electron microscope, methodologies and paths of inquiry, along with the allocation of private and public funds for medical and materials research, would very likely have taken on different patterns, greatly altering the fields of science as we know them today.

This episode provides only one look into the history of the electron microscope; the development of the technology did not end with Ruska's work in the 1930s. After the first experimental models were built in Germany based on Ruska's designs, commercial development followed in the United States, Europe, and

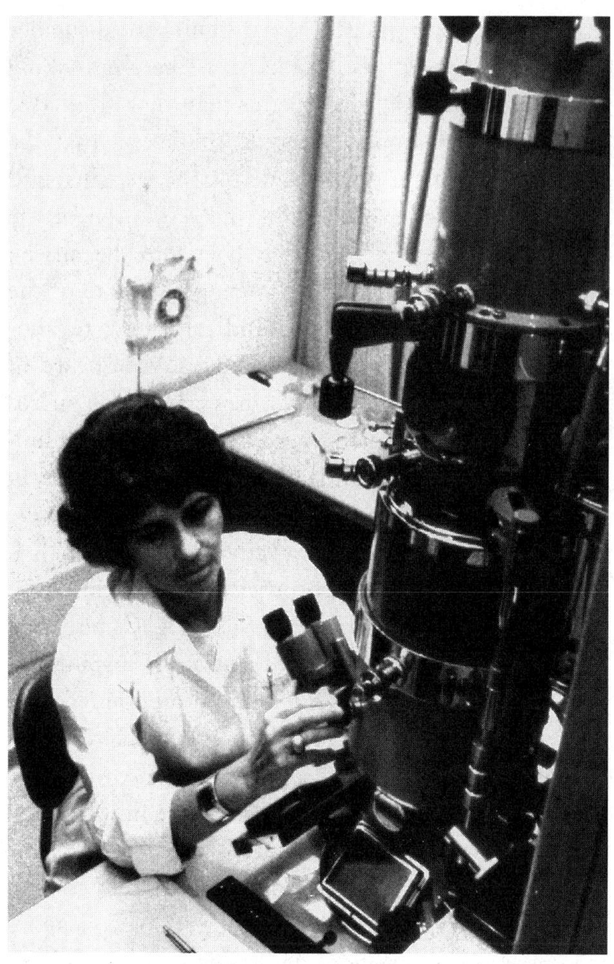

Scientist using an electron microscope (courtesy National Library of Medicine).

Japan. In the 1940s and 50s, the Radio Corporation of America (RCA) developed and marketed several electron microscopes and became a leader in the industry. Reflective of broader trends in the electronics industry, however, Japanese firms later eclipsed American dominance by the 1970s and continue to lead in the commercial manufacture of electron microscopes.

New developments in electron focusing combined, most recently, with adaptations of computer technology to enhance imaging in so-called "scanning" and "tunneling" electron microscopes have led to ongoing improvements in the power of these instruments. Today, far surpassing the resolving power of ordinary light microscopes, state-of-the-art machines are capable of magnifications on the order of 200,000 times an object's original size, bringing researchers to the threshold of seeing individual atoms.—G.K.

See also ELECTRON; MICROSCOPE, OPTICAL

Further Reading: Gregory C. Kunkle, "Technology in the Seamless Web: Success and Failure in the History of the Electron Microscope," *Technology and Culture*, vol. 36, no. 1 (Jan. 1995).

Microscope, Optical

Although lenses for eyeglasses were invented in the 14th century, and low-power magnifying glasses were known at an even earlier date, centuries passed before lenses were used to highly magnify objects far away and close at hand. The first instrument to use two lenses was the telescope, and a few years later the same principle was used for the first compound microscopes (so called because they combined two lenses, one called the *objective* lens, the other the *eyepiece* lens). Credit for its invention is usually assigned to two Dutch eyeglass-makers, Hans Janssen and his son Zacharias Janssen (1580–1638), who built a microscope in the last decade of the 16th century, but other lens grinders undoubtedly hit upon the idea at about the same time.

Although microscopes used the same optical principles as telescopes, they were harder to make, for their lenses were smaller and more difficult to grind. Effective use of a microscope also required some means of bringing it into focus, and it was necessary to have a means of illuminating the object under study. Early microscopes were mounted in a vertical position, usually supported by a tripod that also held the subject being observed. Microscopes that could be inclined from the vertical soon appeared , but it was not until the end of the 18th century that these become more popular than the vertically mounted kinds. Changes in focus were effected by sliding a tube that mounted one lens relative to the tube that held the other. Focusing mechanisms employing rack-and-pinion gears appeared in the late 17th century but did not become common until much later.

While microscopes benefited from some improvements in their mechanical layout, they continued to suffer from two serious optical defects: spherical aberration and chromatic aberration. Spherical aberration occurs because rays of light passing through the periphery of a lens are bent more acutely than those passing near the center of the lens. As a result, there is no single point of focus and the image is blurred. This problem could be partially countered by blocking the peripheral rays through the use of a small-diameter diaphragm placed behind the lens, but at a cost of diminishing the amount of light cast on the subject. Chromatic aberration is caused by light of different wavelengths being refracted at different angles. This results in different colors finding their foci at different points relative to the lens, and manifests itself as fringes of color ringing the object being studied.

Both spherical and chromatic aberration tended to be intensified when multiple lenses were employed. Consequently, some of the best work of early microscopy was done with single-lens instruments. The most notable maker and user of single-lens microscopes was Anton Leeuwenhoek (1632–1723), a Dutch linen draper. Leeuwenhoek made hundreds of microscopes that typically consisted of a double-convex lens he had ground himself that was mounted between two small plates about 5 cm (2 in.) long and 2.5 cm (1 in.) wide. The lenses typically had a focal length of 6 to 7 mm (0.24–0.28 in.) and were capable of magnifying an object as much as 275 times, and perhaps considerably more. They were difficult to use, however, for their short focal length mandated mounting the object to be examined very close to the lens, and then bringing everything right up to the eye. Beginning in the 1660s and continuing to his death in 1723, Leeuwenhoek observed a great number of things, including red blood cells, spermatozoa, insects, yeast, and capillaries. He also discovered protozoa, and identified the three major types of bacteria: bacilli, cocci, and spirilla. Although no one ever surpassed Leeuwenhoek in the breadth and quality of their observations, a great deal of useful scientific work continued to be done with single-lens microscopes, more in fact than was done with compound microscopes for many decades.

This is not to say that nothing important was discovered with compound microscopes. Some of the most significant explorations of the microscopic world were conducted in England by Robert Hooke (1635–1703) using a compound microscope built by Christopher Cock, a London instrument-maker. The results of Hooke's efforts were recorded in a highly influential book, *Micrographia* (1665), in which were presented renditions of magnified insects, plant parts, the point of a pin, crystals, and other common objects. Hooke's writings also are notable for the first biological use of the word *cell*, which Hooke employed to describe the inner structure of cork. In so doing, Hooke calculated that the number of cells in a cubic inch of cork would total more than 1,200,000,000, ". . . a thing most incredible, did not our microscope assure us of it by ocular demonstration."

At about the same time that Hooke was describing

A 17th-century microscope (courtesy National Library of Medicine).

cells and other minute entities, Marcello Malpighi (1628–1694) in Italy was using the microscope to unveil a basic physiological process. In Malpighi's time, the idea that the blood circulates was generally accepted, but many questions remained regarding the connection of arteries and veins, as well as how blood flowed from the former to the latter. In 1661, Malpighi answered these questions when he used a microscope to examine the lungs of a frog, finding that they consisted of many air sacs covered by tiny capillaries that linked small arteries with small veins.

These pioneering efforts were not followed up, and during much of the succeeding century the microscope was used primarily as an entertaining diversion for nonscientists. The decline of microscopy was due in part to religiously based opposition to prying into things that God had intentionally hidden from humankind, but even secular philosophers and physicians objected to microscopic studies that did not support their conceptions of physiology and disease processes. The limited uses to which the microscope was put retarded its development, and for more than a century it underwent modest mechanical improvements, such as better focusing mechanisms, but nothing was done to improve its optical qualities.

The inability to develop better lenses also stemmed from an adherence to the optical theories of Isaac Newton, who seemingly had proved that refracting lenses invariably produced chromatic aberration. Yet as early as the 1730s, a lawyer and amateur microscopist named Chester Hall built a microscope with achromatic lenses by combining a convex lens made from crown glass with a concave one made from a glass with a high lead content, inaccurately known as *flint glass*. During the second half of the 18th century, mathematically based studies of optics coupled with experimentation using lenses made from different kinds of glass led to the production of low-power achromatic microscopes. By the 1820s, microscopes of greater power were being built by using several lens pairs, which were combined largely through trial and error to produce microscopes with little chromatic or spherical aberration. This empirical approach was supplanted a decade later by methods based on the optical theories devised by J. J. Lister (1786–1869) in England. Development continued, and by the end of the 19th century many microscopes had optical qualities that have not been bettered today.

Substantial improvements in optical quality reestablished the microscope's role as a key scientific instrument. From the 1830s onward, microscopic investigations resulted in fundamental biological discoveries, beginning with the observations of Matthias Schleiden (1804–1881) and Theodor Schwann (1810–1881) in Germany, which established that the cell was the basic unit of living beings. The microscope was also essential to the discovery that bacteria and other microorganisms are responsible for the transmission of many diseases. In the 20th century, some of the functions of the optical microscope were taken over by the electron microscope and other instruments, but it still remains an indispensable tool for scientific inquiry and medical practice.

See also BACTERIA; BLOOD, CIRCULATION OF; CELL; EYEGLASSES; GERM THEORY OF DISEASE; GLASS; MICROSCOPE, ELECTRON; TELESCOPE, REFLECTING; TELESCOPE, REFRACTING

Further Reading: S. Bradbury, *The Evolution of the Microscope*, 1967.

Microwave Communications

Microwaves comprise the part of the electromagnetic spectrum that encompasses wavelengths from about 0.3 cm to 30 cm and frequencies from 1 to 100 GHz. This locates them between radio waves at one end and infrared radiation on the other. The existence of the electromagnetic spectrum was predicted in 1864 by James Clerk Maxwell's (1831–1879) theoretical formulations. Radio waves were first generated and detected by Heinrich Hertz (1857–1894) in 1888, and the commercial development of radio soon ensued, but little was done with shorter-length waves. Since it was assumed to be of no commercial or military value, this part of the spectrum was given over to radio hobbyists, who began to explore the potential of short waves in the early 1920s. In 1931, André Clavier demonstrated the potential value of even shorter waves, i.e., microwaves, when he used them for a transmission across the English Channel. In the years that followed, the key impetus for the use of microwaves came from the development of radar. Due in large measure to what had been learned in the course of developing radar, microwaves began to find other applications after World War II. One such application was the microwave oven, but the most significant was the use of microwaves for radio, television, and telephone communication.

Microwaves are sent and received in much the same way as ordinary radio signals. The waves are generated at a particular frequency by an oscillator and are sent along a tubular waveguide, a device invented by G. C. Southworth in the 1920s. The waves are then converted into a plane wave by a transmitting antenna. Microwaves can be transmitted using amplitude modulation, frequency modulation, or phase-shift modulation. They are then detected at their destination by a receiving antenna and a silicon diode that is mounted in a waveguide.

Microwaves are especially useful for long-distance telephony since they can be sent along a large range of frequencies, each of which can carry a signal. The microwave band provides about 100 times as much frequency space as the entire radio spectrum below it—everything from AM and FM broadcasting to citizen's band radio. Moreover, the greater directivity of microwaves means that microwave frequencies can be used many times in the same area without their interfering with one another.

The advantages of microwaves for long-distance telephony were realized in the early 1950s, when the Bell System installed a network of microwave relay stations that connected the east and west coasts of the United States. In the 1960s, terrestrial microwave links began to be supplemented with orbiting satellite links. With the development of microwave relay links on the ground and in space, microwave transmission was expected to be the dominant technology for long-distance telephone communication. This presumed ascendancy of this technology had economic implications. Not only was microwave transmission effective in a technical sense, it also held out the promise of the deconcentration of the telephone industry, for the entire network no longer had to be hard-wired together, as had been the case since the inception of commercial telephone service. The prospect of a decentralized, satellite-based telephone system using microwave transmission was in fact one of the motivating factors in the court-ordered breakup of the Bell System in 1983.

However, just as this was occurring, technological change was once again altering the technical foundation of long-distance telephony, for the development of fiber-optic cables held out the prospect of a more powerful and efficient means of transmitting telephone messages. The spectrum of visible light has 1,000 times the width of the radio spectrum, which gives fiber-optics–based communications systems a vast channel capacity. In consequence, fiber-optic cables have in many cases supplanted satellite-based microwave transmissions. However, microwaves will continue to be employed for satellite-based communications technologies like cellular telephones, as well as for communications between the Earth and space.

See also AMPLITUDE MODULATION; COMMUNICATIONS SATELLITES; FIBER-OPTIC COMMUNICATIONS; FREQUENCY MODULATION; MAXWELL'S EQUATIONS; OVEN, MICROWAVE; RADAR; RADIO; TELEPHONE SYSTEMS, CELLULAR

Microwave Oven—see Oven, Microwave

Military Revolution in Early Modern Europe

The "military revolution" refers to the technological and organizational innovations that enabled Europe to replace Asia as the world's dominant military power between the Renaissance and Industrial Revolution. During the late Middle Ages, Asian armies routinely crushed European forces, as demonstrated by the collapse of the Crusades, the Mongolian invasion of Central Europe, and the Turkish conquest of the Balkans. The success of the Ottoman Turks in particular offered a powerful indictment of the superiority of Asian infantry and cavalry tactics, gunpowder weaponry, command hierarchies, and logistical support over the feudal armies of the West. Yet the military might of western Asia paled in comparison to the power of eastern Asia. The Ming Dynasty of China in the 15th century and the Mughal Empire of India in the 16th century each employed large standing armies armed with sophisticated weaponry and centralized bureaucracies. Nevertheless, by the late 18th century a revolution had occurred: European powers were routinely and decisively defeating Asian armies, as demonstrated by the Russia's conquest of the Crimea, the British East India Company's conquest of Bengal, and the French invasion of Egypt. China's turn at military humiliation would come with the First Opium War (1839–1842). This transfer of military superiority was the result of Western flexibility and Eastern rigidity with regard to technical and organizational changes. The motivation of Europeans to invest continuously in naval, siege, and field warfare innovations during the military revolution was a direct response to their interminable political conflicts. Yet the efficacy of these innovations was aided by their interaction with contemporaneous advances in science and technology.

Illustrating this process was the rise of Western naval supremacy during the 16th century. Especially critical was the Portuguese work of the 15th century under Prince Henry the Navigator and King John II. The Portuguese developed oceangoing vessels that relied on inanimate power for both propulsion and defense, and astronomical science for navigation. The result was the employment of the light and maneuverable caravel, and the heavy, fortresslike carrack for ocean voyages. By the early 16th century, these vessels employed both lanteen and square sails, and were armed with muzzle-loading artillery. Their navigators used the compass, quadrant, and tables of solar declination to determine latitude, as well as a Ptolemeic mapping system to chart their course. Equally significant was the carrack's ability to function simultaneously as a commercial and military vessel.

The Portuguese used their naval innovations to control the coast of Africa and enter the Indian Ocean by 1494. Their initial probe into Chinese waters, however, was decisively crushed in 1522 by the gunships of the Ming Dynasty. Chinese naval power was demonstrated during the early 15th century when Admiral Cheng Ho's fleets of war junks dominated the Indian Ocean. Changes in political priorities rather than technical conservatism led the Ming Dynasty to abandon its commitment to naval expansion. This left a partial vacuum in the Indian Ocean that the Portuguese quickly exploited.

Western naval rivalries stimulated the innovation of increasingly powerful warships during the late 16th century, including the oar- and sail-propelled *galeasse* that Hapsburgs used to crush the Turks at Lepanto in 1571, and the sleek galleon that the English used to deflect the Spanish Armada in 1588. Such naval innovations accelerated during the 17th century with the use of increasingly specialized naval vessels, including bomb ketches for offshore bombardments, frigates for long-range privateering, heavy warships with multiple gun decks for concentrated engagements, and the *flutte* for economical transportation. Consistent funding of scientific education and research also became a standard naval strategy in 17th-century Europe. The Royal Observatory founded by Charles II and the Paris Academy of Science founded by Colbert and Louis XIV are the most direct examples. The political demand for a practical technique to measure longitude, in fact, motivated much of the astronomical and horological research conducted during the 17th and 18th centuries.

On land, the Ming and Qing Dynasties of China, as well as the Mughal and Maratha Empires of India built enormous fortresses that were virtually impermeable to heavy siege artillery. They routinely employed gunpowder weaponry in their active defense as well. The Ottoman Empire, on the other hand, excelled in assaulting fortresses. Their siege of Constantinople in 1453 was a brilliant example of coordinated artillery, naval, and infantry action, while their sieges of Rhodes in 1522 and of Cambria in 1669 demonstrated a mastery of mining attacks. In terms of developing a comprehensive system of siege warfare, however, Western siege armies were outclassing their Asian counterparts by the early 16th century. Although Europeans had used large-caliber bombards to both assault and defend fortified positions during the second half of the 14th century, their enormous weight rendered them difficult to transport, while their stone projectiles made them difficult to supply. Towards the end of the Hundred Years' War (1338–1453), the French developed smaller caliber guns with higher muzzle velocities and placed them on stable carriages for greater mobility. The employment of corned gunpowder and iron shot further increased such artillery power. Thus armed, the French reduced all British strongholds in France except Calais between 1450–51, and crushed English field armies at Formigny and Castillon. Armed with such artillery, the Spanish reduced the Moslem fortresses in Granada to wrap up the *Reconquista* by 1492.

Although fortification designers did respond to artillery changes during the 15th century, it was in Italy during the Hapsburg-Valios Wars that effective countermeasures emerged with the *trace italienne*. Beginning with the Spanish and Portuguese in the early 16th century, and continuing with the Dutch, French, and British during the 17th and 18th centuries, the *trace italienne* allowed Europeans to maintain commercially viable strongholds around the world.

Equally significant for the European military ascendancy was the development of rational approaches to assaulting fortresses. The Dutch forces led by Maurice of Nassau (1567–1625) and Simon Stevin (1548–1620) began using formal geometrical considerations to construct siege lines around Spanish-held fortresses in the 1590s. Sébastian Vauban (1633– 1707), the invincible French military engineer, optimized such a system during the wars of Louis XIV so that a commander could calculate precisely the time and resources needed to reduce a fortress with minimum casualties. Vauban's system relied on fortified artillery emplacements and carefully surveyed trenches that employed both concentric and zigzag geometries. This maximized the siege army's power, while minimizing the threat of enemy counterfire, sorties, and relieving armies. In contrast, Turkish siege lines were confused sprawls. This helps explain why the Ottoman thrusts into Western Europe after their conquest of Hungary in 1526 were frustrated by the growing network of *trace italienne* fortifications.

Another central element in the military revolution was the transition from small decentralized armies focused around feudal cavalry forces to disciplined national armies dominated by infantry and artillery firepower. This transition began during the 14th century with the devastation that French and Hapsburg knights suffered from the English longbow and Swiss pike, respectively. The vast training needed to use the longbow effectively, however, led to the crossbow's becoming the dominant missile weapon for Western infantry forces during the 15th century, followed by the harquebus or matchlock during the 16th century. The vulnerability of such forces to unexpected cavalry assaults, however, presented the problem of integrating them with pikemen to provide mutual support. An initial solution came from the Spanish in the early 16th century with their geometrical *tercio* formation. It involved a rectangular body of pikemen with harquebusiers at its corners. The Spanish ability to discipline and coordinate their infantry to fight in such an integrated formation rendered them virtually invincible in 16th-century field warfare, as demonstrated in the conquest of the Aztec and Inca Empires, the Battle of Pavia (1525), and the field actions of the Dutch Revolt. Nevertheless, such infantry innovations hardly gave the West a decisive advantage over Asian military armies. Europeans, after all, did not dare engage the Ottoman Turks in a large-scale battle for most of the 16th and 17th centuries.

By the late 17th and early 18th centuries, however, the strength of Asian field armies was in decline. Western European field armies were routinely employing innovations in military technology that gave them significant advantages. From a utilization perspective, this involved the drilling of troops to load and fire their firearms in precise sequences, as well as in choreographed maneuvers. Furthermore, this exercise occurred not only during the training phase but also continuously to maintain the discipline and morale of experienced troops. As initiated by Maurice of Nassau during the Dutch Revolt and developed by Gustavus Adolphus during the Thirty Years' War (1618–1648), such drill enabled infantry units to concentrate their fire in devastating volleys even under terrifying combat conditions. Although expensive in terms of time and the use of junior officers, the system of continuous drill routinely executed the transformation of society's dregs into disciplined soldiers, capable of functioning as interchangeable parts in a deadly social machine. In the late 17th century, the effectiveness of these soldiers was extended by their being armed with flintlock muskets and socket bayonets. The musket generated faster and more reliable firing rates than the harquebus, while the bayonet reduced the vulnerability of soldiers while they were loading.

The growing strength of Western field warfare in the 18th century also depended on artillery innovations. While the basic smoothbore-artillery design of the 15th century remained, a series of artillery reforms created both powerful and maneuverable field artillery systems. This began with Gustavus Adolphus's introduction of the three-pounder regimental artillery piece into the Swedish army during the 1620s. The trend accelerated during the mid-18th century with the artillery reforms of Austria and France that furnished the first heavy field guns that could be moved routinely in combat. Equally significant was the way such 18th-century artillery was used. Following the ballistics research conducted during the War of the Austrian Succession, the killing efficiency of Western field artillery improved significantly when directed by officers trained in Newtonian science.—B.D.S.

See also BALLISTICS; BOW; COMPASS, MAGNETIC; CROSSBOW; GUNPOWDER; LONGITUDE, DETERMINATION OF; PTOLEMAIC SYSTEM; SHIPS, SAILING; TRACE ITALIENNE

Milk, Condensed

Cow's milk is about 87 percent water. Removing much of that water reduces the bulk, making milk easier to transport and store. Condensed or evaporated milk, as this product is known, was invented by Gail Borden (1801–1874), who began to experiment with the condensation of milk after previously working as a farmer, newspaper publisher, surveyor, and teacher. Before experimenting with the condensation of milk, Borden had invented a biscuit that contained condensed meat. It won an award at the Britain's Great Exposition in 1851 and was taken on a voyage to the Arctic, but it failed to achieve commercial success.

Borden was a deeply religious man, and one of his motivations for developing condensed milk was to prevent the illnesses that often struck children who had been nourished with the milk of sick cows. The key to Borden's technique was heating milk in a vacuum pan that prevented the entrance of air from the outside, a process that had previously been used by a New York Shaker community for the condensation of fruit juices. Borden applied for a patent to cover this process in 1853, but the patent examiner was not convinced of its novelty. Only after two scientists attested to its uniqueness was a patent granted that year.

As is usually the case, the gaining of a patent was not followed by immediate commercial success. Borden went into partnership with Jeremiah Milbank, a New York banker and food wholesaler, but financial returns were meager until the Civil War. The need to provision large armies during this conflict resulted in a massive demand for Borden's condensed milk, to the great benefit of Borden's personal finances.

While the Civil War was raging in America, Louis Pasteur (1822–1895) in France was making the epochal discovery that certain bacteria were responsible for the spoilage of beer, wine, and many foods. This discovery led directly to the process of pasteurization, the heating of a food or beverage to destroy harmful microorganisms. Condensed milk was a safer product than the whole milk of the time, because the heat used in the condensation process pasteurized the milk. As is often the case, the technology was effective even though the reasons that it worked were not understood at first.

Condensed milk continues to play an important part in the diet of many people, especially infants. As Borden had hoped, the purity of condensed milk has prevented a great deal of sickness. At the same time, however, bottle feeding with some form of cow's milk still remains an inferior substitute for breast feeding in a number of ways.

See also BOTTLE FEEDING; GERM THEORY OF DISEASE; PASTEURIZATION; PATENTS

Milling Machine

The milling machine, or miller, is a metal-cutting machine tool that is often described as the opposite of the lathe. On a lathe, the workpiece rotates in place and the cutter moves along its length. On a milling machine, the cutter rotates and the workpiece moves past it. The lathe cuts cylindrical forms; the milling machine cuts flat forms.

English and French gear-cutting devices and other special-purpose equipment for making clock parts displayed the operating characteristics of the milling machine as early as the late 17th century. The milling machine as a general-purpose machine tool, however, is generally recognized to have taken shape around 1818 in the firearms shops of Connecticut. The armsmakers did not base the machine on European clockmaking practice but more likely developed it as an improvement on planers and shapers. Planers and shapers cut flat surfaces in metal with a single-point tool that encountered the workpiece in a reciprocating motion. The milling machine, with its multitoothed rotary cutter, could produce a broad, flat surface in a single pass.

The arms shops of Eli Whitney (1765–1825) in New Haven, and Simeon North in Middletown built the earliest positively identified millers. They had the same basic features. Both stood on benchtops rather than on the floor. The spindle (the rotating shaft that carried the cutter) was placed horizontally. The workpiece was bolted to a small table that ran longitudinally under the cutter. Later depictions also show the workpiece held in a vise bolted to the table. The table ran in only one direction, left to right from the operator's perspective. Neither machine had built-in means to adjust the relationship between tool and work vertically or forward and backward (the motion known as *cross-feed*).

Whitney and North were both engaged in arms contracts for the United States government and belonged to a competitive but nonetheless tightly knit technical community that also included the government armories. Within a few years of the machine's first appearance, the equipment at the government's Harpers Ferry Armory included some form of miller. The Springfield Armory probably had similar machinery.

Although advances were made in their convenience of use and precision of output, milling machines did not depart substantially from the established format for some 40 years. The first millers to feature vertical adjustment appeared in the 1830s at a textile machine producer in Massachusetts. An apprentice from that shop, Frederick W. Howe, later developed the idea further when he went to work at the seminally innovative arms factory of Robbins & Lawrence in Windsor, Vt. The Vermont machines of the 1840s had crude vertical adjustment and hand-cranked cross-feed, as well as spindle-bracing and stiffer construction overall to enable heavier and more precise cutting. Veterans of Robbins & Lawrence carried the design throughout the New England arms sector, notably to Hartford, Conn., where the firm of Pratt & Whitney improved on it. A lead screw replaced the awkward rack-and-pinion method of table feed, and the floor-standing machine offered superior rigidity. Known as the Lincoln miller, it was the first milling machine offered for general sale (in 1855), rather than being produced for the internal use of a shop making guns or textile machinery. Though sufficiently adaptable and reliable to remain in use into the 20th century, the Lincoln milling machine is more significant as the preeminent product of the nascent machine tool industry.

The miller reached its modern form in 1861 as a joint effort by Frederick Howe and Joseph R. Brown. Howe was serving as superintendent of a rifle factory in Providence, R.I., and he commissioned from Brown's nearby instrument shop a machine to make the drill bits used in arms production. Howe's long experience with millers and Brown's astute problem solving finally resolved the difficulty of accurate vertical adjustment by mounting the entire table on a column and devising a mechanism to move it up and down on ways attached to the column face. Because of the built-in capacity to adjust the work in all three planes, Brown named the machine the Universal Milling Machine. It launched Brown's firm, later known as Brown & Sharpe, as a machine tool producer. This firm, along with Pratt & Whitney, and a slightly later entrant, Cincinnati Milling Machine, contended for technical and market supremacy into the middle of the 20th century.

By the end of the 1860s, vertical-spindle milling machines also became commonplace. They allowed cuts to be made on the top and sides of the workpiece without having to reset it. The vertical spindle also extended the range of millers in both directions. They could more easily handle heavy castings than the horizontal-spindle design, and the ease of setting up a cut made them the preferred option for fine toolmaking work. Horizontal-spindle millers remained the workhorse of high-volume production for many subsequent decades.

Until the middle of the 20th century, these basic milling machine variations underwent constant modification as the machine tool producers leapfrogged each other in a series of technical improvements. Belt-and-pulley drive gave way to all-geared work feeding. Constant-speed drives with incremental adjustment through change gears permitted the close matching of machine speed to work requirements. Vertical spindles gained the ability to feed downward into a workpiece, a critical capability in the demanding job of making the dies for metal-forming operations. Compound feeding, or moving the workpiece in two planes at the same time, allowed the precise cutting of curves and contours. Automobile manufacturers and other mass producers ordered specially configured milling machines for single-purpose uses such as machining cylinder heads.

The line of mechanical development begun by the Connecticut gunmakers was superseded by computer technology during the Korean War. Researchers at the Massachusetts Institute of Technology applied digital methods to describe the desired shape and guide the machine to produce it. Known as numerical control, this application of electronic technology became the formative influence for further innovations in cutting flat and contoured surfaces in metal.—M.W.R.

See also LATHE; MACHINE TOOLS, NUMERICALLY CONTROLLED; METAL STAMPING

Further Reading: Robert S. Woodbury, *Studies in the History of Machine Tools*, 1972.

Mining, Surface

Unlike underground mining, where operations take place beneath the earth's surface, in surface mining (known as *open-cast mining* in Great Britain) the overburden of soil and rocks is completely removed from the surface, allowing the extraction of coal or other minerals. There are two basic forms of surface mining. In the first, the materials to be extracted are dug from a pit. This method is primarily used for the extraction of copper ore, as well as sand, gravel, and stone, but it is also used when coal is found in very thick beds or where the seams are steeply sloped. In the second form of surface mining, commonly known as *strip mining*, the minerals are extracted from a series of cuts. In flat countryside or rolling hills, these cuts take the form of parallel trenches. On steeper hills and the sides of mountains, the cuts form a series of contour bands. Both of these types of strip mining are used extensively for the extraction of coal.

People have used surface mining for thousands of years in order to get at materials that lay close to the Earth's surface. In the late 19th century, surface mining began to be used for deposits well below the earth's surface. This was made possible by the development of heavy machinery that allowed the extraction of large volumes of minerals and the overburden that had covered them. At first, steam was the dominant source of power for these operations. Steam shovels dug out massive quantities of materials (often after they had been loosened by blasting) and loaded them into railroad cars pulled by steam locomotives. Today, electricity and diesel engines provide the massive amounts of power needed for surface mining.

Surface mining uses some of the largest machines ever built. Giant scrapers and bulldozers first remove much of the overburden. Coal, ore, or other minerals are then extracted by draglines, huge cable-operated buckets that are suspended from fixed booms. The booms of draglines may stand taller than a 20-story building, and each dragline bucket may hold as much as 168 m^3 (100 yd^3) of material. Alternatively, mines may use crawler-mounted bucketwheel excavators that are capable of digging up 3,800 m^3 (5,000 yd^3) of material in a single hour. The trucks used for hauling the excavated materials are some of the world's biggest wheeled vehicles, some of them capable of carrying loads of 91 metric tons (100 tons).

The immense quantity of material extracted and moved by these machines is what makes surface mining economically feasible, for the material that is extracted is often covered by so much overburden that vast quantities of earth have to be extracted in order to get at it. At the same time, however, the scale on which surface mining is conducted often poses a severe threat to the environment. The excavations and the spoil left behind scar the terrain, disrupt the habitats of wildlife, pollute lakes and streams, cause floods and landslides, and lead to further erosion. In the past, mining operators simply abandoned a worked-over area, leaving local inhabitants to cope with the consequences. In recent years, however, the governments of many nations have required the reclamation of land after it has been mined. In the United States, the environmental consequences of above-ground mining were addressed by the 1977 Surface Mining Control and Reclamation Act. The act established an Office of Surface Mining as an agency of the Department of the Interior and charged it with the drafting of regulations covering the conduct of mining. Among other matters, these regulations stipulate that mined areas have to be returned to their approximate original contours wherever possible. Spoil piles and depressions cannot be left at the site, and the disturbed areas must have their wildlife habitats restored wherever this is practical. The actual administration of these

Surface mining for coal, Germany (courtesy World Coal Institute).

provisions is the responsibility of the government of the state in which the mining occurs. The constitutionality of the legislation was challenged by mine operators in Virginia and Indiana, who claimed that it deprived them of property without compensation, but it was upheld by the U.S. Supreme Court.

The environmental consequences of surface mining have generated a great deal of justifiable criticism. At the same time, however, supporters of surface mining point out that it is used for the extraction of more than 60 percent of the coal used in the United States. Consequently, significant limitations on surface mining would have serious consequences for energy production. Surface mining is also safer than underground mining, and it is considerably more productive. The average daily output of a worker in a surface mine is approximately 28 metric tons (31 tons); the average production of an underground miner is about 10 metric tons (11 tons). However, some of the superior productivity of surface mining is offset by the costs of reclaiming the land after it has been mined.

See also COAL; EXPLOSIVES; ENGINE, DIESEL; EROSION; LOCOMOTIVE, STEAM; MINING, UNDERGROUND

Mining, Underground

Underground mining is employed when deposits of coal or minerals lie too far beneath the earth's surface to be extracted through other means. Miners have worked underground for thousands of years. Mines for the extraction of copper began to be worked in what is now Serbia around 4500 B.C.E., while underground flint mines in Western Europe go back to at least 4000

B.C.E. Archeologists have even recovered the skeleton of a flint miner in Belgium who was killed in a cave-in around 3500 B.C.E.; the antlers that he used for a pick were found lying next to him.

Early mines were simple excavations, but by the 3d-millennium B.C.E. the inhabitants of the Sinai Peninsula were using an elaborate network of shafts and galleries for the mining of copper. In China, shafts and galleries were dug for the mining of coal at considerable depths; one ancient mine in Hebei province has a vertical shaft that extends 46 m (150 ft) into the ground. Even more impressive were some of the mines worked by the Romans. One mine in Roman-controlled Spain extracted silver-bearing lead from galleries that were 1,070 m (3,500 ft) long and 200 m (650 ft) deep.

The extent of underground mining diminished during the early Middle Ages, although the basic techniques of the Romans continued to be employed. By the middle of the 15th century, mining underwent a revival that was stimulated by increased demand and technological advance. Of particular importance was the development of ways to overcome one of the most vexing problems of underground mining, the removal of water. To this end, mining engineers devised a variety of pumping devices, many of them quite elaborate. Some mines used a chain of treadmill-powered wheels that lifted water towards the surface by passing it from wheel to wheel. Other water-removal methods used force pumps or suction pumps. In addition to stimulating the advance of hydraulic technologies, underground mines also gave rise to the world's first railroads.

The development of technologies for underground mining was not the exclusive province of practically minded miners and engineers. Men of science such as Galileo and Newton also turned their attention to ventilation, pumping, and mineralogy. At the same time, many mining personnel had to have a working knowledge of chemistry, metallurgy, geology, hydraulics, and civil and mechanical engineering. Much of that knowledge was codified by Georg Bauer (usually known by his latinized name, Agricola) in one of the great technological handbooks of all time, *De Re Metallica*, which was posthumously published in 1556. Improve-

A rotary cutter working a coal face in Britain (courtesy World Coal Institute).

ments in underground mining technology increased the output of coal and mineral ores, but equally important, many of them were used to improve productivity in other sectors of the economy.

In the centuries that followed, the output of underground mines expanded substantially, but this was largely the result of increases in scale rather than the introduction of new technologies. However, in the 18th century, mining received a major boost with the invention of the steam engine. The early history of the steam engine is closely associated with mining, for one of its most important early applications was pumping water out of mines. In addition, by the early 19th century, steam locomotives were being used to pull trains in the vicinity of coal mines, and steam-powered ventilation fans came into use around 1830. At about this time the hazards of underground mines were considerably reduced by the introduction of the safety lamp, which lessened the likelihood of underground explosions.

At this time, the actual extraction of coal, ores, and other minerals was still done with simple tools and human muscle, for steam was difficult to employ underground. Mining began to be mechanized in the mid-19th century through the use of compressed air to power circular cutters and drills. Towards the end of the 19th century, electricity began to be used for illumination and power, although many underground mines were not electrified until well into the 20th century.

Today, underground mining is most often used for the extraction of coal. Underground mines are of three general types, the choice of type being determined by the depth of the deposit and the nature of the surrounding terrain. A *drift mine* uses a horizontal entry from a hillside to gain access underground. A *slope mine* uses a inclined shaft that extends from the surface to the deposit; it is used when the deposit lies fairly close to the surface. A *shaft mine* allows access to deep deposits through the use of a vertical shaft from which galleries branch out. Shaft mines can go to considerable depths, as much as 3 km (10,000 ft).

Once underground, the extraction of a deposit can be done through two basic techniques: the room-and-pillar method or the longwall method. In room-and-pillar mining, a portion of the deposit is removed, leaving a space 6 to 9 m (20–30 ft) wide known as a *room*. The rest of the deposit and the associated rock are left in place to serve as pillars that support the roof covering the areas being worked. After the coal has been extracted from the room, the pillars can also be extracted through a process known as *retreat mining*, which allows the roof to collapse as the pillars are removed.

Room-and-pillar mining can employ conventional means of extracting the coal from the face—cutting, scraping, drilling, and blasting—or it can use continuous-mining techniques. Continuous mining uses electrically powered cutters capable of ripping large quantities of coal, salt, or soft minerals from the face. The extracted deposit is then loaded into a scraper conveyer by rotating disks or gathering arms. The conveyer takes the coal to a shuttle car, from which it is transferred to a conveyer belt that takes it to the surface. The use of mechanized loading in both room-and-pillar and continuous mining has resulted in a substantial savings of labor. In the early 1920s, virtually all of the coal mined in the United States was loaded by hand, and even in 1950, 30 percent was still hand loaded. Today, mechanical loaders have almost completely taken over this task.

In the United States, about 30 percent of the coal is extracted through the use of the second basic method of underground coal extraction, longwall mining. This technology was originally developed in Europe; it did not take hold in the United States until about 1960. It is now used in about 100 coal mining operations in the United States. In longwall mining, a cutting head mounted on an armored face conveyor moves back and forth across a coal seam as much as 240 m (800 ft) wide and up to 2,130 m (7,000 ft) long. The cutting takes place under temporary roof supports (known as *shields*) that advance as the seam is cut, leaving behind collapsed roof material known as *gob*.

In the early 1950s, longwall coal mining was intensively studied by a group of British industrial sociologists. Their research showed that the introduction of longwall mining affected worker autonomy and the cohesion of work groups, making it necessary to restructure the work environment so that work groups rather than individual miners were treated as the fundamental units. This entailed making the workers responsible for the allocation shifts and jobs, and paying bonuses to the group rather than to individuals. This research was also the beginning of the "sociotechnical" school of organizational analysis, which requires that attention be paid to the social as well as the technical aspects of a job, especially when a new method of production is introduced.

See also AIR PRESSURE; DRILLS, PNEUMATIC; LOCOMOTIVE, STEAM; MINING, SURFACE; PUMPS; RAILROAD; SAFETY LAMP; STEAM ENGINE

Mir Space Station

Beginning in 1971, a series of Salyut space station missions demonstrated that there were no insurmountable technical, physiological, or psychological barriers to long stays in space. Following the success of the Salyut series, on Feb. 20, 1986, the Soviet space program launched the next generation of space station, Mir ("peace"). On Mar. 13, 1986, two Soviet cosmonauts were launched on a voyage that would put them aboard Mir. This initial crew ended up being very well traveled in the course of their 125 days in space. Leaving Mir, they docked with and then boarded the Salyut 7 space station on May 6, 1986, rejoined Mir on June 25, 1986, and finally returned to Earth on July 16, 1986.

Although considerably smaller than the U.S. Skylab, Mir is one of the largest objects ever orbited, with a length of 13.1 m (43 ft), a maximum diameter of 4.15 m (13.6 ft), and a weight of 21,000 kg (23 tons). Attached at right angles to the main body are 76 square meters of solar panels; together they produce 9 kW of power. Orbiting at an altitude of approximately 350 km (217 mi), Mir is subject to a slight atmospheric drag that, if unchecked, would eventually cause it to fall back to Earth. Counteracting this tendency are gyrostabilizers that provide a reference point for two rockets that maintain Mir's orbit by firing when necessary. Where later versions of Salyut had only two docking ports, Mir has six. At each end, aligned along the craft's central axis, there are two main ports, along with four auxiliary ports perpendicular to the main ports. A supply ship or other spacecraft initially docks with one of the two main ports and then is transferred to one of the auxiliary ports via its onboard robot control arm. These ports can receive special lab modules such as a workshop module for the in-space production of materials, an observatory module for astrophysics, and a laboratory module for biological research.

Mir is resupplied by manned spacecraft and unmanned "Progress" supply craft. It was hoped that Mir would eventually be served by reusable Buran ("snowstorm") space vehicles. Buran had a marked resemblance to the U.S. Space Shuttle, except that it did not have its own rocket motor. The Buran made one unmanned test flight on Nov. 15, 1988, but the program succumbed to the financial problems that accompanied the collapse of the Soviet Union. It is now a forlorn display in a Moscow park. The post-Soviet era was the scene of other difficulties for the space program, as when a temporary inability to provide rocket motors for a Soyuz spacecraft left three cosmonauts stuck in Mir for 2 months beyond the planned mission.

Problems such as these should not obscure the fact that the Russian space station program has been unmatched by any other country. Except for a 6-month period in 1989, Mir has been continuously occupied since the first cosmonauts boarded it in 1986, and it has been the site of record-breaking stays in space, the most recent being the women's record of 169 days set by U.S. astronaut Shannon Lucid in 1996. A major goal of the Mir program has been to find out more about human adaptation to conditions in space, especially weightlessness. It has been learned, for example, that 6 hours of vigorous exercise each day for a month prior to returning to Earth mitigates some of the problems associated with a return to Earth's gravity. Mir's exercise facilities include a stationary bicycle and a treadmill, along with suits that stimulate blood circulation.

By early 1998, Mir had been home to 62 people from more than a dozen countries. United States involvement in the Mir program began in 1991, when George Bush and Mikhail Gorbachev agreed to a number of visits by U.S. astronauts. The first of these occurred in 1995, when the U.S. Space Shuttle docked with Mir from February 3 to February 11. U.S. astronaut Norm Thagard remained on Mir until he was picked up in the course of another Shuttle-Mir rendezvous that occurred from June 27 to July 7 of that year.

This and subsequent stays aboard Mir by U.S. astronauts can be seen as the first phase of cooperation by the United States and Russia in the construction and operation of the international space station that will supplant Mir. Many technical, political. and economic obstacles stand in the way of this project, but when it is completed it will have benefited greatly from the knowledge acquired in the course of the Mir program.

See also SKYLAB; SOYUZ AND SALYUT; SPACE SHUTTLE; SPACE STATION

Missile, Intercontinental Ballistic (ICBM)

An intercontinental ballistic missile (ICBM) is a nuclear-armed, rocket-powered guided missile with a range of more than about 5,000 km (3,000 mi). It can carry a single warhead or a number of separate warheads that may be independently targeted. The missile is guided only during a few minutes of powered flight, after which the warhead or warheads coast ballistically to the target(s). Because an ICBM is guided for only a short time, a small error can produce a wildly inaccurate trajectory; an error of 1/1,000 of a second in terminating the rockets' thrust may cause a target to be missed by several hundred meters.

First deployed around 1960 by the United States and the Soviet Union, the ICBM became the mainstay of their nuclear deterrents. Because of the great expense of developing and deploying large nuclear forces, the ICBM has remained a superweapon confined to only a handful of great powers. In addition to the United States and the successor states of the Soviet Union (soon only Russia will have ICBMs), the sole nation with land-based ICBMs is China. Great Britain and France also possess submarine-launched ballistic missiles (SLBMs), a close relative of the ICBM, with intercontinental range.

The exotic concept of a long-range ballistic missile was first discussed in the 1920s, but only after the Germans deployed the V-2 rocket in 1944 and the United States dropped the atomic bomb on Japan in 1945 did a nuclear-armed ICBM become imaginable. Numerous technical obstacles stood in the way, however. Although the United States and the Soviet Union, and to a lesser extent Great Britain and France, quickly absorbed German rocket technology and personnel, liquid-fuel rocket engines were still too small to boost a missile over such huge distances. The early atomic bombs were also much too heavy and the guidance challenges were extremely daunting: The V-2 was very inaccurate even at its short range. The United States therefore concentrated on manned bombers and cruise missiles. Stalin's Soviet Union, however, lacked effective bombers and desperately needed a delivery system for its atomic bomb, first tested in 1949. Under the leadership of the engineer Sergei Pavlovich Korolev, the Soviet Army energetically pushed liquid-fuel rocket development. In 1953, Korolev's ICBM, the R-7, was approved. Designed to accommodate the massive early Soviet thermonuclear weapon, it was huge, with 20 engines grouped in a core vehicle and four boosters that dropped off in flight. It was first successfully tested in August 1957 and then used to launch the world's first artificial satellite, Sputnik, in October of that year.

With Cold War tensions rising and a hot war taking place in Korea, the U.S. Air Force (USAF) began to fund ICBM development in the early 1950s. But the Atlas program, contracted to Convair (later General Dynamics), only became a high priority in 1954, when it became clear that the Soviet Union might soon threaten the United States with ICBMs. In 1955, President Eisenhower made Atlas a crash program; the Sputnik shock accelerated funding of ballistic missiles still further. The Atlas was tested in 1957, achieved intercontinental range in 1958, and was first deployed in unprotected sites in 1959–60. Since it used supercold liquid oxygen, as did the Soviet R-7, the Atlas required hours to prepare for launch. The R-7 was even more unwieldy, and only a handful were ever deployed. Both missiles were quickly superseded by a second generation with multiple stages and storable, hypergolic (self-igniting) liquid propellants, such as nitrogen tetroxide and hydrazine, and then by a third generation with massive solid-propellant motors that allowed virtually instant launch. Basing modes on both sides quickly evolved toward hardened underground silos strong enough to survive all but direct nuclear hits.

The Soviets were the first to develop a submarine-launched ballistic missile, but it was the American Polaris missile that pioneered large solid-propellant motors. Initiated by the U.S. Navy in 1956, the Polaris program became operational in 1960. Although the first missiles had a range of only 2,250 km (1,400 mi), this program paved the way for the USAF's Minuteman solid-fuel ICBM. With initial basing occurring in 1962, a total of 1,000 Minutemen eventually were deployed. The Soviet Union lagged in large solid rocket technology and did not deploy its first such ICBM until 1969. The Soviets eventually built over 1,600 ICBMs, many with storable liquid propellants. China tested its first ballistic missile of intercontinental range using these propellants in 1975.

During the 1970s and 1980s, the superpowers focused on improving the accuracy of inertial guidance systems, resulting in missiles that had a reasonable probability of coming within a few hundred yards of their target. At the same time, the number of warheads per missile was increased, in some cases to 10 or more multiple independently targeted reentry vehicles per missile. Before these improvements were made, an ICBM was useful only for retaliatory strikes; as such, it was the basis of a policy of "mutually assured destruction." But more accurate ICBMs equipped with MIRVs seemed to make possible a successful counterforce strategy against the other's nuclear forces possible. This, in turn, raised fears that one side might launch a first strike against the other's missile installations, crippling their ability to retaliate. Nuclear arms limitation agreements, followed by the end of the Cold War and actual reductions have greatly eased tensions since, but the United States and Russia will each retain hundreds of ICBMs for many years to come.—M.J.N.

See also ATOMIC BOMB; HYDROGEN BOMB; MISSILES, CRUISE; MISSILES, GUIDED; MISSILES, MULTIPLE INDEPENDENTLY TARGETED REENTRY VEHICLES (MIRVs); MISSILES, SUBMARINE-LAUNCHED; ROCKETS, LIQUID-PROPELLANT; ROCKETS, SOLID-PROPELLANT; SPUTNIK; SALT I AND SALT II; STRATEGIC ARMS REDUCTION TALKS (START)

Missiles, Cruise

Cruise missiles are unmanned aircraft that are able to deliver both nuclear and conventional warheads. Unlike guided missiles, cruise missiles have wings to supply lift, and unlike ballistic missiles they can be controlled throughout their flight. Cruise missiles can be powered by turbojets, turbofans, ramjets, and rockets. The type of powerplant used does much to determine the speed a cruise missile attains. Some can attain speeds up to Mach 3 (three times the speed of sound), while others travel as slowly as 100 kph (62 mph). Range is also variable; some cruise missiles are designed to travel only 20 km (12.4 mi), while others have a range of 3,000 km (1,864 mi). Cruise missiles can be launched from the ground (GLCMs), from aircraft (ALCMs), and from ships and submarines (SLCMs). Cruise missiles are used for a variety of military missions. Some are intended for use against ground troops, others against ships; still others are strategic weapons intended to deliver a nuclear blow.

The ancestor of the modern cruise missile was the German V-1. Powered by a ramjet motor that propelled it up to speeds of 650 kph (400 mph), the V-1 became operational in June 1944. About 20,000 "buzz bombs," as they were sometimes called by the Allies, were launched by the Germans. Although many of them were shot down or failed to reach their targets, they still accounted for nearly 11,000 deaths and 28,000 injuries. The United States deployed some second-generation cruise missiles, such as the Air Force's Snark and the Navy's Regulus II, in the 1950s and 1960s. Spurred by improvements in guidance systems, cruise missiles became an important part of the arsenals of many nations in the 1980s. More than 70 countries now possess antiship cruise missiles such as the French Exocet and the American Harpoon. Their accuracy gives these relatively inexpensive weapons the ability to sink or at least incapacitate large, expensive ships, as was demonstrated in the 1982 Falklands War. The accuracy of cruise missiles also makes them highly effective weapons against land targets. The standard assessment of missile accuracy is the circular error probable (CEP); this indicates the maximum distance from a target that half of the missiles or their warheads will land. Although the information is classified, it is thought that the U.S. Tomahawk cruise missile has a CEP of no more than 6 m (19.7 ft). In other words, in the event of an attack of Tomahawk missiles, half of them would fall within a circle with a radius of 6 m.

One of the reasons for this high level of accuracy has been the development of gyroscope-based inertial guidance systems that keep the missiles on their course. However, inertial guidance systems produce some deviation from an intended course, so for even greater precision, advanced cruise missiles also use a

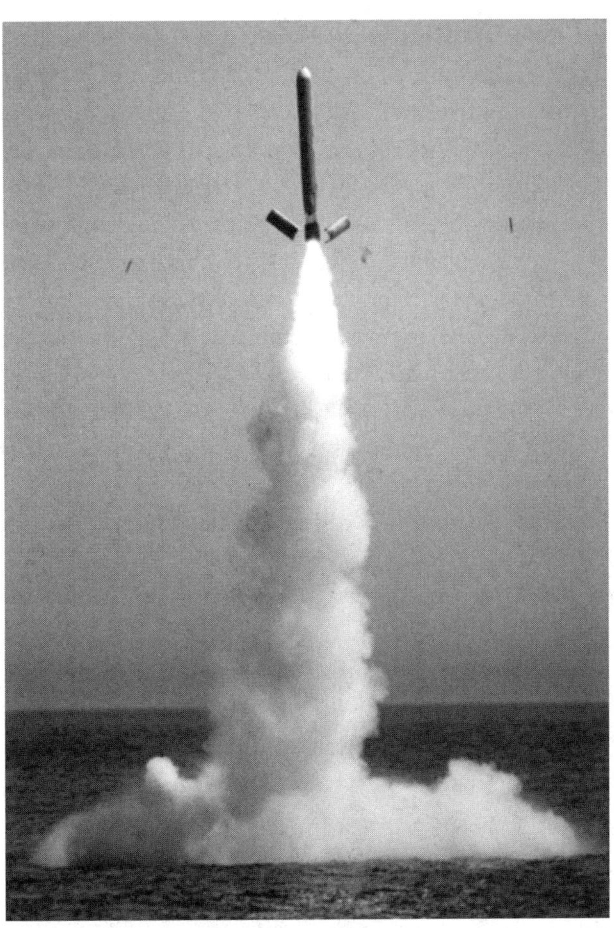

A submarine-launched Tomahawk missile. A pair of short wings and a tail assembly fold out shortly after launch (courtesy U.S. Naval Institute).

guidance technology known as Digital Terrain Comparison (TERCOM). This system uses onboard radar to digitally map the area the missile is flying over; the maps so generated are then compared with digitized maps stored in the missile's computer. Even more advanced guidance systems use a satellite-based global positioning system to provide highly accurate guidance to a target. Even without some of the guidance systems now in use, cruise missiles convincingly demonstrated their accuracy in the 1991 Gulf War against Iraq; of the 288 Tomahawk missiles launched, about 85 percent are thought to have struck their intended targets.

In addition to being accurate, cruise missiles also have the advantage of being difficult to track and shoot down, in large measure due to their ability to fly at very low altitudes. American Tomahawks fly above the ground at an altitude of 30 to 100 m (98–328 ft), depending on the terrain they are flying over. Radar usually cannot find them at these altitudes, for their signature is obscured by buildings, hills, and other obstacles. To make matters even more difficult, many cruise missiles make use of stealth technology, further diminishing their observability by radar and infrared sensors.

Even if they are spotted, cruise missiles are difficult to shoot down. Surface-to-air missiles have a chance of destroying their target only when the latter approaches from a high altitude. They are therefore of little use against ground-hugging cruise missiles. Other defenses, such as lasers or particle-beam accelerators, are not likely to be much better. The elaborate and massively expensive missile defenses envisaged by the Strategic Defense Initiative (the Star Wars program) would be of little value in the event of a cruise missile attack.

Cruise missiles create many difficulties for arms control. They are easy to produce or acquire because they are relatively inexpensive and generally use technologies that are readily available. For example, inertial guidance systems are no longer an exotic technology; the ones used for cruise missiles are essentially the same as the ones used for small private aircraft. Because they make modest financial and technological demands, cruise missiles have been deployed by military forces throughout the world. Many Third World nations have cruise missiles. Most have been bought from the United States and Europe, but Brazil, North Korea, China, Iraq, Israel, South Africa, and Taiwan have been able to make their own cruise missiles, some of which have entered the world arms trade.

It has been especially difficult to incorporate cruise missiles into arms control agreements. Even when agreements limiting the number of cruise missiles can be hammered out, there still remains the problem of verification. Ground-launched cruise missiles and their launchers do not take up much space and they can be moved about, making it difficult to keep track of them. Even more vexing is the fact that, short of on-site inspection, there is no way to distinguish a cruise missile with a nuclear warhead from a conventionally armed one. Consequently, an arms control agreement that limited the number of cruise missiles with nuclear warheads would be very hard, if not impossible, to enforce.

See also GLOBAL POSITIONING SYSTEM; GYROSCOPE; MISSILES, GUIDED; RADAR; ROCKETS, LIQUID PROPELLANT; ROCKETS, SOLID-PROPELLANT; SALT I AND SALT II; STEALTH TECHNOLOGY; STRATEGIC ARMS REDUCTION TALKS; STRATEGIC DEFENSE INITIATIVE; TURBOFAN

Missiles, Guided

Guided missiles are unmanned, self-propelled, flying weapons with guidance systems and high-explosive or nuclear warheads. Since World War II, they have become so important that our era can be called "the age of the guided missile." Most are guided rockets, but there is also an important subclass of winged missiles using air-breathing turbojet or ramjet engines, most notably cruise missiles. Guided missiles can also be divided into tactical (battlefield/short-range) and strategic (long-range) categories.

This class of weapons emerged in the 20th century because of the convergence of a few key technologies: gyroscopes, flight technology, reaction propulsion (jet and rocket engines), and electronics. In the WWI era, the development of aviation and of gyroscopic autopilots small enough to fit into aircraft made possible the first "flying bombs," the precursors of modern cruise missiles. In 1917–18, the United States financed the construction of two small, unmanned biplanes intended to dive on German targets with high explosives, but it was far more difficult than expected to get them to fly without a pilot present. Similarly unsuccessful experiments were carried out in the interwar period by the British. The first true cruise missile was the V-1 ("Vengeance Weapon 1"), launched by Nazi Germany against London beginning in June 1944. This miniature airplane carried a 900-kg (1-ton) warhead to a range of about 240 km (150 mi), propelled by a pulsejet engine and guided by a simple autopilot. Thousands were launched against Allied cities, with some psychological impact but little military effect. Accuracy was poor, and the V-1 could be shot down by conventional antiaircraft defenses.

The Germans were also the first to highlight the potential of ballistic missiles and other guided rockets. The conception and invention of liquid-fuel rocketry by independent enthusiasts and engineers in the interwar years, with the long-term goal of spaceflight, led to discussions of launching rockets with high-explosive or poison-gas warheads against enemy cities hundreds or thousands of miles away. This exotic concept interested the militaries of only two countries in the 1930s: Nazi Germany and the Soviet Union. But the rocket efforts of the latter were unfocused and were severely disrupted by Stalin's purges of 1937–38, when leading rocket designers were shot or sent to concentration camps. The German Army, on the other hand, created a focused liquid-fuel rocket program, led by General Walter Dornberger (1895–1980) and Dr. Wernher von Braun (1912–1977), that eventually produced the A-4 (V-2) ballistic missile, with a range of about 320 km (200 mi). First launched against Allied cities in September 1944, the V-2 could not be shot down because it arrived supersonically, but it was very expensive, inaccurate, and armed with only a 900-kg (1-ton) high-explosive warhead. It was thus even more spectacular and ineffective than the V-1.

During the war, Germany also produced the first antiship missile and experimented with ground-launched antiaircraft missiles. The latter were guided by gyroscopes and either a manual joy stick operated by a ground observer or an automatic guide beam, but none was ever deployed, mostly because of the inadequacies of existing guidance systems and electronics. At war's end, the United States, the Soviet Union, Great Britain, and France all moved to seize control of German rocket technology and personnel, along with their advanced aeronautical research. The Cold War began soon afterward, which, along with the arrival of nuclear weapons in 1945, produced an explosion of missile research and development. Although the United States absorbed the heart of the German Army rocket group led by von Braun, long-range ballistic missiles were not a high priority until after 1950, and especially after 1954, because of the American advantage in nuclear weapons and manned bombers. Stalin's U.S.S.R., on the other hand, lagged in aircraft technology and desperately needed a means to deliver its atomic bomb to the United States, so it funded ballistic missile technology more energetically. To the shock of the West, the Soviet Union was first to successfully test an ICBM in August 1957, but the United States quickly used its superior technical capability to produce better ICBMs in larger numbers. Both superpowers also deployed shorter-range ballistic missiles in Europe during the late 1950s and early 1960s. Great Britain and France struggled with limited resources to build their own nuclear missiles during these years, but only France ultimately deployed homegrown designs. China followed suit in the 1960s and 1970s.

During the 1950s, the United States and Soviet Union experimented with long-range, nuclear-armed cruise missiles as well, without much success. These large unmanned jet aircraft were unreliable, inaccurate, and vulnerable to air defenses. The United States did deploy some Snark missiles in 1959, but deactivated them within 2 years. Shorter-range Matador, Mace, and sea-based Regulus missiles had a slightly longer military life. The most useful members of this type were air-launched cruise missiles on strategic bombers, such as the Air Force's Hound Dog.

Submarine-launched ballistic missiles (SLBMs) had a much greater impact on the strategic arms race. In this case, the Soviet Union was again first, testing the short-range R-11FM missile from a surfaced submarine in 1955. But it was the American Polaris program that was truly revolutionary. Initiated by the U.S. Navy in 1956, Polaris could be launched underwater from submarine missile tubes, and was the first ballistic missile to employ large solid-rocket motors. This technology, developed in the United States, offered many advantages over liquid-propellant rocketry, with its complicated handling systems and often slow reaction times. Polaris had only intermediate range, but it paved the way for the later Poseidon and Trident SLBMs and for the land-based Minuteman and MX ICBMs. The Soviet Union lagged behind in solid-fuel rocketry until the 1980s, and used storable liquid propellants more extensively, but eventually deployed large solid-fuel missiles too.

After 1945, tactical missiles also underwent rapid development, although few types were militarily effective until the mid-1950s. Following the German lead, antiaircraft missiles were a high priority, especially in view of the need for defense against nuclear-armed bombers. The first American operational surface-to-air missile was Nike Ajax, deployed in 1953, which was followed by the solid-fuel Nike Hercules in 1958 and the smaller, mobile Hawk in 1959, as well as by the U.S. Navy's Terrier (1956) and Talos (1959) for fleet protection. The Soviet development of the German Wasserfall, SA-1 (NATO designation), first saw use in 1954, and the more effective SA-2 appeared in 1957. Except for the Hawk, all these missiles were useful mostly against high-altitude aircraft not taking evasive action. But during the 1960s, guidance capabilities improved markedly, as did mobility. In 1964, the United States deployed Redeye, the first shoulder-launched missile to be carried by an individual soldier for use against low-flying aircraft. That was followed by the Soviet SA-7 in the early 1970s and by the Stinger in the 1980s, which had major impacts on the Vietnam and Afghanistan conflicts, respectively.

Less successful were the ultimate extension of antiaircraft missiles, the ABM (antiballistic missile) for nuclear defense. The United States quickly abandoned its single Safeguard site in the 1970s as ineffective, while the Soviets maintained its Galosh ABMs around Moscow anyway. During the Persian Gulf War, the upgraded Patriot antiaircraft missile was used for the first time as an ABM against short-range Iraqi SCUDs, but the Patriot was less effective than advertised.

Small aircraftborne missiles also became significant in the mid-1950s. The U.S. Navy's AIM-9 Sidewinder was the first effective air-to-air missile with heating-seeking (infrared) homing. Sidewinder was imitated by the Soviet Union (the AA-2 in NATO designation) and by other countries. Larger, longer-range air-to-air missiles that homed on a radar beam from the launching aircraft (semiactive guidance) appeared about the same time, such as the Navy's AIM-7 Sparrow and Soviet AA-1 and AA-3. A panoply of more-and-more sophisticated missiles followed, leading ultimately to "fire-and-forget" missiles like the U.S. AMRAAM with their own radar systems, which do not require the attacking aircraft to expose itself by maintaining continuous radar lock. All major powers have also developed a bewildering array of air, ground, and sea-launched missiles for use against submarines, ships, tanks, ground installations, and radar sites. Guidance techniques include radar, television, and, most recently, laser designation. Especially noteworthy for their effects on the "limited" wars from Vietnam to the Persian Gulf have been the air-launched antiradiation missiles like the U.S. Shrike, which passively home on enemy radars, antiship missiles such as the Soviet Styx and French Exocet, and helicopterborne antitank missiles like the American TOW and Hellfire.

The Persian Gulf War of 1991 highlighted the appearance of small and effective cruise missiles as well, notably the U.S. Navy's sea-launched Tomahawk. Very small, efficient turbojet engines plus the increasing reliability and miniaturization of electronics, computers, and inertial guidance systems since the 1960s made these missiles feasible. But particularly important was the creation of terrain-following radars, which permitted the cruise missile to hug the landscape at very low altitudes. The Tomahawk and its USAF equivalent, the ALCM (Air-Launched Cruise Missile), had their roots in the perfection of McDonnell-Douglas's TERCOM terrain-contour-matching system, in the Navy's search for a new antiship missile, and in a renewed interest in cruise missiles arising from the limits placed on ballistic missiles in the superpower arms control agreements of the 1970s. The ALCM was deployed on B-52 bombers in 1982 and the Tomahawk on ships in 1984. The U.S. Air Force also placed nuclear-armed GLCMs (Ground-Launched Cruise Missiles, variants of the Tomahawk) in Europe beginning in 1983, but these were removed within a few years as a result of the Intermediate Nuclear Force Treaty. The Soviet Union also developed effective cruise missiles in the 1980s.

Guided missiles have decisively reshaped military strategy, tactics, and force structures since 1950. Fortunately, the end of the Cold War has slowed the pace of technological change and has decreased the likelihood of nuclear war, but even if directed energy weapons like lasers become more effective, guided missiles will remain central to virtually all military endeavors.—M.J.N.

See also GYROSCOPE; LASER; MISSILE, INTERCONTINENTAL BALLISTIC; MISSILES, CRUISE; MISSILES, SUBMARINE-LAUNCHED; MISSILES, SURFACE-TO-AIR; PRECISION-GUIDED MUNITIONS; ROCKETS, LIQUID-PROPELLANT; ROCKETS, SOLID-PROPELLANT; SALT I AND SALT II; STRATEGIC ARMS REDUCTION TALKS (START); SUBMARINE, TRIDENT; TURBOJET

Missiles, Submarine-Launched

In 1955, the U.S. Navy initiated the Polaris nuclear submarine program to base ballistic missiles in the oceans in the event of war with the Soviet Union. Early candidates for the missiles included liquid-fueled versions of the army's Jupiter missile, but the navy needed a much smaller missile with a solid fuel for launch from submerged submarines. From 1957 to 1960, the Polaris program proceeded from design to construction, with the USS *George Washington* becoming operational on Nov. 15, 1960. It carried 16 Polaris A1 missiles, each with a range of 2,222.4 km (1,200 nautical miles [n.m.]) and a reentry vehicle containing one W-47 800-kiloton (kt) warhead. The first underwater launch of a live missile occurred on Apr. 14, 1960, off San Clemente Island.

Subsequent versions of the Polaris missiles, including the A2s and A3s, carried up to three warheads of 600 kt each. The A3 missiles were aboard the 41st and last fleet ballistic missile submarine of the series, the USS *Will Rogers*, when it departed on patrol in October 1967. Eventually, most of the Polaris submarines were retrofitted with newer Poseidon (C-3) missiles, each with a 3,860- to 4,830-km (2,400–3,000-mi) range and capable of carrying combinations of up to 14 reentry vehicles and warheads of up to 6.4 megatons (mt).

In 1974, a new ballistic missile program, the Trident program, was started. Trident submarines, called the *Ohio*-class submarines, could each carry up to 24 C-4 missiles. Each of those missiles could carry combinations of up to eight reentry vehicles, with warheads of 800 kt. More important, though, the newer C-4 missiles had a range of 8,149 km (4,400 n.m.) and carried multiple, independently targeted warheads. The U.S. Navy planned to replace those missiles in all submarines during the 1980s with a longer-range missile, the D-5 (Trident II), capable of a 11,112-km (6,000-n.m.) range. However, as the Trident submarines proved more undetectable than even the navy anticipated, it reduced its range requirements on the D-5 to achieve greater warhead capacity and/or megatonnage.

Other nations adopted submarine-launched-ballistic-missile-firing (SLBM-firing) submarines, including France, Great Britain, and the former Soviet Union. In addition, nations such as China and India are considered to have the technical capacity to create such weapons, but apparently have not done so as of this writing. Soviet submarines, including the *Deltas* and the *Typhoons*, proved exceptionally capable platforms, with the Typhoons capable of extremely deep operations. However, the United States far exceeded other nations in its ability to develop long-range and accurate missiles based on submarines with exceptionally long cruising ranges. Insofar as an SLBM is a component of the entire system, the capabilities of the U.S. Polaris/Poseidon fleets and the Trident fleets made the American submarine-based missile force a deadly threat to foreign adversaries.

Submarine-launched missiles are carried vertically in tubes. When the vessel intends to fire, it approaches a shallow depth (a classified number, but less than 30 m [100 ft] below the surface). The submarine opens its missile hatch cover, and the prepared missile is ejected from its tube by a powerful steam and/or gas jet. At the same time, the firing sequence bursts a plastic tube enclosure at the top of the tube. Forced out of the tube by the gas, the missile's first-stage, solid-propellant booster ignites.

Completing the firing sequence is a test for any vessel, which must immediately compensate for the loss of between 29,500 kg (70,000 lb). Modern submarines backflood immediately in order to offset the loss of weight. The entire process to eject the missile and ignite the boosters takes less than a minute, based on Poseidon tests. Although the navy has never publicly admitted to firing an entire complement of Trident missiles at one time, estimates are that an *Ohio*-class submarine can release its entire lethal magazine of 24 missiles in under 23 minutes. Most analysts, however, agree that no submarine could have survived in launch mode for more than 5 minutes before becoming targeted by enemy antisubmarine warfare systems or enemy ballistic missiles.

Modern SLBMs are targeted and guided through the use of the satellite Global Positioning System, which can feed navigational data into the missiles targeting computers in flight. Consequently, the accuracy of American SLBMs was considered to be exceptionally high—between 100 and 300 meters (109–328 yards) radius of the target. In addition to the traditional SLBMs, newer submarines, including the USS *Los Angeles* class and the former Soviet *Oscar* class, also have started to carry cruise missiles. Cruise missiles are air-breathing weapons that fly along a horizontal trajectory, hugging the terrain until they reach their target. Although much slower moving than ballistic missiles, cruise missiles are extremely difficult to detect by traditional upward-looking radars. Cruise missiles can carry nuclear warheads, but the warheads are not nearly as powerful as those carried on the SLBMs.

Recent thaws in the Cold War have led to arguments that many, if not all, of the SLBMs are obsolete. Indeed, the navy has capped its procurement of Trident submarines at 18, and has plans to retrofit the remaining Trident force with D-5 missiles. But submarine-launched cruise missiles were fired effectively in the Gulf War (1991), and the sea-based missile force is still considered important to maintaining deterrence against foreign powers. Likewise, France, Great Britain, and Russia have not mothballed their nuclear submarine fleets.—L.S.

See also GLOBAL POSITIONING SYSTEM; MISSILES, CRUISE; MISSILES, MULTIPLE INDEPENDENTLY TARGETED REENTRY VEHICLES (MIRVS); SUBMARINE, TRIDENT

Further Reading: R. A. Fuhrman, "The Fleet Ballistic Missile System: Polaris to Trident," AIAA Paper 77-355, *The Journal of Spacecraft*, vol. 15, no. 5 (Sep.–Oct. 1978): 265–86.

Missiles, Surface-to-Air

Efforts have been made to shoot down airplanes for as long as aircraft have been used for military purposes. For decades, antiaircraft measures entailed the use of guns of some sort. The terrible destruction wrought by bombing during World War II, which culminated with the dropping of the first atom bombs, made a better means of defense imperative, especially as tensions mounted during the early years of the Cold War. On a more positive note, the rapid development of rockets in the late 1940s and early 1950s created new possibilities for air defense. In 1953, these were realized when the United States installed the first surface-to-air missile system for defense against bombers. This system was built around the Nike-Ajax missile and an extensive early-warning radar network. In 1958, the United States began to deploy an improved version, the Nike-Hercules, which was capable of intercepting bombers at an altitude of 45,750 m (150,000 ft). The Soviet Union followed suit; the capabilities of its surface-to-air missiles were dramatically demonstrated in 1960 when one of them shot down an American U-2 spy plane.

In the years that followed, improvements in rocketry, radar, and especially computers greatly increased the effectiveness of surface-to-air missiles. Today's missiles reach their target through a combination of internal and external guidance. After launch, they and their intended target are tracked by giant phased-array radar units. As a missile approaches its target, its onboard infrared heat detector leads it to the target's jet exhaust.

Although a great deal of effort and expense have gone into the development of surface-to-air missiles, they do not provide certain defense against incoming attackers. A bomber may attempt to defeat an intercepting missile by taking evasive maneuvers, and by throwing out chaff that confuses radar and flares that counteract infrared detectors. These countermeasures have been quite effective. During the Vietnam war, the Soviet-built SAM-2 missiles ringed around Hanoi were able to shoot down only 3 percent of attacking B-52 bombers. An even lower rate of success was achieved by Soviet-built SAM-6 missiles during the 1973 Arab-Israeli conflict. In recent years, the detection and interception of aircraft has been made even more difficult by the development of stealth technology and the ability to fly close to the ground in order to evade radar detection.

While using a missile to shoot down an airplane is hard enough, even more difficult is the interception of other missiles. This is a daunting task, for ballistic missiles approach their targets from a very high altitude at a speed of about 16,000 kph (10,000 mph). Despite the inherent difficulties, in the 1960s both the United States and the Soviet Union began to devote substantial resources to the development of an effective antiballistic missile (ABM). In 1969, the U.S. Senate authorized (by a single vote) the construction of an ABM system, only to have its deployment sharply limited by the first Strategic Arms Limitation Treaty. The U.S. ABM program built on an earlier attempt at missile defense, the Nike-Zeus of the late 1950s, but many doubts remained regarding its feasibility. As a result, many U.S. officials were willing to accept the SALT I provisions. Only one American ABM site was actually built (to protect an ICBM site near Grand Forks, N.D., at a price of $5.5 billion). The Soviet Union built an ABM system in vicinity of Moscow.

In addition to worries about its effectiveness, concerns were also voiced that an ABM system would be militarily and politically destabilizing. At the time of the ABM debate, "mutually assured destruction" had become the operative strategic doctrine. Neither the United States nor the Soviet Union could protect themselves from an all-out nuclear attack, but such an attack would be met by an equally devastating counterblow. An ABM system threatened to upset the equation by promising immunity from a retaliatory strike, allowing an aggressor to launch an initial attack with impunity. According to its critics, if the United States or the Soviet Union installed an ABM system, in a crisis situation the other country might be tempted to launch a preemptive first strike that would do more damage than a retaliatory strike against sites defended by ABMs.

The U.S. ABM system consisted of two different missile types, the Sprint and the Spartan. Both were guided by ground radar, powered by solid-fuel rockets, and contained nuclear warheads. In the event of an attack, the Spartan would be the first to be launched, taking off when a perimeter acquisition radar detected reentry vehicles at a distance of up to 4,000 km (2,500 mi). It would then be guided to its target by radar, destroying it as far way as 600 km (375 mi) from its launch site. Since the interception would take place above the atmosphere where there is no air to transmit a shock wave, the actual destruction would be the result of X rays and neutrons released by a nuclear explosion. Any surviving reentry vehicles would then be destroyed by Sprint missiles.

Critics of the ABM pointed out that even if the Spartan and Sprint missiles worked effectively, the radar system that guided them was vulnerable to enemy attack. It was also argued that an ABM system could be subverted by arming ICBMs with as many as 10 reentry vehicles so that no fewer than 10 ABMs would be necessary for the successful interception of a single incoming missile. The addition of decoy vehicles would make matters even more difficult. Finally, nuclear explosions, some resulting from defensive actions and others set off by the enemy, would seriously interfere with radar guidance of ABMs.

ABMs have never engaged attacking ICBMs. However, during the 1991 Gulf War it was widely reported that U.S. Patriot missiles had achieved a high degree of success in intercepting short-range Scud missiles launched from Iraq. As with a number of claims made during the Gulf War, these reports were not grounded in reality. Although official documents remain classified, it is likely that not a single Scud was actually shot down by a Patriot missile.

Efforts to develop an effective antiballistic-missile capability continue, although many of these center on new and largely untried technologies in the place of surface-to-air missiles. Meanwhile, smaller (in some cases handheld) surface-to-air missiles such as the U.S. Stinger play an important tactical role as battlefield defenses against airplanes and helicopters.

See also ATOMIC BOMB; BOMBING, STRATEGIC; MISSILE, INTERCONTINENTAL BALLISTIC; MISSILES, MULTIPLE INDEPENDENTLY TARGETED REENTRY VEHICLES (MIRVS); PRECISION-GUIDED MUNITIONS; RADAR; ROCKETS, SOLID-PROPELLANT; SALT I AND SALT II; STEALTH TECHNOLOGY; STRATEGIC DEFENSE INITIATIVE

Monoclonal Antibodies

Monoclonal antibodies are immunoglobulins produced by cells that have been cloned from a single precursor cell. Hence, these antibodies are termed *monoclonal antibodies*. In 1975, G. Kohler and C. Milstein took advantage of an antibody-secreting tumor cell line from a mouse to create clones of cells that could be grown in tissue culture and that produced large amounts of a single type of antibody. That technology has made possible the commercial availability of specific antibodies against virtually any molecule of interest. The method for accomplishing this has subsequently transformed many areas of biomedical research and practice.

In order to understand the production of monoclonal antibodies and to appreciate their usefulness, a brief overview of normal antibody production by the immune system might be helpful. The type of lymphocyte that produces antibodies is called a B-lymphocyte, or simply a B-cell. When a B-cell is exposed to an antigen that its antibody can recognize, the cell is induced to replicate and produce a clone of many cells like itself, most of which churn out antibodies. An antigen is a molecule that can induce an immune response, usually by stimulating lymphocytes to proliferate and manufacture specific antibodies. This is the way immunity normally develops. Foreign antigens may enter the body as infectious bacteria or viruses, or they may arise by mutation of body cells, as in certain types of cancer. It is the job of the immune system to attack and destroy these foreign antigens.

Cells of the immune system are unusual among normal body cells, in that they undergo gene rearrangement during development. This allows them to produce the many different types of antibodies required to inactivate the plethora of foreign molecules that the body can be exposed to during a person's lifetime. Because lymphocytes undergo gene rearrangement, the genetic makeup of one immune-competent B-lymphocyte will be slightly different from that of most others. Normally, several genetically different B-cells will react to a single foreign molecule, producing "polyclonal" antibodies that are not identical in their antigen-combining sites, and bind to different sites on the foreign molecule. However, antibodies produced by lymphocytes derived from a single precursor cell will all be identical and will have a single type of antigen-binding site (idiotype). This is the essential feature of monoclonal antibodies.

B-cells usually will not proliferate in the absence of antigenic stimulation. However, antibody-producing B-cells (plasma cells) can sometimes develop into a type of cancer called multiple myeloma or plasmacytoma. Because they are cancer cells, they are not limited in their replication as are most normal cells. This means that they have a kind of immortality. A type of plasmacytoma from mice was used in the initial work by Milstein and Kohler and continues to be the basis of most monoclonal antibody production today. The plasmacytoma cells lack an enzyme involved in the nucleic acid salvage pathway: hypoxanthine-guanosine phosphoribosyl transferase (HGPRT). The cells are induced to fuse with spleen cells from a mouse that has been immunized against an antigen of interest. The fused cells are hybrids of the plasmacytoma cells and the normal antibody-producing mouse spleen cells, and are often termed *hybridomas*. After incubation in a medium that kills the nonfused cells, the surviving hybridomas are diluted and plated out in microtiter wells so that there will be approximately one hybridoma cell per well. Clones of the cells that grow in the wells are screened for antibodies against the molecule of interest, usually by ELISA (enzyme-linked immunosorbent assay). Only a few of the hybridomas (perhaps 1 in 500) will produce antibodies to the antigen of interest. However, once a useful clone has been selected, its cells can be grown in large quantities, and virtually unlimited supplies of very specific antibodies can be made.

Monoclonal antibodies are now routinely used by many biochemists, cell biologists, and pathologists in localizing and identifying tissue molecules. Monoclonal antibodies to most purified biological molecules are available from biotechnology companies, either as standard products or generated to order. They are used routinely in biomedical test kits, e.g., for pregnancy and tumor-antigen detection. An exciting future application of monoclonal antibodies will be in "chemical surgery." Monoclonal antibodies attached to specific poisons, such as cancer chemotherapeutic agents, will be targeted to specific cell types that express unique antigens on their surface. Such reagents should be able to selectively kill antigen-expressing cancer cells without destroying normal body cells. New technologies are evolving for monoclonal antibody production. Genetically engineered bacteria will probably become the future sources of these highly specific reagents.—J.S.

See also CANCER; CHEMOTHERAPY; CLONING; ENZYMES; GENE THERAPY; GENETIC SCREENING

Monoculture

Monoculture is the planting of large expanses of agricultural land with only one variety of crop. In many parts of the world, farmers now cultivate a single variety of wheat, rice, or cotton that has been bred to give high yields. This pattern first emerged in industrialized nations that had strong agricultural research programs. Beginning in the 1950s, agricultural researchers developed new strains of wheat and rice in order to increase crop yields in poor, developing countries. The adoption of the improved varieties has brought substantial gains to the farm sectors of both rich and poor nations, but in recent years questions have been raised regarding the sustainability of cropping systems based on monoculture.

In contrast to monocropping practices, traditional agricultural systems often have been based on a multiplicity of different crops. Instead of being sown to a single crop, fields are planted with many different kinds of crops and crop varieties. In the Andes, for example, no less than 36 varieties of potatoes were observed to be growing in a single field.

This kind of cultivation has a number of advantages. Intercropping different kinds of plants helps to check the growth of weeds. The root systems of different plants go to different depths, so nutrients at all levels of the soil can be used. In similar fashion, crops of different heights can complement one another, allowing farmers to make good use of limited acreage. The cultivation of a number of different crops diminishes water runoff and checks erosion. Most important, the cultivation of many different kinds of plants prevents a single pest or plant disease from wiping out the whole crop. Plant diseases and attacking insects tend to be specific to a particular kind of plant, so the use of several different crops, or varieties of the same crop, can provide the first line of defense against potentially devastating pests and diseases. The maintenance of genetic diversity is also important for protection against future problems with plant diseases. On occasion, potentially disastrous diseases have been arrested by hybridizing the affected plants with varieties that have some natural immunity. But if the practice of monoculture eliminates plant diversity, there will be no resistant plants left for this purpose.

Although the cultivation of a multiplicity of crops confers some important advantages, agricultural trends clearly have gone in the opposite direction. The world's agricultural lands now sustain only about 150 plant species. Only three species provide nearly 60 percent of the calories that human beings get from plants. This pattern has not been confined to the most agriculturally advanced countries. In India, for example, 75 percent of the rice-growing areas are sown to only 10 varieties of rice.

There is no denying the fact that monocropped fields with their scientifically bred crop varieties produce much more abundant harvests than farmlands sown to several different kinds of crops or varieties of the same crop. At the same time, however, farmers, agricultural researchers, and government officials should be aware of the inherent dangers of monoculture, especially when it is taken to the point where all biodiversity has been eliminated.

See also GREEN REVOLUTION

Further Reading: Jules N. Pretty, *Regenerating Agriculture: Politics and Practice for Sustainability and Self-Reliance*, 1995.

Moore's "Law"

The development of the integrated circuit allowed the packaging of vast numbers of individual electronic components on a single silicon chip. When they first emerged in the late 1950s, integrated circuits included only few dozen components, but soon the number of transistors, resistors, and capacitors began to grow at a very rapid rate. In 1964, an electronic engineer named Gordon Moore reflected on the immediate past and noted half-jokingly that in the future, the number of components per chip would likely double every year. His prophecy was fulfilled, although by the late 1970s it took 18 months for doubling to occur, a pace that continued through the mid-1990s.

The continual expansion of the number of electronic components on a chip resulted in the commercial production of integrated circuits containing as many as 6 million transistors. The expansion in the number of components packed on a chip has been paralleled by a steady drop in prices of the components. In 1970, a megabyte of computer memory cost more than a half-million dollars; in the mid-1990s it cost less than $40. The ability to include so many components on a chip has provided a technological foundation for everything from cellular telephones to personal computers, and in so doing has revolutionized communications and information handling.

By the mid-1990s, the pace began to slow as chip manufacturers began to run up against insurmountable physical barriers to the continued increase in the number of electronic components. At the same time, the price of a new, state-of-the-art chip-manufacturing plant reached stratospheric levels, easily exceeding $1 billion. The semiconductor industry was beginning to replicate a common pattern of industrial change: A period of rapid improvement in performance amidst falling prices eventually gives way to a considerably slower pace of development. Refinement and product diversity were beginning to take the place of dramatic performance improvements and plummeting prices. No natural or man-made system can indefinitely sustain the rate of growth exhibited by the semiconductor industry from the 1960s to the 1990s. Moore's "law" proved to be remarkably prescient when it was first formulated, but it can no longer be taken as a guide to the future.

See also CAPACITOR; COMPUTER, PERSONAL; INTEGRATED CIRCUIT; RESISTOR; TELEPHONE SYSTEMS, CELLULAR; TRANSISTOR

Further Reading: G. Dan Hutcheson and Jerry D. Hutcheson, "Technology and Economics in the Semiconductor Industry," *Scientific American*, vol. 274, no. 1 (Jan. 1996): 54–62.

Morphine

The narcotic qualities of opium were known for centuries before its chemical constituents were isolated. In 1805, Friedrich Sertuerner (1783–1841), a German pharmacist, extracted an acid from the juice of the opium poppy and converted it into a crystalline form. Sertuerner found that the effects of this substance were more powerful than those produced by opium itself. He named the substance *principium somniferum*, a reference to its sleep-inducing properties. In 1817, he renamed it *morphium*, after Morpheus, the Greek god of dreams. In the same year, the French chemist Joseph Louis Gay-Lussac (1778–1850) referred to it as morphine, the word that is commonly used today.

Morphine use increased as a result of the invention of the hypodermic syringe in 1853. The powerful analgesic properties of injected morphine led to its widespread use by physicians. The use of morphine increased substantially during the American Civil War, when many soldiers received morphine injections to alleviate pain and to treat dysentery. Many soldiers returned home addicted to morphine. They were not alone in their addiction; many other people, including upstanding citizens like one of the founders of the

Johns Hopkins Medical School, were lifelong morphine addicts. Morphine addiction decreased in the course of the 20th century due to tighter restrictions on its distribution, but also because an even more powerful narcotic, heroin, had become available.

Morphine continues to be used under medical supervision as a painkiller. Given intramuscularly, subcutaneously, or orally in doses of 5 to 20 mg every 4 hours, morphine alleviates the pain associated with injury, surgery, or chronic illness. Strictly speaking, morphine does not "block" pain, but it makes pain more tolerable by affecting the patient's emotional state. Morphine is such an effective pain reliever that it is often used as the standard by which other analgesic drugs are evaluated. Still, it is underprescribed. For example, it is estimated that fewer than 50 percent of cancer patients receive adequate doses of morphine and other pain relievers. Many physicians are concerned that administration of morphine will lead to subsequent addiction, even though this rarely occurs with patients who have not had a previous drug problem.

As with all other drugs, users of morphine can build up a tolerance to it; one morphine addict reportedly used 5 grams of morphine daily, more than 1,000 times greater than a normal therapeutic dose. Even when taken in lower doses, morphine can produce a number of side effects: drowsiness, mood swings, depressed respiratory functions, and constipation. Withdrawal from prolonged morphine use is a difficult and painful process characterized by hyperventilation, restlessness, and confusion.

See also HEROIN; OPIUM

Mortising and Tenoning Machines

Mortises and tenons are the nuts and bolts of woodworking, essential for joining separate pieces of wood to create rigid frames in houses, windows, doors, and furniture. They are of various sizes, shapes, and positions, most fairly simple but some extraordinarily complex.

The mortise is a hollowed space of appropriate size to receive a rectangular shaped tenon. To make a mortise by hand, a woodworker used a marking gauge to scribe a rectangle that defined the area that would form the sides and ends of the mortise. He then bored a hole the same width as the mortise in the center of this space. If the mortise was large, additional holes were drilled to remove most of the wood. In the nineteenth century, house carpenters cut mortises with portable rotary and oscillating boring machines that functioned much like an egg beater, but with two handles. Once the hole or holes were bored, chisels were used to finish the sides and square the corners of the mortise.

A steam-powered mortising machine was employed very successfully at the Portsmouth Dockyards in 1804 as one of the special-purpose tools built to form pulleys used by the British Navy. Later, the railroads and wagonmakers employed large, special purpose mortising machines to meet the great demands for rolling stock and wagon wheels. However, most machines in the early nineteenth century were manually powered with a foot pedal or hand lever. Often these machines were mounted on interior posts that supported buildings, the post serving as the frame for the machine. Typically, the mechanism brought the chisel to the workpiece, although a few moved the piece to the tool. The tools were of three types. Rotary cutting tools resembled drill bits and functioned much like a modern router. Solid chisels with squared ends functioned with a reciprocating action, the chisel being reversed in order to square the opposite ends of the mortise. Both types were capable of an oscillating action that moved the cutting tool across the portion to be mortised so that the workpiece could be firmly secured. The third type combined a rotary bit with a reciprocating hollow chisel, the rotary action removing most of the mortise while the chisel squared the ends in a single stroke.

By the middle of the 19th century, high-speed machines run by overhead belt drives were very common—and very dangerous. Since they remained inexpensive and simple to repair, mortising machines were common in most small woodworking shops, especially in factories that produced furniture, windows, and doors. These machines benefited greatly from improvements in tool steel since they cut cleaner mortises and operated for longer periods without the need to sharpen the tool.

Tenoning machines followed the same general line of development. To cut a simple tenon, a woodworker used a square and marking gauge to identify the shape of the tenon. Once marked, the form of the tenon was cut with a saw and then squared and smoothed with chisels or planes. The end of the tenon was chamfered so that it could be inserted easily into the mortise. The time and skill required depended on the nature of the work. In house carpentry, many of the joints were rough, but in fine furniture the process was accomplished with great care and accuracy. In truly complex work, joints were often esoteric and were produced by hand, not by machine.

Successful tenoning machines for simple tasks were first produced in the United States by Joseph Fay in 1831. The machine operated manually, cutting tenons in pieces that were 2-by-12 inches (5 cm by 30.5 cm) or less. One lever drove two cutters against a workpiece secured to a table, while a second lever lowered the cutters against the piece and then raised them to allow for a return. The motion was repeated with each thrust, cutting in a manner just like that of a rabbet plane. Furniture-makers; sash, door, and blind manufacturers; and carriage and railroad car shops used these machines to produce large numbers of identical parts. By the 1850s, machines could tenon two sides of a piece to produce standard parts for doors and windows.

Power-driven tenoning machines consisted of two basic types. Circular-saw tenoners used arrangements of circular saws working at right angles to each other, one set of saws cutting the shoulder of the tenon, the other cutting its length. Unlike the manually operated machine, the wood moved on a carriage past the cutters rather than the cutters moving past the wood. Circular-saw tenoners were capable of cutting large timbers.

The most common method for cutting tenons involved a rotary action. In such machines, two cutter heads with multiple knives revolved around a spindle. As the piece moved toward the cutters, they cut a square or rectangular tenon equal to the distance between the cutter heads. This machine operated much like a planing machine. This same principle also produced circular tenons by securing two cutters in a hollow frame that could be adjusted for cutting tenons of various diameters.—J.C.B.

See also MASS PRODUCTION; PLANING MACHINE; STEEL, ALLOYED

Motion Pictures, Early

The technology that made possible the projection and exhibition of photographed moving images emerged about 100 years ago as in Europe and North America a diverse group of engineers, scientists, inventors, and eccentrics at about the same time created cameras and projectors capable of photographing and displaying motion pictures. Yet this collection of inventions is not equivalent to "The Movies." Cinema as we know it is an economically, socially, and culturally defined mass entertainment. A host of technologies made it possible, but they did not determine its content or the uses to which it was put.

An interest in projected entertainments was not solely a Western one. For example (puppet) shadow theater is common to many traditions including Turkey, Java, India, and China. The technological roots of moving pictures can be traced to the *camera obscura*—a roomsize pinhole camera that projects an upside-down view of an exterior scene on an opposite interior wall. The principles of this device were first recorded in the 11th century in Egypt by Ibn al Haitam. Another forerunner of the movies, the magic lantern—essentially a slide projector—was in use in Europe as early as 1646. By the 1790s, Étienne Gaspard Robertson shocked and confounded Parisian audiences with his *Fantasmagorie*. He used magic lanterns to rear-project images of the dead onto smoke in the darkened chapel of an abandoned monastery.

The principles of photography were well established by Joseph Nicéphore Niepce (1765–1833), Louis-Jacques Daguerre (1789–1851), and William Henry Fox Talbot (1800–1877) by the 1830s. The 19th century also gave rise to a seemingly endless stream of Greek- and Latin-named inventions designed to simulate the appearance of motion in a series of still drawings and, eventually, photographs. This illusion of motion from a series of still images was originally conceptualized as based on a physiological phenomenon known as "persistence of vision," the passive accumulation of images on the retina. Today, the psychology of perception identifies the "motion" of motion pictures as a consequence of "short-range apparent motion." Our perception of short-range apparent motion—for example, the appearance of continuous movement when a light is flashed sequentially between closely spaced dots—is believed to involve a cognitive mechanism identical to our apprehension of real motion. It is thought that both short-range apparent motion and real motion stimulate low-level neural motion detectors in a similar fashion. As viewers we process the motion in a motion picture in the same way as we process motion in the real world.

The precursors to the motion picture included a vast array of devices: the Thaumatrope, Phenakistoscope, Zoetrope, Zoescope, Stroboscope, Phatascope, Phantasmascope, Phenakistiscope, Choreutoscope, Praxinoscope, Sterothaumatrope, Chromascope, Eidotrope, Kinestiscope, Tachyscope, Phosmatrope, and Zoopraxiscope. The idea of moving pictures had seized the popular imagination, and many inventors focused their efforts on creating and perfecting apparatus for showing pictures that moved.

One of the first was the Thaumatrope, a simple circle attached to twisted strings. On one side, for example, there was an image of a bird, and on the other a cage. Pulling on the strings spun the circle, creating the image of the bird in the cage. The Phenakistiscope, the first device to animate a series of drawings, is considered the most important invention of precinema technology. It was created almost simultaneously by Joseph Plateau (1801–1833) in Brussels and Simon Stampfer in Vienna in 1832. The Phenakistiscope (from the Greek words meaning *to trick* and *to look*) consisted of a slotted disk spun before a mirror. On the backside of the disk was a series of drawings, each one only slightly changed in position from the next. The slots served as shutters, and peering at the images reflected in the mirror resulted in the perception of their apparent movement.

The Zoetrope, the "wheel of life" invented by William Horner in 1834 was among the most popular of the early animated amusements. It consisted of an upright, slotted, topless drum in which the viewer placed a circular strip of drawings. As the drum spun, a viewer looking through the slots would see the drawings as a simple animation. These strips of drawings anticipated the frames in strips of film, each with a single image, slightly varied. Jules Etienne Marey, a French physiologist, was interested in scientifically analyzing motion. His photographic gun was capable of recording the motion of a bird in flight. His techniques included multiple exposures on a rotating dry plate (1882) as well as superimposing motion on a single stationary plate. Edweard Muybridge (1830–1904) definitively captured the process of human and animal locomotion using banks of stationary cameras. His Zoopraxinscope projected these photographs using the principles of the Phenakistiscope. In 1893, at the Chicago World's Fair, Muybridge exhibited 20,000 of his still images in the Zoopraxographical Hall.

Designed as parlor amusements, children's toys, or scientific experiments, the content of the technology reflected the social and cultural values of its times. Although Victorian inventors assumed that technology would lead inevitably to social improvement, the Viviscope, for example, included animated drawings of a "country-darkie" eating watermelon, perhaps the first but certainly not the last time that a motion picture presented a racist caricature.

Scientists and engineers in Great Britain, Germany, France, and the United States independently designed components of the motion picture camera and projector: a flexible transparent film base, photo emulsions with a fast exposure time, a suitable transport mechanism and, a reliable shutter. Thomas Edison (1847–1931) and William Kennedy Laurie Dickson in the United States, and August (1862–1954) and Louis (1864–1948) Lumière in France, are generally credited as the inventors of motion pictures. On Dec. 28, 1895, the Lumière brothers' Cinematographe projected 10 short films in the basement of the Grand Cafe in Paris to a paying audience of 35. Edison originally had Dickson design his Kinectoscope as a peep show; it wasn't until Apr. 23, 1896, that Edison's Vitascope (a projector invented by Jenkins and Armat) premiered at Koster and Bial's Music Hall in New York, screening scenes of ocean waves, dancing girls, prizefighters, and excerpts from the Broadway hit *A Milk White Flag*.

Early movies were usually incorporated into vaudeville performances as one of nine 10- to 20-minute acts that were generally unrelated in form and content. This format remained the primary form of distribution until the emergence of nickelodeons (storefront theaters) in 1906. *Scientific American* claimed that by 1910 there were 20,000 nickelodeons in Northern cities. This was the entertainment of urban workers, immigrants, and the poor. Short, silent films were relatively inexpensive and accessible to illiterate, newly arrived immigrants, who with their children soon comprised a majority in many urban areas. The movies served as models for Americanization, teaching new customs, values, and habits of consumption. The rise of the movies in America also corresponded with the decline of "Victorian" social norms; it was an era that produced new roles for women, changes in morality and sexual behavior, and new patterns of work and consumption. The movies reflected these changes, but at the same time they were becoming an increasingly influential force in the shaping of contemporary culture.

New technologies are often put to uses unimagined by their creators. In the case of early movies, entrepreneurs originally sold films by the foot, regardless of content, for they were interested in selling machines, not entertainment. This situation changed early in the 20th century. From about 1904 until the present, the fictional narrative has been the most prominent form of motion picture. All the subsequent technical improvement of motion pictures: sound, color, wide screen, video, and "virtual reality" were not the result of a search for scientific truths but were driven by commercial imperatives: creating and fulfilling audience demand for ever more compelling images of fantasy and escape.

The history of motion pictures is in large measure the history of an industrial, capital-dependent, consumer-driven, dream factory. Perhaps now that video cameras are nearly as available as pens, the future will offer modes of personal expression too rarely seen in the first century of motion pictures.—M.F.

See also MOTION PICTURES, SOUND IN; MOTION PICTURES, TECHNICOLOR; MOTION PICTURES, WIDE-SCREEN; PHOTOGRAPHY, EARLY; ROLL FILM; VIDEOCASSETTE RECORDER; VIRTUAL REALITY

Motion Pictures, 3-D

Seeing in depth is a function of human binocular vision. Each eye sees an object from a slightly different perspective. The combined view creates the appearance of depth and dimensionality. The search for three dimensional representation, like the invention of motion pictures themselves, predates the invention of photography. Around the year 1600, Giovanni Battista della Porta (1537–1602) created pairs of 3-D drawings, although it is not certain how or even if he intended viewers to combine the pictures. In 1838, Charles Wheatsone (1802–1875), a British physicist, used a mirror-based apparatus that imparted a three-dimensional appearance to specially produced drawings by restricting the vision of each eye to a single drawing. The parlor stereoscope used photos shot from two slightly different perspectives (about 6.3 cm [2.5 in] apart) that were viewed through special lenses to create a popular Victorian amusement. The stereoscope was shaped like a pair of binoculars. The photos were arranged on a sliding armature to allow for focusing. Distinct images were presented to each eye, thereby replicating natural vision. By mounting a paddlewheel behind a stereoscope, Coleman Sellers was able to present handheld 3-D moving images in 1861. As the paddlewheel turned, a series of sequential images gave the appearance of 3-D motion.

Commercial applications of 3-D in the 1950s required viewers to wear red and green glasses with polarizing lenses that had been developed by Edwin Land (1909–1981) between 1928 and 1932. Each eye viewed an image shot from a slightly different angle. The first 3-D movie, *Bwana Devil,* premiered on Nov. 27, 1952. A total of 38 3-D films were made in 1953–54, including *Kiss Me Kate* and Alfred Hitchcock's *Dial M for Murder*. The 3-D movie represented an effort by the motion picture industry to combat the loss of audiences to commercial television. In this, it failed. Cumbersome glasses, headaches, and the generally poor production values of 3-D films caused the novelty to fade rapidly. In the words of one advertiser: "Do you want a good movie—or a lion in your lap?"

IMAX is currently experimenting with a 3-D format. Productions have included nature films like *The Last Buffalo* and *Into the Deep*. The first planned IMAX 3-D narrative film is *Wings of Courage*. To view the film, each member of the audience will wear a high-tech version of the old 3-D glasses. A headband places liquid crystal display shutters over each eye and transducers near each ear.

The 3-D movie is part of the historical search for "total cinema." The development of motion pictures has paralleled a search for increased realism—sound, color, motion and 3-D—combined with public spectacle. Today the urge to reproduce our experience of the world has lead to the development of virtual-reality technologies. These interactive, tactile, computer-based systems offer the possibility of individual participation in dimensions unavailable to motion picture spectators.

When viewed historically, it is apparent that novel communications technologies have not readily replaced the old formats. Instead, a process of niche specialization has allowed old and new means of communication like books, telephones, motion pictures, radio, television, and computer technologies to coexist. It remains to be seen whether or not the much ballyhooed information revolution will break down the familiar distinctions among media, and exactly how the development of new audiovisual technologies will affect our understanding of the "movies."—M.F.

See also MOTION PICTURES, TECHNICOLOR; MOTION PICTURES, WIDESCREEN; VIRTUAL REALITY

Motion Pictures, Sound in

Motion pictures and recorded sound both appeared in the last quarter of the 19th century. Combining the two was an obvious step, but many years were to pass before this could be accomplished with a fair degree of success. At the beginning of the 20th century, the movies had achieved a good level of picture quality, while recorded sound, although still leaving something to be desired, was adequate. The main problem was the synchronization of the two so that the sound appeared at the right moments. If the sound and the picture got out of synch, especially when dialogue lines were being spoken, the ensuing scene was disconcerting and sometimes downright ludicrous.

Thomas A. Edison (1847–1931), who had made major contributions to both the phonograph and the motion picture, worked on sound movies for a number of years. In 1913, he unveiled his second "Kinetophone" system (an earlier version appeared briefly in the 1890s). Initial reviews were good, but the complicated equipment rapidly got out of adjustment, ruining the synchronization. A better proposition was devised by Lee de Forest (1873–1961), who had earlier won fame and some fortune as the inventor of the triode vacuum tube. De Forest's system, which he called Phonofilm, used light for recording sound and playing it back. This was accomplished by first using a microphone to convert sounds to electrical signals, much as a telephone does. These signals were amplified and transmitted to a oxide-coated vacuum tube that glowed in accordance with the strength of the electrical signal. Light from the tube then passed through a slit and was photographically recorded near the edge of the movie film, the portion that came to be known as the *sound track*. When the movie was projected, the process was reversed. Light passed through the portion of the film containing the sound track and then fell upon a photoelectric cell. The sound track controlled the intensity of the light, and hence the current produced by the photoelectric cell. This current was then amplified and used to drive the speakers, which in turn replicated the original sound. De Forest started work on Phonofilm in 1920, and after an unsuccessful sojourn in Germany, he presented a Phonofilm program to an American audience in 1923. Major studios expressed little interest, so despite inadequate capitalization de Forest's firm had to go into the production business, making hundreds of short films that were exhibited in specially equipped theaters.

While Photofilm was achieving some modest success, Western Electric and Bell Laboratories, two subsidiaries of American Telephone and Telegraph (AT&T), were experimenting with sound movies. They had already made significant improvements in sound quality through their development of electric recording and electrically actuated speakers. The two were used for a movie sound system that coordinated sound and pictures through the use of projectors and record players powered by synchronous AC motors. The system appeared to be a technical success, but it was a hard sell to movie studios that had grown wary of sound movies and all their imperfections.

They finally found a customer in Warner Brothers, a small studio with large aspirations. Filming the first sound movies presented many difficulties, for every sound—everything from off-stage footsteps to the hissing of the arc lights—showed up on the sound track. It was even necessary to muffle the noise of the cameras' electric motors by enclosing them and their operators in soundproof cabinets. Warner Brothers called their new product Vitaphone, and on Aug. 6, 1926, it was premiered with the screening of a series of musical performances followed by a feature film, *Don Juan*. The actors and actresses remained mute, but the movie featured sound effects like clashing swords and was accompanied by a synchronized musical score that had been specially composed for it.

Two more Vitaphone movies followed, one of them including part of a scene that was rejected by censors in New York. In order to cut the offending segment while retaining the musical accompaniment, studio technicians developed a process known as *duping*, re-recording a sound disk so that a portion could be added or removed. Duping allowed a great deal of flexibility in the creation of a sound track, and it was put to good use in the Warner Brothers 1927 film *The Jazz Singer*, the first movie to include spoken words. Most of the movie's dialogue was rendered through traditional title cards, but also included were 354 words of dialogue, most of them spoken by the movie's star, Al Jolson. A year later, the first all-talking movie, *The Lights of New York*, made its debut.

Other studios began to introduce sound movies that used variants of de Forest's optical system. In 1928, RCA's Photophone appeared. It differed from the de Forest system through its use of a mirror that was vibrated by the sound being recorded. Light beams that bounced off the mirror were then used to create the sound track. The Photophone process produced less background noise than the others, and like the other systems that employed a film sound track, it was inherently more reliable than the disk-based Vitaphone process. By the mid-1930s, Vitaphone was extinct.

By this time, the careers of many silent-film actors and actresses were extinct too. Sound movies demanded good speaking voices, and erstwhile stars with thick accents or unpleasant voices quickly found themselves unemployed. Styles of acting also had to change, as the exaggerated styles of the silent era gave way to less flamboyant modes. Theater organists and pianists lost a steady source of income, while a new occupation, the screenwriter, emerged. Outside the movies, entertainment institutions like vaudeville withered as audiences flocked to lavishly produced musicals. As with all significant technologies, sound movies brought consequences well beyond the technical realm.

Optically based recording systems prevailed for a number of years, but after World War II they were supplanted by technologies that used magnetic tape. In recent years, digital sound tracks using laser-based recording and playback have begun to replace the magnetic ones.

See also MICROPHONE; MOTION PICTURES, EARLY; PHONOGRAPH; TELEPHONE; THERMOIONIC (VACUUM) TUBE

Further Reading: Curt Wohleber, "How the Movies Learned to Talk," *American Heritage of Invention and Technology*, vol. 10, no. 3 (Winter 1995).

Motion Pictures, Technicolor

Experiments with color began with the very earliest development of motion pictures. At the turn of 20th century, the movie pioneers Geroges Melies (1861–1938) and Charles Pathè (1863–1957) used a kind of assembly line to hand-color portions of movie prints. Within a decade, the process was speeded up through the use of stencils that allowed the mechanical application of color. G. A. Smith's Kinemacolor, which appeared in 1909, was the first effective color process. It was based on the alternate exposure of film through red and green filters, and then the projection of the developed film through red and green filters. When the film was screened, the viewer would mentally superimpose the two frames, thereby perceiving a multicolored image. Other films, such as D. W. Griffith's *Birth of a Nation* (1915) and Abel Gance's *Napoleon* (1927), used monochromatically tinted scenes—coloring the film stock before the image was printed on it—to heighten dramatic effect. By the 1920s, 80 percent of American films were tinted or partially tinted.

The most important technology was Technicolor, created through the efforts of Herbert T. Kalmus at the Massachusetts Institute of Technology. The first version of the Technicolor process was successfully demonstrated in 1915. A two-color Technicolor process was used selectively in *Ben Hur* (1925) and *Phantom of the Opera* (1926). The process was expensive, costing up to 30 percent more than a black-and-white production. In the 1930s, the classic three-color dye transfer process appeared. It employed a special camera to make three separate films in green, blue, and red. After being developed, the three films were used to produce a single film, which could then be screened without the need for a special projector. Technicolor was first used in 1932 for a Walt Disney animated short entitled *Flowers and Trees*, while in 1935 *Becky Sharp* became the first feature film shot in

Technicolor. The last movie to use Technicolor film was *The Godfather II*, which was released in 1974.

Through the 1940s, Technicolor films were available only in selected theaters—a marketing strategy later emulated by Cinerama for their widescreen releases. Capitalizing on the spectacle and high production values associated with Technicolor, these films were predominately historical sagas and lavish musicals. Until a 1950 consent degree signed in response to an antitrust suit, Technicolor enjoyed a virtual monopoly in the production of color films. In order to shoot in Technicolor, producers were required to use only Technicolor cameras and operators, and to employ a Technicolor consultant—usually Natalie Kalmus, wife of the company's founder. Subsequently, less-expensive (and less-durable) processes, such as Eastmancolor, which had been introduced in 1952, gradually eroded Technicolor's market share, eventually replacing Technicolor altogether. Since 1976, 96 percent of all American films have been produced in color.

The effect of color in filmmaking is paradoxical. Color mimics our normal vision and ought to give a heightened sense of realism to motion pictures. But even today's color film stocks are only a relatively crude approximation of the range of color perceptions registered by human vision. The historical primacy of black and white, and the absence of color in early newsreels has often tended to make black-and-white footage seem more credible than color. In fact, color has been used expressionistically throughout the history of film. Color describes moods and heightens emotions. It is often used symbolically: to identify characters and states of mind, or both. It is perhaps not an overstatement to assert that color, like so many technical innovations, is more about spectacle than about realism. Color is one more tool used by filmmakers to construct their visions before our eyes.—M.F.

See also MOTION PICTURES, WIDESCREEN

Motion Pictures, Widescreen

The scale of motion picture projection depends on the interrelationship of several factors: the size and aspect ratio of the screen, the gauge of the film, the type of lenses used for filming and projection, and the number of synchronized projectors used. These choices are in turn determined by engineering, marketing, and aesthetic considerations.

Aspect ratio is the width of the screen divided by the height. The classic standard aspect ratio was expressed as 1.33:1. Today most movies are screened as 1.85:1 or 2.35:1 (widescreen). Films shot in these ratios are cropped for television, which retains the classic ratio of 1.33:1. This cropping is accomplished either by removing a third of the image at the sides of the frame, or by "panning and scanning." In this process a technician determines which portion of a given frame should be included. "Letterboxing" creates a band of black above and below the televised film image. This allows the composition as originally photographed to be screened in video.

The larger the film negative, the more resolution. Large film gauges allow greater resolution over a given size of projected image. In the 1890s, film sizes varied from 12 mm to as much as 80 mm until Edison's 35 mm was accepted as the standard. Today, films are screened in a variety of gauges including Super 8 mm, 16 mm and Super 16 mm, 35 mm, 70 mm, and IMAX.

The impetus towards widescreen projection dates from the infancy of cinema. Raoul Grimoin-Sanson patented his multiprojector technique in 1897, just 2 years after the first commercial exhibition of motion pictures. Not to be outdone, the Lumières erected a giant translucent screen, 70 ft wide by 53 ft high, at the Paris Exposition. Films were screened to 25,000 people at a time, half facing one side of the screen and half the other.

Perhaps the most important of the early experiments with widescreen was Abel Gance's 1927 *Napoleon*. Using three cameras and three projectors designed by Andre Debrie, Gance created a triptych 15.2 m by 3.7 m (50 ft by 12 ft). Some shots—a marching army for example—stretched the entire width of the screen. At other times, three distinct pictures were composed, multiplying the aesthetic possibilities.

By 1929, Hollywood was posed to proceed with a number of widescreen formats. But exhibitors who had recently upgraded to sound were unable to capitalize additional equipment as the Depression took hold. Widescreen technologies were put on hold until after World War II, when "free" television began to threaten the film industry's near-monopoly on mass visual entertainment. Dozens of widescreen processes were touted to the public between 1952 and 1954. On Sep. 30, 1952, Cinerama premiered at New York's Broadway Theater. CinemaScope followed in 1953 and Todd -AO in 1955. At the time, it was unclear whether they were destined to be passing novelties like 3-D movies, or as revolutionary as the introduction of sound.

Cinerama was a three-camera, three-projector system displayed on a curved screen with a ratio of 2.62:1. In 1955, *This Is Cinerama* played to nearly 2½ million people in New York, grossing $4.7 million (final grosses exceeded $32 million). Not merely a visual spectacle, the film was seen by its producer, Lowell Thomas, as a critical weapon in the struggle against communism. The first five Cinerama films were American travel films. The American landscape in *This Is Cinerama* concludes with a paean to the American way of life entitled "America the Beautiful."

CinemaScope was a product of the Twentieth Century Fox Studio in 1953. Much less costly to shoot and exhibit than Cinerama, CinemaScope was used in tens of thousands of theaters until 1967. This process relied on an anamorphic lens and a redesign of the 35-mm format to create a widescreen format projected at 2.66:1. Todd-AO premiered with *Oklahoma* in 1955. This 65-mm camera/70-mm projector format offered greater resolution and the sense of spectacle that the public had associated with Cinerama. Targeted at the high end of the market, Todd-AO never played in more than a few hundred theaters.

Today, 70-mm presentations like *Star Wars* and *Apocalypse Now* typically originate on 35-mm film and are enlarged to 70-mm for projection. Innovation in projection technology is once again found in nontheatrical venues. Museums, fairs, and theme parks are sites for IMAX (*image maximization*) presentations, which use film frames 10 times larger than the conventional 35-mm frame. By the end of 1994, there were 119 permanent IMAX theaters with 35 additional theaters planned. These theaters draw on a base of more 100 productions. In addition to the original IMAX format, there is a 48-frames-per-second IMAX HD format and two IMAX 3-D formats

Widescreen technologies promised heightened viewer involvement—immersing spectators and overwhelming them with visual stimulation. The novelty of each technical innovation has for the most part proved to be short-lived. "Bigger is better" has not proved to be a predictor of quality. The most enduring motion pictures have relied on a classical Hollywood formulation—the perennially potent combination of "story and stars."—M.F.

See also MOTION PICTURES, EARLY; MOTION PICTURES, SOUND IN; MOTION PICTURES, 3-D; PHOTOGRAPHY, 35-MM

Further Reading: John Belton, *Widescreen*, 1992.

Motor Scooter

Like a motorcycle, a motor scooter is a two-wheeled powered vehicle; it differs from a motorcycle in having smaller wheels and a floor instead of footpegs. Also, unlike many motorcycles, scooters offer at least minimal weather protection. Some of the first motor scooters were exactly what the name implies, wheeled platforms powered by tiny internal combustion engines. Lacking even a seat, they were intended for travel only over short distances. Other scooters were equipped with seats, but were still intended for short-range journeys. These first scooters enjoyed a brief vogue in the early 1920s, but then disappeared. They had the virtue of simplicity, but they fell between two stools, being less useful than a motorcycle and more expensive than a bicycle.

During World War II, the British built a folding scooterlike vehicle for use by paratroopers; it was sold in modest numbers to the general public after the war. The primary source of the modern motor scooter was Italy. Responding to the desperate need for transportation following World War II, the Vespa ("wasp") made its appearance in late 1946. The Vespa offered low purchase price, easy maneuverability, and enough power to at least keep up with traffic. The Vespa served as an inspiration for many other manufacturers, and in the late 1950s and early 1960s a scooter mania swept over Europe, propelled in part by the postwar baby boom that led to large numbers of young people seeking personal transportation. Many people who might otherwise have bought motorcycles bought scooters instead, resulting in considerable damage to the established motorcycle industry before the scooter boom faded.

Motor scooters continue to be popular in some parts of the world, especially when incomes are high enough for people to afford personal transportation, but not so high as to allow the purchase of automobiles. India, for example, has 12 million scooters and only 3.5 million cars plying its roads. Scooters also have the virtue of low fuel consumption and the ability to thread their way through crowded city streets, giving them a niche in more developed parts of the world as well.

See also MOTORCYCLE

Motor, Electric

The operation of an electric motor is based on the close relationship between electricity and magnetism. An electric current produces a magnetic field, just as a moving magnetic field produces an electric current. An electric motor typically has two major components: a stationary magnet or set of field coils (the stator) and a rotating armature consisting of several lateral segments wound with wire (the rotor). The stator produces a magnetic flux (one equipped with field coils has to be electrically excited for this to happen), as does the rotor when an electric current runs through it. The two components are so designed that magnetized segments of the rotor will be first attracted and then repelled by adjacent portions of the stator. In a direct-current motor, this occurs because the segments of the rotor undergo a change in their magnetic polarity when the current being fed to them is reversed by a commutator attached to the rotor. In an alternating-current motor, the rotor's segments change polarity because the electricity is delivered as three separate out-of-phase waves. In both cases, the rapidly shifting attraction and repulsion of the rotor and stator imparts a rotary motion and produces torque at the motor's output shaft, which is situated on the rotor's axis.

The invention of the electric motor closely followed the realization that an electric current produces a magnetic field. In 1821, a year after Denmark's Hans Christian Oersted (1777–1851) discovered this phenomenon, Michael Faraday (1791–1867) of Great Britain constructed an apparatus that is often described as the first electric motor. It consisted of a bowl of mercury in which a bar magnet rotated vertically, inducing an electric current that moved through an adjacent copper wire, one end of which was immersed in the mercury. Current then traveled to another piece of wire, the end of which was immersed in another bowl of mercury. This bowl contained a fixed bar magnet, so that the interaction of its field with the second wire's induced magnetic field caused the wire to rotate continuously around the magnet.

The subsequent development of electric motors was an international endeavor. In Italy, Salvatore dal Negro built in 1830 an electric motor that produced a reciprocating motion. Another linear motor was built in France by Paul Fromont (1815–1865). Both Fromont and dal Negro used additional mechanical devices to impart a rotary motion. In the United States, Joseph Henry (1797–1878) built an experimental motor in the 1829, as did Charles Grafton Page (1812–1868) and Thomas Davenport (1802–1851) a few years later. In Russia, H. M. Jacobi (1801–1874) powered a paddlewheel boat with a rotary electric motor in 1839; it carried as many as 12 passengers as it ran for several days on the Neva River. In the following year, Scotland's Robert Davidson used eight electromagnets to propel a carriage.

One of the first public demonstrations of an electric motor occurred at the Vienna Exhibition of 1873, when a Gramme generator was used as a motor to operate a pump; its power was supplied by an identical generator. This demonstrated the not-always-obvious fact that an electric motor works just like an electrical generator, except that a generator's output is a motor's input. Movement in a generator produces electric current, whereas electric current causes a motor to move. The close connection between the two is significant because improvements made in generators often could be carried over to motors. One important example of this process was Antonio Pacinotti's (1841–1912) transfer of his previously invented generator armature to an electric motor, which resulted in a more regular delivery of electric current.

For many years, however, the incentive to create better electric motors was limited, for they were of little practical use. The main problem was that the motors lacked an effective, inexpensive source of electricity. For much of the 19th century, batteries were the primary source of electricity, and the energy they supplied was more than 20 times more expensive than the energy provided by contemporary steam engines. Electric motors became an economical source of energy only after improvements in electrical generators made possible a large-scale expansion of electrical supply. Expanding demand was equally important, for the installation of early generators was motivated by the introduction of the first significant commercial application of electric power, arc lighting. Once electrical systems for arc lights were in place, the widespread use of electric motors became an attractive proposition to early power suppliers, for the motors were likely to be used at times and places where electrical demands for lighting were low. As Thomas Edison (1847–1931), always alert to commercial possibilities, observed in his notebook, "Generally poorest district for light, best for power, thus evening up the whole city—note effect of this on investment."

The first industrial motors were generators that used current instead of making it. But their performance was poor; they changed speed when their load changed, and to deliberately change their speed required moving the brushes that supplied the current. In time, motor design began to depart from generator design, and by the 1880s improved electric motors were being used for a variety of purposes, most notably as motive power for electrical railroads. At the other end of the scale, in 1880 Thomas Edison began to market a stencil-cutting pen powered by an electric motor, the first one to be sold in large volume. Electric motors were even beginning to show up in American homes, where they were used to power sewing machines. In 1887, 15 American manufacturers collectively produced 10,000 motors.

All early electric motors ran on direct current. By the end of the 19th century it was evident that alternating current would supplant direct current for most applications. New motor designs were therefore required, for contemporary DC motors were hard to start when fed alternating current, and once they got started they produced a great amount of sparking between the commutator and the brushes. The use of AC for motors also necessitated making fundamental changes in the electrical supply itself. The single-phase alternating current that was first put into use was adequate for lighting, but running electric motors required polyphase current. In the late 1890s, Westinghouse marketed an induction motor invented by Nicola Tesla (1856–1943). Using a rotating field that imparted motion to a squirrel-cage rotor, it served as the standard design for AC motors for many years.

As increasing numbers of motors found their way into industrial settings, they did not simply replace the former prime movers, steam engines and water turbines. Electric motors offered not just a new source of energy but a whole new approach to powering machinery. Before the appearance of industrial electric motors, many enterprises had a central power source that motivated a jumble of overhead shafts, pulleys, belts, clutches, and other contrivances. It was a system with high initial costs and maintenance requirements, substantial power losses, and many occupational hazards. It was also inflexible, as individual machines could not be easily moved from one place to another, nor could the speed at which they operated be easily controlled. In contrast, in a factory run by electricity, each machine could have its own motor, the powering of which required nothing more elaborate than some electrical fittings and a cord of sufficient length. As a result, factories could be laid out in ways that promoted maximum efficiency and flexibility.

The electrically powered factory was seen as the harbinger of a new epoch. Unlike vast, bureaucratically organized factories dependent on a central power source, industrial enterprises of the electric-motor era could be smaller in scale, allowing production to be decentralized and integrated with a variety of lifestyles. For those repelled by the traditional industrial order, the electric motor promised a transformed economy based on clean, efficient, and humane workplaces. That such an industrial utopia failed to emerge demonstrates the perils of excessive enthusiasm for the transformative powers of a new technology. Electric motors have made some work easier and more pleasant, but it should be noted that the sewing machine used by an underpaid sweatshop employee is powered by an electric motor.

The most common type of motor today is the universal motor, so called because it can run on either AC or DC. Universal motors have high starting torque and can run at higher speeds than other kinds of AC motors. First appearing around 1925, literally tens of billions of

Nicola Tesla (courtesy Institute of Electrical Engineers).

them have been used in household appliances, power tools, record players, and many other applications. Much larger motors, usually of the induction type and using alternating current, are essential components of almost all industrial processes. Altogether, the tasks performed by electric motors consume approximately two-thirds of all the electric power generated in the United States. At the same time, the electricity consumed by motors has been used with increasing degrees of efficiency; the best permanent-magnet motors convert into power more than 95 percent of the electricity fed into them. Gains in efficiency have come in large measure through the use of permanent magnets made from rare earths. From the time of their invention in the mid-1960s to 1980, the strength of magnetic fields increased fourfold, and since then it has gone up another 25 percent. Altogether, magnets are 50 times more powerful than they were in 1900.

Most electric motors produce rotary motion, but this is not an essential feature; as was noted above, many of the earliest motors imparted a reciprocating motion. Although overshadowed by rotary motors, many linear induction motors are currently in use. In the future, they may take on an added importance as the source of power for high-speed trains guided by conventional wheels or magnetic levitation.

See also ALTERNATING CURRENT; BATTERY; DIRECT CURRENT; GENERATOR, ELECTRICAL; MAGNETIC LEVITATION VEHICLES; TROLLEY (STREETCAR)

Motorcycle

Once the bicycle was established as a common conveyance, people began to think of adding a source of power to it. At first this meant a steam engine. Although there may have been earlier attempts, it is certain that by 1869 Michel in France and Sylvester Roper in the United States had produced steam-powered velocipedes. In 1884, Copeland equipped a high-wheeler bicycle with a steam engine, and then went on to reportedly produce 200 steam-propelled tricycles in Philadelphia. In France, a few steam-powered tricycles were built in the 1880s by Leon Serpollet and Comte Albert de Dion.

As also happened with automobiles, steam proved to be a dead end as a power source for motorcycles. The future belonged to the four-stroke engine designed by Nicolaus Otto (1832–1891). In 1886, one of Otto's assistants, Gottlieb Daimler (1834–1900), constructed a motorcycle built around one of these engines. A year earlier a tricycle powered by an internal combustion engine had been patented by Edward Butler in England, although an operational machine did not appear until 1887.

As usually happens with an emerging technology, the early years of the motorcycle were characterized by a great deal of experimentation. Even the most basic principles, such as where to put the engine, had not been determined; engines were mounted over the front wheel, atop or even behind the rear wheel, and above and below the rider's feet. In Germany in 1893, the Hildebrand brothers and their assistant Alois Wolfmüller built a motorcycle with what came to be the standard configuration: an engine between the two wheels and fairly close to the ground. This machine, which was built in significant numbers in

France and Germany, used eccentrically mounted rods to transfer power from the engine's two cylinders to the rear wheel. In 1901, the Werner brothers in France introduced a motorcycle that used a belt to drive the rear wheel. This arrangement was employed for many years, even though chain drive had made its appearance early in the century.

By the second decade of the 20th century, enough improvements had been made to the motorcycle that it was widely used for everyday transportation, even by people with little mechanical aptitude. The usefulness and reliability of the motorcycle was underscored by its use in World War I, when both sides employed large numbers of motorcycle-mounted dispatch riders. After the war, postwar European prosperity brought with it an increased demand for individual transportation. Many people who could not afford an automobile rode motorcycles, often with sidecars attached. Such was not the case in the United States, however. Innovative production techniques like the assembly line allowed the Ford Model-T to be sold for a price not much higher than that of a motorcycle. For many years the main buyers of motorcycles in the United States were the police and people likely to be arrested by them.

By the early 1930s, motorcycles had taken on a common configuration: a frame made from tubes or pressed metal, chain drive, a gas tank mounted like a saddle on top, a hand throttle, and in some cases a foot-operated gearshift. Propelling these machines was a wide variety of engines with one, two, or even four cylinders, ranging in size from 100 cc to 1,000 cc. Many smaller motorcycles used two-stroke engines, while four-stroke engines had either overhead or side valves.

The world depression of the 1930s put many manufacturers out of business, and World War II diverted most production to the military. In the late 1940s and early 1950s, the need for cheap transportation in war-ravaged Europe and Japan created a vast market for motorcycles and motor scooters. Firms in Great Britain, Germany, and Italy produced millions of them, but in the late 1950s, the Japanese began to take their place as significant motorcycle manufacturers. By this time, however, in most of the industrial world the motorcycle had become what economists call "an inferior good," something that is sold in diminishing quantities as incomes rise. As people earned more money, automobiles came to be their preferred means of transport. Under these circumstances motorcycles might have died out, but rising affluence also allowed many people to buy them as sport and recreational vehicles. This trend was greatly augmented by the maturing of the postwar baby boom generation, for young people have always been the prime purchasers of motorcycles.

The major beneficiary of these trends was Japan, which by the 1960s had become the world's largest producer of motorcycles. The technical features displayed by Japanese motorcycles were not complete novelties, but Japanese machines must be credited with the popu-

A BMW K-75 sport-touring motorcycle.

larization of electric starting, overhead camshaft engines, disc brakes, and engines with three and four cylinders, along with levels of reliability not usually found in the past. But in the 1980s, demographic and social changes turned against the makers of motorcycles. As the population aged and incomes stagnated, the number of motorcycles on the road declined as only true enthusiasts bought them. At the same time, however, rising incomes in newly industrializing countries are creating new markets for motorcycles, ensuring the future of powered two-wheel transportation.

See also ASSEMBLY LINE; BICYCLE; CHAIN DRIVE; ENGINE, FOUR-STROKE; ENGINE, TWO-STROKE; MOTOR SCOOTER

Mouse

The "mouse," or the graphical user interface device, originated in the early 1960s with Ivan Sutherland of the Massachusetts Institute of Technology and his work on "Sketchpad." Sketchpad was a graphics manipulation package that had several unique features, including a graphical method of input and output, with the output portrayed on a cathode-ray tube (CRT). The technology was devised with a light pen that allowed the user to draw objects that appeared on the screen. Perhaps more important, however, Sketchpad generated a window that split the screen into separate sections displaying different information on different portions of the screen.

Another researcher, Douglas Englebart (1925–), had conducted a review of the light pen and other input technologies, and had contemplated various substitutes without success. Then, while attending a talk in 1964, he recounted that the speaker bored him so much that as he daydreamed he hit upon the idea of the mouse. It contained two small wheels at right angles that measured direction to the left, right, up, and down. That directional information was fed into the computer. On top of the device were two buttons that allowed the user to pause on an item and send a separate signal to the computer. This in essence was what has come to be known as "point and click" technology. A more mature design used a rubber-coated metal ball that protruded from the bottom of a hand-sized controller. Two rollers touching the ball recorded its movements along the x and y axes. As the rollers rotated, encoders made and broke electrical contacts, sending electrical pulses that the computer used to track the mouse. A second mouse design used a light-emitting diode or LED to shine light through holes in the encoders onto photodetectors. As the rollers rotated, the encoders alternately broke and made light beams between the LED and the photodetector.

With any of these technologies, the mouse, combined with the windowed graphical display screen, provided the basics of a graphical user interface system. As work on the mouse unfolded, Xerox Corporation—which specialized in photocopier technology—invested heavily in related research through its Palo Alto Research Center (PARC). Ultimately, Englebart joined PARC, along with several other computer engineers determined to make computers more "user friendly." They focused on translating the numerical and alphabetical commands people gave to computers to a system that would involve symbols. The earliest result of their efforts, the Alto computer, was constructed in 1975. In 1979, Steve Jobs (1955–), the co-creator of Apple Computers, visited PARC and hired one of the Alto designers, Larry Tesler, to design the user interface for the new Apple computer, the Lisa. Although the Lisa was a commercial failure, its successor from Apple, the Macintosh personal computer, was highly successful.

Xerox and PARC did not reap any of the rewards associated with the mouse or the graphic user interface system. But Steve Jobs's Macintosh yielded a stream of technology that was subsequently copied (most notably by Microsoft's popular Windows operating system), and it proved a critical bridge for the introduction of the mouse. After Macintosh computers and Windows operating systems, the future of the mouse was ensured. By 1996, all major operating systems incorporated the mouse. Windows 95 by Microsoft and other new graphical user interface systems cannot be used without a mouse.

Still, some theorists view the mouse as obsolete. PARC's current director, John S. Brown, has argued that the next step will involve computer networks in which people do not have to actively interact; instead, the computers will "read" their intentions. With such technology, the slightest unintended action—even a sneeze—could have substantial consequences!—L.S.

See also LIGHT-EMITTING DIODE; PHOTOCOPIER

Further Reading: Tekla S. Perry and John Voelker, "Of Mice and Menus: Designing the User-Friendly Interface," *IEEE Spectrum* (Sep. 1989): 46–51.

Movies, Drive-In

Although California is thought to be the home of the drive-in culture, the first drive-in movie opened on the other side of the country. In the early 1930s, Richard M. Hollingshead, Jr., set up an outdoor movie theater that used the hood of his car as the support for a projector and the side of his garage as a screen. In 1933, he was granted a U.S. patent for the drive-in movie theater, and in that same year he opened the first example in Camden, N.J. Hollingshead licensed the right to use his invention to other entrepreneurs for a fee of $1,000 plus 5 percent of the gross receipts. California was quick to adopt the novelty of outdoor movies; the second drive-in theater was built on Pico Boulevard in Los Angeles.

Early drive-in theaters set the pattern that was retained for decades: a large screen placed in front of several row of parking spaces that sloped upwards in order to provide good sight lines from the back rows. The viewers, however, lacked individual speakers. The sound emanated from the structure that housed the screen, a feature not always appreciated by those living in the vicinity. Also, since sound travels more slowly than light, viewers in the back perceived an annoying lack of synchronization between the picture and the sound track. The picture also suffered from distortion caused by the need to project upwards, a situation that was eventually remedied by building screens with a slight forward inclination at the top.

The drive-in movie achieved only modest initial success; only 50 of them were in operation in 1941. But they expanded rapidly in the immediate postwar years, and by 1950, 1,700 drive-in theaters served the movie-going public. Drive-in theaters were especially appealing to families with young children, and the postwar baby boom made for vast numbers of the latter. The drive-in was also assured of a steady clientele when the baby boomers entered their teenage years. By this time, the drive-in theater had become notorious as the favored place for romantic encounters, the infamous "passion pits" of popular lore.

With 4,000 screens in use, drive-in theaters achieved their peak level of popularity in the late 1950s. They then went into a decline from which they never recovered; by the mid-1970s only about 850 were in operation. Television took some of their audience, although walk-in theaters were hit harder at first. A decline in the number of children and teenagers took away much of the drive-in's natural audience. The most serious threat to the drive-in's existence was the expansion of suburbia and concomitant rises in land values. Large numbers of drive-in theaters vanished because the land they occupied had become more valuable for other purposes. Many of the drive-ins that remain today are dependent on weekly swap meets for a significant part of their income.

Further Reading: Chester H. Liebs, *Main Street to Miracle Mile: American Roadside Architecture*, 1985.

Mule Spinning

A mule is a multiple-spindle machine for spinning yarn of virtually any fiber (cotton, wool, worsted, asbestos, synthetics, etc.). Invented in England by Samuel Crompton (1753–1827) in the 1770s, it combines features of the spinning jenny and the spinning frame, and thus is a hybrid, or mule (it is also possible that the name is derived from the difficulty of operating it).

The spinning process for cotton and worsted begins when a roving (a long rope of parallel fibers that have been slightly twisted together) from spools or bobbins at the back is drawn out by a series of paired drafting rolls that turn at increasing speeds. In wool spinning, spools of roving turn on friction drums (at a steady speed despite the decreasing diameter of the package) and emerge from pinch rolls without drafting. In both cases, a moving carriage carrying the spindles moves away from the feed, and drafting and spinning take place simultaneously. The rapidly turning spindles (inclined slightly toward the roving) add a twist as the yarn slips off their tips. Because drafting and spinning are happening in the same zone, the yarn is more even than the roving, as twist goes into thin places and draft into thick, evening out the product. As a result, the mule produces better yarn from equal stock than any other machine.

The operator, or mule spinner, controls the outward movement of the carriage and thus the drafting. Once the spinning and drafting finish, the machine stops, the spinner lowers the newly made yarn with faller wires, and winds it onto bobbins (as many as 600) while drawing the carriage in. Years of experience and great skill were formerly required to manage the mule, particularly in controlling the draw and building the "cop" or package of yarn on the spindle or bobbin.

In England, mule spinners hired and trained helpers known as *piecers*, who were in training to become spinners. Highly organized, they represented an important aspect of the management of factory production. In the United States, the scarcity of mule spinners, their great skill, and their craft traditions made them unpopular with mill owners in the mass-production branches of the industry. In both countries, the development of self-acting machinery, which ran without the direct intervention of the spinner, was avidly pursued and was first achieved in 1846. Decades passed as self-actors gained the ability to produce finer counts, first in cotton and, much later, in wool. While one man (and later, in very rare cases, generally in German-owned woolen mills, a woman) could tend two of the automatic machines, the need for skill in tending, adjusting, and repairing the mules had not been eliminated. Consequently, development of alternative styles of spinning evolving from the spinning frame continued, eventually resulting in the production of throstle, ring-, and cap-frames that used much-less-skilled, cheaper, and generally female labor while producing adequate (and later good-quality) yarn, first in cotton and later in wool. Despite the mule's advantages in terms of quality and lower power consumption, its requirement for skilled labor, along with the occupation of greater floor space and the intermittent nature of its production, doomed it in an industrial style devoted to maximizing owners' control by minimizing workers' skill in the United States.—L.G.

See also COTTON; DESKILLING; SPINNING JENNY; SPINNING WHEEL; WOOL

Multiphasic Screening

Multiphasic screening is the administration of a battery of medical tests in order to determine the presence of an incipient chronic disease such as tuberculosis, syphilis, hypertension, thyroid malfunction, and cancer. Since their initiation in the late 1940s and early 1950s, multiphasic screening programs have contributed to the health of some individuals, but considerable controversy still attends their use.

One criticism of multiphasic screening is that any medical test may produce erroneous results. If tests produce a significant number of false negatives, the whole purpose of screening has been defeated, as potentially serious problems escape detection. A false positive, on the other hand, causes needless worry and puts a burden on the medical system as patients seek treatment for nonexistent ailments. In addition to producing some false negatives and false positives, multiphasic screening can also generate results that are ambiguous. The tests used in multiphasic screening have been designed to detect anything that might be medically significant, but there may be no clear demarcation between the normal and the pathological.

Questions have also been raised about the overall efficacy of multiphasic testing, for its role in reducing morbidity and mortality has not been clearly evident. One study conducted in London during the late 1960s showed that screening picked up a large number of ailments (2.3 diseases per person), but most of the disorders not already known by the patients' doctors were insignificant. In most cases, indications of a pathological condition did not lead to new treatments, except for patients with high blood pressures or low hemoglobin counts. In similar fashion, a major clinical trial conducted in the mid-1970s in Salt Lake City, Utah, found that multiphasic screening detected many previously unknown abnormalities, but this knowledge was rarely acted on. Few went back for retesting, and still fewer initiated any treatment for the condition revealed by the screening.

To be an effective component of medical care, multiphasic testing has to be integrated with the diagnostic work performed by individual physicians. But many doctors have been less than enthusiastic about making use of multiphasic screening. Some of their discomfort may stem from a concern that screening transfers some of the physician's traditional tasks to medical technicians. At the same time, however, indifference or even hostility to multiphasic testing may have a sound basis. For multiphasic testing to be effective, the results of the individual tests have to be integrated with the results of diagnostic procedures like physical examinations and taking patients' histories. If the job is taken seriously, it can significantly add to an already heavy workload. Screening is usually undertaken by a public-health agency and not a private medical practice, but it creates expectations that when a disorder is identified, it will then be treated, further adding to the physician's burden.

Finally, multiphasic testing is somewhat at odds with most physicians' conception of their work. The normal mode of medical practice is reactive; patients come in with some problem, and only then are tests done in order to determine the cause of the problem. In contrast, multiphasic testing is often undertaken in the absence of any indications of illness. Consequently, the information it provides is not always considered to be of any great importance.

Further Reading: Stanley Joel Reiser, "The Concept of Screening for Disease," *Millbank Memorial Fund Quarterly*, vol. 54, no. 6 (1978): 403–25.

Multiple Independently Targetable Reentry Vehicles (MIRVs)

Intercontinental ballistic missiles (ICBMs) with nuclear warheads were key weapons of strategic offense during the Cold War (1945–1991) between the United States and the Soviet Union. Ballistic missiles followed the launch of Sputnik by the Soviet Union in 1957, and several types of ICBMs and submarine-launched ballistic missiles (SLBMs) were in operation within the Soviet Union and the United States, as well as in Great Britain and France, by the 1960s. An early ballistic missile operated as follows: Upon launch from an underground silo or submerged submarine, the missile traced an arc up through the atmosphere until its booster burned out, at which point its reentry vehicle (RV)—the "bus" that protects the nuclear warhead as it reenters the Earth's atmosphere—was mechanically released to start its descent to target based on predetermined coordinates programmed into the RV.

While hundreds of ICBMs were being deployed, the American military initiated several antiballistic missile (ABM) programs, including the Nike-Zeus system. American strategists concluded that the Soviets also were preparing an ABM system of their own, and to ensure "deliverability" of the warheads, the RVs either had to be maneuverable or each missile had to carry several RVs. By the early 1960s, the technical community already had worked on RV bus maneuverability with small rocket motors. Years before maneuverable RVs enjoyed support, though, the idea of using multiple warheads had taken hold in strategic thinking.

In the United States, the first multiple warhead missile was a SLBM made for the Polaris submarine. Authorized in September 1960, the Polaris A-3 missile carried three RVs with 200 kilotons (KT, the equivalent of 2,000 lb [907 kg] of explosives) each, but the separation was not adjustable, leading to a "footprint" phenomenon in which the warheads created a distinct pattern when hitting their target. Technically, that RV design was called a MRV, or multiple reentry vehicle, because it lacked the capability of independent targeting. The A-3 became operational in September 1964 for the U.S. Navy. Meanwhile, the U.S. Air Force had started work on the MK. 12, 13, and 14 series, which were true multiple independent reentry vehicles for the new Minuteman II missile. The MIRV bus released its warheads in sets of two, which permitted dual targeting of enemy sites. Traditional measures of accuracy, or CEP (circular error probable), reflected the two-warhead targeting by estimating the outermost limit in which either of the two warheads would land. (In the case of the modern D-5 Trident II missile, for example, the CEP is 122 m [400 ft], meaning that one of the two warheads destined for a particular target would be expected to land within 122 m of that target.)

While the MK. 13 and 14 were never authorized, the MK. 12 received approval in 1963, with General Electric named as the prime contractor. In 1964, the government proceeded with MK. 12 development, then production, for the Minuteman II and for the new Poseidon navy missile. Megatonnage on the warheads varied with target characteristics, distance to target, and number of warheads designated for target, but the trade-off was clear: Any MIRV missile could carry only a fraction of the explosive power contained in a single-warhead missile. (The Soviet SS-18, for example, carried a single 26-megaton warhead—the equivalent of 26 million tons of TNT—while MK. 12 MIRVs carried several 500 KT warheads.)

Final approval for the deployment of MIRVed missiles in the United States occurred in 1965 as part of the planning for the FY1967 budget, at which time both the Minuteman II and the Poseidon were to incorporate the MIRV MK. 12 RVs. A Poseidon test vehicle was fired in late 1966, followed by 1967 flight testing of the RVs. The Department of Defense concluded at that time that it would rely on identical, real RVs rather than decoys to penetrate Soviet defenses, and the navy commenced refitting more than 30 Poseidon subs with the new missiles.

During those planning sessions, the number of RVs per missile varied, depending on whether they were deployed on the Minuteman or the Poseidon, on the perceived needed range of the missile, and on the use of lighter decoys that by this time had supplanted real RVs. Over the next decade, further upgrades occurred with the Minuteman III missiles and development and deployment of the new Trident submarine. The U.S. Air Force and Navy both continued research on an upgraded MIRV, the MK. 500 ("Evader") MARV, or maneuverable reentry vehicle. MARVs were tested but never deployed. Meanwhile, the Trident submarine carried the new Trident I (C-4) missile capable of carrying 10 RVs each. Another upgrade, the Trident II missile, began to be deployed in some submarines in the late 1980s. Originally, the navy intended each missile to carry up to 8 RVs, but the collapse of the Soviet Union permitted navy planners to change the mix of RVs to include large, "silo buster" warheads (known as AORs or "All Out Rounds" of approximately 3,175 kg [7,000 lb]). In any targeting scenario, however, a decrease in accuracy required more megatonnage per warhead, and vice versa. This was significant, for the more maneuvers a RV had to perform, the less accurate it was, hence the need for more explosive power.

As the threat of the Soviet Union diminished and as the Soviet Union itself came to an end, the need for MIRVed missiles has likewise declined (although Russia and some of the former components of the Soviet Union still maintain powerful nuclear arsenals). Instead, recent strategic thinking has emphasized refitting the MIRVed missiles with "silo buster" or conventional warheads for use against threats such as the one posed by Iraq in the 1990–91 Gulf War. Created to penetrate ABM systems that never saw full deployment, the nuclear MIRV outlived its usefulness, and in the post–Cold War era, the technology of multiple independent RVs still may have use, but with conventional warheads.

Policy considerations centering on the MIRV involved criticisms that it accelerated the "arms race" by making each existing (or new) missile into multiple weapons. Consequently, arms control advocates constantly were divided between those who favored limiting launchers and those emphasizing the limitation of warheads. Conversely, the charge that MIRVs were a response to ABM systems led critics to work for the abolition of any antimissile technology. That approach culminated in the ABM Treaty of 1972. Yet efforts to limit the delivery systems proved futile for the most part: Restrictions on launchers were circumvented by expansion of RV numbers, and vice versa. Ultimately, the demise of the Soviet government proved the only successful means of reducing the number of MIRVs.—L.S.

See also MISSILE, INTERCONTINENTAL BALLISTIC; MISSILES, GUIDED; MISSILES, SUBMARINE-LAUNCHED; SPUTNIK; SALT I AND SALT II; SUBMARINE, TRIDENT

Further Reading: Ted Greenwood, *The Making of the MIRV: A Study of Defense Decision Making*, 1988.

Multiplexing

Multiplexing is the technique of combining two or more information signals—voice, data, image, or motion video—for simultaneous transmission over a communications circuit linking two major points. The basic purpose of multiplexing is to improve the economy of a given transmission facility (a coaxial cable, a fiber-optic cable, or a microwave radio channel), thus reducing the costs of installation and operation. All types of information content may be multiplexed: speech, data, images (fax and still photos), and motion video. Different types of information require more or less transmission capacity; for example, a digitally coded network television broadcast signal uses as much capacity as 1,344 digitally coded voice circuits.

The number of signals that can be combined on a single transmission facility depends on the design capacity of the transmitter (multiplexer) at one end and the receiver (demultiplexer) at the other end of the circuit;

high capacities mean more complexity and more expensive equipment. If the facility requires repeaters or regenerators (which amplify or rebuild degenerating signals at various points in a long transmission path), each of those units must be matched to the design of the multiplexing system. Each end of a two-way transmission path requires a combined multiplexer/demultiplexer, called a *muldem* by communications engineers.

Advances in multiplexing technology have helped to drastically reduce the cost per channel in major facilities, such as the transoceanic cables that link continents. For example, the first transatlantic cable for telephone use was installed in 1958 and cost about $6 million per analog channel. By 1993, the cost per digital channel on new fiber-optic cables had dropped to about $4,000, a reduction by a factor of 1,500. If that savings applied to an automobile, your next car would cost less than $10.

The basic concept of multiplexing frequencies was first conceived by Cromwell Fleetwood Varley, who patented the idea in Britain in 1870. His British Patent No. 1,044 described the superpositioning of "tone" signals on direct-current telegraph lines by means of rudimentary filters. Meanwhile, Alexander Graham Bell (1847–1922) was experimenting with his "harmonic telegraph transmitter and a harmonic telegraph receiver." Bell's original intent was to transmit several telegraph channels simultaneously over a single shared wire, using a group of differently pitched vibrating reeds as transmitters, and a group of matched reeds (or tuning forks) as receivers, so each channel passed through the wire as a different frequency. Instead, Bell's invention became the telephone.

The basic requirements for multiplexed voice circuits were known by scientists in 1894, but the devices needed for such systems were simply not available. The tools that finally enabled the building of commercial multiplexers were developed by 1912, but they were primitive and notoriously unstable by modern standards. However, they did the job. The first multiplexing systems used by the Western Union Telegraph Company were installed in 1913, when Western Electric (part of the American Telephone & Telegraph Company) produced five sets of multiplexing equipment to serve Western Union's New York–Boston route. Within 3 years, Western Union had 26 multiplexing circuits in operation and orders for 25 more.

The invention of the triode by Lee de Forest (1873–1961) in 1907, and the subsequent improvements in its stability, performance, and reliability by H. D. Arnold at AT&T and Irving Langmuir (1881–1957) of General Electric, enabled telephone engineers to introduce voice channel multiplexing in 1918. One of the most prominent promoters of telephone multiplexing in the United States was John Stone Stone, who published a seminal paper on "The Practical Aspects of Propagation of High-Frequency Electric Waves Along Wires" in 1912.

Analog signals are multiplexed by "stacking" the relatively narrow bands of voice channel frequencies within a broad band of frequencies, a process called *frequency-division multiplexing* (FDM). Digital signals are usually multiplexed by a process called *time-division multiplexing* (TDM). Each duplex signal is divided into a series of thousands of coded pulses per second, and these are inserted between the pulses of many other signals to form a high-speed stream of coded pulses passing through the cable. In 1986, about 70 percent of the long-distance telephone traffic in the United States still was being handled by analog microwave systems based on frequency-division multiplexing. Less than 5 years later, almost all long-distance telephony employed digital signal streams served by time-division multiplexing.

Fiber-optic cables have far greater capacities for high-speed digital signal transmission. Within a decade after their introduction in about 1980, fiber-optic cables had largely replaced high-capacity copper and microwave circuits in the national and international networks. The first transoceanic fiber-optic cable, TAT-8, was installed in 1988 and transmits at 280 megabits per second over two pairs of fibers (a third pair serves as a standby system), equivalent to about 40,000 multiplexed digital conversations. In this system, a technology known as Enhanced Time Assignment Speech Interpolation (E-TASI) takes statistical advantage of the fact that, on average, each speaker in a telephone conversation talks slightly less than half the time, due to listening or brief pauses in normal conversation. By scanning across many simultaneous conversations, the system selectively connects active speakers while blocking idle speakers.

Subsequent transoceanic fiber-optic cables performed at higher transmission rates, eventually reaching

5 billion bits (5 gigabits) per second of optical transmission. Starting in the mid-1990s, many new long-distance fiber cables were equipped with special optical amplifiers that replaced the electronic signal-conversion systems with short fiber cables in which minuscule amounts of erbium, a rare light-sensitive element, were added. Tiny semiconductor lasers stimulate ions in the erbium, causing them to transfer to the infrared light pulses racing through the amplifier. The ions become photons in the pulses, refreshing the pulse strength. This technique allows cable operators to make future upgrades in the multiplexing performance by replacing only the terminal equipment, not the cable.

In addition to normal fiber-optic communications systems, there are extremely high-speed systems operated by high-volume, long-distance carriers such as AT&T. These special terrestrial and undersea digital cables run at speeds as high as 5 gigabits per second through one active fiber pair, using a process called *wavelength-division multiplexing* (WDM) to transmit simultaneously two separate wavelengths, or "colors," of light in the same fiber, carrying a total of 320,000 circuits.

The capacity of digitally multiplexed voice channels also has been increased by reducing the number of bits per second required to transmit speech signals through a process called *speech compression*. The original transmission capacity required for a digital voice channel was 64 kilobits per second, of which 56 kilobits actually contain the digitized speech, while 8 kilobits transmit the control signals accompanying the call. Speech coding research at Bell Labs resulted in methods for compressing the number of bits needed to characterize certain speech sounds. At the receiving end, the compressed signals are restored by expanding them. The present state of the art has limits, however. At about 8 kilobits, the voice begins to sound metallic, and at 4 kilobits it has a robotlike intonation. Speech-compression technology led to other research in reducing the bits used to digitally transmit images and motion video, so each image (fax or still photo) requires less time to transmit, and more video channels can be squeezed through a given multiplexer.—R.Q.H.

See also FIBER-OPTIC COMMUNICATIONS; LASER; MICROWAVE COMMUNICATIONS; SUBMARINE CABLE; TELEPHONE; THERMOIONIC (VACUUM) TUBE

Music, Electronic

In the 1940s, when the word *electronic* first appears in dictionaries, composers began responding to the new age by seeking an alliance of their art with science and mathematics: Electronic music studios were born. However, if by definition electronic music is made with electronically generated or modified sounds, the first electronic music was actually performed much earlier. The evolution of electronic music is, *a fortiori*, fed by technological advances, and toward the end of the 20th century it had evolved into a digital technology.

A comprehension of electronic music requires some understanding of the nature of sound. Sound is perceived when pressure waves in air cause our eardrums to vibrate. The three predominant perceptual parameters of sound are pitch, loudness, and timbre. Therefore, in order to play music generated electronically, it was necessary to build a machine that would cause pressure waves in air, allowing the performer access to these parameters. Each of these perceptual parameters has a physical analog: Pitch is determined by the frequency of the pressure wave, loudness is determined by the amplitude, and timbre is determined by the spectrum of the wave, or the overtone structure of each sound. These parameters vary over time, and such variations give us the aural cues that allow us to recognize a particular instrument in the orchestra and expressive features of the sound: vibrato, the sharpness of attacks, the "bending" of notes.

The use of electricity to generate music goes back to the last quarter of the 19th century. In 1874, Elisha Gray (1835–1901) used an electromagnet as the basis of an electronic oscillator. The electromagnet controlled a metallic switch that turned a battery on and off, creating an unstable circuit. When the switch was on, the current flowed through the magnet, creating a magnetic field that turned off the switch. The closed switch shut off the current, turning off the magnetic field, so that the switch turned back on. This process repeated itself, turning the switch on and off periodically at a frequency that depended on the inertia of the switch. By using 24 of these switches with different inertias, Gray was able to play two octaves. He also built a primitive loudspeaker to amplify these vibrations; by running an oscillating current through an

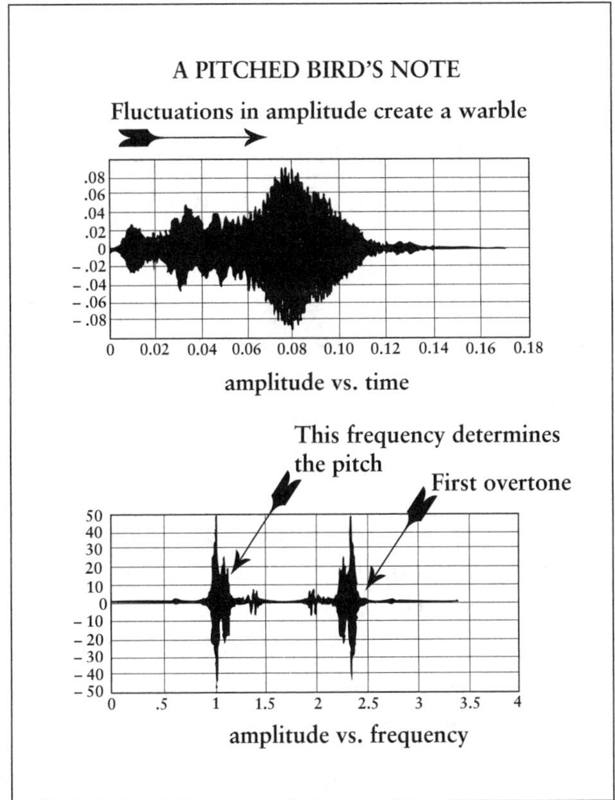

The structure of a bird's note: Loudness is determined by the amplitude; pitch is determined by the frequency; timbre is determined by the overtone.

electromagnet, another metallic piece was activated at an amplitude proportional to the current.

In 1896, Thaddeus Cahill patented a device that generated tones using a series of electrical generators mounted on axles. There were 12 axles, each of which produced a tone on the chromatic scale. The circuitry was activated by a keyboard, and the signals were then passed to a series of acoustic horns. There were enough generators to produce the primary overtones of each note as well, so that the relative amplitude of these could be controlled, generating sounds that imitated orchestral instruments. The device was gigantic; it weighed 200 tons and cost $200,000, but it caused a sensation in the United States as well as Europe and was a breakthrough in that it granted the performer access to all three primary sound parameters.

At the turn of the century, two new approaches to the electronic generation of sound appeared: the current arc oscillator, responsible for a whining sound that emanated from London streetlights around 1900, and the vacuum-tube triode amplifier valve. William Duddell attached a keyboard to a voltage-controlled "singing arc" to create a novelty instrument, while vacuum-tube technology spawned an entire family of musical instruments. The first of these were monophonic, i.e., they played only one melodic output at a time. The most well known were the Theramin (1924), the Ondes Martenot (1928), and the Trautwein (1931). The Theramin was unique in that it did not use a keyboard to control the frequencies of the oscillation but had instead two capacitive detectors: a vertical rod and a horizontal loop, which controlled pitch and amplitude, respectively. The performer's proximity to either device altered the electrical fields they generated, allowing control of these two sound parameters. Live performances on the Theramin and the Ondes Martenot can still be heard occasionally, and several fine recordings are available. The next generation of vacuum-tube instruments were polyphonic; they could play several notes at once, sounding chords and harmonic progressions. The Givelet (1929) was the first of these; with one oscillator per key, it required 200 tubes. Soon (1935) the Hammond organ gained popularity due to its more economical method of tone generation. It used rotating disks in a magnetic field, much like the Telharmonium, and boasted many different orchestral sounds.

After World War II, the tape recorder became an essential part of the electronic-music compositional process. Sounds of all sorts, naturally produced by instruments or machines, such as railroad engines, rubber bands, or natural physical systems like the ocean or the human heart, were recorded and then processed electronically. This type of music became known as *musique concrète*: concrete sounds processed and collaged into a musical piece. Instruments were invented to cope with the enormous tedium of gluing together many tiny bits of magnetic tape, each holding a fraction of a second of sound. One of these was the *Phonogène* invented by Pierre Schaeffer, which had 12 capstans, each with a different diameter, so that a loop of tape could be played back at the 12 different speeds corresponding to successive half-step transpositions. A keyboard allowed the player to select one of the capstans or double the motor's speed, transposing taped sounds up one octave.

The invention of the transistor in the 1950s initi-

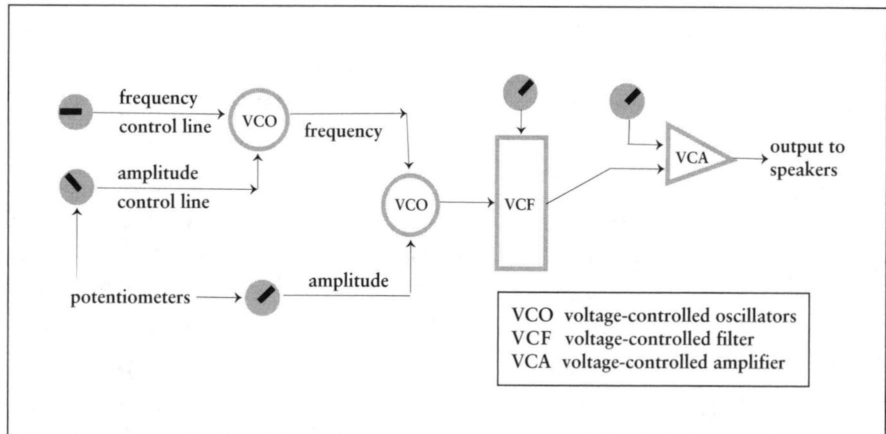

Schemetic of a voltage-controlled synthesizer module (illustration by the author).

ated a major leap forward for the world of electronic music. Transistors were tiny and cheap, and they could be put together into modules, each unit being controlled by some set of voltage characteristics. The voltage-controlled oscillators led to the simultaneous invention of the first two synthesizers in 1966: Robert Moog's Moog Synthesizer and Don Buchla's Buchla Electronic Music System. By the end of the decade, several manufacturers were vying for a market that they correctly saw as taking over the music industry.

The main ingredient of the voltage-controlled synthesizer is a number of oscillators, each with frequency and amplitude input controls. The oscillator can be configured to create several different waveforms, for example, triangle, square, and pulse, in addition to sine waves. The output of these oscillators can be fed through filters, added or subtracted from each other ("ring modulation"), or fed to the inputs of another oscillator. This linking of oscillators is known as *frequency modulation* or *amplitude modulation.* Eventually, analog logic devices were added to the synthesizers to perform signal detection functions such as envelope detection, or vocoding (the application of the timbral or spectral characteristics of one signal onto another, causing a spoken voice, for example, to sound like a violin).

The voltage-controlled synthesizer had a relatively long and healthy existence, with many metamorphoses along the way, until integrated circuit chips took over the role of the transistor. The "analog" synthesizer, with waves generated by electrical current through transistors, was gradually replaced by digital instruments, which output stored waveform patterns from tables or recordings. There are still some musicians who prefer the "organic" sound of the analog instruments, and revivals, such as the Oberheim OBL-MX, are currently on the market.—A.R.

See also AMPLITUDE MODULATION; COMPUTER MUSIC; ELECTROMAGNET; FREQUENCY MODULATION; GENERATOR, ELECTRICAL; INTEGRATED CIRCUIT; MICROPROCESSOR; TAPE RECORDER; THERMOIONIC (VACUUM) TUBE; TRANSISTOR

Musket

The principal early modern military firearm, the musket is a large-caliber, smoothbore, muzzle-loaded shoulder weapon. The word is generally considered to derive from the Italian *moschetto* ("sparrowhawk"), or less commonly, *mosca*, ("fly"). Both etymologies suggest the weapon's origin as a small piece of artillery rather than a large hand weapon, for such ordnance was traditionally named for winged creatures.

The musket was originally introduced into military service by the Spanish in the mid-16th century. It first supplemented, then supplanted, the harquebus (or arquebus) as the principal infantry firearm, due largely to its superior ability to penetrate armor. The musket was distinguished from the harquebus mostly by size, having a larger bore—generally 10 to 12 lead balls to the pound, each with a diameter of 18 to 20 mm (0.7–0.8 in.) rather than the harquebus's 16 to 17 balls with a diameter of 16 to 17 mm (0.6–0.7 in.)—and a longer barrel, on the order of 1.2 m (4 ft). To this end, the first muskets were normally fired from a standing position using a forked rest for support. Firing from a prone position was uncommon due to the difficulties engendered by muzzle loading and using matchlock ignition (a mechanical arrangement for thrusting a smoldering match into the weapon's priming pan to ignite the powder charge).

The proliferation of the musket greatly affected the tactics and organization of early modern European armies, as well as the forces they encountered in the course of imperial expansion and the development of seaborne trade. Mid-16th-century infantry tactics were based on squares of pikemen, supported by harquebusiers and musketeers firing more or less individually. A century later, the emphasis had shifted to musketeers disposed in linear formation to exploit the capabilities of their weapons; soldiers fired in unison, their ranks kept solid by rigid drill and unquestioning subordination to their officers. The pike continued to be used in reduced numbers for pursuit and to deter cavalry charges.

Improved muskets continued to displace the pike throughout the 17th century. As the weapon's rate of fire quadrupled from the old standard of about one round every 2 minutes, soldiers using them became less vulnerable to shock action. This increase was partly due to mechanical improvements. Muskets had become light enough that their rests could be dispensed with, greatly simplifying handling. Furthermore, the matchlock was first improved, then was gradually replaced by a variety of flintlock ignition systems. No less important to the increased effectiveness of the musket as an infantry weapon was improved drill, with excess motions eliminated and the remaining ones incessantly practiced.

The musket completely superseded the pike shortly after 1700 through the addition of the bayonet. Prior to this, the musket had been of little use at close quarters, although its heft and robustness made it a fairly serviceable club. The bayonet began in the mid-17th century as a dagger with a round, tapered grip suitable for plugging into the muzzle of the musket, thus turning the weapon into a makeshift spear. This rather clumsy arrangement was soon replaced by the socket bayonet. This was a more specialized device with a tubular butt and an offset blade that could be kept fixed to the weapon while loading and firing. This innovation allowed the same troops to deliver both missile and shock action for the first time since the Roman legionaries, and thus radically simplified the structure and employment of 18th-century infantry, while at the same time enhancing their effectiveness.

The adoption of flintlock ignition and the socket bayonet marked the maturity of the musket as an infantry weapon; as such, its characteristics were one of the major determinants of land warfare until circa 1850. The smoothbore musket was not a weapon suited to sharpshooters; accuracy fell off substantially beyond 75 m (250 ft). To compensate for the inherent inaccuracy of the musket, armies relied on the concentrated fire of linear close-order formations. This in turn required the high levels of subordination produced by the ferocious discipline characteristic of European armies of the period. The weapon's rate of fire was largely a function of efficient and relentlessly repeated drill. Throughout the 18th century, and indeed through the first third of the 19th century, the musket's design was remarkably stable. Mechanical refinements were limited to minor, incremental innovations. Such alterations included improved lockwork, the Prussian introduction of the double-ended iron ramrod (allowing more vigorous loading with less wasted motion) and the conical touch-hole to facilitate priming, and a general lightening of the weapon.

The sum of these developments over almost a century raised the rate of fire of the individual infantrymen to four and—under exceptional circumstances—even five rounds per minute. This allowed the decrease of the infantry line's depth from the six to eight ranks of the 17th century to the two or three of the Napoleonic Wars, with a concomitant increase in frontage and thus in combat power. This performance was not to be significantly bettered until the introduction of fulminate cap-based percussion ignition to European military weapons in the 1830s, which also brought better performance in adverse weather than did the flintlock, with its exposed priming charge. This, in turn, was followed at midcentury by the revolutionary application of rifling to the musket via the Minié-Delvigny system, at which point the rifle musket's development becomes subsumed by that of the rifle.—B.H.

See also RIFLE

Further Reading: Geoffrey Parker, *The Military Revolution*, 1988.

Nails

A nail driven into a piece of wood is held by the fibers it displaces as it penetrates the wood. Although other kinds of fasteners may hold tighter and look better, nailing is easy, fast, and cheap. In the past, however, nailing was a fairly expensive proposition, for the cost of making nails limited their use. Much labor was required because the nails had to be forged one at a time. This entailed cutting a taper on a red-hot iron rod with a hammer and a triangular "hardy" or "hack iron." The nail was then inserted into the opening of a "nail header" and the blunt end flattened to form the nail's head. Nailmaking was a standard activity for most blacksmiths, but it was also done by many farmers during the slack winter months.

In the late 18th century, more sophisticated devices began to be used to form nails from iron plates. These hand- or foot-powered machines used a screw-driven knife to shear off the nail. Its head was then formed by a lever-operated hammer. Automatic headers came into use in the 1830s, but nailmaking still required a lot of hand labor to manipulate the plates that were cut into nails. Automatic feeds for nailing machines did not appear until the late 19th century.

At first, carpenters and joiners were reluctant to use machine-made nails because they thought that the shape of these nails reduced their holding power. Machine-made nails tapered on two sides only, whereas hand-forged nails tapered on all four sides. But mechanization cut the price of nails to such an extent that it became increasingly difficult to justify the use of handmade nails. Cheap machine-made nails in turn helped to effect a transformation in construction technology. Instead of building a structure around heavy beams that were held together with wooden pegs and elaborate mortise-and-tenon joints, the new method was based on the use of relatively light structural members held together by cheap, machine-made nails. The combination of dimensionally smaller lumber and cheap nails substantially reduced the cost of construction in money and time.

One remnant of traditional practices still in operation today is the convention of sizing nails in "pennies." The pennies refer to the cost of 100 hand-forged nails a long time ago. Thus, 10 cents once bought 100 10-penny (abbreviated as 10d) nails, while 2 cents bought 100 2d nails. A 10d nail is 7.6 cm (3 in.) in length; a 2d nail is 2.5 cm (1 in.) long.

Nails are available in a wide variety of sizes and shapes. Nails designed for furniture or cabinetmaking, known as *finish nails*, have very small heads that can be sunk below the surface; the small depression is then filled in, leaving a smooth surface. There are nails specially designed to be driven into nonwood materials such as masonry or gypsum board. Some nails have greater gripping power because they are equipped with spiral threads, annular rings, or special coatings. Nails are usually made from steel, often galvanized (i.e., coated with zinc) for corrosion resistance. Brass nails are sometimes used for particular applications such as upholstery.

See also CONSTRUCTION, BALLOON-FRAME

Napalm

Since early times, warriors have used fire as a weapon. Burning oil, "Greek fire," and heated solid shot have all been utilized. Throughout history, incendiary weapons have generated controversy. Branded as terror weapons, their legitimacy has frequently been questioned. Modern flame weapons were first used during World War I, when Germans attacked the British and French with flamethrowers (*Flammenwerfers*) in 1915. Soon, troops on both sides were using flamethrowers with petroleum-based fuels as a means of inflicting casualties in close-quarters trench fighting along the western front.

Early in World War II, German and Japanese troops employed flamethrowers against enemy troops and flammable structures. The United States rushed to develop flamethrowers and incendiary aerial munitions on the eve of its entry into the war. Gas and diesel oil mixtures were tried, but were unsatisfactory; the flame stream burned itself up on its flight from the flame gun to the target. A more viscous fuel was needed, one that would burn more slowly and could deliver more fire on the target. Early experiments with fuel thickeners composed of soap, raw rubber, or crankcase oil, some of which dated back to World War I, were unsuccessful.

Early field experiments were only marginally successful. Tests demonstrated the need for a flame fuel thickener that would retard the rate at which the flame fuel burned after it was expelled under pressure from the nozzle of the flame gun directed by the soldier carrying a flamethrower. Similarly, experiments carried out by American airmen dropping fuel tanks filled with highly flammable aviation fuel were relatively ineffective as incendiary weapons because the fuel burned too quickly.

In late 1941, an American civilian advisory group, the National Defense Research Committee (NDRC), experimented with various compounds built around aluminum hydrochloride which, when combined with gasoline, formed a jellylike substance. They also found a rubber-based thickener called IM (isobutyl methacrylate). It formed a substance that had the consistency of applesauce when added to gasoline. Unfortunately, the Japanese conquest of Malaya cut off most of the rubber supplies on which wholesale use of IM would depend.

In tests conducted by the U.S. Army Air Force (USAAF), both mixtures functioned well as a filler for incendiary bombs, but tended to break down into their original components after being combined, especially when being handled during shipment. These compounds also failed as flamethrower fuels because temperature or humidity variations caused unwanted chemical reactions in the gel. The average soldier, working under less than ideal conditions in the battle area, might easily blend the fuel incorrectly.

A different group of scientists, under the direction of Dr. Louis Fieser of the NDRC, provided the Army's Chemical Warfare Service (CWS) with the substance it needed to produce a much more efficient filler for bombs and flamethrowers. Fieser's group blended the fatty soaps extracted from aluminum napthanate and aluminum palmitate. The chemical reaction that resulted formed a compound that, like IM, had the look and consistency of applesauce. When dried, this material could be separated into marble-sized balls and packed in tightly sealed containers. On being mixed with gasoline, the chemicals in the compound would cause a thickening of the gasoline, thus lowering its burning flash point.

In a series of USAAF tests, the compound produced impressive fires when dropped and the CWS was convinced that it had found a suitable fuel thickener. Pressed for a name for his discovery, Dr. Fieser elected to name it by using letters from the first syllables of the two main elements: It was called "napalm."

The evolution of the fire bomb as a close air support weapon is a story of imagination and improvisation. American fighter pilots began the practice of jettisoning partially filled external fuel tanks during missions over France prior to D-day in 1944. The dropped fuel was then ignited by firing tracer bullets into the area where it had landed. The addition of napalm thickening was a natural second step. The thickener caused the jellied gasoline to stick to surfaces and to burn more slowly, thereby making the blaze more damaging to human beings and flammable material.

Like their contemporaries in Europe, pilots in the Southwest and Central Pacific commands began experimenting with napalm in drop tanks early in 1944. Many ground commanders requested air attacks, maintaining that flame attacks by fighter aircraft carrying large amounts of napalm were more effective than ground troops armed with flamethrowers.

The Marine Corps, long a leader in close air support for its ground forces, put air-delivered napalm to use during the assault on Iwo Jima in February 1945 and again on Okinawa in April–May 1945. A number of Marine Corps units cited the bombs in their after-action reports, stating that they often caused Japanese troops to break from cover in an attempt to evade the napalm blaze. Once in the open, the running enemy could be killed with small-arms fire. Similar success was reported by USAAF fighters flying in support of ground troops in the Philippine Islands. Selected enemy strong points were blanketed with napalm bombs dropped by the aircraft that flew over the target area in waves. Success led to continued employment of napalm by American air units for the rest the war in the Pacific.

Five years later, the United States and its United Nations allies would again employ napalm, this time in Korea. Most of the flame weapons used in the Korean conflict were World War II–vintage incendiary bombs and flamethrowers taken out of storage. Tactics varied little from those employed during the last year of fighting in the Pacific and proved to be successful in assaulting dug-in communist troops.

The Army and Marine Corps in Vietnam used both portable and vehicle-mounted flamethrowers. In the Mekong Delta, the U.S. Navy employed flamethrowers mounted on armored, shallow draft boats. Despite some successful use of flamethrowers, the heat, humidity, and inhospitable terrain with which ground forces had to contend discouraged their widespread use. Instead, the commanders of ground units relied on the dropping of napalm by tactical air support to destroy enemy forces with which they were engaged. The use of napalm was heavily criticized by the foreign press and by antiwar activists in the United States and elsewhere. But troops in contact regarded it highly, not only for its casualty-producing effects but also because it burned away vegetation concealing enemy positions.

The Vietnam War saw the last tactical employment of napalm by the United States until January 1991. Then, U.S. Marine airmen tapped stocks of obsolescent napalm bombs in order to bomb Iraqi forces. Napalm is available for use today in a number of countries. However, it has been relegated to the rear rank of munitions. Just as today's "smart bombs" have largely replaced iron bombs for the destruction of precision targets, so too have sophisticated fuel-air explosives, and antipersonnel, armor-defeating munitions replaced napalm as the first choice to aircrews charged with the elimination of enemy ground forces.—J.M.

See also GREEK FIRE; PRECISION-GUIDED MUNITIONS

National Academy of Science

The National Academy of Science (NAS) was established by Congress in 1863 to serve as an official advisor to government and the public at large on matters of science and technology. It is analogous to Britain's Royal Society for the Promotion of Natural Knowledge (founded in 1660 during the reign of Charles II), France's *Academie Des Sciences* (founded in 1666 by Jean-Baptiste Colbert), and Russia's *Akademiya Nauk* (founded in 1724 by Peter the Great). The NAS is a nongovernmental organization with a self-perpetuating membership of highly accomplished scientists and engineers. New members are nominated by existing members and voted on by the full membership. There are currently 1,800 American members and 300 foreign associates. Over 120 members have won Nobel Prizes. The membership is divided into 25 sections based on academic specialties ranging from astronomy to animal sciences, and includes the social as well as the natural sciences. An elected president heads the NAS.

In 1964, the National Academy of Engineering (NAE) was formed as an associated organ of the NAS. The NAE has over 1,700 members and 150 foreign associates. Its members frequently work closely with members of the NAS on related projects and reports. For example, the Water Science and Technology Board and the Computer Science and Technology Board are made up of equal number of scientists and engineers. In recent years these boards have published a number of highly visible reports that have served as key references for important environmental laws such as Superfund and the Clean Water Act. They have also shaped Congressional debate on topics such as the commercial and legal consequences of rapidly changing information technologies. In 1970, the Institute of Medicine was chartered as an additional associated organization of the NAS. It follows the mandate of the 1863 NAS charter to provide government with advice about matters pertaining to public health and medical research.

In 1916, the academy established the National Research Council (NRC) to coordinate the many activities of scientists and engineers in universities, industry, and government who had been enlisted to do NAS work. In 1950, the NRC and the NAS were administratively joined. The NRC coordinates the work of 6,000 volunteers who serve on 600 NAS committees that address scientific issues of public importance and publish reports that are frequently central to ongoing policy deliberations. These committees are composed of NAS members and nonmembers, and they serve to acquaint member scientists with complex public-policy issues, while at the same time connecting public officials and civic activists to relevant scientific expertise. These committees produced over 200 reports on issues of public concern in 1995 alone. The primary scientific research publication of the NAS is the *Proceedings of the National Academy of Sciences*.

As noted by the president of the NAS, the organization has four major functions. First, it serves to validate scientific excellence by thrusting NAS members into the public arena and asking them to lead by example. Careful nomination and election of NAS members is therefore of considerable public importance. Second, the NAS is charged with maintaining the vitality of the scientific enterprise: upholding the highest standards for research and encouraging talented young people to enter scientific fields and mentoring them effectively in their early years of training and research. Third, NAS seeks to apply scientific knowledge to public policy. Here, the NAS/NRC committee system plays the key role by bringing scientists, concerned civil leaders, and policymakers together to address issues of public concern, and ensuring that the latest scientific knowledge is applied in responsible ways. Finally, at a time when U.S. legislators from both parties have proposed substantial cuts in federal budgets for research and development, the NAS has dedicated increased attention to educating the Congress and the public about the importance supporting scientific research and training. This public-relations undertaking has been fostered by the development of a site on the World Wide Web and the electronic publishing of numerous full-length NAS reports. The NAS has also launched a new program entitled "Public Understanding of Science" and has created a new Center for Science, Mathematics, and Engineering Education.—T.I.

See also INTERNET; NOBEL PRIZE; SCIENTIFIC SOCIETY

National Aeronautics and Space Administration (NASA)

Motivated by a concern that the United States was falling behind other nations in aviation technology, the federal government created the National Advisory Committee for Aeronautics (NACA) in 1915. In 1920, NACA got its own facility, the Langley Memorial Aeronautical Laboratory in Virginia. In 1939, a second center, the Ames Aeronautical Laboratory, was established at Moffett Field in California, followed a year later by an engine research facility, the Lewis Flight Propulsion Laboratory, in Cleveland, Ohio.

The research done at NACA facilities made major contributions to aircraft design in the 1920s and 1930s, especially in regard to the promotion of faster, more efficient flight through streamlining. In the late 1940s, NACA was involved with successful efforts to fly faster than the speed of sound. Yet all these accomplishments were eclipsed by the Soviet Union's orbiting of an artificial satellite on Oct. 4, 1957. Determined not to lose "the space race," the United States endeavored to strengthen its own space program. On July 29, 1958, President Eisenhower signed into law the National Aeronautics and Space Act, and on Oct. 1, 1958, NACA became the National Aeronautics and Space Administration (NASA), the agency that assumed overall responsibility for the U.S. space program. In addition to the former NACA facilities, two other centers of space technology were brought into the fold: the Redstone Arsenal in Huntsville, Ala. (subsequently renamed the George C. Marshall Space Flight Center), and the Jet Propulsion Laboratory in Pasadena, Calif. NASA also acquired the staff of the Navy's ill-fated *Vanguard* program. In 1959, the Goddard Space Flight Center was established in Greenbelt, Md. Originally intended to be the center for all space operations, its responsibilities were eventually limited to unmanned programs. In 1962, with the manned program becoming the centerpiece of the U.S. space program, NASA's operational center for human spaceflight moved from Langley to a new site near Houston, Tex. In 1973, this facility, which had been known simply as the Manned Space Flight Center, became the Lyndon B. Johnson Space Center. In 1963, construction began on what was to become the John F. Kennedy Space Center on Merritt Island at Cape Canaveral, Fla., the first launch occurring in 1967.

During this period, NASA continued to involve itself with aeronautical work that ranged from the reduction of hydroplaning when aircraft landed on wet runways to managing the X-15 program for research into high-altitude, high-speed flight. But spaceflight was NASA's main occupation. In addition to launching numerous unmanned space probes, NASA was deeply involved with human spaceflight. Beginning with Project Mercury, moving through Project Gemini, and culminating with the moon program, Project Apollo, NASA's efforts produced resounding successes. These triumphs did not come cheaply, however; by the 1965 fiscal year, NASA was spending more than $5 billion annually—4.4 percent of the federal budget and 0.8 percent of the U.S. gross national product.

NASA's successes were not achieved by simply throwing money at technical problems. It had a skilled staff that was highly motivated to solve the myriad technical problems connected with spaceflight. During its early years, NASA's personnel did a great deal of construction and testing in house, but the immensity of the projects, especially the ones involving manned spaceflight, necessitated the use of outside contractors. As a result, NASA personnel had to develop managerial as well as technical capabilities. At the same time, however, the widespread use of contracting tended to deprive NASA personnel of the hands-on involvement that characterized the early years of the space program.

Although NASA successfully responded to President John F. Kennedy's call to put a man on the moon by the end of the 1960s, triumphant space missions did not translate into long-term public support. By 1973, less than 3 years after Apollo 17 had made the last manned trip to the moon, for every American who wanted to see spending on space increased, there were eight who wanted it cut. By this time NASA was feeling the effects of financial stringency. After reaching a high point in the mid-1960s, NASA had to go through a long period of stagnant or declining budgets. While budgets were being pinched, NASA also was subjected to more political oversight. The combined effects of budgetary squeezes, political interference, and the diminished glamour of the space program made NASA a less-attractive place to work. As a result, the average age of agency personnel increased as fewer new employees entered the organization. All of these factors combined to make NASA an organization that had become overly concerned with procedures, more risk averse, and less flexible—in short, more bureaucratic.

Beginning in the late 1970s, much of NASA's energies were devoted to the Space Shuttle program. Tragically, the explosion of the *Challenger* shortly after launch revealed serious organizational weaknesses within NASA. A few years later, a major optical flaw in the Hubble Space Telescope again called into question NASA's ability to administer large projects. NASA is currently engaged in planning for a space megaproject for the 21st century, the space station *Freedom*. According to some observers, NASA will have to revive the organizational environment of the early years of the space program if this project is to be successful.

See also APOLLO PROJECT; *Challenger* DISASTER; GEMINI PROJECT; HUBBLE SPACE TELESCOPE; MERCURY PROJECT; SPACE PROBE; SPACE SHUTTLE; SPACE STATION; SPUTNIK; STREAMLINING; SUPERSONIC FLIGHT

Further Reading: Howard E. McCurdy, *Inside NASA: High Technology and Organizational Change in the U.S. Space Program*, 1993.

National Institute of Standards and Technology

In 1901, the United States Congress established the National Bureau of Standards as an agency of the Department of Commerce. The nation's industries were becoming more technologically sophisticated, and there was a need to create standards that could be used by all firms within a particular industry. In the years following its creation, the National Bureau of Standards (NBS) took on a variety of tasks. To cite a few examples, it established standards for lighting and electrical power usage, developed uniform temperature measurements for different metals in their molten states, and engaged in studies of corrosion.

During the 1980s, many American political and economic leaders were deeply concerned about the state of American technology. There was a common fear that the United States was falling behind Japan and Western Europe in the creation, development, and application of new technologies. One response to this concern was the passage by the U.S. Congress of the Omnibus Trade and

Competitiveness Act in 1988. One part of this legislation gave the National Bureau of Standards a significant role in the development and promotion of new productive technologies. In recognition of these expanded responsibilities, the Bureau was renamed the National Institute of Standards and Technology (NIST). In the mid-1990s, it had 3,300 staff members working at its two main centers in Gaithersburg, Md., and Boulder, Colo.

The National Institute of Standards and Technology funds and organizes four major programs. The first of these is the Advanced Technology Program (ATP). Begun in 1990, this program awards funds to individual firms and industry consortia to foster research and development on new technologies. NIST does not provide all or even most of the total funding; the recipients usually pay more than half the R&D costs incurred. Applicants are evaluated according to several criteria: (1) the proposed project's potential to produce a significant economic impact, (2) the extent to which it embodies cutting-edge technology that is technically and scientifically sound, (3) the strength of the commitment by the firm or consortium, and (4) the likelihood that participation in the program will make a real difference.

NIST also organizes and partially funds the Manufacturing Extension Partnership. In this program, NIST provides technical assistance for small and medium-sized firms. Working in conjunction with state and local governments that provide a major portion of funds, MEP is somewhat similar to the agricultural extension services that have brought improved agricultural technologies to the nation's farmers. In 1992, MEP's funding came to $70 million.

The third NIST program harks back to the original charge of NBS: the setting of technical and scientific standards. NIST's Laboratory Research and Services Division is charged with the development and delivery of test measurement techniques for R&D, production, and marketing. The seven laboratories in the division do basic scientific research in many fields: electronics and electrical engineering, chemical science and technology, information technology, materials science and engineering, manufacturing engineering, physics, and building and fire research.

Finally, NIST administers the Baldrige National Quality Award. Named after Secretary of Commerce Malcolm Baldrige, the award has been given since 1988 to small businesses, manufacturing firms, and providers of services. The selection of award winners takes into account a number of criteria that affect and indicate an enterprise's performance: leadership, information and analysis, strategic planning, human resource development and management, process management, business results, and customer focus and satisfaction.

The expanded duties of NIST reflect the belief that government agencies can make a positive contribution to technological advance and economic growth. In the 1990s, many members of Congress wanted to reduce government involvement in the economy. Consequently, political support for the National Institute of Standards and Technology has diminished somewhat since the time of its creation.

See also DEPARTMENT OF AGRICULTURE, U.S.; RESEARCH AND DEVELOPMENT; STANDARDIZATION

National Institutes of Health

The National Institutes of Health (NIH) is the United States government agency that supports and undertakes biomedical research in order to acquire new knowledge to help prevent, detect, diagnose, and treat disease and disability. Its origins can be found in the Laboratory of Hygiene, which was founded as a one-room government laboratory in 1887, with a budget of $300. Currently, NIH is one of eight agencies within the Public Health Service, which in turn is housed within the Department of Health and Human Services. The NIH is made of 24 separate institutes, centers, and divisions that include the National Cancer Institute; the National Heart, Blood, and Lung Institute; the National Institute on Aging; and the National Institute on Environmental Health Sciences. The agency headquarters and the majority of the intramural research is housed in 70 buildings on 300 acres in Bethesda, Md. Supported exclusively by tax dollars, the 1995 NIH budget was $11.2 billion. Of that budget, 81 percent went to supporting extramural research and training at 1,700 universities and research institutions across the United States and in a number of foreign countries; 11 percent of the budget went to support intramural research at NIH laboratories. The remaining 8 percent of the budget was absorbed by administrative costs. The agency is headed by a director appointed by the President, but each institute also has its own leadership and operates with considerable independence. The NIH has 19,000 employees, 4,500 of whom have professional or research doctorates.

The NIH is far and away the largest supporter of biomedical research in the United States. It solicits grant applications from established researchers, and assesses and ranks them through a process of peer review. The NIH receives approximately 36,000 applications each year and, at any one time, supports about 35,000 research projects. This considerable American investment in biomedical research has produced extraordinary results. NIH funding has supported the work of 83 Nobel laureates, 4 of whom have done their work in NIH laboratories.

The NIH has also been a major supporter of predoctoral and postdoctoral training; many of the current generation of leading American biomedical researchers were supported by NIH training fellowships. In an era of tighter budgetary constraints, concerns have been raised about adequacy of funds for training and the effects of fiscal stringency on future generations of biomedical research scientists. The NIH is also charged with the active dissemination of new knowledge. It maintains the National Library of Medicine, which is the largest source of medical information in the United States.

NIH's mission to generate new knowledge in the fight against life-threatening and debilitating diseases has resulted in a division among supported projects between clinical and basic research. While it has varied modestly over time, approximately two-thirds of the NIH research budget has gone to clinical work and one-third to basic research. The agency's priorities regarding which health issues or specific diseases to address with what levels of funding have varied over time. However, the process of peer review by established research scientists has undoubtedly insulated the NIH from the public politics of perceived health crises that could result in wide swings in research priorities. At the same time, however, the NIH has responded to new health issues or the emergence of new diseases both by the creation of new institutes such as the National Cancer Institute and the National Institute of Environmental Health Sciences and by the reallocation of funding within existing institutes. Currently, the NIH defines its priority diseases as cancer, heart disease, stroke, blindness, arthritis, diabetes, kidney disease, Alzheimer's disease, communication disorders, mental illness, drug abuse and alcoholism, and AIDS. It also identifies the health of infants and children, women and minorities, and the aging process as health issues of special importance.—T. I.

See also Nobel Prize

National Science Foundation

The National Science Foundation (NSF) was founded in 1950 following the passage by the U.S. Congress of the National Science Foundation Act. Deliberations regarding the creation of the NSF were part of a larger national debate about the appropriate role for government in the support of basic and applied science during and immediately after World War II. During the war, scientists, industry representatives, government officials, and populist lawmakers sought to enhance the government's ability to harness science and technology resources to the cause of rapid and effective military mobilization. Discussion was broadened after the war by a report, *Science—The Endless Frontier*, authored by Vannevar Bush (1890–1974), the director of the Office Scientific Research and Development, which was submitted to President Truman in 1945. The report's principal themes were the importance of basic research to the achievement of national goals—security, prosperity, health—and the need to develop indigenous scientific and technical potential as opposed to importing it from abroad. With time, it became clear that federal support for basic research was crucial if this national capability was to be realized.

The early architects envisioned the National Science Foundation as the lead agency in this national effort, but President Truman vetoed the first NSF bill passed by Congress in 1947. By the time the NSF Act became law in 1950, other government agencies concerned with Cold War security, such as the Office of Naval Research, the Atomic Energy Commission, and the Department of Defense, had already won sizable basic research allocations. The NSF that emerged after 1950 never took on the role of lead government research agency, and its budgets consistently reflected this more modest role. Its share of funding for basic research in the total national research and development budget was 7 percent in 1960, and the budget has never exceeded 10 percent. However, unlike the defense-related agencies and the National Institutes of Health, which oriented their research towards particular applied missions, NSF focused on basic research and training in the natural sciences through the 1950s and early 1960s. Congress broadened the role of the agency with an amendment to the NSF Act in 1968, authorizing it to fund applied as well as basic research and to support the social sciences in addition to the natural sciences and engineering.

NSF's mission, as spelled out in the NSF Act, is to promote the progress of science; to advance the national health, prosperity and welfare; and to secure the national defense. It is led by a director and a National Science Foundation board of 24 members, each of whom is appointed by the President and approved by the Senate. One deputy director and eight assistant directors administer the diverse range of programs. From its founding, NSF has allocated the largest portion of its funds to individual scientific investigators who submit research proposals for discrete projects. Each proposal is evaluated on its merits through a process of peer review. NSF also supports a large program of graduate fellowships in the natural and social sciences and in engineering. Much smaller allocations of funds have gone to the construction of scientific facilities, institutional support, group projects, and, more recently, the development of research centers. With a budget of about $3 billion in the mid-1990s, NSF makes about 20,000 research and fellowship awards each year. Unlike the NIH, NSF has no intramural research program or staff; it relies entirely on universities, colleges, and other research institutions for the conduct of its supported research.

The NSF also fosters interchange of scientific information between scientists in the United States and other countries, supports the development of computers and computer applications, promotes activities that encourage the application of new knowledge, recommends national policies for basic research and scientific education, and supports activities designed to increase the number of women, minorities, and other underrepresented groups in science and technology.—T.I.

See also DEPARTMENT OF DEFENSE, U.S.; NATIONAL INSTITUTES OF HEALTH; PEER REVIEW

Natural Gas

Natural gas is 85 to 95 percent methane (CH_4), accompanied by nitrogen, carbon dioxide, helium, and hydrogen sulfide. When it lies in a reservoir that contains petroleum, it usually includes some heavier hydrocarbons, such as propane, ethane, heptane, and butane. Depending on the mix, natural gas has an energy density ranging from 29 to 39 megajoules (6.9–9.3 calories) per cubic meter. Natural gas that is used commercially is odorless; the "gas smell" associated with it is provided by an additive, ethyl mercaptan (HSC_2H_5).

Drilling for natural gas goes back more than 2,000 years. Wells were sunk in China's Sichuan province for the extraction of natural gas as early as the 3d-century B.C.E. Although the gas was used for domestic lighting and cooking (after having been transported through bamboo pipelines), its primary application was for the evaporation of brine for making salt. The first gas well in the United States was sunk in Fredonia, N.Y., in 1821, nearly 4 decades before the drilling of the first oil well. It went down to a depth of 8.2 m (27 ft). In recent years, some gas wells have been sunk more than 9,150 m (30,000 ft) into the ground.

Petroleum deposits always include some gas (associated gas). Gas may also exist by itself, particularly in deep deposits (nonassociated gas). During the early days of the petroleum industry, gas was useful because it forced oil out of its reservoir and up the borehole, but it had little value as a fuel. In the absence of compressors and pipelines, there was no way to get the gas to where it could be used, so it was flared off at the source. For many decades, gas had been widely used for illumination, but the gas used was made from coal, while naturally occurring gas went to waste.

Pipelines for the transportation of gas began to be built in the late 19th century. In 1891, a 193-km (120-mi) gas pipeline supplied Chicago with gas produced in Indiana. In the 1930s, a 1,600-km (1,000-mi pipeline ran from Texas to Chicago. Gas pipelines began to be built in earnest in North America after 1945 and in Europe after 1960, stimulating the widespread use of natural gas. When pipeline transportation is impossible, as when gas has to be transported overseas, it is liquefied and shipped in special containers that keep it at –162°C (–260°F), at which point it occupies a volume 1/600 of its gaseous state.

Governmental agencies have long been involved with the production and transport of gas. In 1912, gas pipelines were ruled to be common carriers, and hence under the control of the Interstate Commerce Commission. In 1954, the U.S. Supreme Court upheld the 1938 Natural Gas Act, paving the way for the control of natural gas production by the Federal Power Commission (FPC). Under the FPC, the price of natural gas shipped across state lines was controlled, whereas intrastate gas was not controlled. As a result, when energy prices increased as a result of actions by the Organization of Petroleum Exporting Countries (OPEC) in the 1970s, gas tended to remain in the state where it was produced, causing significant economic dislocations. In 1978, prices were partially decontrolled; they were completely decontrolled in 1989. Predictions that the price of gas would skyrocket proved erroneous, for higher prices stimulated increased exploration and production, bringing supply and demand back into equilibrium.

Gas is in many ways the most desirable of fossil fuels. It burns cleanly, requires no on-site storage facilities for domestic and industrial uses, produces no ash, and requires little in the way of pollution control equipment. However, emissions of unburned methane can be problem, for it is a much more powerful greenhouse gas than carbon dioxide. The clean-burning quality of natural gas has made it especially attractive as a fuel for cars and trucks. But natural-gas–using vehicles have higher initial costs, require more space for the fuel tank, and give less mileage between refuelings.

In addition to being an important industrial fuel, natural gas is the most important feedstock and source of energy for the production of chemical fertilizers.

With natural gas likely to remain in high and perhaps increasing demand, there is a natural concern about the extent of gas reserves. Estimating the reserves of any natural resource is a difficult exercise fraught with many sources of potential error. More than 50 billion cubic feet of natural gas is produced in the United States each day, accounting for 27 percent of U.S. domestic energy consumption. At the present rate of extraction, there are more than 50 years of natural gas reserves in the United States, accounting for only about 7 percent of the world's proven reserves. There seems little likelihood of serious shortages of natural gas, although political events can always cause disruptions.

See also CLIMATE CHANGE; DRILLING, PERCUSSION; DRILLING, ROTARY; FERTILIZER, CHEMICAL; FUELS, FOSSIL; HABER PROCESS; LIGHTING, GAS; RESOURCE DEPLETION

Natural Selection

The modern theory of biological evolution began in 1858 with the simultaneous publication of articles by Charles Darwin (1809–1882) and Alfred Russel Wallace (1823–1913) in the *Journal of the Linnean Society.* Darwin's "On the Tendency of Species to Form Varieties" and Wallace's "On the Perpetuation of Varieties and Species by Means of Natural Selection" both insisted that changes occurred more or less randomly in organisms. Furthermore, those changes that made it less likely that the organism would survive through its reproductive cycle would be damped out as other organisms whose changes gave them some advantage in surviving through their reproductive cycle produced greater numbers of offspring. If this process were carried out in an environment that slowly changed and in which the existing species were originally almost perfectly adapted to the initial environment, some new varieties would be better adapted to the new environmental conditions and would thus gradually replace the original populations. Over long periods of time, the successful organisms would come to differ so much from the original populations that they would be recognized as entirely new species. Wallace termed this process *natural selection*, and his terminology prevailed over Darwin's initial "struggle for life." Indeed, Darwin popularized natural selection in his 1859 treatise *On the Origin of Species by Natural Selection or the Preservation of Favoured Races in the Struggle for Life.*

Today, almost all biologists believe that the most important mechanism producing changing populations of living beings is natural selection. However, Darwin himself suggested that sexual selection may increase the reproductive rates of some varieties even where the number of "competitors" that survive through reproductive age is equal or greater. More recently, population biologists have suggested that some survival-neutral changes may be propagated simply because of random population fluctuations, giving rise to new kinds of organisms by what is termed *genetic drift*.—R.O.

See also EVOLUTION; EVOLUTION, PUCTUATIONAL MODEL OF

Navigation

Finding out where you are and where you are going while in unfamiliar territory has always been a challenge. Whether traveling on the high seas, through the air, or across featureless terrain, people had to be able to navigate accurately if they were to get to their destinations in a reasonable amount of time. On many occasions, the ability to navigate has been the difference between life and death.

A fundamental method of navigation is known as *dead reckoning*. This entails plotting one's course by taking into account the speed and direction of travel. In centuries past, speed was determined by throwing overboard a floating object and then observing the time that elapsed as a known length of the ship passed by it. Alternatively, a line was attached to the object, and note was taken of how long it took for a known length of line to pay out. Often, the line had knots at regular intervals to facilitate this measurement; this is the origin of a *knot*, or nautical mile per hour, where 1 nautical mile = 1.15 mi = 1.85 km. The term *dead reckoning* probably is derived from the fact that the floating object was "dead in the water" as the ship passed by it. Today, speed and distance usually are measured by mechanical logs that are towed behind ships. In an airplane, these are measured by instruments connected to a pitot tube. The direction of travel is ascertained through the use of a magnetic compass or a gyrocompass.

Of course, dead reckoning is useful only if one's position is already known. Position is given in terms of two coordinates: latitude and longitude. Latitude is the distance, measured in degrees, of one's position north or south of the equator. It is relatively easy to determine, because the position of the sun and stars varies according to one's latitude. Finding latitude is facilitated by the use of an instrument known as a *sextant*. The sextant determines latitude by measuring the angle (in degrees of arc) between the horizon and a star, or the angle between the horizon and the sun at high noon (in navigational terms, this angle is known as an *altitude*). The sextant is so named because it has an arc of 60 degrees of a circle (i.e., one-sixth of a circle). Similar instruments with greater or lesser arcs are frequently called sextants as well.

The sextant was developed independently in the early 1730s by an American, Thomas Godfrey, and an Englishman, John Hadley (1682–1744). Their devices replaced an earlier instrument known as a *cross staff*, which in turn had replaced the *astrolabe*. Hadley, who had previously worked on an early reflecting telescope, equipped his sextant with a mirror that allowed simultaneous viewing of the horizon and the sun. The angle between the two could then be taken from a scale inscribed around the instrument's edge. The scale encompassed only 40 degrees, so the instrument really was a quadrant. In 1757, the arc was extended to 60 degrees, giving rise to the true sextant. Modern sextants fall into two categories, marine sextants and air sextants. In the latter instrument, a built-in artificial horizon takes the place of the Earth's horizon. Air sextants are usually equipped with periscopes that allow observation of celestial bodies without the need for the aircraft to be fitted with a transparent dome for sighting.

Longitude is more difficult to determine. The first successful means of determining longitude was made possible by John Harrison's invention in the 18th century of a highly accurate timepiece that could be carried on a ship. The navigational methods used from the late 18th century onwards have relied on some version of a sextant, an accurate timepiece (or at least one that gains or loses time at a known rate), and numerical tables collected in nautical almanacs.

The first step in finding one's position is to determine the "projection" of a selected star on the surface of the Earth; this is a point that lies on an imaginary line extending from the star to the center of the Earth. Finding position is done by using a sextant to determine the star's altitude and noting the exact time at which the observation was made. The navigator can then refer to a table in a nautical almanac that gives the latitude and longitude of a point projected by a particular star when the altitude of that star at a particular time is known. The next step is to find the distance from that point to the location of the navigator. This is done by taking star's altitude expressed in minutes of arc (60 minutes = 1 degree) and subtracting it from 5,400. This gives the distance in miles between the point and the navigator. The navigator then draws a circle on a chart (i.e., a map used for navigation), using the point as the center and the distance as the radius. The procedure is then repeated using another star as a reference, and another circle is drawn. The intersection of these circles marks the latitude and longitude of the observer.

A similar procedure can be used with the sun serving as the target and using two measurements separated by an interval of time. This introduces complications when drawing the intersecting lines (due to the movement of the vessel between observations), but readings of the sun may be required when the stars are not visible due to the time of day or the presence of a cloud cover.

The procedure outlined above is not used for actual navigation, for a circle drawn on a flat map would be something quite different if it were transposed onto a spherical surface. Instead of drawing two circles, it is necessary to draw two straight lines that represent small portions of these circles; the intersection of these lines then marks the navigator's position. The technique used for plotting these "position lines" is given in any practical guide to navigation.

In recent decades, navigating by the sun and stars has been supplemented and in some cases supplanted by new navigational technologies. In inertial navigation, onboard sensors indicate the position, heading, and speed of a craft (typically a submarine or long-range aircraft). Gyroscopes sense angular motions of the craft, and accelerometers sense changes in its speed. These data are mathematically processed to provide the information used to keep the craft on its proper course.

Many ships and aircraft follow a course set by synchronized radio signals sent out by paired transmitters located at known points. Measurements of the different arrival times of the signals from the two transmitters then give the difference in distances from the stations. From 1964 onwards, orbiting satellites have played an important role in navigation. Ships and aircraft have used the Doppler effect to determine the relative velocity between them and the satellite. When this is combined with known rates of change in distances between the two—the precise orbit of the satellite and the speed of the ship or airplane—the position of the latter can be known with a good degree of accuracy. In recent years, orbiting satellites have been the basis of global positioning systems (GPS) that provide an easy-to-use and highly accurate method of navigation.

See also ASTROLABE; COMPASS; DOPPLER EFFECT; GLOBAL POSITIONING SYSTEM; GYROSCOPE; LONGITUDE, DETERMINATION OF; TELESCOPE, REFLECTING

Further Reading: Leonard Gray, *How to Navigate Today*, 6th ed., 1986.

Neanderthals

The Neanderthals were a specialized race of early *Homo sapiens* that inhabited Europe and western Asia (the Middle East) from about 150,000 to 35,000 years ago. When the first specimens were discovered, in the latter half of the 19th century, prominent scientists discounted them as skeletons of modern human beings with bone diseases. Eventually, however, the association with extinct animal species indicated that the Neanderthals were a form of prehuman. Considerable data have since been gathered, advancing our understanding of Neanderthal morphology and behavior.

Early members of this group, commonly classified as *Homo sapiens neanderthalensis*, resembled their *H. erectus* predecessors but had an expanded brain; cranial capacities reached 1,500 cc, slightly larger than the modern average of 1,400 cc. Later specimens exhibit the extreme traits usually attributed to the Neanderthals. Although large, the skull is extremely low, with large, arching brow ridges giving way to a sloping forehead. The back of the skull protrudes in an "occipital bun." The face, particularly the nasal portion, is pulled forward, and the nasal aperture is large. The anterior teeth are also large, and specimens often show considerable wear on the incisors. The lower portion of the skull slopes down and back, with no prominent chin. The skeletal anatomy reflects an emphasis on strength. The body is short and stocky, with bones designed to support massive muscles. The bones of the hands are nearly twice as broad as modern human counterparts.

It has been proposed that this morphology is the result of adaptation to a glacial environment. Beginning approximately 75,000 years ago, the climate in Europe changed considerably as ice sheets advanced southward. The short, stocky body of the Neanderthal would retain heat more efficiently. These muscular individuals would have been better able to withstand a seminomadic lifestyle with many hours of hunting and gathering. It has even been suggested that the large nose would more effectively warm and moisten the cold, dry air. On the other hand, western Asian populations, represented by fossils from Israel and Iraq, exhibit similar characteristics, despite their exposure to a milder climate.

Behaviorally, the Neanderthals made clear advances over their ancestors. Their tool culture, the Mousterian industry, is more varied and complex. It includes tools made by the Levallois technique, which requires preshaping a core to a desired configuration, allowing the removal of many long, thin, sharp flakes. These flakes were then fashioned into knives, scrapers, burins, and spear points that were hafted onto wooden handles. Such tools were undoubtedly used in cooperative hunting and butchery, as evidenced by fossilized bones of wild horses, oxen, bison, and deer present at Neanderthal sites. Hides from prey animals would have enhanced the Neanderthals' ability to survive, along with the controlled use of fire and the exploitation of cave shelters.

Neanderthal culture and sociality go beyond complex tools. Specimens found at Shanidar cave in Iraq indicate that individuals who were elderly or infirm were supported by the group, rather than left behind. The "Old Man of Shanidar," with his crippled leg, missing arm, and useless eye, may have contributed to the group's survival as a wise man, or even a shaman. Other sites have yielded skeletons apparently buried in shallow graves, often bent into a fetal position, and occasionally in association with tools or flower pollens. Although these data continue to raise debate, many have concluded that the Neanderthals not only buried their dead but also provided them with grave goods, perhaps for use in an afterlife; some graves were apparently adorned with flowers. It is therefore possible that the Neanderthals, rather than being the clumsy cavemen of popular imagination, were highly evolved hominids with complex social, cultural, and ritual lives.—L.E.M.

See also EVOLUTION, HUMAN; *Homo erectus*

Neolithic Agricultural Revolution

According to physical anthropologists, human beings—that is, creatures of the genus *Homo*—have been around for 2 to 2.5 million years. For the great majority of this stretch of time, people supported themselves through gathering and hunting. The expression is often rendered as "hunting and gathering," but gathering deserves first billing, for in most places the bulk of the food was obtained through gathering rather than hunting. It was not always an easy life, but contrary to common belief, most people did not live on the edge of starvation. Food supplies and the number of people to be fed were kept in balance because human beings took positive steps to limit their numbers. Through unfortunate practices like infanticide and more benign ones like prolonged lactation, population levels were kept stable, thereby preventing the overshoot-and-crash scenario described by Thomas Malthus (1766–1834) in his *Essay on the Principle of Population.*

Beginning about 10,000 years ago, human beings began to gain their sustenance through agriculture. Instead of dining on what nature had already provided, people began to cultivate plants for future consumption. This change is sometime called the Neolithic agricultural revolution. The term Neolithic refers to the new stone age, a period in which human beings created more efficient stone tools through grinding instead of flaking, as well as clay pots and baskets. It has been called a revolution because farming changed far more than the means of subsistence; some of the basic patterns of human existence were altered. According to some historical schemas, there have been only two true revolutions while people have been on Earth: the Neolithic agricultural revolution and the Industrial Revolution that began in the 18th century.

Neolithic agricultural revolutions occurred independently in a number of distinct regions of the world: the Middle East (10,000 years ago), what is now China (7,000 years ago), and Mesoamerica (5,000 years ago). The reasons for the shift to agriculture are not at all clear, and much debate still surrounds the issue. It was formerly thought that human beings turned to agriculture when the Earth's climate became drier, reducing the availability of game and wild plants. This is certainly a plausible hypothesis, but it does not mesh with what we now know about past climatological changes. There is no evidence that changes in regional climates immediately predated the invention of agriculture. Another climatological hypothesis states that a recession of the globe's ice caps at the end of the Pleistocene era led to the extinction of large game animals and created conditions conducive to farming. There is no question that the ice caps did retreat and that the climate became warmer and drier, but the same phenomenon had occurred in the past without triggering the emergence of agriculture. Another theory based on biological changes speculates that gathering practices forced a shift to farming. According to this theory, the harvesting of wild wheat left behind grains that clung tightly to the plant. Over time, these grains predominated, and wheat lost its ability to propagate itself. Consequently, people had to plant the wheat if they were to get a harvest. This is an ingenious theory, but it does not explain the planting of nongrain crops like peas and lentils, which commenced at about the same time as the cultivation of wheat.

One of the great advantages of agriculture is that it allows the support of many more people in a given area of land. Whereas the typical gatherer-hunter economy needs 10 square kilometers for every person, nonirrigated farming requires only 0.5 square kilometers per person, and irrigated farming needs only 0.1 square kilometers. The ability of agriculture to feed more people than gathering and hunting has naturally led to the speculation that the Neolithic agricultural revolution was motivated by the growth of human populations. But this theory is not completely satisfying, for it is not certain that the human population had grown prior to the invention of agriculture, or if it did, why this happened after many generations of stability. It may be that in some parts of the world, at least, the cumulative effects of human population may have stressed the environment to a point where the returns to gathering and hunting were substantially diminished. Other population-based explanations posit that the foundation of settled communities impelled the development of agriculture. From this standpoint, agriculture was a consequence of social development rather than a cause. This theory has gained a number of adherents recently, although it begs the question of why stable communities came into being when they did.

Efforts to explain why an agricultural revolution occurred will always be unconvincing if it is assumed that taking up farming was accompanied by a complete displacement of gathering and hunting. In fact, there is evidence that agriculture coexisted with gathering and hunting for many centuries. The Neolithic agricultural revolution is therefore not a true revolution if the term *revolution* is meant to imply a clean break with the past. It would also be a mistake to view the invention and development of agriculture as an unalloyed blessing for humanity. In terms of the amount of energy derived from food crops relative to the energy put into their cultivation, agriculture is usually inferior to gathering and hunting, especially the latter. This is because most agriculturists devote considerably more time to farming than foragers spend on gathering and hunting. The main advantage of agriculture, as noted above, is that it allows more people to subsist on a given amount of land. These people may not be better fed and may have to work harder than gatherer-hunters, but there are more of them. Whether or not this can be viewed as progress is a matter that can be endlessly debated.

Although agriculture did not represent a major step forward in per-person productivity, it did allow a sedentary way of life. Some early agriculturists combined crop cultivation with seminomadism through the practice of what has been labeled slash-and-burn (or swidden) agriculture. This practice, which still endures in some parts of the world, entails cutting down trees and other vegetation, burning them, and then planting a crop. After the harvest, the farmer abandons the field, allowing it to revert to its wild state. After the passage of a number of years, the land is ready for another round of slashing and burning. Although this form of agriculture can support population densities 10 times higher than those of foraging economies, it still requires a lot of land, much more than is needed for sedentary agriculture.

In addition to its ability to support large populations, sedentary agriculture made possible the establishment of more complex societies. Increases in population size and density stimulated the development of specialized activities. Unlike the small, undifferentiated, transient communities of gatherers and hunters, agricultural societies allowed the development of specialized craftsmen, administrators, artists, and religious practitioners. Organizing militarily for protection or conquest also was more easily accomplished when people lived in dense communities.

Many of the farming techniques that originated with the Neolithic agricultural revolution persisted for centuries. Successful cultivation required the regular application of fertilizers, the movement of water in the case of arid croplands, the use of simple tools, and, above all, a great deal of human effort. Over the centuries, farming became more productive as improvements were made in the breeding of plants and animals, the use of new organic fertilizers, and the development of improved farm tools. But it was not until the 19th century that agriculture began to depart in a fundamental way from its Neolithic origins.

See also AGRICULTURAL REVOLUTIONS IN INDUSTRIAL AMERICA; AGRICULTURE, IRRIGATED; EVOLUTION, HUMAN; FERTILIZERS, CHEMICAL; FERTILIZERS, ORGANIC; INDUSTRIAL REVOLUTION; IRRIGATION, CENTER-PIVOT; LEGUMES; POTTERY; SELECTIVE BREEDING OF ANIMALS

Further Reading: David Rindos, *The Origin of Agriculture: An Ecological Perspective*, 1984.

Neon

Neon is an inert gas (i.e., since it has eight electrons in its valence shell, it does not combine with other atoms to form chemical compounds). Neon was discovered in England by William Ramsay and Morris W. Travers in 1898 while they were isolating rare gases through the fractional distillation of liquefied air. The discovery of neon gas came at a propitious time, as many experimenters were running electrical discharges through various gases in order to produce light. Meanwhile, Georges Claude (1870–1960), a French chemist, was working with liquid air in the hope of coming up with a cheaper means of extracting oxygen. His experiments produced sizable quantities of rare gases, neon among them. In 1909, he discovered that neon produced a red glow when stimulated by an electric charge. Another inert gas, argon, was found to produce a blue glow. Claude also discovered that more colors could be obtained by coating the inside of the tube with different substances.

In 1910, Claude demonstrated the possibilities of neon by using it to illuminate a sign in Paris. Now working intensively on neon lighting, in 1915 Claude received a patent for corrosion-resistant electrodes used as electrical terminals for neon lights. Claude's inventive abilities were coupled with the business acumen of one of his associates, Jacques Fonseque, who made signage the focus of the infant neon industry. Claude's patent gave him a firm grip on the industry, and during the 1920s his firm franchised neon to signmakers throughout the world. The first application of a neon sign in the United States consisted of two signs installed at a Packard automobile agency in Los Angeles in 1923; its brilliance caused traffic to stop and crowds to gather. The 1930s may be considered the golden age of neon. It was much in evidence at the Chicago Century of Progress Exposition (1933–34), the Paris Exposition (1937), and the New York World's Fair (1939), while commercial enterprises, movie theaters especially, used neon to generate visual interest and excitement.

The effective use of neon requires craft skills of a high order. The fabrication of a neon sign or other object begins with a design that is rendered on an asbestos pad. Glass tubing is heated to the correct temperature and then bent to the intended configuration, occasionally being rested on the asbestos pad in order to check congruence with the design. After the tubing has been bent to the proper shape, electrodes are inserted, and the tube is partially evacuated with a vacuum pump. An electrical current is then used for a "bombarding" process that removes impurities from the inside of the tube. The tube is then evacuated further, and neon is admitted to the tube. The tube is then sealed off and "aged" by passing an electric current through it in order to stabilize the gas with the vacuum.

In the 1930s, these craft skills were partially bypassed as large numbers of standardized signs were manufactured to advertise beer and other products. After World War II, electric signage deteriorated further, as fluorescent tubes in plastic shadow boxes offered a mode of advertising cheaper than neon. These shabby signs evoked a growing revulsion to electrical signs in general, and neon was unfairly blamed for contributing to visual blight.

While neon has not regained its place as the premier material for illuminated signs, in recent years it has gained increasing recognition as a medium for artistic expression. Neon is especially suited to public art, as exemplified by the dazzling display at the United Airlines terminal at O'Hare Airport in Chicago or as an architectural enhancement. Unfortunately, while neon has many aesthetic possibilities, its use is currently limited by its undeserved popular image as a tawdry advertising medium.

See also CHEMICAL BONDING; LIGHTS, FLUORESCENT; MASS PRODUCTION

Further Reading: Rudi Stern, *Let There Be Neon*, 1979.

Net Fishing, Commercial

Fishnets are the basis of commercial fishing. Although their basic design has changed little since prehistoric times, in the last 50 years the use of synthetic materials in conjunction with advanced maritime technology has increased the yield of fish to such an extent that the depletion of fish stocks has become a major problem in some parts of the world. The United Nations Food and Agriculture Organization, in cooperation with the major maritime nations, now monitors commercial fishing in waters beyond national jurisdictional boundaries where it has successfully secured moratoriums on certain types of net fishing. While experimentation continues with electric current as a means of catching fish in bulk, nets continue to be the primary means of harvesting fish.

Throughout the world, many varieties of nets are used to catch fish. They may be divided into three basic types: gill, pound, and pass. "Gill nets," so-called because they catch fish by their gills, are used to catch scattered congregations of fish over large expanses of water. They are suspended vertically beneath the surface by cables and weights, at varying depths down to about 75 m (246 ft). These nets can be used in any body of water regardless of bottom conditions and force of current. At sea, it is common to join individual nets in an array several kilometers long and allow them to drift as they are carried along by a vessel. In coastal waters, it is common practice to anchor and buoy them. In shallow waters, nets are attached to poles set in the bottom. In rivers and estuaries, drift gill nets are allowed to float downstream behind boats to which they are fastened, a practice similar to that employed at sea.

"Pound" or "trap nets" catch fish by entanglement, and are used to ensnare fish during their annual migrations along coasts and up rivers. These nets have one or two chambers of varying mesh size to confine the catch, while "wing nets," which extend from surface to seabed, guide the fish to an entry device. The device allows the fish to enter the pound or trap but bars their escape. In deep water, the nets are supported by a system of buoys and anchors. In shallow water, the nets are often attached to pilings. The catch is removed by pump, through the use of much smaller nets, or by hand.

"Pass" or "surrounding nets," which capture fish by encircling them, include seines and trawls. Seines are suspended vertically. Drag seines occupy the full depth of the water. Circular or purse seines, which are closed from the bottom by a cable, occupy a layer of water close to the surface. Danish seines are situated close to the bottom and are used in conjunction with traction guide ropes that drive fish into the path the net is moving. A trawl is a wide-mouthed, funnel-shaped net that is towed horizontally near the bottom or at midwater by a specialized vessel known as a *trawler*. The net's mouth is held open by hydrodynamic kites known as *otter boards*. Fish are trapped when the net is hauled to the surface.

Archeological evidence indicates that nets were in use in 2500 B.C.E. in Southern Europe by Upper Paleolithic hunter-gatherers and even earlier in Southwestern Asia. Ancient nets were made from cord fashioned from a wide variety of plant fibers and animal tissues. Later, cotton, hemp, manila, flax, and sisal became standard net components. In the 20th century, those materials were largely supplanted by nylon, polyester, polypropylene, and polyethylene. These are all rotproof synthetics that given rise to concerns about "ghost fishing." This occurs when lost or abandoned nets, particularly gill drift nets, continue to trap fish and in some instances sea mammals. The extent of ghost fishing and its effect on sea life is a matter of considerable of controversy between the fishing industry and environmental groups. The issue was partially resolved in the early 1990s by a United Nations–sponsored moratorium on large-scale, high-seas drift net fishing.

Nets are composed of meshes (measures) or squares that are formed by knotting or knitting. Modern net fishing was given a strong impetus in the 19th century by the development of netmaking machines and the transition from sailing ships to steam-powered vessels. Following World War I, diesel-engine-powered ships made possible the use of larger, heavier nets. With World War II came synthetic, rotproof materials and higher-powered vessels. Unfortunately, in many parts of the world these technological successes have been followed by the depletion of the fish stock; this in turn has necessitated the use of even more advanced technologies to catch the fish that remain.—A.L.

See also COTTON; ENGINE, DIESEL; NYLON; POLYETHYLENE; POLYPROPYLENE; STEAMBOAT

Further Reading: Andres von Brandt, *Fish Catching Methods of the World*, 1964.

Neuron

Virtually all animal behaviors are organized by the coordinated activity of neurons, cells specialized to perform communication functions in nervous systems. Understanding the physiology of behavior thus requires an understanding of the biology of neurons and their interactions in neural circuits. These investigations were given an important theoretical foundation when, near the close of the 19th century, researchers concluded that individual neurons remain separate in neural circuits rather than fuse in syncytia (like muscle cells, for example), as previously believed. This discovery supported the idea that specific regions of the human brain perform specific information-processing functions. The neuron doctrine and the concept of localized functions encouraged people to think that the cellular basis of brain function could be understood by studying neurons and their interaction in defined circuits.

The importance of neurons and nervous systems in directing thought and behavior was not always obvious. Aristotle (384–322 B.C.E.) argued that the human brain served to cool the blood in order to regulate the temperature of the heart, where intellectual functions were performed. The Roman physician, Galen (130–200 C.E.) strongly rejected Aristotle's "cardiocentric" notion and asserted the primacy of the brain in intellectual, sensory, and motor function, but he thought that the brain communicated with the body by means of "spirits" conducted along hollow nerves. The person most credited with establishing the electrical nature of nervous communication remains Luigi Galvani (1737–1798), who, in 1786, noticed that frog leg muscles twitched either when he applied electrical stimuli directly to the muscle tissue or to the nerve attached to that muscle. Approximately 60 years later, the German physiologist Emil DuBois-Reymond (1818–1896) showed that nerve cells and muscle fibers produce their own regenerative electrical signals.

For most of the 19th century, the most widely accepted idea of how electricity was conducted through the nervous system reflected the view that the cellular elements must be fused together in a syncytium. One of the major proponents for this "anastomosis theory" was Camillo Golgi (1843–1926), who developed the use of silver nitrate as a stain capable of revealing the nerve cell body and its tangle of processes. Nerve cells are highly variable in structure but generally consist of a cell body from which extends a single long process called an *axon* and an arborization of processes individually called *dendrites*. Ironically, it was Golgi's technique that was later used to prove the anastomosis theory wrong. The Spanish neuroanatomist and histologist Santiago Ramón y Cajal (1852–1934) applied Golgi's silver staining technique without finding evidence of cell fusion in nervous systems and concluded that nerve cells remain independent. He thus became the major proponent for the neuron doctrine.

Once the electrical nature of neuronal communication and the neuron doctrine were established, the central problems became those of explaining how neuronal electrical signals are generated and conducted from one neuron to the next. Development of the oscilloscope and the intracellular micropipette electrode around the time of World War II allowed researchers to record electrical activity across the membrane of individual neurons. The potential difference across the membrane of living, quiescent neurons typically falls between –40 and –120 mV, with the inside of the cell more negative than the outside. This is called the *resting potential*, and it results from the differential distribution of diffusible ions (typically sodium and potassium) across the neuronal membrane. This delicate equilibrium sets the stage for development of the basic electrical signal of neural communication, the *action potential*.

The mechanism of the action potential was discovered in the early 1950s by Alan Hodgkin (1914–) and Andrew Huxley (1917–), who made intracellular recordings of action potentials in the unusually large axons of neurons in the squid. They observed that sufficiently strong electrical stimuli elicited rapid changes in voltage across the neuronal membrane, bringing the cell to a positive voltage inside before the normal negative resting potential was restored. They also showed that these voltage changes are produced by the coordinated flux of sodium and potassium ions across the membrane. Once initiated, the action potential travels the length of the axon without decreasing in amplitude.

Given the neuron doctrine, signals must move from one neuron to another across small gaps termed *synapses* by Charles Sherrington (1857–1952). Signals are transmitted across some synapses by direct electrical conduction, but most synapses use a chemical mechanism. In the 1960s, Bernard Katz and his colleagues found that when an action potential invades the end of the axon of a chemical synapse, it sharply increases the probability of release of neurotransmitter molecules from small membrane-bound packets stored in the axon terminal. These molecules diffuse across the synaptic gap and bind to specific receptor molecules on the target cell's surface. This binding triggers events in the postsynaptic cell, leading to a change in that cell's membrane potential.

Through synaptic connections, neurons form circuits. The complexity of neural circuits varies enormously and plays a major role in determining the complexity of information processing by the nervous system. The simplest circuit, such as that associated with the knee-jerk reflex, consists of a sensory neuron that synapses with a motor neuron that in turn connects to a muscle. Most neural circuits, however, have interneurons interposed between the sensory and motor cells. The human brain contains on the order of 100 billion neurons, with each neuron participating in 10,000 synapses on average, indicating the possibility of unimaginably complex circuits. The pattern of connectivity in a neural circuit may be genetically programmed, but it often can be varied under the influence of experience (learning) or the modulatory effects of various chemicals such as hormones. This flexibility of neural circuits creates possibilities for adapting behavior to changing circumstances.

A variety of approaches have been taken in trying to unravel the neural organization of animal behavior. Recordings from individual neurons in simple circuits have shown that the key features of behaviors can be explained entirely on the basis of the interaction of neurons acting in definable circuits. However, single-cell recordings have been less useful for explaining the neural organization of complex behaviors. In these cases, extracellular electrical recordings from large numbers of neurons at a time, as in electroencephalograms, electrical stimulation of the nervous system, ablation experiments in which parts of the nervous system are removed, and recent advances in neural imaging have proved helpful. Positron Emission Tomography (PET) and adaptations of Magnetic Resonance Imaging (MRI), for example, reveal metabolic activity in specific regions of the brain that correlates with electrical activity. These new imaging methods produce spectacular evidence in human beings that particular mental or behavioral tasks are accompanied by coordinated activity in clusters of definable brain regions. These studies confirm the belief of a century ago that localized regions of the brain, hence localized neural circuits, perform specific functions. Complex behaviors are built from the coordinated activity of many such specialized regions working in series and in parallel.

Current techniques allow investigation either at the level of the single cell or at the level of very large numbers of neurons. Few techniques allow exploration of the broad midrange. Consequently, modeling studies are becoming increasingly important. Information gathered from the single-cell and many-cell levels can be used by investigators to inform the creation of computer models or artificial neural networks from which hypotheses regarding mechanisms of brain function can be generated. The testing of these hypotheses and the development of new methods for analyzing complex information processing by the brain will extend our understanding of how neurons work together to generate thought and behavior.—N.C.

See also CELL; ELECTROENCEPHALOGRAM; HORMONES; MAGNETIC RESONANCE IMAGING; OSCILLOSCOPE

Further Reading: Eric R. Kandel and James H. Schwartz, *Principles of Neural Science*, 1985.

Neutron

In 1911, Ernest Rutherford (1871–1937) demonstrated that atoms were not irreducible units of matter but were comprised of even more elementary particles, protons and electrons. The concept of the nuclear atom came in response to a number of experimental findings, but a number of puzzles remained unsolved. One of the biggest was the discrepancy between the atomic weights and atomic numbers of most elements. Atomic number is defined as the number of protons in an atom's nucleus; atomic weight (or more properly, atomic mass) is the sum of the masses of all particles comprising the atom plus the binding energy that holds together the nucleus. Atomic weight is based on a scale that assigns an atomic weight of 12.000 to the most common form of carbon and expresses the atomic weights of all other elements as ratios of this weight (in the past, oxygen, with an assigned atomic weight of 16, was commonly used as the standard). In this schema, an individual proton has an atomic weight of about 1, while each electron has a weight (mass) that is so small as to be negligible for most purposes.

On the basis of these conventions, an element's atomic number should be almost the same as its atomic weight (the latter being a bit larger due to the presence of electrons). This, however, is almost never the case, the sole exception being hydrogen. In order to resolve this anomaly, some scientists postulated that the nucleus contained electrons and additional protons, the charge of which was neutralized by the electrons. But as more was learned about protons and electrons, it became apparent that electrons were too large to fit in the atom's nucleus. Moreover, as quantum theory demonstrated that atomic particles have wavelike properties, it was apparent that the required energy level was impossibly high for electrons, given the short wavelengths that quantum theory stipulated.

Another possibility existed, that the nucleus of an atom contained a particle with the mass of a proton but carrying no charge. This idea had many adherents, but it lacked experimental validation. A particle with a neutral charge was difficult to find because it would not ionize atoms, and ionization was the chief means of detecting a subatomic particle. Proof of the neutron's existence came as a consequence of research in a different area of nuclear physics, the transmutation of one element into another by bombarding the nuclei of light elements such as beryllium with alpha particles. Transmutation experiments conducted in Germany by Walther Bothe (1891–1957) and in France by Irene Joiliet-Curie (1897–1956) and Frederic Joliet-Curie (1900–1958) produced powerful radiation that was capable of penetrating several inches of lead. The intensity of the radiation was indicated by the substantial ionization it produced. Moreover, when the radiation passed through thin sheets of material containing hydrogen (paraffin, for example), it ionized gas at an increased rate. Clearly, something in the rays was causing protons to be stripped from hydrogen atoms and propelled through space.

Further experimentation done by James Chadwick (1891–1974) produced similar results. Chadwick, however, produced the correct interpretation: The alpha particles were dislodging neutral particles from the nucleus of the beryllium, and these in turn were dislodging protons from the paraffin molecules. Chadwick also reasoned that the protons were being set into motion by a particle with a similar mass, for energy transfer in collisions is greatest when the colliding objects have equal masses. Equally important, the penetrating power of the particles could be explained by their lack of charge because their motion was not affected by the electrical fields of the atoms. By measuring the velocities of protons as they were ejected from several materials and then applying well-established collision theory, Chadwick was able to determine the neutron's mass, finding it to be quite close to the mass of a proton.

The discovery of the neutron explained the discrepancies between atomic number and atomic weight. The fact that all elements other than hydrogen had atomic weights greater than their atomic numbers was the result of the presence of neutrons in their nuclei. For example, nitrogen's seven protons give it an atomic number of 7, while its protons and neutrons are together responsible for virtually all of the element's atomic weight of 14.0067. The existence of the neutron also cleared up another long-standing problem: the existence of isotopes, alternative forms of elements that had identical chemical properties but different atomic weights. Isotopes could now be explained by the presence of greater or lesser numbers of neutrons in their nuclei.

For decades following its discovery, it was generally believed that the neutron was a unique, fundamental particle. But in the early 1960s, Robert Hofstadter (1915-1990), an American physicist, found that both protons and neutrons had the same core of positively charged matter that was surrounded by two shells of mesons. In the case of the neutron, one of the mesons was negatively charged, offsetting the positively charged matter and resulting in a total charge of zero.

In addition to providing a better understanding of the structure of the atom, the discovery of the neutron created many new research opportunities. Because they lack a charge, neutrons are far more effective than charged particles in initiating nuclear reactions.

See also ATOMIC NUMBER; ATOMIC WEIGHT; ELECTRON; ISOTOPE; NUCLEAR FISSION; PARTICLE ACCELERATORS; PROTON; QUANTUM THEORY AND QUANTUM MECHANICS

Newtonian Mechanics—see Mechanics, Newtonian

Nipkow Disk

The basic principle underlying television is the conversion of light into an electrical signal that is then converted back to light by the TV receiver. The electrical signals do not create the TV picture instantaneously; rather, the screen is scanned by an electron beam that sequentially lights up approximately 250,000 picture elements on the screen, creating a new image 30 times a second. The basic idea behind this process goes back to 1884, well before electronic technology had reached a point where it could be used for the transmission and reproduction of visual images. It was at this time that the German physicist Paul Friedrich Nipkow (1860–1940) attempted to devise a means of scanning a scene so that it could be visually represented as a series of parallel lines of different intensities.

The basis of Nipkow's technology was a thin disk that rapidly rotated on a shaft. Along the disk were a series of holes arranged in a spiral pattern. The rotating disk was placed between a lens that concentrated light from the subject and a photoelectric cell. The latter was made from selenium, which in 1873 had been discovered to produce a photoelectric effect; that is, its electrical conductivity increased when light struck it. As the disk rotated, light from a particular part of the subject passed through a hole and hit the photoelectric cell. In one rotation of the disk, the scene was completely scanned in a sequential order.

It was Nipkow's hope that the electrical currents sequentially triggered by light passing through the disk's holes could be used to produce an image corresponding to the subject. Nipkow never quite succeeded in achieving this, largely due to selenium's slow electrical response to changes in light intensity and the impossibility of amplifying the signal, given the technology of that time. But in the 1920s, the basic idea was taken up in the United States by the Bell Telephone Laboratories and in Britain by John Logie Baird (1888–1946). Baird's system was used in 1930 when the British Broadcasting Corporation began to transmit the first regular television programming. This pioneering effort proved to be a false start however. Although 20,000 television sets were in use in the United Kingdom by the end of the decade, mechanically based television was fatally defective. Picture resolution was poor due to the small number of scanned lines on the screen, while the production of TV programs was made difficult by the need for very high levels of illumination. Further development of television required the use of all-electronic technologies.

See also TELEVISION

Further Reading: John R. Pierce, *Electrons, Waves, and Messages*, 1956.

Nitrogen Cycle

In 1815, the armies of Napoleon had been vanquished, and the new government of defeated France faced a serious food shortage. Hoping to improve the situation, the French government asked the eminent physiologist François Magendie (1783–1855) to chair a commission that would determine if the gelatinous extract of meat could serve as a nourishing food. The commission eventually disbanded, but Magendie continued with his research into nutrition. By conducting experiments with dogs, in 1816 he found that foods containing nitrogen were an essential part of the dogs' diet. In the 1830s,

Gerardus Mulder, a Dutch chemist, identified a class of nitrogen-containing organic substances that he called *proteins*. In 1842, the German chemist Justus von Liebig (1803–1873) found that the consumption of foods containing proteins was essential to the growth of bodily tissues.

It therefore came to be understood that nitrogen was an integral part of proper nutrition, but it was not known how nitrogen made its way into the plants and animals that were consumed for food. By the end of the 18th century, it was understood that nitrogen was a major component of the air, and Horace Bénédict de Saussure (1740–1799) had discovered that plants could not take in atmospheric nitrogen; rather, they obtained nitrogen from the soil through some unexplained process.

In the mid-19th century, the French chemist Jean-Baptiste Boussingault (1802–1887) demonstrated that plants did not need nitrogen derived from organic sources; nitrates or ammonium salts served just as well. Boussingault also made the crucial discovery that legumes had the ability to add nitrogen compounds to nitrogen-poor soils. He further found that animals also cannot absorb nitrogen from the air; they had to derive the nitrogen in their tissues from the food they ate.

The mechanisms through which nitrogen was absorbed by legumes began to be ascertained in 1862, when Louis Pasteur (1822–1895) put forth the suggestion that microorganisms might play a crucial role in the process. In the 1880s, the French chemist Pierre Berthelot (1827–1907) provided empirical evidence for this hypothesis when he grew nitrogen-fixing plants in soil that had been sterilized; under these circumstances, no nitrogen was added to the soil. It is now known that nitrogen-fixing bacteria live in nodules on the roots of legumous plants. There, they convert atmospheric nitrogen into ammonia (NH_3). Nitrifying bacteria then convert the ammonia into nitrates, which are then taken up by growing plants.

Some nitrogen is also found in the soil and is returned to the soil through a self-perpetuating process, hence the term *nitrogen cycle*. Nitrates already existing in the soil are taken up by plants and incorporated into them. Some of these plants are eaten by animals, which also assimilate nitrogen. When these plants and animals die, nitrifying bacteria convert nitrogen-containing compounds into nitrites, which in turn are converted into nitrates by another kind of nitrifying bacteria. These compounds are returned to the soil, where they can again be assimilated by growing plants.

Farm lore holds that crops grow especially well after a powerful thunderstorm. This belief actually has a scientific foundation; the electrical discharges of a thunderstorm convert nitrogen in the air into oxides of nitrogen. These oxides move into the soil, where they are converted into nitrates that aid in the growth of plants.

See also AMMONIA; COMPOSTING; FERTILIZER, CHEMICAL; LEGUMES

Nobel Prize

The Nobel Prize was established by Alfred Nobel (1833–1896), the inventor of dynamite. When he died in 1896, he left 33 million Swedish kroner (at the time the equivalent of $9 million) to support the awarding of prizes in a number of scientific and humanistic fields. Along with conferring prestige, it was expected that the cash awards would provide financial support for further work.

Five prizes are awarded annually: chemistry, physics, medicine or physiology (which has come to include biochemistry, bacteriology, virology, genetics, molecular biology, and biophysics), literature, and economics (added in 1969). A prize for advancing the cause of world peace is granted on an irregular basis. The prizes in the sciences are distributed by the Royal Swedish Academy of Sciences and the Royal Caroline Medico-Surgical Institute. The first prizes were awarded on Dec. 10, 1901, the fifth anniversary of Nobel's death. December 10 continues to be the date on which the prizes are awarded in a lavish ceremony hosted by the king of Sweden.

The Nobel Prize does not cover all the sciences; fields such as geology and oceanography are not eligible. Even the subdiscipline of astrophysics was informally excluded from consideration for the physics prize for many decades. There is also a strong tendency for the prize to be awarded for empirically grounded discoveries rather than for advances in pure theory. Albert Einstein (1879–1955), for example, was awarded the 1921 prize in physics primarily for his investigation of the photoelectric effect rather than for his theories of general and special relativity.

The Nobel Prize is universally recognized as the highest honor that can be received in the sciences and the other fields for which it is awarded. Not only does the prize distinguish its recipients, it also adds luster to the institutions with which they are or ever have been affiliated. The tally of Nobel Prizes received is also taken as a gauge of an entire nation's scientific prowess, even though the assigning of nationality to individual laureates can be rather arbitrary. Some laureates, for example, were born in one country, educated in another, and did their prize-winning research in yet another.

The determination of prize winners in each field is the responsibility of a five-member committee that invites nominations, investigates candidates, and makes the final selection. Although there is no set formula for determining winners, their work has been original and of substantial importance in almost all instances. Publications of prize winners have been cited frequently by other scientists, and in many cases the laureates made contributions in a number of areas. Since the selection process is secret, it is impossible to say if other criteria, such as age or nationality, have been factors, but there is some likelihood that this has occurred from time to time.

The prize is never awarded posthumously, which means that some deserving work gets passed over. In any given year, the prize can be shared among no more than three scientists. As a result, some important scientific discoveries have been excluded for consideration because too many people were involved in them. The restriction of the prize to three recipients also means that some worthy collaborators do not get the recognition they deserve. Despite these limitations, the majority of the prizes have been awarded for collaborative rather than individual work. Still, with science becoming an increasingly collaborative activity, the difficulties of recognizing collaborative efforts present serious dilemmas for the committees charged with the awarding of Nobel Prizes.

Nobel laureates do not come from a representative cross section of the population. Rather, they have been disproportionately drawn from middle- and upper-middle-class families. The majority have come from mainline Protestant backgrounds, although Jews are overrepresented relative to their numbers in the general population. Conversely, the percentage of Catholic laureates is smaller than their percentage in the population as a whole. In the United States, the vast majority of laureates received their graduate training at a few elite universities, and they began to publish their research early in their career. Many of them were mentored by scientists who were themselves Nobel laureates, or would receive the prize later in the their careers.

Although the Nobel Prize has enhanced the prestige of individual scientists and science as a whole, it has had some negative consequences as well. Many laureates have complained that winning the prize made them into celebrities, which in turn diminished the time they had for their current research. Many laureates also have been troubled that the prize was awarded for research that they did not consider their best, or that too many years separated the prize-winning research and the award of the prize. Laureates may also be concerned that one or more of their collaborators did not also receive the prize. It is sometimes argued that competitive prizes like the Nobel Prize attenuate the cooperation necessary for the advancement of science, but there is no hard evidence in support of this. All in all, the Nobel Prize has been beneficial to science, even though its overall impact probably has been modest.

See also EXPLOSIVES

Further Reading: Harriet Zuckerman, *The Scientific Elite: Nobel Laureates in the United States*, 1977.

"Normal Accidents"

Although technological advance has brought many benefits, it has also expanded the danger of catastrophic events such as oil spills, the venting of toxic chemicals, and the release of radioactive materials. There are two ways of looking at these occurrences. Seen from one perspective, they are aberrations; although the consequences are tragic, they can be viewed as unique events not likely to be repeated. Indeed, the lessons learned from each accident will help to prevent more accidents of this sort from happening. Other analysts are far more pessimistic. From their perspective, accidents are inherent to the functioning of large-scale organizations and the technologically sophisticated systems they employ. Far from being aberrations, they are a normal part of modern life.

The concept of "normal accidents" was developed by sociologist Charles Perrow in a book of the same name. According to Perrow, many organizations have some features that make accidents virtually inevitable. The first of these is "interactive complexity." Organizations are made up of separate components—equipment, procedures, and individuals—that interact with one another in a predictable, familiar manner. But along with predictable interactions are more complex situations: unplanned, unforeseen events where something that happens in one part of the organization produces unintended consequences in another part. To take a hypothetical example, a building's air conditioning breaks down, which causes an employee to open a window in order to get some fresh air. At this point a bird flies in, and while the employees are chasing the bird, one of them bumps against a button that sets off an alarm. This in turn causes a worker in another building to activate a sprinkler system that ruins dozens of important blueprints that had been spread out in the boss's office. This may seem like a far-fetched sequence of events, but real disasters such as the nuclear accident at Three Mile Island were set off by series of events that seem equally improbable.

"Interactive complexity" is not inherently hazardous; in most cases an unexpected event will have no ramifications beyond its immediate setting. But when an organization has the property of "tight coupling," problems in one place can quickly spread to other places. Tight coupling means that the various parts of the organization are closely connected. This happens when interrelations between the different parts have no alternative modes; there is only one way to accomplish a given task. Tight coupling also occurs when processes are constrained by rigid time limits, so that there is no time to find alternative paths. In general, there is very little slack in the system. Due to tight coupling, a problem occurring in one part of the organization may not be easily confined; it may cause a multiplicity of new problems to cascade through the whole organization.

Efforts to improve the safety of a system by adding redundant components (like an emergency cooling line at a chemical plant) may not improve the situation, for the extra components may themselves increase interactive complexity. Also, a profusion of redundant parts and subsystems may make the system more difficult to understand when something does go wrong. And finally, as can be the case with safety measures in general, the addition of more safety features may cause operators to take greater risks than they did before the features were installed.

Other organizational characteristics intensify the tendency for things to go awry from time to time. Organizations usually have multiple and even conflicting goals. Safety may be articulated as a major goal, but other goals, such as meeting a production quota or making a profit, may override concerns about safety. Second, the hierarchical structure typical of most organizations may impede appropriate responses to problems when they occur. Decentralized authority is necessary if problems are to be effectively addressed when they first emerge, but tightly coupled organizations with a high degree of complex interactivity require centralized decision making and its complement, the tendency of lower-level functionaries to remain immobilized until they receive orders from above. Finally, learning from past mistakes can be inhibited by organizational realities. Learning is possible only when there is a clear understanding of what has happened. But an ingrained organizational culture may produce biased interpretations of the event, especially when the situation is shrouded with ambiguity. This tendency is even more pronounced when individual and departmental interests are at stake, for a successful learning experience may have to include the assignment of blame, and this is likely to be resisted by the person or group on the receiving end.

Students of organizations do not unanimously agree with Perrow's formulations. Some studies seem to show that a high degree of reliability can be built into inherently risky procedures, such as landing airplanes on the flight decks of aircraft carriers. Even so, the theory of normal accidents demonstrates the shortsightedness of simply assuming that "it can't happen here."

See also BHOPOL; CHERNOBYL; OIL SPILLS

Further Reading: Charles Perrow, *Normal Accidents : Living with High-Risk Technologies*, 1984.

Normal Curve

Whenever we look at people, places, or things, we see enormous variability and variety. On many occasions we will find it necessary to describe and summarize the ways in which our observations are similar and dissimilar. One important conceptual tool for accomplishing these is the normal curve.

The illustration below shows a hypothetical distribution. In generating this distribution, we asked a computer statistics program to give us a normal distribution with a mean of 22 and a standard deviation of 1.09. As was noted elsewhere, the mean provides us with a value that is the most typical score, and the standard deviation gives us a value that is the most typical or average deviation from the mean. Thus, for this hypothetical distribution, the most typical value is 22, and the most typical deviation is 1.09.

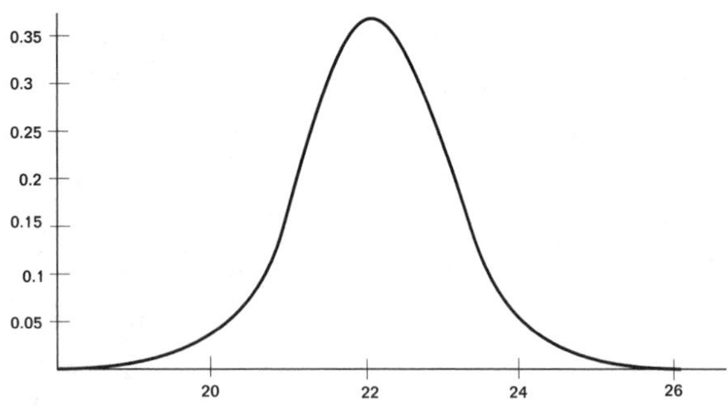

In the figure, the horizontal axis presents the values of the hypothetical variable; the vertical axis presents of the number of observations at each point on the horizontal. Note that this distribution has one peak and that the area to the left of the mean equals the area to the right of the mean. In fact, if you were to fold this distribution at the mean, the two areas would neatly overlap each other. Note also that the mean value, 22, is also the modal value. That is, the most popular value (i.e., most frequently occurring) is 22. Finally, note also that the median also equals 22. That is, the value 22 splits the distribution in half, and thus 50 percent of the observations lie below 22, and 50 percent lie above.

This is an important property of any normal distribution: The mode is equal to the median. Imagine further for a moment that we removed a value close to the mean, say one value of 23, and at the same time add a value far above the mean, say 28. The median and the mode would stay the same, but the mean would be moved above the value of 22. Put another way, the mean is sensitive to extreme values and is pulled in the direction of such values. This is because the mean is the sum of all observations divided by the number of observations.

We now want to transform this distribution into what is called a "standard normal distribution." We do this by taking each raw score, subtracting the mean from it, and dividing the result by the standard deviation or

$$\text{Standard score } Z = \frac{x - \bar{x}}{sd}$$

where Z is the standard score or Z score, x is the individual raw score, x is the mean, and sd is the standard deviation of the distribution. Such a standardized normal distribution always has a mean of 0 and a standard deviation of 1, regardless of the units of the original distribution. Such a distribution is presented in the figure below.

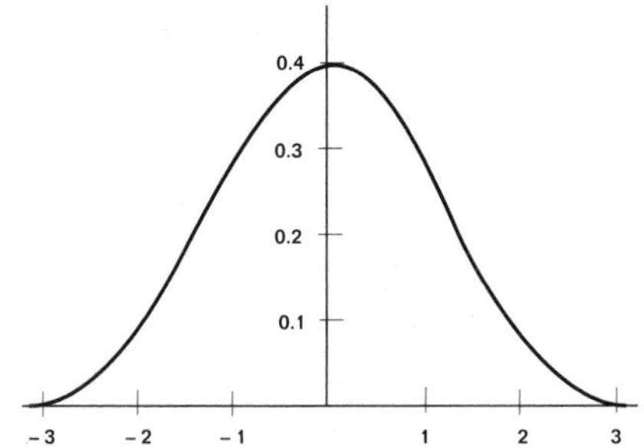

Some properties of the standard normal distribution make it an important tool in statistics. First, as you can see, the distribution is symmetrical around the mean. Second, the area under the curve equals unity or 1.0.

We can use these properties of the standard normal curve to make some interesting observations. First, it can be shown that the area between the mean and one standard deviation above the mean equals 34.10 percent of the total area under the curve. Given the symmetry of the curve, this allows us to conclude that the area within plus or minus one standard deviation of the curve equals 68.20 percent of the total area under the curve. Alternatively, we can conclude that 31.80 percent of the area under the standard normal curve lies beyond plus or minus one standard deviation of the mean. If we move out to two standard deviations from the mean, the area between plus and minus two standard deviations is 95.4 percent of the area under the curve.

We can use these properties of the normal curve to make some simple inferences. First, let use define *probability* as the relative frequency of the event we are interested in divided by the total number of events that can occur. Thus, in flipping a fair coin, we expect that heads will occur 50 percent of the time because we expect in a finite number of flips, say 10,000, that we will get 5,000 heads. Put another way, we can say that the chance that we will get a head in flipping a coin is .50.

Now consider a class in which scores are normally distributed, and the mean score on an exam is 83 and the standard deviation is 3. If we randomly select a student from the class list, what is the chance that the student will have a score greater than 90? To get the answer, we must first turn our raw score into a Z score and utilize the normal curve. The Z score here is

$$Z = \frac{90 - 83}{3} = 2.33$$

In graphic terms, look at the standard normal distribution above and estimate where a Z score of 2.33 would fall. Our question then becomes: What is the area of the curve beyond a Z score of 2.33? Put another way, we are asking for the relative frequency of scores beyond $Z = 2.33$ to all scores in the distribution. Solving such a problem requires calculus, but fortunately statisticians have created tables of such areas that can be found in any standard introductory statistics book. In this case, reference to such a table indicates that area beyond a Z score of 2.33 is 0.01. Thus, the probability of randomly selecting a student with a score of 90 or greater is .01 or 10 percent.

In effect, we have used the properties of the normal curve to reach this probability statement. This demonstrates both the utility of the normal curve and the nature of statistical inference. It should be noted that the normal curve is only one distribution used by the statistician in making statistical inferences.—J.D.S.

See also CALCULUS, DIFFERENTIAL AND INTEGRAL; STATISTICS, DESCRIPTIVE; STATISTICAL INFERENCE

Further Reading: T. H. Wonnacott and R. J. Wonnacott, *Introductory Statistics for Business and Economics*, 4th ed., 1990.

Not-Invented-Here Syndrome

Innovations, be they scientific, technological, or otherwise, often expose the innovators to risk. People tend to be risk averse, which helps to explain why resistance to new things often crops up. One of its manifestations is the "Not-Invented-Here" Syndrome, or NIH for short. People and organizations are reluctant to adopt new things, and they are especially disinclined to adopt inventions that come from the outside. As Henry Ford II lamented, in the automobile company that bore his name,

> There's too much NIH—not invented here. . . . Lots of times a guy brings something in, and unless the improvement is rather dramatic, nothing happens. The status quo is a hell of a lot easier than making changes.

Psychological resistance to unfamiliar things is one source of the NIH syndrome, but it is not the only one. Compared to the ones coming from the outside, internally generated innovations are more likely to be congruent with established organizational structures and activities. Moreover, an internally generated innovation will probably have a "champion" within the organization who has a financial, career, or psychological stake in the success of the innovation. Such a person often plays a vital role in overcoming internal resistance to change.

Although manifestations of the NIH syndrome are understandable, efforts to overcome it may be essential to the continued vitality of an organization. A dynamic organization can ill afford to ignore useful inventions, processes, and ideas just because they were developed elsewhere. An organization that chooses to resist or ignore potentially useful new things because of their source may eventually find that the preservation of security and pride have come at a considerable price. While its competitors have been moving ahead, it has successfully maintained comfortable roles, routines, and structures, right up to the day that it declares bankruptcy.

N Rays

The discovery of X rays in 1895 generated a great deal of excitement in the scientific world. X rays marked the beginning of a new era in physics, as three other mysterious emanations were discovered in rapid succession:

alpha rays, beta rays, and gamma rays. And since these phenomena had come onto the scene with no prior anticipation, it stood to reason that other mysterious rays were waiting to be discovered.

In 1903, a French physicist named René Blondlot (1849–1930) announced that he had discovered a new form of radiation. Blondlot came upon his discovery in the course of conducting research on X rays. While attempting to determine if X rays could be polarized by electrically charged plates, he came up with the idea that polarization might be detected by placing an electrical spark in the plane of an X-ray beam. Under these circumstances, the presence of polarized X rays would be indicated by an increased brightness of the spark. To his delight, Blondlot observed just such an effect. But he also found that the X-ray beam was bent when it passed through a glass prism, which was a direct contradiction of all previous experimental evidence. He therefore concluded that the spark was intensified by a hitherto unknown form of radiation, which he called N rays in honor of his home institution, the University of Nancy.

Blondlot went on to devise a number of experiments aimed at gaining a greater understanding of N rays. In this effort he was joined by a number of other scientists, some of whom found N-ray emanations from a number of unexpected sources, including the human body. By the middle of 1904, the French Academy of Science's official journal had published more than 50 articles on N rays.

Some scientists, however, remained skeptical. One of them was R. W. Wood (1868–1955), a physics professor at Johns Hopkins University. After failing to reproduce Blondlot's results, Wood journeyed to Nancy to see things firsthand. Blondlot and his associates put on a number of demonstrations, but Wood remained unconvinced, for the observed effects seemed to require a fair degree of subjectivity on the part of the observer. Another experiment proved to be much more convincing, but not in the way that Blondlot would have wanted. In this experiment, N rays were passed through an aluminum prism (it had been determined that N rays readily passed through this material), which refracted the rays and caused them to spread out in a distinct spectrum. The experimenters duly noted the expected results—even though Wood had surreptitiously removed the prism while they had been working in the darkened room.

The publication of Wood's report in the scientific journal *Nature* effectively ended N-ray research outside France, but Blondlot persisted. Improved equipment seemed to give less-ambiguous results, but Blondlot was still forced to argue that the perceptive abilities of the observer were crucial to the observation of the effects of N rays. This was a very insecure position for a scientist to take, for two essential elements of the methodology of physical science are: (1) the recording of data in a manner that is independent of the observer, and (2) the replication of experimental results by someone other than the person who did the experiment first. Blondlot was not a charlatan, and he did not consciously attempt to dupe the scientific world, but because he failed to realize the limitations of his experimental technique, his name will always be associated with one of the great fiascoes of 20th-century physics.

See also ALPHA PARTICLES; BETA PARTICLES; GAMMA RAYS; LIGHT, POLARIZED; X RAYS

Further Reading: Irving M. Klotz, "The N-Ray Affair," *Scientific American*, vol. 242, no. 5 (May 1980): 168–75.

Nuclear Fission

The nuclei of most atoms are stable. However, the atoms of radioactive elements have nuclei that will occasionally split into two parts, which then emit neutrons and release energy. This is an uncommon event, however. Scientific experiments and technological applications that require fission reactions have to induce them by bombarding atomic nuclei with neutrons. This process is the basis of atomic bombs and nuclear reactors. Fissioning is associated with a nearly instantaneous release of energy in the former and the controlled generation of energy in the latter. In both cases, enormous amounts of energy can be produced. The fissioning of the nucleus of a single uranium atom yields about 3.2×10^{-11} joule. In contrast, when one carbon atom combines with oxygen during combustion, the energy yield is 6.4×10^{-19} joule. This means that for a given amount of mass, the fissioning of uranium produces about 2,500,000 times more energy than the combustion of carbon.

The fissioning of atomic nuclei began during the second decade of the 20th century. In 1919, Ernest

Rutherford (1871–1937) reported experiments undertaken 2 years earlier in which he had bombarded nitrogen atoms with alpha particles. This caused a hydrogen nucleus (i.e., a proton) to be split off from the nitrogen nucleus. It was an infrequent event, however, as only one alpha particle in about 300,000 found its target. Rutherford went on to detach protons from the nuclei of most of the lighter elements, but the next major step was taken by John Douglas Cockcroft (1897–1967) and Ernest T. S. Walton (1903–1995), who used protons in the place of alpha particles. Since protons carry an electrical charge, they could be accelerated by the application of a changing electrical field. Although the protons were lighter than alpha particles, their greater velocity made them considerably more energetic. In 1932, Cockcroft and Walton used accelerated protons to split individual lithium atoms into pairs of alpha particles. Meanwhile in the United States, the ability to fission nuclei was considerably advanced by the invention of the cyclotron, which by 1939 was producing deuterons (nuclei of deuterium or "heavy hydrogen") with energies of 19 million electron-volts.

Important as they were for the advance of pure science, it was not apparent that these laboratory demonstrations would ever lead to practical applications. One of the skeptics was Rutherford, who saw little likelihood that the energy locked in the atomic nuclei could ever be controlled; in 1933 he called the whole idea "moonshine." In 1938, a few years after Rutherford's dismissive comment, two German chemists, Otto Hahn (1879–1968) and Fritz Strassmann (1902–1980), produced the element barium by bombarding uranium with neutrons. At first, scientists could not explain how uranium had produced barium, which is much lighter than uranium. The crucial insight came a year later, when the Austrian physicist Lise Meitner (1878–1968) and her nephew Otto Frisch (1904–1979) showed that Hahn and Strassman had produced the first artificial fissioning of an element. A uranium nucleus had split into two nearly equal fragments, a barium nucleus and a krypton nucleus. In addition, two neutrons had been emitted. Most importantly, it was evident that matter had not been conserved: The mass of the neutrons plus the mass of the uranium nucleus was greater than the mass of the barium nucleus, the mass of the krypton nucleus, and the mass of two electrons. The mass did not disappear; it had been converted into energy in accordance with Einstein's famous equation $E = mc^2$, where E is an abbreviation for energy, m for mass, and c for the speed of light. Since light travels at a speed of approximately 300,000 km per second, the equation implies that a small loss of mass can release a vast amount of energy.

Einstein fully understood that fissioning might lead to the production of bombs with immense explosive power. In 1939, he wrote a letter to President Franklin Roosevelt, urging him to commit the United States to the development of a fission (atomic) bomb. Einstein felt that such a project was scientifically feasible, and he was concerned that scientists in Nazi Germany might be working on an atomic bomb. Roosevelt took action, and in 1940 the Manhattan Project was underway. A group of scientists under the leadership of Enrico Fermi (1901–1954) was assigned responsibility for taking the first steps. Fermi's group developed the first nuclear reactor under a squash court at the University of Chicago. On Dec. 2, 1942, the reactor produced the first artificially created chain reaction. A chain reaction occurs when the number of fission events per unit time remains constant. This means that each fission produces one neutron that initiates another fission instead of being captured by some nonfissionable material.

The possibility of inducing a chain reaction had become evident in April 1939, when two French physicists, Irène Joliot-Curie (1897–1956) and her husband Frédéric (1900–1958) found that when a uranium nucleus underwent fission, more neutrons were released than had been required to initiate the fissioning process in the first place. This meant that the excess neutrons were available for the fissioning of other nuclei. In that same year, Niels Bohr (1885–1962) and J. A. Wheeler showed that uranium-235, a natural isotope of uranium, was more easily fissioned than ordinary uranium or uranium-238, a common isotope. This discovery set off intensive efforts to separate uranium-235 from uranium ore. Achieving this required the solution of many scientific and technological problems. The design and construction of an atomic bomb posed further difficulties. After an immense outpouring of money and effort, the first atomic explosion took place at the Manhattan Project's Alamogordo, N.Mex., test site, where on July 16, 1945, a bomb equivalent to

20,000 tons of TNT was detonated. The power of fission had been convincingly—and terrifyingly—demonstrated.

See also ALPHA PARTICLES; ATOMIC BOMB; CYCLOTRON; ENERGY, MEASURES OF; ISOTOPE; NEUTRON; NUCLEAR REACTOR; RADIOACTIVITY AND RADIATION; RELATIVITY, SPECIAL THEORY OF

Nuclear Reactor

A nuclear reactor is a device that produces a vast amount of energy from a small amount of fuel. It generates energy in the form of heat from a process called *nuclear fission*. Nuclear fission is the splitting of the nuclei of heavy atoms, usually uranium or plutonium. An atomic bomb gets its destructive power from uncontrolled fission. However, in a nuclear reactor the fission is kept under control. Even when the rate of fission gets out of control, which occurred at Chernobyl, it is not possible for a nuclear reactor to sustain a nuclear explosion. It cannot happen because the geometry of the fissionable nuclei in a nuclear reactor, even when radically altered through fuel meltdown, prevents an explosion. However, accidents at Chernobyl and Three Mile Island have demonstrated that serious accidents can occur at nuclear reactors.

The primary use for nuclear reactors is to serve as heat sources for the generation of electricity. The heat produces steam for powering turbines, and the turbines run the electrical generators. The first significant application of nuclear power was the power plant of the U.S. submarine *Nautilus*, which was launched in 1956. Commercial production of electricity by a nuclear power also began in 1956, when Britain's 4.2-mW Calder Hall station went into operation. The first U.S. commercial nuclear-power station was opened in Shippingport, Pa., in 1957. Today, the United States has approximately 100 electrical generating units that use nuclear reactors as their heat source, and 20 percent of the electricity used in the United States is generated from nuclear reactors. No country has as many reactors as the United States, although several countries generate a higher percentage of their electricity from nuclear reactors. Nuclear reactors are also used to produce material for nuclear weapons, although with the end of the Cold War, this use has decreased. In addition, nuclear reactors are used to produce radioactive isotopes that are used in medicine, food preservation, and various industrial tasks.

Nuclear reactors vary in size and design. Most of them have the following basic parts: core, moderator, coolant, control rods, pressure vessel, and various safety systems. The core is the central part of the reactor and contains the nuclear fuel; this is where the fission process takes place. The fuel used by most nuclear reactors is a mixture of two isotopes of uranium, U-235 and U-238. Natural uranium consists of 99.3 percent U-238 and 0.7 percent U-235. Only U-235 is fissionable: Consequently, to be used as nuclear fuel, natural uranium must be "enriched" to increase the percentage of U-235.

U-235 will capture only a "slow" neutron, whereas the neutrons released by the fission process are "fast" (high energy) neutrons. Therefore, to have a chain reaction that sustains the fission process, the fast neutrons must be moderated (slowed down). Most moderators are based on graphite, water, or heavy water. Heavy water is water in which the hydrogen atoms have been replaced by an isotope of hydrogen called deuterium. Moderators must have the capability to slow down neutrons without capturing them. If a moderator did not reduce the speed of the neutrons, many of them would be captured by U-238 nuclei, which do not fission.

The coolant carries the heat produced by fission out of the reactor. It transfers the heat to other systems of the nuclear facility, where it is usually employed for the generation of electricity. The coolant also controls the temperature of the core and prevents it from overheating. Various substances, including gases, liquids, and liquid metals, are used as a coolant. In some reactors, the coolant also serves as a moderator. This is often the case in water-cooled reactors.

The control rods regulate the temperature and/or power level of the reactor. The control rods include boron or cadmium materials that can capture (absorb) neutrons without being altered by them. To start up a reactor, the control rods are withdrawn slowly by control rod drive mechanisms. Conversely, during a controlled shutdown of the reactor the control rods are inserted into the core. If an accident condition is sensed by one of the safety systems, the control rods are inserted into the core very quickly (less than 1 second). It is amusing to consider that when the first chain reaction took place in 1942, the control rods were suspended over the reactor core by a rope. A workman with an ax was stationed close by, ready to sever the rope in the event of an emergency.

Reactor cores that produce significant amounts of power (i.e., power reactors) are contained inside of pressure vessels. Low-power reactors, which are generally used for research, are sometimes contained in open pools. The pressure vessels are usually made of steel, and they also contain the coolant, moderator, and control rods. Pressure vessels also serve as a radiation shield. However, all power reactors require additional shielding to protect personnel and equipment.

The various safety systems shut the reactor core down when necessary, cool it, and keep it in a safe condition. The safety system associated with the control rods have already been mentioned. Some reactors also have other neutron-absorbing materials that can be inserted into the core in an emergency. Most reactors have emergency cooling systems to ensure that the core is kept cool, thereby preventing meltdown of the core. In the event of the release of a large amount of radioactivity, most reactors are provided with a containment building to prevent the escape of that radioactivity to the surrounding environment. These containment buildings are constructed and tested to have very low leakage rates. At Three Mile Island, the reactors were situated inside containment buildings, and very little radioactivity was released as a result of the accident. In contrast, at Chernobyl, no containment buildings were installed, and most of the radioactivity was released to the environment, contaminating large portions of Europe to varying degrees.

Reactors are generally classified by the type of coolant they use. Here we will follow the conventional reactor typology: pressurized-water reactor (PWR), boiling-water reactor (BWR), gas-cooled reactor (GCR), pressurized heavy-water reactor (PHWR), light-water graphite reactor (LWGR), and the liquid metal reactor (LMR). Over 50 percent of the operating power plants in the world are PWRs. PWRs use enriched uranium and are cooled and moderated by light water (i.e., normal water; the term is used to differentiate it from "heavy water," deuterium oxide). In a PWR the coolant is maintained at a pressure that prevents film boiling in the core. After serving this purpose, the coolant is transported to a heat exchanger that causes water in a separate water system to boil and drive a steam turbine.

A boiling-water reactor (BWR) also uses enriched uranium and is cooled and moderated by light water. However, in a BWR, film boiling occurs in the core, and the steam goes directly to the steam turbine to generate electricity. All of the commercial nuclear reactors operating in the United States are either PWRs or BWRs.

Gas-cooled reactors (GCRs) can use natural or enriched uranium and are moderated by graphite and cooled by carbon dioxide. Most of these reactors are located in the United Kingdom. This technology was developed in Britain after World War II when enriched uranium was not available in that country.

Pressurized heavy-water reactors (PHWRs) also use natural uranium or slightly enriched uranium. They are cooled and moderated by heavy water. However, unlike PWRs and BWRs, the water used as a coolant is not the same water used as a moderator. Most of the work on these reactors was done in Canada, and most of the ones in operation are called CANDU (Canadian Deuterium Uranium) reactors. Canada, like the United Kingdom, did not have access to enriched uranium after World War II. After much development work, the first CANDU reactor went into operation in 1963.

Light-water graphite reactors (LWGRs) use enriched uranium, are moderated by graphite, and are cooled by light water. The only commercial reactors of this type are located in the former Soviet Union. The ill-fated Chernobyl reactors are of this type. These reactors are not as inherently safe as other types of reactors. However, improvements have been made since the accident at Chernobyl.

Finally, liquid metal reactors (LMRs) use enriched uranium or plutonium, have no moderator, and are cooled by sodium. This technology was originally developed in the United States, but France is now the leader in this technology. These reactors are sometimes called *breeder reactors* or *fast breeder reactors*. They are called breeder reactors because after a fuel cycle they end up with more fissionable material than when they started. This occurs because the reactor contains a large amount of natural uranium. After the U-238 absorbs a fast neutron, it becomes Pu-239, which is fissionable. Some of the Pu-239 will fission during that fuel cycle,

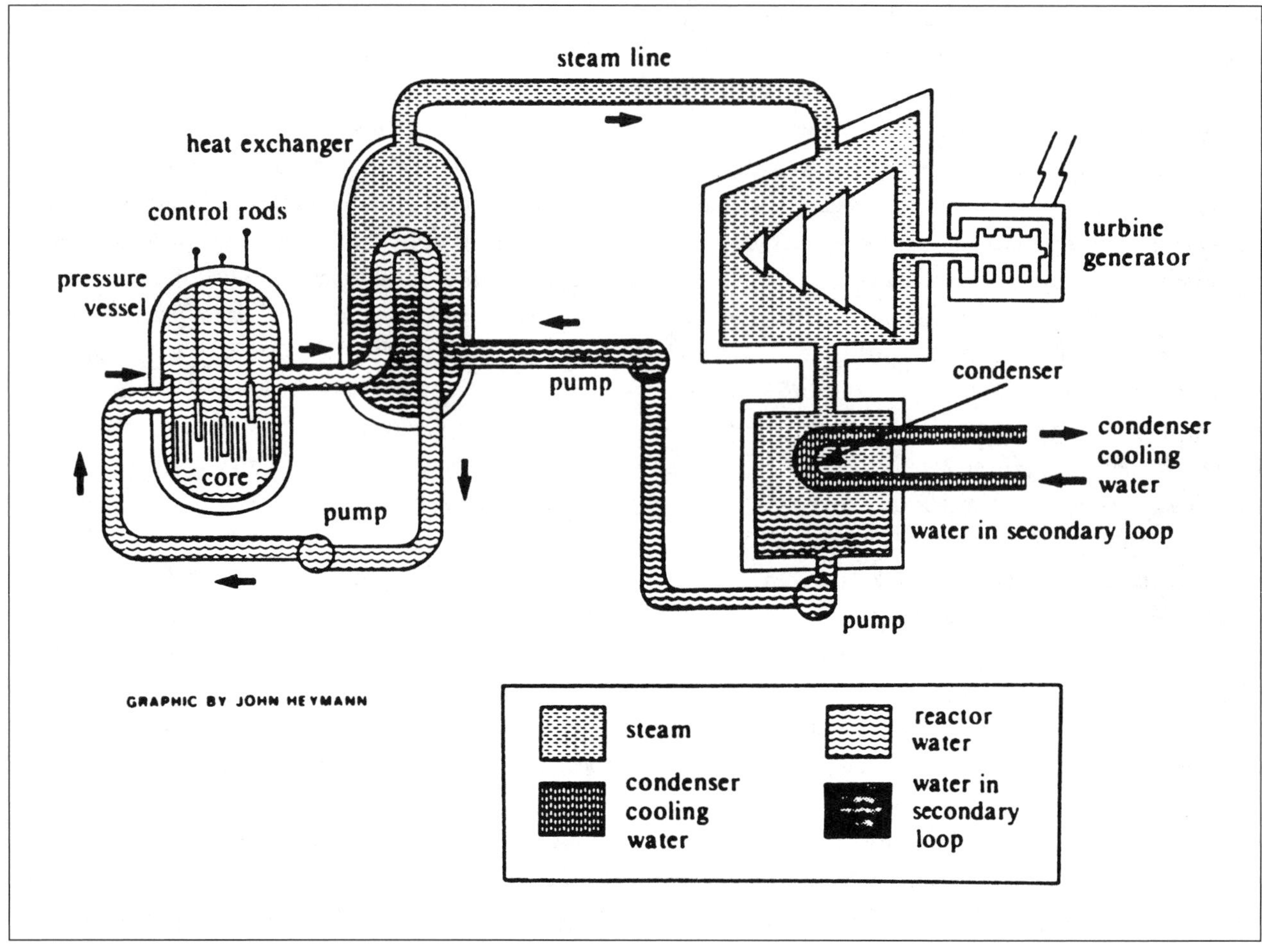

Schematic diagram of a light-water reactor. Nuclear reactions in the core heat water that travels to the heat exchanger, where water in the secondary loop is converted to steam. The steam is then used to run a turbine that powers an electrical generator (from S. Hilgartner, R. C. Bell, and R. O'Conner, *Nukespeak*, 1982, p. 114).

but most will not. It therefore has to be reprocessed in order to be used as fuel. France led the way with the development of breeder reactor technology, but in 1992 the government decided to decommission the world's largest fast breeder reactor, the 1200-mW Superphénix. The United States has largely abandoned this technology due to increased concerns with the proliferation of plutonium, and decreased concern with the availability of enriched uranium.

Using nuclear reactors to generate electricity is still controversial; in particular, the disposal of nuclear wastes remains a concern for many. The economics of nuclear power are also questionable. In 1954 Lewis Strauss, the chairman of the U.S. Atomic Energy Commission, declared that electricity from nuclear power plants would be "too cheap to meter." Future events were not kind to this prediction, and given present fossil fuel prices, nuclear power remains the more expensive way of generating electricity. However, concerns about global warming brought on by the combustion of fossil fuels may make nuclear power more attractive in the future. There is also the possibility that the future generation of nuclear power will make use of fusion rather than fission reactions, but that prospect is still a long way off.

In any event, today's nuclear reactors are a significant source of electrical energy in many parts of the world. France, to take the most prominent example, derives 75 percent of its electrical power from fission. On a smaller scale, nuclear reactors are necessary for the production of radioactive isotopes, which are of great importance to medicine.—D.K.

See also ATOMIC BOMB; CANCER; CHERNOBYL; CLIMATE CHANGE; FUSION ENERGY; IRRADIATED FOOD; ISOTOPE; NUCLEAR FISSION; NUCLEAR SUBMARINE; NUCLEAR WASTE DISPOSAL; RADIOACTIVE TRACERS; THREE MILE ISLAND; TURBINE, STEAM

Further Reading: Richard Rhodes, *Nuclear Renewal*, 1993.

Nuclear-Waste Disposal

The disposal of waste from nuclear facilities has been one of the major issues affecting the development of nuclear-power plants. The problem is long-term, for some of the wastes from nuclear facilities will be radioactive for thousands of years. Over a long period of time, it is feared, radioactive wastes will seep through the ground and affect water supplies. But this is not an inevitable outcome; there are solutions to these problems, and these will be discussed below.

Nuclear wastes are an inevitable outcome of producing energy through a nuclear reaction. When fission occurs, the uranium nucleus splits into two smaller nuclei, and neutrons are emitted. The two smaller nuclei are usually radioactive. The neutrons may cause another fission, may be absorbed by a nucleus in the reactor's fuel supply, or may be absorbed by a nucleus outside the fuel. Absorption inside the fuel by a uranium-238 nucleus leads to the formation of plutonium-239. Other isotopes of plutonium also form inside the spent fuel. The spent fuel consists of uranium, plutonium, and various radioactive fission fragments. Many of the fission fragments will remain radioactive for hundreds of years before they decay to form stable nuclei. However, some of the plutonium isotopes will remain radioactive for thousands of years. In the United States, the spent fuel is called *high-level waste*. The remaining radioactive material is somewhat misleadingly termed *low-level waste*. The storage plans for the two classes of waste are quite different. Most countries with radioactive waste have different storage plans for the spent fuel and the remaining waste.

When neutrons are absorbed by nuclei outside the fuel, they often form radioactive nuclei. Most of the radioactivity in low-level waste comes from these activated nuclei. Usually there will be some leaking fuel elements so that low-level waste may also contain some fission fragments. However, uranium or plutonium cannot escape from the fuel unless a fuel meltdown occurs, and in that event all affected waste would be classified as high level. The volume of low-level waste is much greater than the volume of high-level waste because the low-level waste consists of a great deal of material such as rags, clothing, and tools, which are only slightly radioactive and may be rather bulky. The low-level waste also may contain material such as filters and ion exchange resins, which are highly radioactive. It is important to note that the term *low level* is used to distinguish these wastes from high-level wastes from spent fuel; it does not indicate the intensity of the radiation level. Some of the low-level waste has a very high radiation level, and the material may remain radioactive for hundreds of years. All nuclear waste unrelated to a reactor, such as waste from nuclear medicine, is classified as low-level waste.

At the time of this writing in the late 1990s, the United States was storing all the spent fuel from commercial nuclear power plants in on-site pools. This storage meets all safety requirements, but it is not adequate for long-term storage because of the manpower requirements to monitor the pools and keep the spent fuel covered with water. The federal government is responsible for the development of long-term storage for nuclear waste. One favored location is Yucca Mountain in Nevada. This is a site that has been geologically silent for millions of years. In addition, the groundwater level is very deep in this location. These two criteria—no recent earthquakes or volcanic activity and long distance from water sources—are important for siting any nuclear-waste disposal facility. In addition, the facility must be distant from population centers, as is the case with Yucca Mountain. Yet for all its potential advantages, the project has run into stiff opposition from the legislature and citizens of Nevada. In 1992, the U.S. Department of Energy reported that the site would not be ready until 2010, and another study indicated that its costs might be prohibitively high.

Yucca Mountain is hardly exceptional; no matter how stringent the requirements, there has been political opposition to nuclear-waste disposal facilities. Much of it is driven by the NIMBY (Not In My Back Yard) syndrome. Some of the opposition may be the result of misinformation, but the unwillingness to receive nuclear wastes will never be resolved by the gathering of more precise data, for there will always be different answers to the question "How safe is safe enough?" Meanwhile, the delays caused by local opposition do not pose a safety concern, as temporary facilities operate at adequate levels of safety. However, if the spent-fuel pools at nuclear-power plants become full and there are no interim storage facilities, a shutdown of these plants would inevitably follow.

Spent fuel has to be processed before it is deposited in a long-term storage facility. A process that has been used by the Department of Energy for spent fuel from the nuclear-weapons program and by some other countries is called *vitrification*. The spent fuel is processed to remove the radioactive nuclei, and the radioactive nuclei are then mixed with molten glass. When the glass cools down, the radioactive nuclei are encased in the glass. Since glass is highly impermeable to water, the radioactive particles remain encased in the glass even if exposed to water. One alternative to vitrification has been developed by Sweden. Rather than processing the spent fuel to remove the radioactive nuclei, complete fuel rods are encased in copper. The copper helps remove heat so that the fuel rods do not melt, and copper is very resistant to corrosion. Whatever the method, the encapsulated wastes have to be buried in geologically inactive structures to ensure full safety.

Low-level waste is generally stored in sealed 55-gallon (208-liter) drums. If any liquid is present in the waste, it is removed or solidified before storage. One solidification technique entails mixing the liquid with cement in a drum, resulting in a solid block of cement inside the drum. Unlike spent-fuel repositories that are or will be underground, low-level waste facilities are generally above ground. These facilities present a number of problems: Some of the low-level waste remains radioactive for hundreds of years; the drums are subject to corrosion; and cement and other solidification materials, unlike glass, are permeable to water. For these reasons, each low-level waste facility is located at some distance from groundwater and is usually lined to prevent seepage from the facility. Depending on location, some low-level waste sites have or will have monitoring wells surrounding the site to test the groundwater for radioactivity.

Locating a site for a low-level waste facility is subject to the same political opposition as the site for high-level waste. All of the existing and planned low-level facilities operating in the United States face opposition. Most power plants are storing some or all of their low-level waste onsite. This does not pose a problem for most power plants because they have adequate space. However, the lack of low-level storage may eventually affect hospitals and institutions performing medical research.

Although the storage of nuclear waste is not 100 percent safe, nuclear waste can be stored with a reasonable degree of safely. However, political opposition is strong, especially at the local level, and the nuclear establishment has to convince the public that the benefits of nuclear power outweigh the potential hazards of nuclear wastes.—D.K.

See also ISOTOPE; NUCLEAR FISSION; NUCLEAR REACTOR; RADIOACTIVITY AND RADIATION; TRACERS, RADIOACTIVE

Nuclear Winter

The enormous destructive power of nuclear weapons has been evident since the detonation of the first atomic bombs in 1945. A nuclear war would be marked by widespread blast damage, uncontrollable fires, and massive doses of lethal radiation. In the 1980s, another horrifying possibility was presented: the threat of a nuclear winter. According to this scenario, a nuclear war would put so much smoke and dust into the troposphere (the layer of atmosphere that extends 10 to 16 km [6–10 mi] above the Earth's surface) that most of the sun's radiation would never reach the ground. The result would be a radical cooling of the northern hemisphere; temperatures in the middle of the North American continent might never get above –25°C (–13°F) for months on end. Many people would die from the cold, and many others would die from starvation following massive crop failures. Even a limited nuclear war involving the detonation of no more than 100 megatons could produce subfreezing temperatures that would persist for months.

The possibility of a nuclear winter was first raised

in 1983 by Richard Turco, Brian Toon, Thomas Ackerman, James Pollack, and Carl Sagan. At the core of their analysis was the expectation that urban areas, the primary targets in an all-out nuclear war, would provide enormous amounts of combustible material that would be ignited and then injected into the air. The authors also postulated that the dust raised by ground-level nuclear explosions would produce smaller particles of dust than had been previously assumed; these tiny particles would then rise several kilometers above the Earth's surface. The projected volumes of smoke and dust were then incorporated into a computer model of global atmospheric circulation that indicated where this material would eventually come to rest.

As might be expected, the nuclear-winter hypothesis sparked a considerable amount of public debate. Coming as it did in a period of heightened tensions between the United States and the Soviet Union, the threat of a nuclear winter helped to mobilize support for a unilateral nuclear freeze in the United States. On the other hand, for those with a more hawkish disposition, the prospect of a nuclear winter reinforced a belief in the importance of a strong military deterrent as the best means of preventing nuclear war. In some quarters, the nuclear-winter hypothesis also strengthened a commitment to developing an advanced antiballistic missile system.

The nuclear-winter hypothesis has been criticized for its reliance on computer modeling. As the debate over global warming has indicated, weather patterns on Earth are extraordinarily complex, and no computer is powerful enough to take into account all of the variables that influence it. The specification of the variables to put into the model is also an inexact process. Critics of the nuclear-winter scenario argue that its authors have been excessively pessimistic in their assumptions, while others are of the opinion that they are too optimistic.

More recent efforts to assess the likelihood of a nuclear winter have not been conclusive. One trial involving the use of an improved computer model concluded that a nuclear war would result in a "nuclear fall," with temperatures dropping significantly, but not to the extent predicted by the original model. The 1991 Gulf War seemed to offer a real-world test of the likelihood of a nuclear winter when Iraq set fire to Kuwaiti oil fields, but the smoke that was produced did not rise high enough into the atmosphere to allow any firm conclusions to be drawn.

It is to be hoped that the nuclear-winter hypothesis will always be an academic concern. The definitive test of the hypothesis would be a nuclear war, and no sane person wants to see that happen.

See also ATOMIC BOMB; CLIMATE CHANGE; HYDROGEN BOMB; MATHEMATICAL MODELING; STRATEGIC DEFENSE INITIATIVE

Further Reading: Richard P. Turco et al., "Nuclear Winter: Global Consequences of Multiple Nuclear Explosions," *Science*, vol. 222 (1983): 1283–92.

Numerals, Arabic

The Arabic numeral system has been described as one of humanity's greatest inventions. Its power lies its use of zero as a place holder, which makes it possible to do extensive arithmetical calculations without the use of special counting devices. Although the zero-based system spread to Europe from the Arab world, this was not its place of origin. In the 3d-century B.C.E., the Babylonians were using a character to indicate an empty column in the middle of a number, e.g., 904, but it was not used to convert an integer into a larger number, e.g., 900. Zero was used as a place holder in 9th-century India, although it is possible that the practice was derived from Indochina or from the earlier Babylonian notation. In China, zero first appeared in print in 1247, but a blank space had been used for a place holder for many centuries prior to this date.

Whatever the ultimate source, India was an excellent locale for the cultivation of mathematics, for three of its major religions—Jainism, Hinduism, and Buddhism—all considered the study of arithmetic as an essential part of preparation for the priesthood. When the Arabs learned of the numerical system used in India, it did not happen by accident; the Caliph Al-Mamun had established in Baghdad a combination university, library, and translation bureau known as the House of Wisdom. It was there that a scholar named Al-Khowarizmi came into contact with Indian astronomers who were familiar with the use of zero as a place holder. He put that knowledge to work when he wrote his book, *Arithmetic*. It was translated into Latin around 1120, the author's name giving rise to the term *algorism* as a method of calculation using Indian-Arabic numerals, and eventually the modern term *algorithm*.

At this time, Europe was an intellectual backwater, far removed from the mathematical accomplishments of earlier centuries. The Crusades and trading relationships brought many Europeans into contact with Arab culture, and by the end of the 10th century the Indian-Arabic notation began to appear in European documents, albeit very infrequently at first. The first systematic attempt to use these numerals was mounted by Leonardo of Pisa (1175–1250?), commonly known as Fibonacci. Pisa was one of Europe's great commercial centers. At the age of 12, Fibonacci joined his father, who at the time was the head of a customs house on the coast of North Africa. There Fibonacci learned the Arabic system, which in 1202 was presented in his *The Book of the Abacus*. This had nothing to do with the counting device also known as an *abacus*; in those days, the term was synonymous with arithmetic. The book had a limited initial impact, for circulation was limited in the days before the printing press, and in any event its content and presentation were too advanced for most people.

Two other books helped to popularize the system: *The Poem of Algorism* by Alexander De Villa Dei, and *Common Algorism* by John of Halifax (also known as Sacrobosco). Both of them, based on Al-Khowarizmi, were published in 1220 and 1250, respectively. Their intended audience were the students of the emerging European universities, and in fact they were little more than lecture notes.

Although it greatly facilitated calculation, the new system met with considerable resistance. Many found it hard to conceptualize how the numeral 0 had no value itself but had the capacity to change another numeral into a much larger value, as when 7 became 70,000. There was also considerable concern about the ease with which numbers could be altered, and well into the 16th century Chinese-Indian-Arabic numerals were avoided in contracts and bank drafts. As with many fundamental innovations, the system spread slowly and amidst considerable resistance.

See also ABACUS

Nutritional Guidelines

Nutritional guidelines are used to indicate the necessary components of a proper diet. Efforts by the U.S. federal government to produce and disseminate nutrition standards began in 1943, when the National Research Council published a set of Recommended Dietary Allowances (RDAs) for foods deemed essential to good health. The date is significant, for the impetus for developing RDAs came from wartime food rationing for civilians and the need to provide adequate diets for men and women in the armed forces. The RDAs proved to be so useful that they were continued after World War II. They are modified, usually slightly, about every 5 years.

RDAs cover 17 essential elements of nutrition: 10 vitamins, 6 minerals, and 1 protein. The RDA roster also includes a listing of "safe and adequate" amounts for 12 additional nutrients. A separate set of RDAs for total caloric intake is also provided. These lists provide different RDAs for men and women, as well as RDAs for specific ages, heights, and weights.

On the basis of these lists, the U.S. Food and Drug Administration draws up a separate set of guidelines known as Recommended Daily Allowances (abbreviated as U.S. RDAs). These provide only a single standard for each nutrient, one that is intended to meet the nutritional needs of all people, irrespective of sex and age.

Although RDAs are based on nutritional research, their compilation sometimes involves political considerations. In 1985, the NRC's Committee on Dietary Allowances had prepared a new set of RDAs that, among other things, reduced the allowances for vitamins A and C. The publication of the new standards was blocked by another component, the Food and Nutrition Board, a separate body that was empowered to review revisions of RDAs. Although they did not take issue with the research results that led to the new recommendations, members of the board feared that lowered requirements might lead to reduced funding for federal school lunch and food stamp programs.

RDAs are the basis for food plans that specify the daily consumption of particular food groups. The best known of these, the U.S. Department of Agriculture's Pyramid Guide to Daily Food Choices, originated as the Basic Four Food Group Plan in 1956 and has been modified several times since then. It specifies that a proper daily diet for an adult should include 2 to 3 servings of milk or milk products; 2 to 3 servings of meat, fish, or poultry, or the equivalent in eggs, cheese, or legumes; 2 to 4 servings of fruits; 3 to 5 servings of vegetables; and 6 to 11 servings of grain products such as bread, pasta, and cereal. Dietary recommendations from both governmental and nongovernmental sources also suggest that people can lower the risk of cancer by consuming fiber and complex carbohydrates (starches) and reducing the consumption of fats.

Nutritional guidelines are the basis for food labels that indicate the percentage of various RDAs provided by a typical serving of the product. It would of course require a very large label to indicate the product's contribution to the nutrition of every category of consumer. Consequently, product labels state the product's share of RDAs for four groups at most: infants, children under 4, pregnant and lactating women, and everybody else. The majority of labels do not even go this far; they provide the product's RDA contribution for only one or two of these groups.

See also CHEESE; FOOD AND DRUG ADMINISTRATION, U.S.; LEGUMES; NATIONAL SCIENCE FOUNDATION; VITAMINS

Further Reading: U.S. Department of Agriculture, *Home and Garden Bulletin*, no. 253-1 through 253-8, 1993.

Nylon

One of the first of the artificial polymers, nylon is still one of the most widely used today. Nylon is one of the most successful examples of research sponsored by a large corporation; it is the product of individual genius, but its invention and subsequent commercial success also required the collaborative efforts of many scientists, engineers, and technicians. The financial resources necessary for a project of this scope were provided by E. I. DuPont de Nemours (more familiarly known simply as DuPont), a chemical firm whose origins lay in the manufacture of gunpowder.

In 1928, DuPont hired Wallace Carothers (1896–1937), at that time a chemistry instructor at Harvard, to conduct research of his own choosing. With the strong encouragement of DuPont's research director, Carothers turned his scientific interest toward artificial polymers, and in 1929, a year after entering DuPont, he published an important paper on condensation polymerization, a hitherto neglected area of polymer science and technology. Aided by Julian Hill, another DuPont chemist, Carothers' initial efforts centered on the creation of a polyester by reacting carboxylic acid with alcohols. The resulting compound could be drawn into fibers, but these had a low melting point and were adversely affected by water. The next step was to react acids and amines to form polyamides. These could not be converted into useful fibers, however, and Carothers largely abandoned the effort to produce polymer fibers by mid-1933.

By this time, the Depression had put a financial squeeze on America's industrial corporations, and DuPont was disinclined to devote substantial funds to research that promised few financial rewards. In 1934, Carothers resumed work on polymer fibers after some prodding by a new research director more interested in commercially oriented research than the "pure" variety. Success came on May 23 of that year, when Donald Coffman, an assistant to Carothers, used an aminononanoic ester to make 4 grams (.14 oz) of polymer. On the following day, it was drawn out to form fibers. This substance can be identified as the first of the family of polymers collectively known as *nylon*. The kind of nylon that DuPont eventually produced in commercial quantities was first synthesized on Feb. 28, 1935. It came to be known as Nylon 6-6 (or 66), after the number of carbon atoms in the dibasic acid and diamine that were its starting point.

DuPont's management saw the potential of the new material, and they decided at an early date that women's hosiery would be the primary application. Nylon had many of the properties of the silk that had been used for stockings, while having the advantage of being much more durable. The focus on hosiery considerably simplified research and marketing, but a great deal of development work still lay ahead. As is usually the case, the scaling up of a laboratory procedure into a commercial operation required the solution of many technical problems. Numerous research teams were engaged in the development of melt spinning (melting the nylon to a viscous fluid and then drawing filaments through the tiny apertures of a spinneret), the large-scale manufacture of intermediate chemicals, and controlling the polymer's molecular weight by stopping the reaction at just the right moment. The knitting of the stockings on existing machines also caused many problems. One of nylon's deficiencies compared to silk was that it lacked a coating that protected the fibers as they were processed. An artificial coating had to be developed to make up for this shortcoming.

On Oct. 27, 1938, DuPont used the New York World's Fair as the setting for its public announcement of nylon. A year and a half went by before nylon stockings were put on sale, but they met with immediate acceptance. When the United States entered World War II, all nylon production was diverted to military uses: Parachutes, glider tow ropes, and cords for airplane tires were the main applications. Peace and the resumption of sales of nylon hosiery led to minor riots in some places, as people fought over limited initial supplies of the stockings. Having captured most of the market for women's stockings, DuPont then promoted the use of nylon for fishing line, brush bristles, molded and extruded parts, and fabrics. Nylon has been one of the most outstanding examples of profits derived from industrial research and development. In subsequent years, industrial R&D at DuPont produced several other successful synthetic fibers, most notably orlon and dacron.

See also POLYMERS; RESEARCH AND DEVELOPMENT

Further Reading: David Hounshell and John K. Smith, *Science and Corporate Strategy: DuPont R&D, 1902–1980*, 1989.

Index

C

D

E

H

J

K

L

M

O

P

S

W